25 YEARS OF P53 RESEARCH

25 Years of p53 Research

Edited by

PIERRE HAINAUT

International Agency for Research on Cancer, World Health Organization,
Lyon, France

and

KLAS G. WIMAN

Department of Oncology-Pathology, Cancer Center Karolinska, Karolinska Institute,
Stockholm, Sweden

 Springer

A C.I.P. Catalogue record for this book is available from the Library of Congress.

ISBN-10 1-4020-2920-9 (HB) Springer Dordrecht, Berlin, Heidelberg, New York
ISBN-10 1-4020-2922-5 (e-book) Springer Dordrecht, Berlin, Heidelberg, New York
ISBN-13 978-1-4020-2920-2 (HB) Springer Dordrecht, Berlin, Heidelberg, New York
ISBN-13 978-1-4020-2922-6 (e-book) Springer Dordrecht, Berlin, Heidelberg, New York

Published by Springer,
P.O. Box 17, 3300 AA Dordrecht, The Netherlands.

Printed on acid-free paper

Printed in the Netherlands.

TABLE OF CONTENTS

FOREWORD

p53 was discovered independently by several researchers* in 1979 as a cellular protein that forms a complex with the SV40 large T protein. The 25 years that have passed since this discovery represent an extraordinary scientific journey with a major impact on molecular biology and cancer research. p53 has changed face from obscure oncogene to key tumor suppressor gene with potentially great clinical impact. Yet the p53 field has not followed a straight course. In retrospect, it is possible to discern at least three phases of p53 research during the first 25 years.

The first phase of p53 research, 1979-1988, can be characterized as a search for a p53 identity. The observation by Lane, Levine and others that the SV40 large T oncoprotein binds p53 in SV40-transformed cells indicated a role of p53 in tumor development. This notion was supported by the finding that many tumor cells overexpressed p53. The identification of complexes between p53 and other protein encoded by DNA tumor viruses, i.e. the adenovirus E1B 55K protein and the human papilloma virus E6 protein, showed that SV40 large T was not a special case, as different viruses had apparently evolved proteins to target one and the same cellular protein and presumably, cellular function. However, that function remained elusive. In these early years, p53 was viewed as an obscure oncogene.

Two observations during this period were important indications of very significant advances that were to follow later: the demonstration by Maltzman and Czyzyk in 1984 that p53 is induced by DNA damage and the identification of p53 gene alterations in virus-induced leukemias and tumor cell lines independently by Benchimol and Rotter 1984-85.

The second phase, 1988-1994, brought a major paradigm shift that profoundly changed the current understanding of p53, and, beyond, of the molecular biology of cancer (Figure 1). This was triggered both by experiments in vitro and studies of human tumor samples. Data from the

---

*The four publications that sparkled the field in 1979 were: Lane and Crawford, Nature 278: 261-263 on 15 March; Linzer and Levine, Cell 17: 43-52, May issue; De Leo, Jay, Appella, Dubois, Law and Old, PNAS 76: 2420-2424, May issue; Kress, May E, Cassingena and May P, J Virol 31: 472-483, August issue.

groups of Levine and Oren demonstrated that wild type p53 cDNA clones were in fact able to suppress transformation of rodent cells in culture, while point mutant version of p53 were transforming. Moreover, work by Vogelstein and colleagues revealed frequent point mutations in p53 in colorectal carcinomas and other tumors. In most cases, one p53 allele was mutated whereas the other allele was lost. This was in complete agreement with the classical tumor suppressor paradigm that had emerged from studies of the retinoblastoma gene, the prototype tumor suppressor gene. In addition, inherited p53 mutations were found in the familial Li-Fraumeni cancer syndrome. All these findings led to the recognition of p53 as a key tumor suppressor.

In the following years, the induction of p53 in response to DNA damage was further substantiated and p53 was dubbed "guardian of the genome". DNA sequencing revealed mutagen fingerprints in p53 in human tumors, connecting exposure to certain DNA-damaging agents with specific p53 mutations. Studies of the effects of wild type p53 on human tumor cells showed that p53 could curb cell cycle progression and induce cell death by apoptosis. The generation of p53 null mice in 1992 and the demonstration of their dramatically increased tumor incidence was a solid proof of p53's tumor suppressor function. Meanwhile, molecular studies revealed that p53 was a transcription factor with DNA binding specificity. The first p53 target gene to be discovered was the cell cycle inhibitor p21/WAF1 in 1993, linking p53 with cell cycle arrest.

The crystal structure of the p53 core domain bound to DNA published by Pavletich and colleagues in 1994 was an important milestone that has allowed a deeper molecular understanding of the effect of p53 point mutations and amino acid substitutions in the p53 core domain and a classification of p53 mutations according to their structural consequences.

The third phase of p53 research, beginning in 1995 and still ongoing, can be labelled "Mastering complexity en route to clinical applications". The identification of Mdm2 as a p53 antagonist and critical regulator of p53 in 1997 provided a better understanding of p53 degradation in the proteasome. Mdm2 is itself inhibited by ARF, a protein that is induced in response to oncogenic stress. Overexpression of Mdm2 in a fraction of tumors that carry wild type p53 can functionally inactivate p53.

The complexity has increased substantially with the identification of two p53-related genes, p73 and p63, that share extensive structural homology with p53 and can transactivate similar sets of target genes. Both p73 and p63 are expressed as numerous isoforms, some of which lack the N-terminal transactivation domain and therefore do not activate transcription. More recent work has also demonstrated the existence of N-terminally truncated p53 isoforms, increasing complexity even further.

After 25 years of research, the p53 field is reaching a turning point where we can identify perspectives for applications in the clinic and in public health. The aim of this book is to catch the essential lessons of 25 years of research and to identify the major paths towards applications.

Most importantly, what should we expect from the next 25 years of p53 research? First, there is undoubtedly still a lot to be learned about the different p53 family members and their isoforms, with respect to their role in diverse cellular proceses such as growth suppression, apoptosis, senescene, development, differentiation, and DNA repair, and probably other processes as well. Second, we may hope for the exploitation of the impressive amount of information about p53 status in human tumors to improve cancer diagnosis and prognosis. This should also allow the development of tailored therapy regimens according to mutant p53 status. In addition, the accumulated information on mutagenesis and cancer aetiology should be useful for cancer prevention. Finally, we will hopefully see novel efficient anti-cancer drugs that are based on targeting p53 in cancer cells.

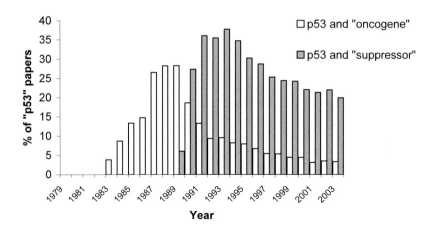

Figure -1. Years of paradigm shift: proportion of p53 papers with the keywords "oncogene" or "tumor suppressor gene" per publication year

This graph, compiled from PubMed search, captures the change in perception of the role of p53 during years 1988-1992, from an "oncogene" to a "tumor suppressor gene".

Klas G. Wiman and Pierre Hainaut
December 2004

Chapter 1

THE FIRST TWENTY-FIVE YEARS OF P53 RESEARCH

Harlan Robins*, Gabriela Alexe*, Sandra Harris[#] and A. J. Levine*[#]
Institute for Advanced Study and the Cancer Institute of New Jersey#

SETTING THE STAGE

During the 1960s, the field of cancer research lacked clear direction. Several facts appeared to be well-established and correct, but the relationships among these observations were not apparent. Fifty years of research had demonstrated that viruses with both DNA and RNA genomes could cause cancer in animals. Over the next 45 years six new viruses were to be discovered that were able to initiate cancers in humans (Epstein-Barr Virus, Human T-Cell Leukemia Virus, Hepatitis B and C Viruses, Kaposi Sarcoma Virus and the Papilloma Viruses) (McKinnel et al., 1998). It was equally clear from the perspective of the 1960s that certain chemicals, when applied to animals, were able to initiate cancers (Yamagawa et al., 1918). Chemical carcinogenesis was a field both separate and distinct (both in the experiments one did and the experimentalists who did them) from viral carcinogenesis and very few scientists thought to find a common ground between concepts generated in each field. Thirdly, the study of mouse genetics demonstrated that some cancers were clearly inherited and these observations confirmed many prior publications that suggested a role for cancer causing genes in humans and other animals (DeOme, 1965). Finally epidemiologists, studying a variety of important variables that predispose humans to developing cancers, had made the very striking observation that the rates of cancer incidence increase exponentially with age and begin to rise dramatically by the fifth and sixth decade of life (Miller, 1991). While these four observations were all accepted facts the relationship between

P. Hainaut and K. G. Wiman (eds.), 25 Years of p53 Research, 1-25.
© 2005 *Springer. Printed in the Netherlands*

these concepts was not clear and researchers who studied viruses hardly ever discussed chemicals and those who thought about genes and viruses didn't know what to make of aging as an important variable. Literally researchers from each of these fields, virology, chemical carcinogenesis, genetics and epidemiology never got together to discuss these issues.

Things began to change when it became clear that some RNA tumor viruses packaged an extra gene in their genomes and this gene could cause the cancer (Kawai and Hanafusa, 1971; Bader, 1972). The cancer causing gene, or oncogene, was shown to be derived from the hosts chromosome and when sequenced contained mutations that activated the oncogene so that it behaved as a dominant mutation giving rise to the cancer in cells infected by the retrovirus (Stehelin et al., 1976). These studies united the concepts for the role of viruses and genes and chemicals that could cause mutations in selected genes, be they from the host or a virus carrying a host gene. The power of this explanation and the unity it gave to three of the four observations discussed above, kept everyone from focusing on two additional observations that contradicted the oncogene dogma. First, somatic cell genetics were employed to fuse a cancer cell with a normal cell in culture. The resultant hybrid, unlike the cancer cell, no longer made tumors in isogenic animals, suggesting that the gene(s) that made the cell tumorigenic was (were) recessive to the normal allele from the normal cell (Jonasson et al., 1977). When these hybrid cells occasionally did produce tumors in animals, those tumors contained cells that had lost some chromosomes. This was interpreted as the loss of genes that prevented cancer formation. The idea that humans and animals have genes that prevent cancer formation or reverse the oncogene phenotype was novel. Independently, in 1971 A. Knudson hypothesized that two independent mutations in the same gene, later called the retinoblastoma gene, could give rise to a childhood cancer of the eye (Knudson, 1971). Knudson noted that the same tumor, a retinoblastoma, had two very different presentations in young children. Some children developed this tumor within the first year after birth, and these children had bilateral tumors in both eyes and as many as three or more tumors per eye. Other children developed these tumors over several years after birth, and these children had unilateral tumors and only one tumor in one eye. He went on to explain these observations using a common hypothesis that unified both classes of tumors. The tumors that presented at a very young age were due to an inherited mutant allele followed by a spontaneous mutation in the other allele (resulting in multiple tumors in both eyes at a very early age). The other class of tumors were due to a rare event of two independent mutations one in each of the two alleles in a cell (resulting in a single tumor in one eye at a later age). This idea suggested a different type of gene than an oncogene (by definition a dominant gene) was involved in the origins of cancer and it was variously called an anti-oncogene, recessive oncogene or a tumor suppressor gene. The idea that several oncogenes and tumor suppressor genes must sustain mutations in the same cell to give rise to a cancer, and that any one mutation is necessary but not sufficient to produce a cancer came to be appreciated, understood and demonstrated at some time in the future (Knudson and Strong, 1972; Land et al., 1983). This concept would explain why cancer was usually a disease of the elderly (it took a long time to accumulate many mutations in the same cell) and that the rate of cancer formation would rise exponentially with the age of the population. This unification of four very diverse observations into a single hypothesis for the origins of cancers in humans gave the field some confidence that these ideas might be correct.

HOW DO THE DNA TUMOR VIRUSES CAUSE CANCERS?

The small DNA tumor viruses, discovered in the mid 1950's, were quickly tested to determine if they too carried oncogenes that were derived from cellular DNA sequences. From this it became clear that the small DNA tumor viruses (SV40, polyoma, the adenoviruses and later the papillomaviruses) encoded their own genes (not the cellular genes) that caused the cancer and therefore were termed viral oncogenes. When these viruses were inoculated into a host animal, the animal would develop a tumor at the site of injection after a long latent period. It was most common that the infectious virus disappeared and that a single cell (a clone) developed into a tumor with the viral DNA integrated into a cellular chromosome. This DNA was differentially expressed and viral m-RNA made one or a few viral proteins in the tumor cells. These proteins were recognized as foreign by the host's immune system that responded by making antibodies against the viral encoded proteins. Thus, these viral oncogene products were termed tumor antigens. These antibodies were then employed to demonstrate that the viral proteins were common to all tumors made by that virus, were different when different tumor viruses were employed to initiate the tumors and that the tumor antigens were most commonly encoded by the viral genomes. An extensive genetic analysis with these tumor viruses provided strong evidence that one or two viral encoded genes were required to cause these tumors and the products of these genes were most often the tumor antigens. In every case the viral tumor antigens were also required for an efficient replication of the virus. For SV40 the proteins were called the large T-antigen and the small t-antigen, the adenoviruses encoded the E1A proteins and the E1B proteins (E1B-58K and

E1B-19K) and for the papilloma viruses the E6 and E7 proteins. Mutations in these viral oncogenes resulted in the inability to form a tumor in animals.

The next question that came under study was how did the viral tumor antigens act to initiate tumors in animals or transform cells in culture? It was in the pursuit of this question that several groups uncovered the p53 protein. In 1979 David Lane and Lionel Crawford (Lane and Crawford, 1979) demonstrated that the immunoprecipitation of the SV40 T-antigen also detected a second protein of 53,000 molecular weight, called p53. They could show that the dilution of their tumor antisera always produced the same ratio of T-antigen and p53 which demonstrated that there was a T-antigen –p53 complex in the cell extract (it is unlikely that the antibodies to two different proteins were in equal concentrations). The SV40 T-antigen bound to the p53 protein in the cell. At the same time Daniel Linzer and Arnold Levine (Linzer and Levine, 1979) employed antisera from animals bearing SV40 induced tumors to detect both p53 and the viral T-antigen in SV40 transformed cells. Antibodies in this sera also immunoprecipitated the p53 protein from teratocarcinoma cells in the absence of the SV40 T-antigen. The peptide maps of the p53 proteins from the SV40 transformed cells and the teratocarcinoma cells were identical. These results demonstrated that the p53 protein was a cellular protein, animals bearing SV40 induced tumors also made antibodies against the p53 protein, and monoclonal antibodies to the SV40 T-antigen co-immunoprecipitated the p53 protein demonstrating the T-antigen p53 complex (Linzer and Levine, 1979). The concentration of the p53 protein in SV40 transformed cells was much greater then in normal cells in culture. In teratocarcinoma cells in culture p53 was higher in its concentration than in normal cells but lower than in SV40 transformed cells [13]. The presence of an SV40 T-antigen – p53 complex and the higher levels of p53 in transformed cells suggested that p53 might act as a transforming gene product or oncogene. At the very least the presence of antibodies directed against the p53 protein demonstrated that it was a tumor antigen. In SV40 transformed cells that contained a temperature-sensitive mutation in the SV40 T-antigen gene, shifting to the non-permissive temperature inactivated T-antigen function, made the cell revert to a non-transformed phenotype, and drastically lowered the levels of p53 in the cell (Linzer et al., 1979). This demonstrated that T-antigen really did control p53 levels in a cell. At this time, Llyod Old and his colleagues (DeLeo et al., 1979) demonstrated that animals immunized with spontaneous transformed and tumorigenic cells also made antibodies to the p53 protein and so it was clear that the p53 protein could well be called a tumor antigen in its own right. At a later time L. Crawford and his colleagues showed that some humans with cancers made antibodies directed against the p53 protein (Crawford et al., 1984).

The generality of these observations received a big boost when it was shown that an adenovirus tumor antigen, the E1B-58k protein, which was quite distinct from the SV40 T-antigen, bound to the p53 protein in adenovirus transformed cells (Kao et al., 1990; Sarnow et al., 1982). Similarly a human papilloma virus oncogene product, the E6 protein, bound to the p53 protein in cells derived from human tumors caused by this virus (Werness et al., 1990). Thus three distinct tumor virus groups encoding diverse proteins evolved a mechanism to complex with the same cellular protein, the p53 protein. About this same time the retinoblastoma (Rb) gene was identified and cloned (Friend et al., 1986). The Rb protein, the product of a tumor suppressor gene, was shown to bind to the adenovirus E1A gene product (Whyte et al., 1989) the SV40 T-antigen (DeCapricio et al., 1988) and the papilloma E7 protein (Munger et al., 1989). Thus three different tumor viruses encoded oncogene products that bound to the cellular proteins p53 and Rb. The real meaning of these observations was only poorly understood until the functions of the p53 protein and the Rb protein were elucidated, but they made everyone feel confident that they were on the right track.

CLONING THE P53 GENE: IS IT AN ONCOGENE OR A TUMOR SUPPRESSOR GENE?

The cloning of the p53 c-DNA and gene were carried out by several groups from a wide variety of cellular sources including both transformed and normal cells (Crawford et al., 1984; Beinz et al., 1984; Oren et al., 1983; Pennica et al., 1984). Once these c-DNA and genomic clones were in hand the biological activities of these clones were tested. The fact that the SV40 T-antigen regulated and increased the levels of the p53 protein made most think that p53 was an oncogene whose over-expression (mutant or not?) resulted in transforming the cell. At the time there were two assays for testing an oncogene, one group of oncogenes was able to immortalize cells in culture but not change other properties of these cells (E1A, Myc) while other oncogenes could fully transform immortalized cells (E1B, Ras) but could not transform non-immortalized cells in culture unless myc or E1A were added as well (Land et al., 1983). Very quickly three groups demonstrated that the p53 c-DNA clones were like myc or E1A and could immortalizes cells or could fully transform cells when added to the Ras oncogene clone (Eliyahu et al., 1984; Jenkins et al., 1984; Parada et al., 1984). p53 was declared an oncogene. Moshe Oren's laboratory had a genomic clone of p53 and A Levine's group had a c-DNA clone of p53 and they exchanged these clones for further experiments. There were two complicating aspects to these results; first the c-DNA from the Levine laboratory did not immortalize cells in culture nor did it transform cells along with the ras oncogene. The Levine laboratory could repeat the results of M. Oren's showing that p53 was an oncogene when they used his clones but could not reproduce these results when the Levine c-DNAs were employed. Second, the amino acid sequence of the Oren and Levine clone differed at codon 135 (a valine and alanine difference). In a series of experiments several things became clear; 1. the wild type p53 c-DNA does not transform cells (Hinds et al., 1989), 2. mutant p53 c-DNAs or mutant genomic clones are commonly found in cells that are grown in culture, in fact p53 mutations are commonly selected for as cells adapt to long term culture conditions (Harvey and Levine, 1991), and 3. a mutant p53 c-DNA or genomic clone can act in a dominant negative fashion (the p53 protein is a tetramer and faulty subunits will inactivate the wild type p53 function) and transform cells (Eliyahu et al., 1988; Kraiss et al., 1988; Finlay et al., 1989). The Levine group went on to show that the wild type p53 c-DNA and its protein can actively inhibit oncogenes from transforming cells in culture (Finlay et al., 1989). In fact a very similar observation had been observed in murine erythroleukemia cells transformed with a retrovirus containing an oncogene (Munroe et al., 1988; Ben David et al., 1988) where the integration of the viral DNA disrupted the p53 gene function in these cancer cells. Thus p53 was behaving as a tumor suppressor gene in all of these assays. These conclusions were independently demonstrated by Vogelstein and his colleagues when they sequenced three human colon carcinomas and showed that p53 mutations were found in their p53 genes and the other allele was lost or reduced to homozygosity (Baker et al., 1989; Nigro et al., 1989). This is the hallmark of a tumor suppressor gene.

Thus three different approaches all led to the conclusion that the p53 gene product acted as a tumor suppressor protein and that the viral oncogene products bound to the p53 protein must therefore inactivate it. Mutations in both p53 alleles were selected for in non-viral induced cancers. Adding back the wild type p53 gene to a cancerous cell in culture killed the cell or blocked the action of oncogenes. At this time there were two examples of tumor suppressor genes (retinoblastoma and p53) and the field turned its attention to elucidating the functions of these genes and their products.

THE FUNCTIONS OF THE P53 GENE AND THE DOWNSTREAM PROGRAM

One of the first clues about the function of the p53 protein came from the observation that it bound to DNA and that tight binding to DNA occurred in a sequence specific fashion (Funk et al., 1992; Zauberman et al., 1993; el-Deiry et al., 1992). Steinmeyer and Deppert (1988) selected for DNA sequences that would bind to the p53 protein even at low concentrations and sequence analysis of these DNA's gave a consensus for the optimal DNA binding sequence: RRRCWWGYYY where R is a purine, W is A or T and Y is a pyrimidine. A core fragment of the p53 protein containing its DNA binding domain was co-crystallized with this consensus oligonucleotide and the protein was found to make strong contacts with the C and G residues and weaker contacts with the other sequences (Cho et al., 1993). At this time a series of experiments demonstrated that as a result of this DNA binding p53 behaved as a transcription factor (Fields et al., 1990; Raycroft et al., 1990; Kern et al., 1991). This set off a search for the genes regulated by the p53 protein and these target genes are discussed further later in the chapter.

At about this same time the p53 protein in cells was shown to bind to another protein and temperature sensitive mutants of the p53 protein regulated the levels of this interacting protein (Momand et al., 1992). The purification and sequencing of this p53 binding protein identified it as the MDM-2 protein (Momand et al., 1992) which had recently been shown to be an oncogene in mouse cells (Fakharzadeh et al., 1991). The MDM-2 protein was found to bind to the p53 protein and block its ability to act as a transcription factor (Momand et al., 1992) and the MDM-2 gene in humans, called HDM-2, was shown to be amplified and over-expressed in some human sarcomas (Oliner et al., 1992). Furthermore the MDM-2 gene was shown to be transcriptionally regulated by the p53 protein, containing a number of DNA sequences in the first intron of the gene related to the p53 DNA consensus sequence (Zauberman et al., 1995). This meant that p53 and MDM-2 formed an autoregulatory loop where increased p53 activity increased MDM-2 levels which in turn decreased p53 activity resulting in declining MDM-2 levels (Piksley and Lane, 1993; Wu et al., 1993). This forms a failsafe mechanism to prevent p53 activity from getting too high in a cell. Subsequently it was shown that MDM-2 is an E3 ubiquitin ligase that transfers ubiquitin to p53 resulting in its degradation (Honda et al., 1997). This type of circutry between p53 and MDM-2 means that the levels and activities of these proteins in a cell oscillate out of phase with each other over time (Bar-Or et al., 2000) and this has been shown to be the case in single cell experiments (Lahav et al., 2004). This relationship between p53 and its negative regulator MDM-2 can be disrupted in several different ways; 1. The p53 gene can be mutated so that the cell doesn't make MDM-2 proteins (present in 50-55% of cancers), 2. the MDM-2 gene can be amplified so it blocks p53 functions (found in 30% of sarcomas), 3 .p53 protein modifications (phosporylations) can occur in or near the p53-MDM-2 binding sites (Unger et al., 1999; Lin et al., 1994) and disrupt this protein-

protein interaction as is the case after p53 activation in response to the appropriate signals, 4. MDM-2 can be inactivated by the ARF protein or by interaction with some ribosomal proteins (Zhang et al., 2003; Lohurm et al., 2003).

Another gene regulated by the p53 transcription factor is the p21/ Waf-1/ Cip-1 gene (el-Deiry et al., 1993). This gene contains perfect p53 DNA binding consensus sites that regulate it by the activation of the p53 protein. One of the functions of the p21 protein is to bind to the cyclin E cdk-2 protein kinase that must act in late G1 of the cell cycle and block its activity (Harper et al., 1995; Xiong et al., 1993). This is in part the reason why p53 activation can lead to cell cycle arrest in G1. Similarly the 13-3-3 sigma gene is regulated by p53 and this protein binds to the CDC-25 protein, keeping it in the cytoplasm where it is unable to function as a nuclear phosphatase thus permitting cells to go from G2 to M phase (Draetta and Eckstein, 1997; Taylor and Stark, 2001). This contributes to a G2-M block that is sometimes observed after p53 activation. Thus some of the downstream genes regulated by p53 contribute to a cell cycle arrest. Another set of p53 responsive genes promotes apoptosis in a cell by helping to activate the release of cytochrome c from the mitochondria (bax, noxa, perp, etc.) and contributing via the production of APAF-1 (Rozenfeld-Granot et al., 2002) to the activation of caspase 9 and 3 followed by apoptosis. p53 also activates a second apoptotic pathway increasing the levels of the Fas ligand and the KILLER DR receptor in the caspase 8 and 3 pathway (Sheikh et al., 1998). Thus a second major p53 response is programmed cell death. p53 also regulates some genes that participate in DNA repair reactions in the cell (p53R2 an alternative ribonucleotide reductase subunit) and a set of gene products that produce secreted proteins after a p53 response (thrombospondin, maspin, inhibitors of plasminogen activators). These gene products may alter the extracellular matix and could impact upon the regulation of cell division, metastasis, angiogenesis, or other functions. Figure 1 depicts these pathways and see also the recent review by Nakamura (2004).

Among the more interesting aspects of the p53 inducible and regulated pathway is the elaborate negative feedback loops that are formed by three p53 regulated genes and their products. First is the p53 -MDM-2 feedback loop that has been discussed above. p53 also regulates the Cyclin G gene that makes a protein that combines with the PP2A phosphatase and removes a phosphate residue from the MDM-2 protein (Okamoto et al., 2002) thus increasing the MDM-2 activity and lowering p53 levels in a cell. A mouse with the Cyclin G gene knocked out is viable but has higher constitutive p53 levels in its cells (Jensen et al., 2003). The phosphate group removed from

MDM-2 by Cyclin G -PP2A is added to MDM-2 by one of several cyclin-cdk kinases suggesting a link to cell cycle events.

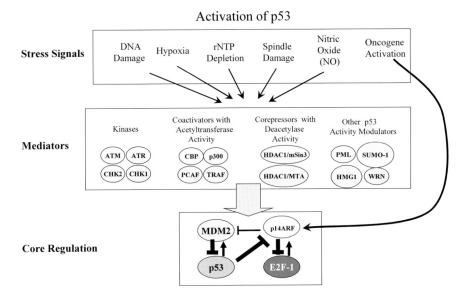

Figure 1a. Downstream of p53: known transcriptional response of the activated p53 protein.

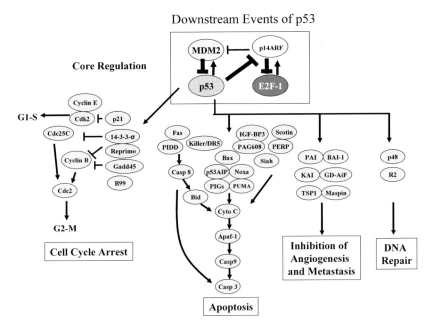

Figure 1b. Downstream of p53: known transcriptional response of the activated p53 protein.

Another negative regulator of p53 that is encoded by a p53 responsive gene is Wip-1 (Fiscella et al., 1997). Wip-1 is a phosphatase that acts upon MAP kinase that in turn can phosphorylate the p53 protein at two sites resulting in its increased activity as a transcription factor. The dephosphorylation of MAP kinase by Wip-1 lowers MAP kinase activity and reduces p53 activity (Takekawa et al., 2000). Thus MDM-2, Wip-1 and Cyclin G are all p53 regulated genes that in turn negatively regulate p53 activity or levels and both MDM-2 (Taubert et al., 2003) and Wip-1 (Bulavin et al., 2002; Sinclair et al., 2003) genes are found to be amplified in selected cancers. Figure 2 summarizes this negative feedback loop and network.

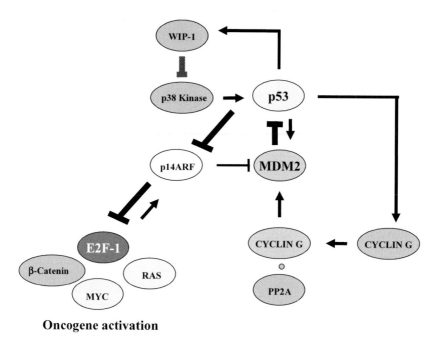

Figure 2. Negative feedback loops that control p53 activity in the cells.

To date some 30-35 genes have been shown to contain the p53 responsive elements in their DNA and, by one criteria or another, have been shown to be regulated by p53 either in a positive or negative fashion. Hoh and her colleagues (Hoh et al., 2002) have formulated an algorithm that scans the human or mouse genome for p53 responsive DNA elements or sequences adjacent to genes that may be regulated by the p53 protein (see http://linkage.rockefeller.edu/p53). They identified in the mouse and human genome 16 genes that had excellent p53 responsive elements and tested these genes for their transcriptional regulation after p53 activation in cells in

ERRATUM

25 Years of p53 Research

P. Hainaut and K.G. Wiman (eds.)

© 2005, ISBN 1-4020-2920-9

Contrary to the information provided in the Table of Contents, Chapter 19: Novel Approaches to p53-Based Therapy: ONYX-015, which starts on p. 421, was written only by Frank McCormick and not co-written by David A. Wood.

culture (Hoh et al., 2002). To date 12 of those genes have been shown to increase or decrease their abundance after p53 activation. There is some cell or tissue type specificity in some of these responses and this has been observed in mice as well. A survey of the p53 responsive DNA sequence elements in many p53 regulated genes demonstrates that there are always two RRRCWWGYYY palindromes separated by a 0-21 base pair spacer of any sequence and a good deal of sequence degeneracy is permitted in these sites. An oligonucleotide chip analysis of genes up- or down-regulated after p53 levels rise in a cell identified a number of genes that have p53 responsive elements (as in Figure1) and many genes whose m-RNA levels change but don't have recognizable p53 responsive elements (Zhao et al., 2000). This suggests that their may well be a program of gene activation or repression begun by p53 regulated genes that is no longer dependent upon p53 for its activity and that among the p53 regulated genes might well be transcription factors that carry out this program. Figure 3 shows the kinetics of mRNA levels (increased or decreased) for a series of genes in lymphoblastoid cell lines, as detected by Affymetrix chips after exposure to gamma irradiation, a known activator of the p53 pathway. Using the data provided by Jen and Cheung, (2003) but utilizing different clustering algorithms, we addressed the problem of identifying among 126 IR-responsive genes in common between 3 Gy and 10 Gy exposure, clusters of genes which are highly correlated in their temporal expression patterns at the two doses (Figure 3 and http://www.csb.ias.edu/Research/clusters.htm). As in Jen and Cheung our analysis reveals a complex program of gene expression in these cells after p53 activation. A similar program was observed in carcinoma cells undergoing a p53 response (Robison et al., 2003).

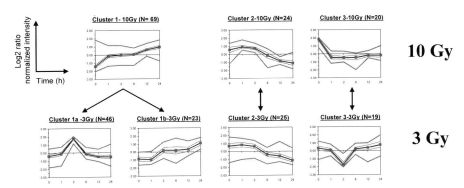

Figure 3. This analysis was performed to better identify clusters of genes whose expression following IR insult display similar temporal patterns. The data that consisted of Affymetrix chip analysis of gene expression in lymphoblastoid cells, was provided by Jen and Cheung [78] (see also http://www.csb.ias.edu/Research/clusters.htm for details). The consensus

profiles show a highly coordinated IR response of genes after 3 Gy and 10 Gy IR exposure: (i) more than 90% of the genes in two of the four 3Gy profiles (Cluster 3Gy-1a and Cluster 3Gy-1b) display similar temporal patterns after 10 Gy exposure (in Cluster 10Gy-1), and (ii) more than 85% of the genes in each of the remaining two 3Gy profiles (Cluster 3Gy-2 and Cluster 3Gy-3) display similar temporal expression patterns after 10 Gy radiation (in Cluster 10Gy-2, and in Cluster 10Gy-3, respectively). Each profile identified in this analysis is stable to data perturbation with white noise, has high homogeneity, and has a very low likelihood to occur by chance. Additionally, Jen and Cheung identified 16 profiles of smaller size in response to 3Gy, and 10 Gy radiation exposure.

THE ACTIVATION OF P53 AND THE UPSTREAM PROGRAM

The p53 protein is synthesized in most cells in the body and has a very short half-life of 6-30 minutes depending upon the cell or tissue type. Under these circumstances there is very little p53 regulation of p53 responsive genes. A variety of stress signals will activate the p53 protein so that there is an enhanced transcription of the p53 responsive genes. Activation is associated with and is caused by protein modifications of p53 (phosphorylation and acetylation). This in turn results in an increased half-life and increased concentrations of the p53 protein (Maltzman and Czyzyk, 1984; Price and Calderwood, 1993; Maki and Howley, 1997). While activation of the p53 protein was first carried out using temperature-sensitive mutants of the p53 gene, the first demonstration by a real physiological stress that activated p53 was by Maltzman and Czyzyk (1984) who showed that UV light damage increased the level of the p53 protein in cells. This was roundly ignored by the field for quite a while until Kastan (Kastan et al., 1991) and others demonstrated that a wide variety of DNA damaging agents producing very different DNA lesions, each can activate the p53 response (Huang et al., 1996). UV damage involves the formation of thymine dimers, gamma radiation results in single or double stranded breaks in DNA, alkylating agents often react with guanine residues producing alkylated G-residues and each of these lesions has a distinct set of repair activities in the cell. Associated with this repair process are a series of enzymatic activities; protein kinases, histone acetylases, and possibly histone methylases, sumo ligases, or other such activities, that recognize the type of DNA damage and modify the p53 protein, signaling to it the existence of that type of damage. Although it has been difficult to assign a specific kinase to a specific signal, the available evidence suggests that the ATM kinase (Canman et al., 1998; Banin et al., 1998) and a CHK kinase (Zhao et al., 2002; Gatei et al., 2003) may well play a role in the single and double strand break stress signals and the ATR kinase could be involved in UV damage recognition (Unsal-Kacmaz et al., 2002). The patterns of phosphorylation after such stress signals have been intensively studied and it appears that different kinases may yield different combinations or patterns of phosphorylation on the p53 protein. UV damage or gamma radiation produces distinct p53 transcriptional responses as examined by oligonucleotide arrays (Zhao et al., 2000). These data suggest that the p53 protein integrates the input signals from different stresses and responds accordingly with a distinct transcriptional output. Thus, removal of each of these lesions might be by a different mechanism in cells. There are also sets of p53 responsive genes that are always transcribed in a p53 response independently of the type of input stress or the cell type under study. These genes include p21, MDM-2, Gadd-45, 14-3-3 sigma, and Cyclin G. Today we recognize a variety of stress signals that activate the p53 pathway such as DNA damage, hypoxia, spindle poisons, the size of the ribonucleotide triphosphate pools in a cell, NO signaling, cold shock, denatured or altered proteins and even some oncogene mutations result in enhanced p53 activity in a cell (reviewed recently by Nakamura (2004). Thus the p53 protein is modified by a wide variety of stress signals or alarms, it then processes this information using a protein modification code yet to be elucidated and responds by activating a transcriptional program (Figure 3) resulting in either cell cycle arrest, apoptosis, DNA repair, the modification of the cellular matrix and communication with neighboring cells.

Clearly these upstream signals that communicate with p53 can mobilize a large number of enzymatic functions such as kinases and phosphatases, histone acetylase complexes and histone deacetylases, PML bodies, helicases, ubiquitin ligases, etc. which all may play a positive or a negative role in modulating p53 activity and p53 responses (see Figure 1). The modified p53 protein must then enter into the transcriptional machinery of the cell, which may respond to the protein modification code, and promote p53 interactions with other proteins to enhance the rate of transcription of selected genes. Together these upstream inputs and their downstream responses create a highly regulated network that responds to the stress signals (Figures 1, 3).

INACTIVATION OF THE P53 PATHWAY IN CANCERS

The p53 pathway is composed of hundreds of genes and many of them will have single nucleotide polymorphisms that impact upon the efficiency of p53 function. Thus we can expect that genetic difference between people will contribute to the cellular and molecular responses to stresses and this may well impact upon the age of onset of cancers, the incidence of cancers and the responses to therapy that results in DNA damage. A small number of families inherit one p53 mutant allele in the germ line. These families have the Li-Fraumeni syndrome with early age of onset of cancers, and in some cases multiple independent cancers. The penitrance of the p53 mutant allele is almost 100% in people and the tumors frequently, but not always, reduce to homozygosity for the mutant allele. The tumors are commonly sarcomas, but breast, colon and several other types of cancers are observed (Frebourg et al., 1992; Malkin et al., 1990). The mouse with no p53 alleles develops thymic lymphomas at a young age and these tumors are not observed in humans. The heterozygous mouse frequently develops sarcomas and this is similar to the human spectrum of tumors (Jacks et al., 1994). About 50 percent of all cancers have p53 somatic mutations in both p53 alleles (Hollstein et al., 1991). A few cancer types do not usually have any p53 mutations (teratocarcinomas), and others (melanomas, some leukemias) have a very low frequency of p53 mutations (about 10% of the time) (Hollstein et al., 1991; Drexler et al., 2000). In teratocarcinomas, which are germ cell tumors, the p53 protein is not functional (Lutzker and Levine, 1996) but has a wild type DNA sequence. Because it is not functional there is no selection pressure to inactivate it via mutation. The p53 protein can be activated and kill the cell by apoptosis after DNA damage. Interestingly teratocarcinomas respond very well to chemotherapy and are cured most of the time. Similarly leukemias rarely have p53 mutations and respond very well to chemotherapy. When they relapse these tumors often, but not always, now harbor p53 mutations.

The International Agency for Cancer Research (IARC) maintains a database of p53 mutations (http://www.iarc.fr/p53) that encompasses 18,585 examples of somatic and 1114 of germ line mutations. Of the somatic mutations 82% are point mutations whereas 18% are insertions, deletions or more complex rearrangements. When the entire p53 c-DNA or gene is sequenced the great majority of the mutations are located in the DNA binding or core domain of the p53 gene and protein. Because of this most researchers have sequenced only exons 5-8 encompassing codons 108 to 298 which is the DNA binding domain. Possibly because of this bias 94% of all point mutations in the p53 gene have been localized to codons 100-310. About 35% of these point mutations are localized in six hot spots in the gene at codons 175, 245, 248, 249, 273, and 282. When tested, the proteins with these p53 mutations fail to bind to the p53 DNA response element efficiently (Bullock et al., 2000; Kern et al., 1991; Epstein et al., 1998) and fail to transcribe p53-regulated genes. The hot spot mutations correspond to amino acid residues that make contact with the nucleotides in the p53 response element. This suggests that mutations resulting in a loss of function of the

DNA binding and transcription factor properties of the protein are being selected. A mutational analysis of the amino terminal region of the p53 protein, where the transactivation domain resides, indicates that at least two independent point mutations are required to inactivate the transcriptional activity of this domain (Lin et al., 1995) so it is clearly easier and more common to obtain point mutations in the DNA binding domain.

Some other tumor suppressor genes, such as the APC gene and the p16 gene, are commonly inactivated by point mutations that result in stop codons that lead to a loss of function. In the p53 database both missense and nonsense mutations are found in many cancers. The missense proteins are not transcriptionally functional and so no MDM-2 protein is made in most cells. Due to the decreased abundance of the negative regulator protein, mutant p53 protein is found in much larger amounts in cancer cells than in normal cells with the wild type p53 protein (Hinds et al., 1990). When the missense protein is produced in large amounts in a cell along with wild type p53 protein both wild type and missense proteins are synthesized and enter into a tetrameric protein complex that is inactive because of the mutant or faulty subunits. Thus p53 mutant c-DNA clones show a dominant loss of function phenotype and can transform cells in culture (Hinds et al., 1989). It is not at all clear that this has any functional significance in vivo or in tumors. However many of the missense mutations in the p53 gene have been shown to have a potential gain of function phenotype (Blandino et al., 1999; Dittmer et al., 1993). When a p53 missense c-DNA clone is added to a cell that is normal but has no p53 gene, the cell can grow more rapidly, become more tumorigenic when inoculated into animals and can gain a drug resistant phenotype. These experiments have been carried out by a number of different research laboratories, they appear to be quite reproducible and even show some allele specific phenotypes, all of which suggests that the missense p53 mutations generated in cancers could have a gain of function phenotype. If this was the case then one could understand why missense mutations would occur more frequently in the p53 gene in cancers than nonsense mutations that are true loss of function mutations.

One of the questions never answered properly is whether or not missense p53 mutations are selected for over and above the frequency of nonsense mutations or neutral mutations observed in the database. The IARC database was employed to ask this question for all possible point mutations in codons 100-310 that could lead to a missense mutation or a nonsense mutation. The way to accomplish this is by comparing the number of ways a mutation in any base in a codon can result in a missense mutation or a nonsense mutation with the number of times this has occurred in cancers with p53 mutations. The necessary assumption is that the database contains a large enough set of mutations that differences in the point mutation rates are balanced out. In addition it is important to consider the set of mutations that were selected, above the noise, by using the silent mutations as background noise. The frequency of silent mutations (a nucleotide change that does not result in an amino acid change) in the database suggests a background level of unselected mutations such that a mutation should occur in four or more separate tumors so as to be above this background level. Then the mutation is clearly selected for some property. In Table 1 we carry out this exercise for mutations that are found in 4 or more tumors. Although it is not shown, increasing this cutoff has very little effect on the results. In column A is shown the number of different nucleotide changes that have led to a missense, nonsense, or silent mutation, respectively, and were represented in the database as least 4 times. The total number of tumors in the database that have the set of nucleotide changes described in column A is represented in column B. Finally, in column C, we calculate the average occurrence of each type of mutation (column B divided by column A). The surprising result of this analysis is that when we correct for the number of positions in which these mutation types could occur, missense mutations occur at about the same frequency or ratio (B/A) as nonsense mutations.

Table 1. Comparison of missense, nonsense and silent mutations in p53. Numbers correspond to mutations that are found in 4 or more tumors in the IARC TP53 database. The total number of tumors in the database that have the set of nucleotide changes described in column A is represented in column B. Column C: average occurrence of each type of mutation (column B divided by column A).

	A	B	C
MISSENSE	529	12296	23,2
NONSENSE	50	1145	22,9
SILENT	76	445	5,9

We have added the analysis in column A because different bases in different sequence contexts have different rates of mutation as is shown in Table 2. By far the most common mutation in the p53 gene is a C to T change in the dinucleotide CpG (see table 2) whether or not the mutation results in a nonsense, missense or silent amino acid change. For nonsense mutations C to T and G to T changes make up 75% of the mutations observed. For missense mutations C to T, G to A, and G to T changes make up 63% of the mutations in the database. For silent mutations the C to T and G to A changes make up 62% of the mutations found. With these base pair biases in the rate of mutations taken into account, it does not appear that missense mutations or nonsense mutations are preferentially selected in the cancers. Recently Yang et al. (2003) applied a variety of mutation rate models to this same p53 database and also concluded that selection for missense and nonsense mutations is about equal in the DNA binding region.

Table 2. Frequency of mutation types in p53 IARC database. Breakdown of nonsense, missense and silent mutations in the IARC TP53 database according to the nature of base change. The total number and percentage of each base change is shown.

Type	nonsense				missense				silent			
	mutation	tot number	percent	CpG	mutation	tot number	percent	CpG	mutation	tot number	percent	CpG
	A->C	0	0		A->C	243	2		A->C	7	0.8	
	A->G	0	0		A->G	1365	10		A->G	37	4	
	A->T	51	4		A->T	412	3		A->T	9	1	
	C->A	72	5		C->A	353	3		C->A	54	7	
	C->G	58	4		C->G	501	4		C->G	35	4	
	C->T	712	52.5	480	C->T	2549	19	1580	C->T	316	38.3	24
	G->A	103	8		G->A	4198	31.2	2533	G->A	201	24.3	31
	G->C	0	0		G->C	737	5		G->C	24	3	
	G->T	299	22.5		G->T	1790	13.3		G->T	25	3	
	T->A	36	3		T->A	398	3		T->A	14	2	
	T->C	0	0		T->C	492	4		T->C	86	12.1	
	T->G	7	0.5		T->G	397	3		T->G	18	2	
Sum		1338				13435				826		

These data contradict the gain of function hypothesis and suggests three possible explanations for the contradiction. 1. The gain of function phenotypes observed with missense p53 mutations are observed in cell culture and animal models but are not operative in human cancers, 2. There are times that cancers select for nonsense mutations and others where missense mutations that have a gain of function are selected for by a tumor. Here the genetic background and the nature of the oncogene and tumor suppressor gene mutations could influence whether a gain of function mutation is selected. Even the cell or tissue type of the tumor could influence this. We could use the IARC database to look at mutations from individual tumors to see if they maintain the equality of selection between missense and nonsense mutations. However, specific mutation rates are modified so strongly by particular carcinogens that we would no longer trust that the large database could smooth out the varying rates. 3. Something is fundamentally wrong with the interpretation of the gain of function experiments (the mutant p53 still acts as a transactivator that changes the cell, it still binds to DNA and alters the cell) but these properties are not important in cancers and are not selected for by tumors. This gain of function hypothesis remains one of the unresolved problems in the field.

The IARC database does not permit one to examine the frequency of p53 mutations in one or more cancers because data are not presented for the total number of tumors where the p53 gene has been sequenced so as to know what percentage of these tumors had p53 mutations. Based upon a large number of studies in the literature it would be conservative to claim that about 30% of all tumors examined contained p53 mutations (likely to be >50%). If we accept a 30% cutoff then the IARC database contains about 18585 mutations (out of an estimated 62,000 tumors sequenced) of which 826 are silent mutations. These are usually considered neutral mutations that are not selected for or against in a tumor or in the evolution of an organism. Because this may be as pure an estimate of a mutation frequency in the absence of selection (not a rate as we know nothing of the number of cell divisions) in tumor cells in vivo, we attempted to estimate this number. There are 826 silent mutations in about 62,000 tumors (a third of which have p53 mutations) which implies that 1.3% of tumors have silent mutations. 268 of 630 nucleotides in codons 100-310 can yield silent mutations. Therefore, the estimate for the neutral mutation rate per nucleotide in the tumor is .013/268 = 5×10^{-5}. An estimate of the frequency of germ line changes per base pair is commonly about 3×10^{-8} or about 1,000 times lower than the silent mutation frequency suggested from this analysis. There are many estimates and assumptions in this calculation, but it does suggest something that was intuitively thought to be correct, namely that the spontaneous mutation frequency (as measured using only neutral mutations) in a cancer is about 1,000 fold higher than in a normal cell. Thus it implies that mutator gene phenotypes are involved in raising the frequency of mutations in cancer.

The possibility that these silent mutations are actually polymorphisms and not somatic mutations can be ruled out. There are 6 known SNPs in the coding region of p53 (also 1 in the promoter region and 12 in introns). Of these 6 SNPs, 4 are silent and 2 change amino acids. Only one of the 4 silent SNPs (Arginine 213) is in the region of the gene being studied (codons 100-310). This change could account for at most 2 of the 826 silent mutations.

Finally it should be pointed out that mutations that do not alter an amino acid in the protein may not be selectively silent. Such changes could alter RNA folding, the rate of RNA processing or the rate of translation. RNA–protein interactions might fail when there is a change in the structure or sequence of the m-RNA brought about by a so called neutral mutation. While we don't usually think of these changes as critical to function, these ideas have been poorly tested.

There are a number of other mutations in the p53 pathway that alter p53 functions. As reviewed in Figure 2 there are four negative and one positive feedback loops for p53 regulation. Thus MDM-2 and WIP-1 gene amplifications in sarcomas and breast cancers respectively reduce p53 activity (Taubert et al., 2003; Bulavin et al., 2002). Cyclin G over-expression also reduces p53 function and cyclin G knockout mice have more p53 protein (Jensen et al., 2003). The AKT kinase in the IGF- PI3K-PTEN pathway has been shown to phosphorylate the MDM 2 protein resulting in the movement of MDM-2 into the nucleus where it more effectively degrades the p53 protein (Ashcroft et al., 2002; Mayo et al., 2001). Several other signal transduction pathways produce transcription factors that enhance ARF synthesis or activity, which in turn inhibits MDM-2 and positively regulates p53. Beta catenin-TCF-4 (the product of the WNT-APC pathway), E2F-1 (the product of the Cyclin D-Rb pathway), MYC, RAS and p38 MAPK which acts via the ETSandAP-1 transcription factors, are all examples of interconnections between several signal transduction pathways and p53. Thus we are beginning to understand not only the p53 pathway but also its many connections to signal transduction pathways that play central roles in the origins of cancers. It will now be up to the new field of systems biology to construct these pathways, model them and make clear predictions which can be tested in experiments and ultimately shown to benefit cancer patients with predictive and prognostic outcomes.

After 25 years of research with the p53 gene and its protein we have built an infrastructure upon which to extend our detailed understanding of its functions. The p53 gene and its protein are not essential for life (i.e. the knock out mouse is born alive) but it is quite clear that it is essential for life to faithfully reproduce itself. The p53 gene in worms and flies protects the germ line from stress and mutations. In the mouse and human these functions still operate effectively but the role of p53 has also been adapted to faithful cellular reproduction of somatic cells, as vertebrates regenerate their tissues. Responses to stress that disrupt our homeostatic mechanisms, cause mutations that impact information transfer, and result in pathogenic outcomes, are the business of the p53 pathway. We need to understand this business better.

REFERENCES

Ashcroft M., Ludwig R.L., Woods D.B., Copeland T.D., Weber H.O., Macrae E.J., Vousden K.H. Phosphorylation of HDM2 by Akt. Oncogene, 2002. 21: 1955-1962.

Bader J. Temperature-dependent transformation of cells infected with a mutant of Bryan Rous sarcoma virus. Journal of Virology, 1972. 10: 267-276.

Baker S.J., Fearon E.R., Nigro J.M., Hamilton S.R., Preisinger A.C., Jessup J.M., vanTuinen P., Ledbetter D.H., Barker D.F., Nakamura Y. et al., Chromosome 17 deletions and p53 gene mutations in colorectal carcinomas. Science, 1989. 244: 217-221.

Banin S., Moyal L., Sheih S., Taya Y., Anderson C.W., Chessa L., Smorodinsky N.I., Prives C., Reiss Y., Shiloh Y., Ziv Y. Enhanced phosphorylation of p53 by ATM in response to DNA damage. Science, 1998. 281: 1674-1677.

Bar-Or L.R., Maya R., Segel L.A., Alon U., Levine A.J., Oren M. Generation of oscillations by the p53-Mdm2 feedback loop: a theoretical and experimental study. Proceedings of the National Academy of Sciences USA, 2000. 97: 11250-11255.

Beinz B., Zakut-Houri R., Givol D., Oren M. Analysis of the gene coding for the murine cellular tumour antigen p53. EMBO J, 1984. 3: 2179-2183.

Ben David Y., Prideaux V.R., Chow V., Benchimol S., Bernstein A. Inactivation of the p53 oncogene by internal deletion or retroviral integration in erythroleukemic cell lines induced by Friend leukemia virus. Oncogene, 1988. 3: 179-185.

Blandino G., Levine A.J., Oren M. Mutant p53 gain of function: differential effects of different p53 mutants on resistance of cultured cells to chemotherapy. Oncogene, 1999. 18: 477-485.

Bulavin D.V., Demidov O.N., Saito S., Kauraniemi P., Phillips C., Amundson S.A., Ambrosino C., Sauter G., Nebreda A.R., Anderson C.W., Kallioniemi-Fornace A.J. Jr., Appella E. Amplification of PPM1D in human tumors abrogates p53 tumor-suppressor activity. Nature Genetics, 2002. 31: 210-215.

Bullock A.N., Henkel J., Fersht A.R. Quantitative analysis of residual folding and DNA binding in mutant p53 core domain: definition of mutant states for rescue in cancer therapy. Oncogene, 2000. 19: 1256.

Canman C.E., Lim D.S., Cimprich K.A., Taya Y., Tamai K., Sakaguchi K., Appella E., Kastan M.D., Siliciano J.D. Activation of the ATM kinase by ionizing radiation and phosphorylation of p53. Science, 1998. 281: 1677-1679.

Cho Y., Gorina S., Jeffrey P.D., Pavletich N.P. Crystal structure of a p53 tumor suppressor-DNA complex: understanding tumorigenic mutations. Science, 1994. 265: 334-335.

Crawford L.V., Pim D.C., Lamb P. The cellular protein p53 in human tumors. Mol. Biol. Med., 1984. 2: 261-272.

DeCapricio J.A., Ludlow J.W., Figge J., Shew J.Y., Huang C.M., Lee W.H, Marsillo E., Paucha E., Livingston D.M. SV40 large tumor antigen forma a specific complex with the product of the retinoblastoma susceptibility gene. Cell, 1988. 54: 275-283.

DeLeo A.B., Jay G., Appella E., Dubois G.C., Law L.W., Old L.J, Detection of a transformation-related antigen in chemically-induced sarcomas and other transformed cells of the mouse. Proceedings of the National Academy of Sciences USA, 1979. 76: 2420-2424.

DeOme K. Formal discussion of: multiple factors in mouse mammary tumorigenesis. Cancer Research, 1965. 25: 1348-1351.

Dittmer D., Pati S., Zambetti G., Chu S., Teresky A.K., Moore M., Finlay C., Levine A.J. p53 gain of function mutations. Nature Genetics, 1993. 4: 42-46.

Draetta G., Eckstein J. Cdc25 protein phosphatases in cell proliferation. Biochim. Biophys Acta., 1997. 1332: M53-M63.

Drexler H.G., Fombonne S., Matsuo Y., Hu Z.B., Hamaguchi H., Uphoff C. p53 alterations in human leukemia-lymphoma cell lines: in vitro artifact or prerequisite for cell immortalization? Leukemia, 2000. 14: 198-206.

el-Deiry W.S., Kern S.E., Pietenpol J.A., Kinzler K.W., Vogelstein B. Definition of a consensus binding site for p53. Nature Genetics, 1992. 1: 45-49.

el-Deiry W.S., Tokino T., Velculescu V.E., Levy D.B., Parsons R., Tent J.M., Lin D., Mercer W.E., Kinzler K.W., Vogelstein B. Waf1, a potential mediator of p53 tumor suppression. Cell, 1993. 75: 817-825.

Eliyahu D., Raz A., Gruss P., Givol D., Oren M., Levine A.J. Participation of p53 cellular tumour antigen in transformation of normal embryonic cells. Nature, 1984. 312: 646-649.

Eliyahu D., Goldfinger N., Pinhasi-Kimhi O., Shaulsky G., Skurnik Y., Arai N., Rotter V., Oren M. Meth A fibrosarcoma cells express two transforming mutant p53 species. Oncogene, 1988. 3:. 313-321.

Epstein C.B., Attiyeh E.F., Hobson D.A., Silver A.L., Broach J.R., Levine A.J. p53 mutations isolated in yeast based on loss of transcription factor activity: similarities and differences from p53 mutations detected in human tumors. Oncogene, 1998. 16: 2115-2122.

Fakharzadeh S.S., Trusko S.P., George D.L. Tumorigenic potential associated with enhanced expression of a gene that is amplified in a mouse tumor cell line. EMBO J., 1991. 10: 1565-1569.

Fields S., Jang S.K. Presence of a potent transcription activating sequence in the p53 protein. Science, 1990. 249: 1046-1049.

Finlay C.A., Hinds P.W., Tan T.H., Eliyahu D., Oren M., Levine A.J. Activating mutations for transformation by p53 produce a gene product that forms an hsc70-p53 complex with an altered half-life. Molecular and Cellular Biology, 1988. 8: 531-539.

Finlay C.A., Hinds P.W., Levine A.J. The p53 proto-oncogene can act as a suppressor of transformation. Cell, 1989. 57: 1083-1093.

Fiscella M., Zhang H., Fan S., Sakaguchi K., Shen S., Mercer W.E., Vande Woude G.F., O'Conner P.M., Appella E. Wip1, a novel human protein phosphatase that is induced in response to ionizing radiation in a p53-dependent manner. Proceedings of the National Academy of Sciences USA, 1997. 94: 6048-6053.

Frebourg T., Friend S.H. Cancer risks from germline p53 mutations. Journal of Clinical Investigation, 1992. 90: 1637-1641.

Friend S.H., Bernards R., Rogelj S., Weinberg R.A., Rapaport J.M., Albert D.M., Dryja T.P, A human DNA segment with properties of the gene that predisposes to retinoblastoma and osteosarcoma. Nature, 1986. 323: 643-646.

Funk W.D., Pak D.T., Karas R.H., Wright W.E., Shay J.W. A transcriptionally active DNA-binding site for human p53 protein complexes. Molecular and Cellular Biology, 1992. 12: 2866-2871.

Gatei M., Sloper K., Sorensen C., Syljuasen R., Falck J., Hobson K., Savage K., Zhou B.B., Bartek J., Khanna K.K. Ataxia-telangiectasia-mutated (ATM) and NBS1-dependent phosphorylation of Chk1 on Ser-317 in response to ionizing radiation. Journal of Biological Chemistry, 2003. 278: 14806-14811.

Harper J.W., Elledge S.J., Keyomarsi K., Dynlacht B., Tsai L.H., Zhang P., Dobrowolski S., Bai C., Connell-Crowley L., Swindell E. et al., Inhibition of cyclin-dependent kinases by p21. Mol Cel Biol, 1995. 6:. 387-400.

Harvey D., Levine A.J. p53 alteration is a common event in the spontaneous immortalization of primary Balb/c murine embryo fibroblasts. Genes and Development, 1991. 5: 2375-2385.

Hinds P., Finlay C., Levine A.J. Mutation is required to activate the p53 gene for cooperation with the ras oncogene and transformation. Journal of Virology, 1989. 63: 739-746.

Hinds P.W., Finlay C., Quartin R.S., Baker S.J., Fearon E.R., Vogelstein B., Levine A.J. Mutant p53 DNA clones from human colon carcinomas cooperate with ras in transforming primary rat cells: a comparison of the "hot spot" mutant phenotypes. Cell Growth and Differentiation, 1990. 1: 571-580.

Hoh J., Jin S., Parrado T., Edington J., Levine A.J., Ott J. The p53MH algorithm and its application in detecting p53-responsive genes. Proceedings of the National Academy of Sciences USA, 2002. 99: 8467-8472.

Hollstein M., Sidransky D., Vogelstein B., Harris C.C. p53 mutations in human cancers. Science, 1991. 253: 49-53.

Honda R., Tanaka H., Yasuda H. Oncoprotein MDM2 is a ubiquitin ligase E3 for tumor suppressor p53. FEBS Lett., 1997. 420: 25-27.

Huang T.S., Kuo M.L., Shew J.Y., Chou Y.W., Yang W.K., Distinct p53-mediated G1/S checkpoint responses in two NIH3T3 subclone cells following treatment with DNA-damaging agents. Oncogene, 1996. 13: 625-632.

Jacks T., Remington L., Williams B.O., Schmitt E.M., Halachmi S., Bronson R.T., Weinberg R.A. Tumor spectrum analysis in p53- deficient mice. Current Biology, 1994. 4: 1-7.

Jen K.Y., Cheung V.G. Transcriptional response of lymphoblastoid cells to ionizing radiation. Genome Research, 2003. 13: 2092-2100.

Jenkins J.R., Rudge K., Currie G.A. Cellular immortalization by a cDNA clone encoding the transformation-associated phosphoprotein p53. Nature, 1984. 312: 651-654.

Jensen M.R., Factor V.M., Fantozzi A., Helin K., Huh C.G. Thorgeirsson S. Reduced hepatic tumor incidence in cyclin G1-deficient mice. Hepatology, 2003. 37: 862-870.

Jonasson J., Povey S., Harris H. The analysis of malignancy by cell fusion: VII. Cytogenetic analysis of hybrids between malignant and diploid cells and of tumors derived from them. Journal of Cell Science, 1977. 24: 217-254.

Kao C.C., Yew P.R., Berk A.J. Domains required for in vitro association between the cellular p53 and the adenovirus 2 E1B 55K proteins. Virology, 1990. 179: 806-814.

Kastan M.B., Onyekwere O., Sidransky D., Vogelstein B., Craig R.W. Participation of p53 protein in the cellular response to DNA damage. Cancer Research, 1991. 51: 6304-6311.

Kawai S., Hanafusa H. The effects of reciprocal changes in temperature on the transformed state of cells infected with a Rous sarcoma virus mutant. Virology, 1971. 46: 470-479.

Kern S.E., Kinzler K.W., Baker S.J., Nigro J.M., Rotter V., Levine A.J. Friedman P., Prives C., Vogelstein B. Mutant p53 proteins bind DNA abnormally in vitro. Oncogene, 1991. 6: 131-136.

Kern S.E., Kinzler K.W., Bruskin A., Jarosz D., Friedman P., Prives C., Vogelstein B. Identification of p53 as a sequence-specific DNA-binding protein. Science, 1991. 252: 1708-1711.

Knudson A.G. Mutation and cancer: statistical study of retinoblastoma. Proceedings of the National Academy of Sciences USA, 1971. 68: 820-823.

Knudson A.G., Strong L.C. Mutation and cancer: A model for Wilm's tumor of the kidney. Journal of the National Cancer Institute, 1972. 48: 313-324.

Kraiss S., Quaiser A., Oren M., Montenarh M. Oligomerization of oncoprotein p53. Journal of Virology, 1988. 62: 4737-4744.

Lahav G., Rosenfeld N., Sigal A., Geva-Zatorsky N., Levine A.J., Elowitz M.B., Alon U. Dynamics of the p53-Mdm2 feedback loop in individual cells. Nature Genetics, 2004. 36: 147-150.

Land D.P., Parada L.F., Weinberg R.A. Cellular oncogenes and multistep carcinogenesis. Science, 1983. 222: 771-778.

Lane D.P., Crawford L.V. T antigen is bound to a host protein in SV40-transformed cells. Nature, 1979. 278: 261-263.

Lin J., Chen J., Elenbaas B., Levine A.J, Several hydrophobic amino acids in the p53 amino-terminal domain are required for transcriptional activation, binding to mdm-2 and the adenovirus E1B 55 kd protein. Genes and Development, 1994. 8: 1235-1246.

Lin J., Teresky A.K., Levine A.J. Two critical hydrophobic amino acids in the N-terminal domain of the p53 protein are required for the gain of function phenotypes of human p53 mutants. Oncogene, 1995. 10: 2387-2390.

Linzer D.I., Levine A.J. Characterization of a 54K dalton cellular SV40 tumor antigen present in SV40-transformed cells and uninfected embryonaly carcinoma cells. Cell, 1979. 17: 43-52.

Linzer D.I., Maltzman W., Levine A.J. The SV40 A gene product is required for the production of a 54,000 MW cellular tumor antigen. Virology, 1979. 98: 308-318.

Lohurm M.A., Ludwig R.., Kubbutat M.H., Hanlon M., Vousden K.H. Regulation of HDM2 activity by the ribosomal protein L11. Cancer Cell, 2003. 3:. 577-587.

Lutzker S.G., Levine A.J. A functionally inactive p53 protein in teratocarcinoma cells is activated by either DNA damage or cellular differentiation. Nature Medicine, 1996. 2:. 804-810.

Maki C.G., Howley P.M. Ubiquitination of p53 and p21 is differentially affected by ionizing and UV irradiation. Molecular and Cellular Biology, 1997. 17: 355-363.

Malkin D., Li F.P., Strong L.C., Fraumeni J.F. Jr., Nelson C.E., Kim D.H., Kassel J., Gryka M., Bischoff F.Z. et al., Germ line p53 mutations in a familial syndrome of breast cancer, sarcoma and other neoplasms. Science, 1990. 250: 1233-1238.

Maltzman W., Czyzyk L. UV irradiation stimulates levels of p53 cellular tumor antigen in nontransformed mouse cells. Molecular and Cellular Biology, 1984. 4: p. 1689-1694.

Mayo L.D., Donner D.B. A phosphatidylinositol 3-kinase/Akt pathway promotes translocation of Mdm2 from the cytoplasm to the nucleus. Proceedings of the National Academy of Sciences USA, 2001. 98: 11598-11603.

McKinnell R.G., Parchment R., Perantoni A.O., Pierce G.B. The biological basis of cancer. 1998; University of Cambridge Press.

Miller R. Gerontology as oncology. Research on aging as the key to the understanding of cancer. Cancer, 1991. 68: 2496-2501.

Momand J., Zambetti G.P., Olson D.C., George D., Levine A.J., The mdm-2 oncogene product forms a complex with the p53 protein and inhibits p53-mediated transactivation. Cell, 1992. 69: 1237-1245.

Munger K., Werness B.A., Dyson N., Phelps W.C., Howley P.M. Complex formation of human papillomavirus E7 proteins with the retinoblastoma tumpr suppressor gene product. Eur. Mol. Biol. Organ. J., 1989. 8: 4099-4105.

Munroe D.G., Rovinsky B., Bernstein A., Benchimol S. Loss of a highly conserved domain of p53 as a result of gene deletion during Friend virus-induced erythroleukemia. Oncogene, 1988. 2: 621-624.

Nakamura Y. Isolation of p53-target genes and their functional analysis. Cancer Sci., 2004. 95:. 7-11.

Nigro J.M., Baker S.J., Preisinger A.C., Jessup J.M., Hostetter R. Cleary K., Bigner S.H., Davidson N., Baylin S., Devilee P., Glover T., Collins F.S., Weston A., Modali R., Harris C.C., Vogelstein B.. Mutations in the p53 gene occur in diverse human tumor types. Nature, 1989. 342: 705-708.

Okamoto K., Hongyun L., Jensen M.R., Zhang T., Taya Y., Thorgeirsson S.S., Prives C. Cyclin G recruits PP2A to dephosphorylate Mdm2. Mol Cell, 2002. 9: 761-771.

Oliner J.D., Kinzler K.W., Meltzer P.S., George D.L., Vogelstein B. Amplification of a gene encoding a p53-associated protein in human sarcomas. Nature, 1992. 358: 15-16.

Oren M., Levine A.J. Molecular cloning of a cDNA specific for the murine p53 cellular tumor antigen. Proceedings of the National Academy of Sciences USA, 1983. 80: 56-59.

Parada L.F., Land F., Weinberg R.A., Wolf D., Rotter V. Cooperation between gene encoding p53 tumor antigen and ras in cellular transformation. Nature, 1984. 312: 649-651.

Pennica D., Goeddel D.V., Hayflick J.S., Reich N.C., Anderson C.W., Levine A.J., The amino acid sequence of murine p53 determined from a c-DNA clone. Virology, 1984. 134: 477-482.

Pisksley S.M., Lane D.P. The p53-mdm2 autoregulatory feedback loop: a paradigm for the regulation of gr.owth control by p53? Bioessays, 1993. 15:. 689-690.

Price B.D., Calderwood S.K. Increased sequence-specific p53-DNA binding activity after DNA damage is attenuated by phorbol esters. Oncogene, 1993. 8:. 3055-3062.

Raycroft L., Wu H.Y., Lozano G. Transcriptional activation by wild-type but not transforming mutants of the p53 anti-oncogene. Science, 1990. 249: 1049-1051.

Robinson M., Jiang P., Cui J., Wang Y., Swaroop M., Madore S., Lawrence T.S., Sun Y. Global genechip profiling to identify genes responsive to p53-induced growth arrest and apoptosis in human lung carcinoma cells. Cancer Biol. Ther., 2003. 2: 406-415.

Rozenfeld-Granot G., Krishnamurthy J., Kannan K., Toren A., Amarigliti N., Givol D., Rechavi G. A positive feedback mechanism in the transcriptional activation of Apaf-1 by p53 and the co-activator Zac-1. Oncogene, 2002. 21: 1469-1476.

Sarnow P., Ho Y.S., Williams J., Levine A.J. Adenovirus E1b-58kd tumor antigen and SV40 large tumor antigen are physically associtaed with the same 54kd cellular protein in transformed cells. Cell, 1982. 28: 387-394.

Sheikh M.S., Burns T.F., Huang Y., Wu G.S., Amundson S., Brooks K.S., Fornace A.J. Jr, el-Deiry W.S. p53-dependent and -independent regulation of the death receptor KILLER/DR5 gene expression in response to genotoxic stress and tumor necrosis factor alpha. Cancer Research, 1998. 58: 1593-1598.

Sinclair C.S., Rowley M., Naderi S., Couch F.J. The 17q23 amplicon and breast cancer. Breast Cancer Res. Treat., 2003. 78: 313-322.

Stehelin D., Varmus H.E., Bishop J.M., Vogt P.K. DNA related to the transforming gene(s) of avian sarcoma virus is present in normal avian DNA. Nature, 1976. 260: 170-173.

Steinmeyer K., Deppert W. DNA binding properties of murine p53. Oncogene, 1988. 3: 501-507.

Takekawa M., Adachi M., Nakahata A., Nakayama I., Itoh F., Tsukauda H., Taya Y., Imai K. p53-inducible wip1 phosphatase mediates a negative feedback regulation of p38 MAPK-p53 signaling in response to UV radiation. EMBO J, 2000. 19: 6517-6526.

Taubert H., Schuster K., Brinck U., Bartel F., Kappler M., Lautenschalger C., Bache M., Trump C., Schmidt H., Holzhausen H.J., Wurl P., Schlott T. Loss of heterozygosity at 12q14-15 often occurs in stage I soft tissue sarcomas and is associated with MDM2 amplification in tumors at various stages. Mod. Pathol., 2003. 16: 1109-1116.

Taylor W.R., Stark G.R. Regulation of the G2/M transition by p53. Oncogene, 2001. 20: 1803-1815.

Unger T., Juven-Gershon T., Moallem E., Berger M., Voght Sionov R., Lozano G., Oren M., Haupt Y. Critical role for Ser20 of human p53 in the negative regulation of p53 by MDM2. EMBO, 1999. 18: 1805-1814.

Unsal-Kacmaz K., Makhov A.M., Griffith J.D., Sancar A. Preferential binding of ATR protein to UV-damaged DNA. Proceedings of the National Academy of Sciences USA, 2002. 99: 6673-6678.

Werness B.A., Levine A.J., Howley P.M. Association of human papillomavirus types 16 and 18 E6 proteins with p53. Science, 1990. 248: 76-79.

Whyte P., Williamson N.M., Harlow E. Cellular targets for transformation by the adenovirus E1A proteins. Cell, 1989. 56: 67-75.

Wu X.W., Bayle H.J., Olson D., Levine A.J. The p53-mdm-2 autoregulatory feedback loop. Genes and Development, 1993. 7: 1126-1132.

Xiong Y., Hannon G.J., Zhang H., Casso D., Kobayashi R., Beach D. p21 is a universal inhibitor of cyclin kinases. Nature, 1993. 366: 701-704.

Yamagawa K, Ichikawa K., Experimental study of the pathogenesis of carcinoma. Journal of Cancer Research, 1918. 3: 1-29.

Yang Z., Ro S., Rannala B, Likelihood models of somatic mutation and codon substitution in cancer genes. Genetics, 2003. 165: 695-705.

Zauberman A., Barak Y., Ragimov N., Levy N., Oren M. Sequence-specific DNA binding by p53: identification of target sites and lack of binding to p53-MDM2 complexes. EMBO J, 1993. 12: 2799-2808.

Zauberman A., Flusberg D., Haupt Y., Barak Y., Oren M. A functional p53-responsive intronic promoter is contained within the human mdm2 gene. Nucleic Acids Res., 1995. 23:. 2584-2592.

Zhang Y., Wolf G.W., Bhat K., Jin A., Allio T., Burkhart W.A., Xiong Y. Ribosomal protein L11 negatively regulates oncoprotein MDM2 and mediates a p53-dependent ribosomal-stress checkpoint. Molecular and Cellular Biology, 2003. 23: 8902-8912.

Zhao R., Gish K., Murphy M., Yin Y., Notterman D., Hoffman W.H. Tom Mack DH, Levine AJ., Analysis of p53-regulated gene expression patterns using oligonucleotide arrays. Genes and Development, 2000. 14: 981-993.

Zhao H., Watkins J.L., Piwnica-Worms H. Disruption of the checkpoint kinase 1/cell division cycle 25A pathway abrogates ionizing radiation-induced S and G2 checkpoints. Proceedings of the National Academy of Sciences USA, 2002. 99: 14795-14800.

Chapter 2

REGULATION OF P53 DNA BINDING

Kristine McKinney and Carol Prives
Columbia University, Department of Biological Sciences, New York, NY, USA

p53 is one of the most frequently mutated genes in human cancers and, as a result, is also one of the most well-studied genes in the history of cancer research. Although many functions have been ascribed to p53 over the years, one of the first activities to be characterized was the ability to bind DNA sequence-specifically through its central domain (reviewed in Vogelstein & Kinzler, 1992). This domain, also frequently referred to as "the core" due to its protease resistance (Bargonetti et al., 1993; Pavletich et al., 1993), contains the most evolutionarily conserved sequences of the protein, both between p53 proteins from different species and between the different p53 family members, p63 and p73 (reviewed in Yang et al., 2002). This region is also the most frequently mutated domain of p53 in the major forms of human cancer (Hainaut & Hollstein, 2000; Olivier et al., 2002). Consequently much research has focused on understanding this crucial ability as well as its regulation. Indeed, the regulation of p53 DNA binding has generated much debate recently, specifically with regard to the role of the C-terminus.

In addition to a sequence-specific DNA binding domain, p53 also contains a transactivation domain in its N-terminus and can therefore bind and transactivate targets *in vivo* and *in vitro* (reviewed in Vogelstein & Kinzler, 1992). Other well-characterized functional domains are diagrammed in Figure 1. This chapter will focus on what is known about the involvement and regulation of the individual domains of p53 in sequence-specific DNA binding.

P. Hainaut and K.G. Wiman (eds.), 25 Years of p53 Research, 27-51.
© 2005 *Springer. Printed in the Netherlands.*

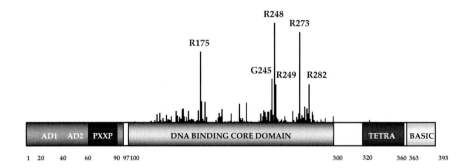

Figure 1. Schematic diagram of p53 functional domains highlighting the clustering of tumor-derived mutations in the sequence-specific DNA binding domain (generated in IARC database October 2003 http://www.iarc.fr/p53). Indicated are the N-terminal activation domains (AD1, AD2), the proline rich region (PXXP), the central core sequence specific DNA binding domain (DNA BINDING CORE), the tetramerization domain (TETRA) and the C-terminal sequence non-specific nucleic acid binding region (BASIC).

TRANSACTIVATION DOMAIN

The first 100 amino acids of p53 contain two transactivation domains, a proline-rich domain and a nuclear export signal. The transactivation domains have been mapped between amino acids 20-40 (Chang et al., 1995; Lin et al., 1994; Unger et al., 1993) and 40-60 (Candau et al., 1997; Venot et al., 1999; Zhu et al., 1998) and are both required for full transactivation ability. Missense mutations in these two domains, however, can have different effects in cells (Chen et al., 1996; Walker & Levine, 1996; Zhu et al., 1999; Zhu et al., 2000). The p53 N-terminus also contains a proline-rich domain (residues 60-90) which contains five copies of the motif "PXXP." This domain has been implicated in regulation of p53-mediated apoptosis (Baptiste et al., 2002; Sakamuro et al., 1997; Venot et al., 1998; Walker & Levine, 1996; Zhu et al., 1999). It is also interesting to note that the proline-rich region is reportedly required for interaction with the nuclear matrix in a DNA-damage dependent manner (Jiang et al., 2001; Okorokov et al., 2002) although the physiological relevance of this association has not yet been adequately addressed. The list of verified and potential p53 target genes is increasing rapidly, as is our understanding of when and in which cell types particular genes are important for the physiological outcomes of p53 and its family members' (reviewed in Harms et al., 2004).

Although its transactivation domains are pivotal to p53 function in cells, there are only a few reports of their direct effects on DNA binding (reviewed in Jayaraman & Prives, 1999). For example, it was demonstrated that antibodies directed at N-terminal epitopes can protect p53 from thermal denaturation and stabilize p53 DNA binding (Friedlander et al., 1996b; Hansen et al., 1996). Therefore it is possible that one way to increase p53 function in cells would be through protein-protein interactions with this intrinsically destabilizing domain. It is interesting to note that interaction with the N-terminal antibody 1801 can stabilize DNA binding of some tumor-derived mutants of p53 even at elevated temperatures (Friedlander et al., 1996b). A potential explanation for this effect may be found in the report that multiple monoclonal antibodies directed at N-terminal epitopes result in a marked reduction in dissociation rate of p53 from DNA *in vitro* (Cain et al., 2000). This is consistent with the observation that the TATA-binding protein (TBP) subunit of the basal transcription factor TFIID, as well as TFIID itself, interacts with the N- and C-termini of p53 and is capable of stimulating p53 recognition of target sequences both in the absence and presence of a TATA box (Chen et al., 1996). Similarly, the adenovirus E1B 55K protein interacts with the N-terminus of p53 and stabilizes its interaction with DNA (Yew et al., 1994).

Residues 80-93 in the poly-proline region have also been reported to comprise a negative autoregulatory domain since deletion of this domain results in an increase in p53 DNA binding (Muller-Tiemann et al., 1998). Furthermore, in the same study a peptide spanning this domain was shown to increase p53 DNA binding when titrated in *trans* in a manner that required the presence of residues 80-93 and 363-393 (another autoinhibitory region that will be discussed below).

The N-terminus of p53 is highly post-translationally modified (reviewed in Appella & Anderson, 2001). Phosphorylation events in this region have been implicated in both the stability of p53 as well as the specificity of target transactivation (Chao et al., 2003; Oda et al., 2000), although only phosphorylation of Ser46 has been shown to affect target specificity at the level of DNA binding. Oda et al. report *in vitro* evidence using nuclear extracts from transiently transfected cells that an Ser46Ala mutation decreases p53 binding to the p53-regulated Apoptosis-Inducing Protein 1 (p53AIP1) target sequence (and not to a p21 binding site). Further work including a DNA binding assay using purified recombinant p53 Ser46Ala to understand if other interacting proteins are required for this effect as well as a chromatin immunoprecipitation experiment to confirm that altered DNA binding specificity is the result of serine 46 phosphorylation *in vivo* would be interesting. Another possible example comes from the Ser18Ala mutant (homologous to Ser15 in human p53) knock-in mouse which shows reduced transactivation of some target genes and an abrogated ability of p53 to induce cell cycle arrest and apoptosis, although there is no data addressing the effect on the DNA binding ability of p53 *per se* (Chao et al., 2003).

SEQUENCE-SPECIFIC DNA BINDING DOMAIN

Sequencing of the p53 locus from over 16,000 human tumors provides striking evidence for the importance of the sequence-specific DNA binding domain for intact tumor suppressor function (Hainaut & Hollstein, 2000; Olivier et al., 2002) (Figure 1). 97% of the mapped mutations cluster in the sequence-specific DNA binding domain (Hainaut & Hollstein, 2000; Olivier et al., 2002) and render the protein inactive for target sequence binding (reviewed in Bullock & Fersht, 2001; Ko & Prives, 1996). This region lies in the center of the protein (approximately amino acids 100-300) and also has the zinc-binding activity required for proper protein folding. Consensus sequences for p53 binding contain two copies of the inverted pentameric sequence $PuPuPuC^A/_T{}^T/_AGPyPyPy$ separated by 0 to 13 base pairs with the 4th C and 7th G being the least variant. Sequences generally conforming to this consensus have been found in many genes that are induced by p53 and have been shown to bind to p53 *in vitro*, to serve as p53 response elements in both reporter transactivation assays *in vivo* and transcription assays *in vitro*. In many cases p53 has also been shown to bind to these sites in chromatin immunoprecipitation (ChIP) assays. Reports using computational genomics and microarray studies suggest that there are 300 - 1600 potential binding sites for p53 in the human genome (Cawley et al., 2004; Hoh et al., 2002). Recently a p53 response element in a microsatellite sequence of the *PIG3* promoter which contains 10-17 tandem repeats of the pentanucleotide sequence $(TGPyCC)_n$ (Contente et al., 2002) was identified. This microsatellite sequence has substantial homology to the classic p53 consensus sequence but differs in the number of repeats and is also the first p53 response element shown to be polymorphic.

p53 is also capable of transrepression of some genes although the mechanisms are less well-understood (reviewed in Oren, 2003). While in some cases repression is caused by target gene products that are induced by p53 such as p21 (Gottifredi et al., 2001; Lohr et al., 2003), other mechanisms involving direct repression have also been described. One study described a large co-repressor complex that could be recruited by p53 via its proline-rich domain (Zilfou et al., 2001). However some *cis*-acting DNA sequences required for p53-transrepression have also been identified. Some of these sequences conform to the consensus p53 binding site while others are essentially permutations of the consensus. The first class of repression target sequences found in the *Bcl-2* and *α-fetoprotein* genes contains p53 binding sites which overlap with the binding site of another more potent activator (Budhram-Mahadeo et al., 1999; Lee K. C. et al., 1999). In these cases, p53 actually transactivates when bound to the promoter, but since its activity is weaker than the transcription factor it has displaced, the target mRNA levels decrease. In a related mechanism, there is a p53 binding site adjacent to an enhancer in the *HBV* gene. p53 represses transcription of HBV in a manner which requires the presence of the enhancer sequence (Ori et al., 1998). When the enhancer is deleted, p53 binds and activates transcription of HBV. Similarly, the *survivin* gene contains a p53 binding site which conforms to the consensus, i.e. two half-sites separated by 3 base pairs and is near an E2F binding site which is required for p53-mediated repression (Hoffman et al., 2002). In contrast, the sequences in the *MDR1* (*multi-drug resistance 1*) gene required for p53-mediated transrepression do not conform to the p53 consensus. Rather, they contain four direct repeats of the canonical quarter site (rather than two copies of two inverted repeats) (Johnson et al., 2001). If the MDR1 promoter sequences are mutated into inverted repeats as in the p53 consensus, p53 becomes an activator of the promoter instead of a repressor. Similar sites were found by sequence analysis in other promoters repressed by p53 and it will be interesting to find out how general this putative p53 repression sequence is. Since p53 binds its consensus sequences in a manner similar to its three-dimensional organization, in other words as a "dimer of dimers," it will be interesting to understand if p53's interaction with a site containing straight repeats is fundamentally different (McLure & Lee, 1998).

Two interesting aspects of the regulation of the core domain are worth mentioning. The first is the role of metal ions and redox (Hainaut & Mann, 2001). Core cysteine residues Cys176 in the L2 loop and Cys238 and Cys242 in the L3 loop along with His179 in Helix 1 coordinate a single zinc molecule in the core. Not surprisingly mutation of the corresponding cysteines in mouse p53 ablates DNA binding (Rainwater et al., 1995) while use of zinc but not other divalent metals at physiological concentrations is required for maintaining the wild-type conformation of p53 (Meplan et al., 2000). This is further supported by observations that divalent metals such as copper (Hainaut et al., 1995; Verhaegh et al., 1997) and cadmium (Meplan et al., 2000) inhibit p53 DNA binding *in vitro* and *in vivo*. The importance of core cysteine residues is also supported by results showing that redox factors Ref1/APE (Gaiddon et al., 1999; Jayaraman et al., 1997; Seo et al., 2002; Ueno et al., 1999) and thioredoxin (Ueno et al., 1999) can affect p53 DNA binding and transactivation activities. Indeed Cys277 is involved in the selenium-requiring Ref1 activation of p53 DNA binding (Seo et al., 2002) and has been shown to play an important role in differential regulation of p53 binding to its target sites (Buzek et al., 2002). It merits pointing out that while redox modulation is key in DNA binding by the core domain, it apparently plays no role in the DNA interactions of the C-terminal sequence non-specific DNA binding domain that is discussed below (Fojta et al., 1999; Parks et al., 1997).

The second feature of this region to be noted is the role in DNA binding of the L1 loop (residues 112-125) which is the sole "cold spot" in the core of human p53 tumor derived mutation. Studies using yeast genetics to identify core residues with altered or increased transactivation ability have yielded an inordinate number of L1 loop mutants (Resnick & Inga, 2003 and references therein). The most well studied of these, Ser121Phe, has been shown to be an unusual mutant in that, in contrast to wild-type p53, it binds well to a single 10-mer of the canonical 20-mer consensus sequence (Saller et al., 1999). Interestingly, Ser121Phe displays greater pro-apoptotic activity than does wild-type p53 along with defects in induction of some p53 targets such as p21 and Mdm2. This is in contrast to other core domain mutants that have been identified that are selectively defective in activating and binding to sites in some pro-apoptotic target genes (Friedlander et al., 1996a; Ludwig et al., 1996). Identification of core domain mutants that preferentially impair arrest or apoptosis would be invaluable for determining the roles of these two outcomes in tumorigenesis in mouse models.

Studies examining the wild-type functions of the core have provided insights into possible modes of regulation of DNA binding. In addition, multiple protein-protein interactions have been mapped to this domain and may directly or indirectly impact the ability to interact sequence-specifically with DNA. On the other hand, studies that have focused on mutant proteins and the restoration of their sequence-specific DNA binding ability have also provided both useful insights to the molecular mechanisms of DNA binding and candidate drug therapies for cancer treatment.

Insights from studies of wild-type p53

Research focused on the DNA binding properties of the central domain have yielded interesting clues as to possible modes of regulation and the molecular mechanism of DNA binding. The isolated core domain is capable of binding target sites in a cooperative manner as a tetramer, even without the aid of the C-terminal tetramerization domain (reviewed in Ko & Prives, 1996) and there is some evidence that protein-protein interactions between core molecules partially mediate this phenomenon (Rippin et al., 2002). The core DNA binding domain can induce ~40-60 degree bends in its target sequences upon binding (Balagurumoorthy et al., 1995; Cherny et al., 1999) and actually prefers sites containing intrinsically flexible sequences (Nagaich et al., 1997a). Furthermore, Nagaich et al. have proposed that a bend in the DNA is actually required to accommodate all four molecules in the tetramer and avoid steric clashes (Nagaich et al., 1999; Nagaich et al., 1997b). Therefore DNA conformation is a potential way to regulate DNA binding through the core domain. Interestingly HMGB1, an abundant nuclear protein capable of bending DNA, has been shown to augment both DNA binding *in vitro* and *in vivo* (Jayaraman et al., 1998), although whether this ability requires the p53 core is not fully understood. Another interesting finding is that when bound to DNA containing more than one target sequence, core domains can interact with each other and bend any intervening DNA into loops (Jackson et al., 1998; Stenger et al., 1994; Zhou & Prives, 2003). Since many p53 target promoters have more than one p53 binding site, it is tempting to speculate that this activity could potentially aid in the formation of transcriptional pre-initiation complexes in the context of genomic DNA.

Protein-protein interactions

Most functions involving the core domain depend on its ability to interact with DNA sequence-specifically, however there are a few noteworthy exceptions. Some interesting protein-protein interactions have been localized to the core domain including the p53 homologs p63 (Yang et al., 1998) and p73 (Jost et al., 1997). Specifically, p73 isoforms have been shown to interact with mutant p53 in a core-dependent manner (Gaiddon et al., 2001; Strano et al., 2000) and tumor-derived mutants of p53 can thereby interfere with p63 and p73 transactivation (Di Como et al., 1999; Gaiddon et al., 2001; Strano et al., 2002). Furthermore in some tumors the expression patterns of various isoforms of p63 are altered in a complex manner that is not yet fully understood (Di Como et al., 2002; Urist et al., 2002) but may reflect a role in the process of oncogenesis. On a related note, Flores et al. demonstrated a requirement for the wild-type p63 and p73 gene families in the p53-dependent induction of target genes in E1A-immortalized mouse embryo fibroblasts (Flores et al., 2002). They were able to show that removal of the p63 and p73 genes resulted in a decreased occupancy of p53 at apoptotic promoters, although the precise function/mechanism of the p53 homologs in this context is still elusive. In addition two proteins termed ASPP1 and ASPP2 (for Apoptosis Stimulating Protein of p53) interact with the p53 core and are capable of directly enhancing the interaction of p53 with pro-apoptotic (but not pro-arrest) target sequences *in vivo* (Samuels-Lev et al., 2001).

The idea that the sequence-specific binding function of p53 is the only activity targeted in oncogenesis could be somewhat over-simplified. It was recently demonstrated that p53 can interact with Bcl-XL and Bcl-2 (Mihara et al., 2003), two anti-apoptotic proteins that reside in the outer mitochondrial membrane and protect its integrity (reviewed in Green & Evan, 2002). It was reported that interaction with these proteins depends on the p53 core domain and that tumor-derived mutants have lost the ability to interact with them but not the ability to localize to mitochondria. Furthermore p53 can relocalize to mitochondria in a stress-dependent manner in cells and induce release of cytochrome c from purified mitochondria *in vitro*. In this scenario, p53 would act in a manner analogous to the related pro-apoptotic BH3-only proteins by sequestering Bcl-XL and Bcl-2 (reviewed in Chipuk & Green, 2004). Furthermore, the p53-Bcl-XL interaction is even stronger than some BH3-only proteins such as Bid (Chipuk et al., 2004). These and other data (reviewed in Baptiste & Prives, 2004) imply that it is possible that the core domain is targeted for mutation during oncogenesis because it not only disrupts the DNA binding ability of p53 but also the ability to directly induce apoptosis through interaction with Bcl-XL and Bcl-2 proteins in the mitochondria. It should be noted that results by Dumont et al. urge determination of the polymorphic status of residue 72 (Dumont et al., 2003) and in general more experiments are required to clarify the regulation, cell-type specificity and physiological relevance of this localization (reviewed in Manfredi, 2003).

Mutant Rescue

Restoring the DNA binding function of p53 mutant proteins which are highly expressed in tumors has long been a prominent goal in the field (reviewed in Bullock & Fersht, 2001). The structure of a single core domain sequence specifically bound to DNA has provided a framework for understanding the precise defects of tumor-derived mutants (Cho et al., 1994). The six most common mutant forms of p53 found in cancer can be classified as either DNA-contact mutants (Arg248 and Arg273) or conformational mutants (Arg175, Gly245, Arg249 and Arg282), however nuclear magnetic resonance data argues that Arg248 mutation also causes significant long-range structural changes (Wong et al., 1999). Many if not all of these so-called "hot-spot" mutations result in less thermodynamically stable proteins. Although highly impaired for DNA binding, many tumor-derived mutant versions of p53 retain some residual, albeit temperature-sensitive, DNA binding capacity *in vitro* and *in vivo* (Di Como & Prives, 1998; Friedlander et al., 1996a; Friedlander et al., 1996b). However, at 37°C it has been shown that several mutants can interfere with wild-type p53 binding to its promoter sites when expressed together at similar levels in the same cells (Willis et al., 2004).

These and other data led to the design of large-scale screens for compounds that could increase p53 mutant sequence-specific DNA binding. Rastinejad et al. isolated a small molecule, termed "CP-31398," on the basis of its ability to restore recognition by the native-state specific antibody PAb1620 and lose recognition by the unfolded mutant conformation-specific antibody PAb240 *in vitro* (reviewed in Bullock & Fersht, 2001). CP-31398 was able not only activate p53-dependent transactivation in mutant-expressing cells but also to slow tumor growth in mice (Foster et al., 1999). Although these results are exciting, the mechanism by which the compound is able to activate p53 appears to involve reduced ubiquitination (Wang W. et al., 2003) rather than stabilization of the folded conformation. More recently another small molecule, named "PRIMA," was isolated by Wiman and Selivanova and colleagues (Bykov et al., 2002). PRIMA is able to restore DNA binding to p53 mutants *in vitro* and can induce apoptosis in cells in a manner that requires both mutant p53 and an intact transactivation domain. Perhaps the most promising result is that intravenous administration of PRIMA to mice harboring mutant p53 tumor xenografts showed tumor shrinkage with no apparent toxicity. There is recent evidence, however, showing that PRIMA-induced p53-dependent apoptosis does not require transcription or protein synthesis and can proceed in a manner that depends on Bax and its translocation (Chipuk et al., 2003).

Using a very different approach which still focused on stabilizing the folded form of mutant proteins, the Fersht group identified a small peptide derived from p53BP2, a protein which interacts with the core domain (Gorina & Pavletich, 1996; Iwabuchi et al., 1994). This peptide is able to preferentially bind the native fold of the core thereby shifting the equilibrium towards a folded state (Friedler et al., 2002). Importantly, this peptide is also displaced by DNA and should therefore allow p53 to bind and transactivate targets. However, it should be noted that because of irreversible denaturation/aggregation, such compounds can only rescue mutant conformation shortly after their biosynthesis (Friedler et al., 2003). Regardless, such "small-molecule chaperones" remain an intriguing class of potential drugs although their efficacy in cells and tumor models remains to be tested.

Yet another strategy aimed at increasing p53 thermostability was employed by Nikolova et al. who re-engineered p53 based on a compilation of evolutionary comparisons. Through the introduction of multiple amino acid substitutions, they were able to maintain DNA binding activity as well as increase the melting temperature of the resulting p53 by 5.6°C (Nikolova et al., 1998). Such mutants may be of use in the ongoing efforts to deliver wild-type p53 to tumors using adenoviral vectors (Wilson, 2002). Although data from phase I clinical trials indicate low toxicity, they also suggest that the more general problem of efficiency and potency of infection by adenoviral vectors is more pressing (Lang et al., 2003; Pagliaro et al., 2003).

Focusing on the DNA contact class of p53 mutations instead, Wieczorek et al. reasoned that secondary mutations which introduced novel interactions between amino acid side chains and the phosphate backbone might act as suppressor mutations. Therefore, they singly mutated several residues in p53 to arginine or lysine. One resulting mutant was in fact able to restore DNA binding ability as well as the ability of p53 to transactivate reporters and suppress colony formation (Wieczorek et al., 1996).

TETRAMERIZATION DOMAIN

The oligomerization domain is located between amino acids 320-360 (reviewed in Chene, 2001). This domain is responsible for the mostly tetrameric state of p53 in solution and along with the basic region comprises the minimal transforming domain. The structure of the tetramerization domain has been determined and forms a "dimer of dimers" which folds/assembles in the same manner. Furthermore p53 binds DNA with the same organization i.e. each dimer binds two contiguous quarter-sites of the target sequence. The second nuclear export signal (NES) also resides in the tetramerization domain (residues 340-351) and may require ubiquitination for efficient usage (Boyd et al., 2000; Geyer et al., 2000; Stommel et al., 1999). In addition, the dominant nuclear localization signal (NLSI, amino acids 316-325) overlaps the tetramerization domain (Shaulsky et al., 1990).

Interestingly, the tetrameric status of p53 requires a larger hydrophobic side chain at residue 341 than 344. An inversion of this relationship results in dimeric p53. The tetramerization domain has been the focus of research aimed at enabling exogenously added p53 to avoid forming inactive co-tetramers with inactive endogenous mutant forms of p53 that are often present at high levels in tumors. Indeed rationally designed mutants of p53 that can form homotetramers yet cannot co-tetramerize with wild-type p53 were shown to maintain intact tumor suppressor activity when the wild-type and mutant forms of p53 were coexpressed in cells (Waterman et al., 1996). In addition, there is some evidence that the tetramerization of p53 can be regulated. Sakaguchi et al. found that phosphorylation of serine 392 could increase the association constant for tetramer formation by 10-fold *in vitro.* Furthermore they showed that while phosphorylation of serine 315 alone had little effect, when combined with Ser392 phosphorylation it counteracted the increased tetramerization induced by Ser392.

BASIC DOMAIN

An extraordinary amount of research has focused on the last 30 amino acids of p53. This highly basic domain is located between amino acids 364-393 and is itself capable of interacting with DNA in a sequence nonspecific fashion (Foord et al., 1991; Wang Y. et al., 1993). Two minor nuclear localization signals (NLSs) reside in this region (Dang & Lee, 1989; Shaulsky et al., 1990) that also contains multiple ubiquitination sites (reviewed in Michael & Oren, 2003), the major site of sumoylation *in vitro* and *in vivo*: lysine 386 (Gostissa et al., 1999; Rodriguez et al., 1999) and many other stress-inducible modification sites including phosphorylation, acetylation and glycosylation (reviewed in Appella & Anderson, 2001).

The extreme C-terminus of p53 has long been a focus of experiments addressing the regulation of sequence-specific DNA binding (reviewed in Ahn & Prives, 2001). Although the interaction of the C-terminus with DNA is not sequence-dependent, there are many reports that it is capable of recognizing a variety of non-B form DNA structures such as single-stranded ends (Bakalkin et al., 1995; Bakalkin et al., 1994; Jayaraman & Prives, 1995), insertion/deletion mismatches (Lee S. et al., 1995), gamma irradiated DNA (Miyashita & Reed, 1995), three- and four-way junctions (Lee S. et al., 1997), stem-loops (Kim E. et al., 1997), recombination intermediates (Dudenhoffer et al., 1998), supercoiled DNA (Mazur et al., 1999; Palecek et al., 2004), gapped DNA (Zotchev et al., 2000), minicircles (McKinney & Prives, 2002) and hemicatenated DNA (Stros et al., 2004). It is interesting to note that in some cases secondary structure actually augments the ability of p53 to recognize its target sequence, for example when binding single-stranded overhangs (Zotchev et al., 2000), minicircular DNA (McKinney & Prives, 2002) supercoiled DNA (Mazur et al., 1999; Palecek et al., 2004) and hemicatenated DNA (Stros et al., 2004) while in other cases it inhibits that ability, for example when binding gapped DNA (Zotchev et al., 2000). In addition, it can recognize and dissociate DNA aggregates (Yakovleva et al., 2001) and has a strand reannealing ability (Bakalkin et al., 1994; Wu et al., 1995) as well as a strand transfer activity (Reed et al., 1995).

It was noted quite early that many different alterations of the C-terminus result in a striking increase in binding to small oligonucleotide probes *in vitro*. First of all, deletion of the last 30 amino acids (generating p53Δ30) results in an increase in p53 sequence-specific DNA binding (Hupp et al., 1992). Secondly, interaction with the monoclonal C-terminal antibody 421 as well as other molecules like *E. coli* dnaK (Halazonetis et al., 1993; Hupp et al., 1992), c-abl (Nie et al., 2000), 14-3-3 (Waterman et al., 1998) or short single strands of DNA (Jayaraman & Prives, 1995) also results in an increased ability to bind targets *in vitro*. Post-translational modifications including phosphorylation by protein kinase C (Takenaka et al., 1995), casein kinase II (Hupp et al., 1992) and acetylation by p300/CBP or pCAF (Gu & Roeder, 1997; Liu et al., 1999; Sakaguchi et al., 1998) can also augment p53's ability to bind DNA sequence-specifically. Even peptides spanning the last 30 amino acids can increase the ability of p53 to bind to DNA sequence-specifically *in vitro* (Hupp et al., 1995; Jayaraman & Prives, 1995; Muller-Tiemann et al., 1998). An early hypothesis proposed that the C-terminus maintains p53 in a conformationally inert "latent" state which could be converted allosterically to an activated state by treatments such as the abovementioned (Halazonetis & Kandil, 1993; Hupp & Lane, 1994). This model was intellectually appealing when extended into cells since many of the post translational modifications of the C-terminus are stress-inducible (Appella & Anderson, 2001).

Further support for the idea that the C-terminus is a negative regulator of core DNA binding came from the identification of cellular proteins that could serve as non-covalent activators of p53 DNA binding in a manner that requires the C-terminus. The multifunctional protein Ref-1/APE-1, which is a redox factor as well as an A/P endonuclease, is capable of increasing p53 DNA binding *in vitro* and transactivation *in vivo* (Jayaraman et al., 1997). Downregulation of endogenous Ref-1 in cells using antisense results in diminished p53 transactivation, indicating a requirement of Ref-1 for full p53 response in cells (Gaiddon et al., 1999). In addition, HMGB1 (HMG-1), an abundant nonhistone chromosomal protein, was identified on the basis of its ability to increase p53 DNA binding *in vitro* (Jayaraman et al., 1998; McKinney & Prives, 2002). Moreover, co-transfection experiments demonstrate that HMGB1 is also capable of increasing p53 transactivation *in vivo*.

Anderson et al. extended these observations in a careful study that dissected the effects of nonspecific DNA on p53 sequence-specific DNA binding in the presence and absence of classic C-terminal stimulators (Anderson et al., 1997). Their data led to a reinterpretation of the C-terminal allosteric hypothesis. Surprisingly, they found that the long, nonspecific DNA that was commonly used as competitor in EMSAs was an inhibitor of p53 DNA binding. Furthermore p53Δ30 was immune to this effect and inhibition of full-length p53 could be blocked by the addition of PAb 421. They therefore concluded that defining the C-terminus as an autoinhibitor of sequence-specific DNA binding was dependent on the inclusion of long nonspecific DNA in the *in vitro* assays used to measure DNA binding. However, since genomic DNA is also long, they speculated that the presence of genomic DNA in cells would similarly inhibit p53 DNA binding and therefore mechanisms must be in place to neutralize the C-terminus upon activation. In support of the Anderson et al. hypothesis, Hoffmann et al.

tested the ability of small, C-terminally derived peptides *in vitro* to bind to DNA sequence-nonspecifically with and without defined post-translational modifications. They found that phosphorylation at the PKC site (Ser378) or the CKII site (Ser392) resulted in a decrease in nonspecific DNA binding activity to small oligonucleotides and modification of both sites rendered the peptides almost completely incapable of interaction with DNA (Hoffmann et al., 1998). Although this proposed mechanism of C-terminal autoinhibition was tested using long, non-specific DNA in *trans* (which is not always the case in cells), these results clearly showed that there was something different about the interaction of the C-terminus with long DNA and was important in paving the way for thinking about this key domain in new ways. Perhaps the most compelling argument against conformational C-terminal-dependent changes was a report by Ayed et al. that there is no difference in the NMR-derived structures of full-length ("latent" form) and p53Δ30 ("active" form) (Ayed et al., 2001).

Data questioning the *in vivo* relevance of the latency hypothesis have since emerged. First, cells inducing a C-terminally deleted form of p53 at physiological levels were reported to have defects in the induction of p21 and apoptosis (Chen et al., 1996). Second, the increase in binding of p53 to target sites in chromatin before and after DNA damage is directly proportional to p53 protein levels (Kaeser & Iggo, 2002). Experiments performed *in vitro* using purified p53 protein have also provided differing conclusions. Espinosa and Emerson demonstrated that p53Δ30 is less able to bind long and/or chromatinized DNA than is full-length p53 (Espinosa & Emerson, 2001). When binding sites are in the context of structured DNA, the C-terminus is actually required for efficient target recognition (Gohler et al., 2002; Mazur et al., 1999; McKinney & Prives, 2002; Palecek et al., 2004). Finally, it has been reported that in the case of some binding sites addition of PAb 421 inhibits rather than stimulates p53 binding to some targets *in vitro* (Resnick-Silverman et al., 1998; Thornborrow & Manfredi, 1999). These data suggest that there may be a positive, promoter specific requirement for the C-terminus, for example the 3' site in the *p21* promoter (which contains two p53 recognition motifs) as well as the binding sites in the *cdc25* and *BAX* promoters.

ChIP experiments to address the role of post-translational modifications in p53 DNA binding regulation in cells, particularly acetylation of the C-terminus, have produced some contradictory results. It is clear that p53 can be acetylated *in vivo* in a stress-dependent manner (Liu et al., 1999; Luo et al., 2004) and that p300/CBP and pCAF can augment p53 transactivation (Avantaggiati et al., 1997; Gu et al., 1997; Scolnick et al., 1997). Furthermore, deacetylases including both HDAC1 (Juan et al., 2000; Luo et al., 2000) and Sir2α (Langley et al., 2002; Luo et al., 2001; Vaziri et al.,

2001) have been demonstrated to deacetylate p53 at lysine 382. Importantly, lower levels of p53 acetylation as a result of increased HDAC1 or Sir2α activity correlate with decreased apoptosis (Luo et al., 2001; Luo et al., 2000). The above data are consistent with physiological regulation of p53 function by acetylase/deacetylase activity but do not address the mechanism by which they act.

Data using structured DNA target sites *in vitro* and chromatin immunoprecipitation (ChIP) experiments *in vivo* to address the mechanistic contribution of acetylation argue that it does not augment p53 DNA binding to its target sites in the context of chromatin (Barlev et al., 2001; Espinosa & Emerson, 2001; Hsu et al., 2004) in contrast to what was previously deduced from *in vitro* studies using small oligonucleotide binding sites (Gu & Roeder, 1997; Sakaguchi et al., 1998). It is interesting to note that one study found that acetylated p53 was enriched at promoters (Luo et al., 2004), although it is not clear whether acetylation of p53 itself is required for increased promoter occupancy or simply the result of the concomitant recruitment and increase in local HAT activity that accompanies transcriptional activation (Hsu et al., 2004). In fact it has been reported that p53 DNA binding can augment its ability to be acetylated by p300 (Dornan et al., 2003). *In vitro* transcription assays using chromatinized templates and some *in vivo* data suggest that acetylation of p53 itself is not required for the coactivation by p300 but histone acetyltransferase activity is required, especially on the tails of histones H3 and H4 (An et al., 2004; Espinosa & Emerson, 2001; Hsu et al., 2004). Somewhat contradictorily, Barlev and colleagues showed that transactivation and apoptosis induction by a hypoacetylated p53 mutant (K320/373/381/382R) transiently overexpressed in cells is compromised relative to wild-type p53 (Barlev et al., 2001). While they do not observe a difference in p21 promoter occupancy of the p53 mutant, they do observe a defect in the recruitment of the CBP histone acetyltransferase (as well as other coactivators such as TRAAP) to the promoter. Although CBP and p300 are highly homologous proteins, in striking contrast with p300, the interaction of CBP and p53 has been shown to depend on the CBP bromodomain and is strongly enhanced by acetylation of p53 at K382 (Mujtaba et al., 2004). Therefore it appears that the reported differences in the precise role of p53 acetylation in the recruitment of coactivators could be the result of surprisingly different regulation of two closely related proteins. Since the C-terminal lysines could be sites of post-translational modifications other than acetylation and could therefore have multiple effects when mutated *in vivo*, further experiments would be useful in resolving this question.

Although recent data strongly urge reconsideration of the idea that the C-terminus is a negative regulator of p53 DNA binding, there are some data in the literature which remain difficult to reconcile. Notably microinjection of antibody 421 into cells results in activation of a p53-specific reporter (Abarzua et al., 1995; Hupp et al., 1995). In addition, there are reports that peptides derived from the C-terminus are capable of activating p53 (wild-type and several tumor-derived mutants) transcription from reporters and apoptosis using co-overexpression (Abarzua et al., 1996; Selivanova et al., 1997). Because the C-terminus of p53 is important for efficient degradation (Kubbutat et al., 1998; Rodriguez et al., 2000) and the site of multiple protein-protein interactions, it remains an untested possibility whether antibody interaction with the C-terminus or an excess of the isolated C-terminus in *trans* could affect the stability of the protein and/or act as a dominant negative. In addition, it was reported that a similar peptide (encompassing residues 361-382) fused to an internalization sequence was capable of inducing apoptosis in a manner that required overexpressed wild-type or a mutated (and thereby stabilized) p53 in breast cancer cell lines (Kim A. L. et al., 1999). Although they show that the peptide does not alter p53 levels upon internalization, it is not clear that there is similar internalization of the peptide in different cell lines. A key experiment would be to use the chromatin immunoprecipitation assay to test directly whether it is the DNA binding activity *per se* that is being affected by these treatments.

SUMMARY

Recent studies regarding the regulation of p53-mediated transcriptional activation suggest that a critical question is what and when other coactivators are recruited to the promoters of p53 target genes. For instance, Kaeser and Iggo found large differences in the type and extent of histone modification on various p53 regulated promoters (Kaeser & Iggo, 2004). In addition, Espinosa and colleagues followed the assembly of transcription initiation complexes on arrest and apoptotic promoters over time and under different stress conditions (Espinosa et al., 2003). Their study shows fundamental differences in the timing and presence of key members of the basal transcriptional machinery including RNA Pol II, TBP and TFIIB. Further, they were able to show that recruiting the proper assortment of factors likely depends both on p53 as well as some *cis-* acting DNA sequences in the promoter. Another recent study using recombinant wild-type and mutated histone-containing chromatin demonstrated stepwise requirements for the histone acetyltransferase p300 as well as two arginine histone methyltransferases PRMT1 and CARM for efficient *in vitro* transactivation from a synthetic promoter containing multiple copies of the p53 response element derived from the *GADD45* promoter (An et al., 2004). It would be interesting to determine if and which of these factors are required for the appropriate regulation of p53 response genes in cells.

One constant theme in p53 research is that the DNA binding activity of p53 is crucial to its function as a transcription factor and as a tumor suppressor. A precise understanding of the mechanism of DNA binding has already generated productive rational drug design *in vitro* and may yet yield a useful cancer therapy. Furthermore, given that p53 is one of the most well-studied transcription factors in the literature and subject to complex regulation, it is likely to continue to serve as a paradigm for mechanisms of specific gene regulation.

ACKNOWLEDGEMENTS

M. Poyurovsky is thanked for help in preparing this review. CP is supported by grants from the National Cancer Institute.

REFERENCES

Abarzua P., LoSardo J. E., Gubler M. L., Neri A. Microinjection of monoclonal antibody PAb421 into human SW480 colorectal carcinoma cells restores the transcription activation function to mutant p53. Cancer Res 1995; 55:3490-4.

Abarzua P., LoSardo J. E., Gubler M. L., Spathis R., Lu Y. A., Felix A., Neri A. Restoration of the transcription activation function to mutant p53 in human cancer cells. Oncogene 1996; 13:2477-82.

Ahn J., Prives C. The C-terminus of p53: the more you learn the less you know. Nat Struct Biol 2001; 8:730-2.

An W., Kim J., Roeder R. G. Ordered Cooperative Functions of PRMT1, p300, and CARM1 in Transcriptional Activation by p53. Cell 2004; 117:735-48.

Anderson M. E., Woelker B., Reed M., Wang P., Tegtmeyer P. Reciprocal interference between the sequence-specific core and nonspecific C-terminal DNA binding domains of p53: implications for regulation. Mol Cell Biol 1997; 17:6255-64.

Appella E., Anderson C. W. Post-translational modifications and activation of p53 by genotoxic stresses. Eur J Biochem 2001; 268:2764-72.

Avantaggiati M. L., Ogryzko V., Gardner K., Giordano A., Levine A. S., Kelly K. Recruitment of p300/CBP in p53-dependent signal pathways. Cell 1997; 89:1175-84.

Ayed A., Mulder F. A., Yi G. S., Lu Y., Kay L. E., Arrowsmith C. H. Latent and active p53 are identical in conformation. Nat Struct Biol 2001; 8:756-60.

Bakalkin G., Selivanova G., Yakovleva T., Kiseleva E., Kashuba E., Magnusson K. P., Szekely L., Klein G., Terenius L., Wiman K. G. p53 binds single-stranded DNA ends through the C-terminal domain and internal DNA segments via the middle domain. Nucleic Acids Res 1995; 23:362-9.

Bakalkin G., Yakovleva T., Selivanova G., Magnusson K. P., Szekely L., Kiseleva E., Klein G., Terenius L., Wiman K. G. p53 binds single-stranded DNA ends and catalyzes DNA renaturation and strand transfer. Proc Natl Acad Sci U S A 1994; 91:413-7.

Balagurumoorthy P., Sakamoto H., Lewis M. S., Zambrano N., Clore G. M., Gronenborn A. M., Appella E., Harrington R. E. Four p53 DNA-binding domain peptides bind natural p53-response elements and bend the DNA. Proc Natl Acad Sci U S A 1995; 92:8591-5.

Baptiste N., Friedlander P., Chen X., Prives C. The proline-rich domain of p53 is required for cooperation with anti-neoplastic agents to promote apoptosis of tumor cells. Oncogene 2002; 21:9-21.

Baptiste N., Prives C. p53 in the cytoplasm: a question of overkill? Cell 2004; 116:487-9.

Bargonetti J., Manfredi J. J., Chen X., Marshak D. R., Prives C. A proteolytic fragment from the central region of p53 has marked sequence-specific DNA-binding activity when generated from wild-type but not from oncogenic mutant p53 protein. Genes Dev 1993; 7:2565-74.

Barlev N. A., Liu L., Chehab N. H., Mansfield K., Harris K. G., Halazonetis T. D., Berger S. L. Acetylation of p53 activates transcription through recruitment of coactivators/histone acetyltransferases. Mol Cell 2001; 8:1243-54.

Boyd S. D., Tsai K. Y., Jacks T. An intact HDM2 RING-finger domain is required for nuclear exclusion of p53. Nat Cell Biol 2000; 2:563-8.

Budhram-Mahadeo V., Morris P. J., Smith M. D., Midgley C. A., Boxer L. M., Latchman D. S. p53 suppresses the activation of the Bcl-2 promoter by the Brn-3a POU family transcription factor. J Biol Chem 1999; 274:15237-44.

Bullock A. N., Fersht A. R. Rescuing the function of mutant p53. Nat Rev Cancer 2001; 1:68-76.

Buzek J., Latonen L., Kurki S., Peltonen K., Laiho M. Redox state of tumor suppressor p53 regulates its sequence-specific DNA binding in DNA-damaged cells by cysteine 277. Nucleic Acids Res 2002; 30:2340-8.

Bykov V. J., Issaeva N., Shilov A., Hultcrantz M., Pugacheva E., Chumakov P., Bergman J., Wiman K. G., Selivanova G. Restoration of the tumor suppressor function to mutant p53 by a low-molecular-weight compound. Nat Med 2002; 8:282-8.

Cain C., Miller S., Ahn J., Prives C. The N terminus of p53 regulates its dissociation from DNA. J Biol Chem 2000; 275:39944-53.

Candau R., Scolnick D. M., Darpino P., Ying C. Y., Halazonetis T. D., Berger S. L. Two tandem and independent sub-activation domains in the amino terminus of p53 require the adaptor complex for activity. Oncogene 1997; 15:807-16.

Cawley S., Bekiranov S., Ng H. H., Kapranov P., Sekinger E. A., Kampa D., Piccolboni A., Sementchenko V., Cheng J., Williams A. J., Wheeler R., Wong B., Drenkow J., Yamanaka M., Patel S., Brubaker S., Tammana H., Helt G., Struhl K., Gingeras T. R. Unbiased mapping of transcription factor binding sites along human chromosomes 21 and 22 points to widespread regulation of noncoding RNAs. Cell 2004; 116:499-509.

Chang J., Kim D. H., Lee S. W., Choi K. Y., Sung Y. C. Transactivation ability of p53 transcriptional activation domain is directly related to the binding affinity to TATA-binding protein. J Biol Chem 1995; 270:25014-9.

Chao C., Hergenhahn M., Kaeser M. D., Wu Z., Saito S., Iggo R., Hollstein M., Appella E., Xu Y. Cell type- and promoter-specific roles of Ser18 phosphorylation in regulating p53 responses. J Biol Chem 2003.

Chen X., Ko L. J., Jayaraman L., Prives C. p53 levels, functional domains, and DNA damage determine the extent of the apoptotic response of tumor cells. Genes Dev 1996; 10:2438-51.

Chene P. The role of tetramerization in p53 function. Oncogene 2001; 20:2611-7.

Cherny D. I., Striker G., Subramaniam V., Jett S. D., Palecek E., Jovin T. M. DNA bending due to specific p53 and p53 core domain-DNA interactions visualized by electron microscopy. J Mol Biol 1999; 294:1015-26.

Chipuk J. E., Green D. R. Cytoplasmic p53: Bax and Forward. Cell Cycle 2004; 3:429-31.

Chipuk J. E., Kuwana T., Bouchier-Hayes L., Droin N. M., Newmeyer D. D., Schuler M., Green D. R. Direct activation of Bax by p53 mediates mitochondrial membrane permeabilization and apoptosis. Science 2004; 303:1010-4.

Chipuk J. E., Maurer U., Green D. R., Schuler M. Pharmacologic activation of p53 elicits Bax-dependent apoptosis in the absence of transcription. Cancer Cell 2003; 4:371-81.

Cho Y., Gorina S., Jeffrey P. D., Pavletich N. P. Crystal structure of a p53 tumor suppressor-DNA complex: understanding tumorigenic mutations. Science 1994; 265:346-55.

Contente A., Dittmer A., Koch M. C., Roth J., Dobbelstein M. A polymorphic microsatellite that mediates induction of PIG3 by p53. Nat Genet 2002; 30:315-20.

Dang C. V., Lee W. M. Nuclear and nucleolar targeting sequences of c-erb-A, c-myb, N-myc, p53, HSP70, and HIV tat proteins. J Biol Chem 1989; 264:18019-23.

Di Como C. J., Gaiddon C., Prives C. p73 function is inhibited by tumor-derived p53 mutants in mammalian cells. Mol Cell Biol 1999; 19:1438-49.

Di Como C. J., Prives C. Human tumor-derived p53 proteins exhibit binding site selectivity and temperature sensitivity for transactivation in a yeast-based assay. Oncogene 1998; 16:2527-39.

Di Como C. J., Urist M. J., Babayan I., Drobnjak M., Hedvat C. V., Teruya-Feldstein J., Pohar K., Hoos A., Cordon-Cardo C. p63 expression profiles in human normal and tumor tissues. Clin Cancer Res 2002; 8:494-501.

Dornan D., Shimizu H., Perkins N. D., Hupp T. R. DNA-dependent acetylation of p53 by the transcription coactivator p300. J Biol Chem 2003; 278:13431-41.

Dudenhoffer C., Rohaly G., Will K., Deppert W., Wiesmuller L. Specific mismatch recognition in heteroduplex intermediates by p53 suggests a role in fidelity control of homologous recombination. Mol Cell Biol 1998; 18:5332-42.

Dumont P., Leu J. I., Della Pietra A. C., 3rd, George D. L., Murphy M. The codon 72 polymorphic variants of p53 have markedly different apoptotic potential. Nat Genet 2003; 33:357-65.

Espinosa J. M., Emerson B. M. Transcriptional regulation by p53 through intrinsic DNA/chromatin binding and site-directed cofactor recruitment. Mol Cell 2001; 8:57-69.

Espinosa J. M., Verdun R. E., Emerson B. M. p53 functions through stress- and promoter-specific recruitment of transcription initiation components before and after DNA damage. Mol Cell 2003; 12:1015-27.

Flores E. R., Tsai K. Y., Crowley D., Sengupta S., Yang A., McKeon F., Jacks T. p63 and p73 are required for p53-dependent apoptosis in response to DNA damage. Nature 2002; 416:560-4.

Fojta M., Kubicarova T., Vojtesek B., Palecek E. Effect of p53 protein redox states on binding to supercoiled and linear DNA. J Biol Chem 1999; 274:25749-55.

Foord O. S., Bhattacharya P., Reich Z., Rotter V. A DNA binding domain is contained in the C-terminus of wild type p53 protein. Nucleic Acids Res 1991; 19:5191-8.

Foster B. A., Coffey H. A., Morin M. J., Rastinejad F. Pharmacological rescue of mutant p53 conformation and function. Science 1999; 286:2507-10.

Friedlander P., Haupt Y., Prives C., Oren M. A mutant p53 that discriminates between p53-responsive genes cannot induce apoptosis. Mol Cell Biol 1996a; 16:4961-71.

Friedlander P., Legros Y., Soussi T., Prives C. Regulation of mutant p53 temperature-sensitive DNA binding. J Biol Chem 1996b; 271:25468-78.

Friedler A., Hansson L. O., Veprintsev D. B., Freund S. M., Rippin T. M., Nikolova P. V., Proctor M. R., Rudiger S., Fersht A. R. A peptide that binds and stabilizes p53 core domain: chaperone strategy for rescue of oncogenic mutants. Proc Natl Acad Sci U S A 2002; 99:937-42.

Friedler A., Veprintsev D. B., Hansson L. O., Fersht A. R. Kinetic instability of p53 core domain mutants: implications for rescue by small molecules. J Biol Chem 2003; 278:24108-12.

Gaiddon C., Lokshin M., Ahn J., Zhang T., Prives C. A subset of tumor-derived mutant forms of p53 down-regulate p63 and p73 through a direct interaction with the p53 core domain. Mol Cell Biol 2001; 21:1874-87.

Gaiddon C., Moorthy N. C., Prives C. Ref-1 regulates the transactivation and pro-apoptotic functions of p53 in vivo. Embo J 1999; 18:5609-21.

Geyer R. K., Yu Z. K., Maki C. G. The MDM2 RING-finger domain is required to promote p53 nuclear export. Nat Cell Biol 2000; 2:569-73.

Gohler T., Reimann M., Cherny D., Walter K., Warnecke G., Kim E., Deppert W. Specific interaction of p53 with target binding sites is determined by DNA conformation and is regulated by the C-terminal domain. J Biol Chem 2002; 277:41192-203.

Gorina S., Pavletich N. P. Structure of the p53 tumor suppressor bound to the ankyrin and SH3 domains of 53BP2. Science 1996; 274:1001-5.

Gostissa M., Hengstermann A., Fogal V., Sandy P., Schwarz S. E., Scheffner M., Del Sal G. Activation of p53 by conjugation to the ubiquitin-like protein SUMO-1. Embo J 1999; 18:6462-71.

Gottifredi V., Karni-Schmidt O., Shieh S. S., Prives C. p53 down-regulates CHK1 through p21 and the retinoblastoma protein. Mol Cell Biol 2001; 21:1066-76.

Green D. R., Evan G. I. A matter of life and death. Cancer Cell 2002; 1:19-30.

Gu W., Roeder R. G. Activation of p53 sequence-specific DNA binding by acetylation of the p53 C-terminal domain. Cell 1997; 90:595-606.

Gu W., Shi X. L., Roeder R. G. Synergistic activation of transcription by CBP and p53. Nature 1997; 387:819-23.

Hainaut P., Hollstein M. p53 and human cancer: the first ten thousand mutations. Adv Cancer Res 2000; 77:81-137.

Hainaut P., Mann K. Zinc binding and redox control of p53 structure and function. Antioxid Redox Signal 2001; 3:611-23.

Hainaut P., Rolley N., Davies M., Milner J. Modulation by copper of p53 conformation and sequence-specific DNA binding: role for Cu(II)/Cu(I) redox mechanism. Oncogene 1995; 10:27-32.

Halazonetis T. D., Davis L. J., Kandil A. N. Wild-type p53 adopts a 'mutant'-like conformation when bound to DNA. Embo J 1993; 12:1021-8.

Halazonetis T. D., Kandil A. N. Conformational shifts propagate from the oligomerization domain of p53 to its tetrameric DNA binding domain and restore DNA binding to select p53 mutants. Embo J 1993; 12:5057-64.

Hansen S., Hupp T. R., Lane D. P. Allosteric regulation of the thermostability and DNA binding activity of human p53 by specific interacting proteins. CRC Cell Transformation Group. J Biol Chem 1996; 271:3917-24.

Harms K., Nozell S., Chen X. The common and distinct target genes of the p53 family transcription factors. Cell Mol Life Sci 2004; 61:822-42.

Hoffman W. H., Biade S., Zilfou J. T., Chen J., Murphy M. Transcriptional repression of the anti-apoptotic survivin gene by wild type p53. J Biol Chem 2002; 277:3247-57.

Hoffmann R., Craik D. J., Pierens G., Bolger R. E., Otvos L., Jr. Phosphorylation of the C-terminal sites of human p53 reduces non-sequence-specific DNA binding as modeled with synthetic peptides. Biochemistry 1998; 37:13755-64.

Hoh J., Jin S., Parrado T., Edington J., Levine A. J., Ott J. The p53MH algorithm and its application in detecting p53-responsive genes. Proc Natl Acad Sci U S A 2002; 99:8467-72.

Hsu C. H., Chang M. D., Tai K. Y., Yang Y. T., Wang P. S., Chen C. J., Wang Y. H., Lee S. C., Wu C. W., Juan L. J. HCMV IE2-mediated inhibition of HAT activity downregulates p53 function. Embo J 2004; 23:2269-80.

Hupp T. R., Lane D. P. Allosteric activation of latent p53 tetramers. Curr Biol 1994; 4:865-75.

Hupp T. R., Meek D. W., Midgley C. A., Lane D. P. Regulation of the specific DNA binding function of p53. Cell 1992; 71:875-86.

Hupp T. R., Sparks A., Lane D. P. Small peptides activate the latent sequence-specific DNA binding function of p53. Cell 1995; 83:237-45.

Iwabuchi K., Bartel P. L., Li B., Marraccino R., Fields S. Two cellular proteins that bind to wild-type but not mutant p53. Proc Natl Acad Sci U S A 1994; 91:6098-102.

Jackson P., Mastrangelo I., Reed M., Tegtmeyer P., Yardley G., Barrett J. Synergistic transcriptional activation of the MCK promoter by p53: tetramers link separated DNA response elements by DNA looping. Oncogene 1998; 16:283-92.

Jayaraman L., Prives C. Activation of p53 sequence-specific DNA binding by short single strands of DNA requires the p53 C-terminus. Cell 1995; 81:1021-9.

Jayaraman L., Moorthy N. C., Murthy K. G., Manley J. L., Bustin M., Prives C. High mobility group protein-1 (HMG-1) is a unique activator of p53. Genes Dev 1998; 12:462-72.

Jayaraman L., Murthy K. G., Zhu C., Curran T., Xanthoudakis S., Prives C. Identification of redox/repair protein Ref-1 as a potent activator of p53. Genes Dev 1997; 11:558-70.

Jayaraman L., Prives C. Covalent and noncovalent modifiers of the p53 protein. Cell Mol Life Sci 1999; 55:76-87.

Jiang M., Axe T., Holgate R., Rubbi C. P., Okorokov A. L., Mee T., Milner J. p53 binds the nuclear matrix in normal cells: binding involves the proline-rich domain of p53 and increases following genotoxic stress. Oncogene 2001; 20:5449-58.

Johnson R. A., Ince T. A., Scotto K. W. Transcriptional repression by p53 through direct binding to a novel DNA element. J Biol Chem 2001; 276:27716-20.

Jost C. A., Marin M. C., Kaelin W. G., Jr. p73 is a simian [correction of human] p53-related protein that can induce apoptosis. Nature 1997; 389:191-4.

Juan L. J., Shia W. J., Chen M. H., Yang W. M., Seto E., Lin Y. S., Wu C. W. Histone deacetylases specifically down-regulate p53-dependent gene activation. J Biol Chem 2000; 275:20436-43.

Kaeser M. D., Iggo R. D. Chromatin immunoprecipitation analysis fails to support the latency model for regulation of p53 DNA binding activity in vivo. Proc Natl Acad Sci U S A 2002; 99:95-100.

Kaeser M. D., Iggo R. D. Promoter-specific p53-dependent histone acetylation following DNA damage. Oncogene 2004; 23:4007-13.

Kim A. L., Raffo A. J., Brandt-Rauf P. W., Pincus M. R., Monaco R., Abarzua P., Fine R. L. Conformational and molecular basis for induction of apoptosis by a p53 C-terminal peptide in human cancer cells. J Biol Chem 1999; 274:34924-31.

Kim E., Albrechtsen N., Deppert W. DNA-conformation is an important determinant of sequence-specific DNA binding by tumor suppressor p53. Oncogene 1997; 15:857-69.

Ko L. J., Prives C. p53: puzzle and paradigm. Genes Dev 1996; 10:1054-72.

Kubbutat M. H., Ludwig R. L., Ashcroft M., Vousden K. H. Regulation of Mdm2-directed degradation by the C terminus of p53. Mol Cell Biol 1998; 18:5690-8.

Lang F. F., Bruner J. M., Fuller G. N., Aldape K., Prados M. D., Chang S., Berger M. S., McDermott M. W., Kunwar S. M., Junck L. R., Chandler W., Zwiebel J. A., Kaplan R. S., Yung W. K. Phase I trial of adenovirus-mediated p53 gene therapy for recurrent glioma: biological and clinical results. J Clin Oncol 2003; 21:2508-18.

Langley E., Pearson M., Faretta M., Bauer U. M., Frye R. A., Minucci S., Pelicci P. G., Kouzarides T. Human SIR2 deacetylates p53 and antagonizes PML/p53-induced cellular senescence. Embo J 2002; 21:2383-96.

Lee K. C., Crowe A. J., Barton M. C. p53-mediated repression of alpha-fetoprotein gene expression by specific DNA binding. Mol Cell Biol 1999; 19:1279-88.

Lee S., Cavallo L., Griffith J. Human p53 binds Holliday junctions strongly and facilitates their cleavage. J Biol Chem 1997; 272:7532-9.

Lee S., Elenbaas B., Levine A., Griffith J. p53 and its 14 kDa C-terminal domain recognize primary DNA damage in the form of insertion/deletion mismatches. Cell 1995; 81:1013-20.

Lin J., Chen J., Elenbaas B., Levine A. J. Several hydrophobic amino acids in the p53 amino-terminal domain are required for transcriptional activation, binding to mdm-2 and the adenovirus 5 E1B 55-kD protein. Genes Dev 1994; 8:1235-46.

Liu L., Scolnick D. M., Trievel R. C., Zhang H. B., Marmorstein R., Halazonetis T. D., Berger S. L. p53 sites acetylated in vitro by PCAF and p300 are acetylated in vivo in response to DNA damage. Mol Cell Biol 1999; 19:1202-9.

Lohr K., Moritz C., Contente A., Dobbelstein M. p21/CDKN1A mediates negative regulation of transcription by p53. J Biol Chem 2003; 278:32507-16.

Ludwig R. L., Bates S., Vousden K. H. Differential activation of target cellular promoters by p53 mutants with impaired apoptotic function. Mol Cell Biol 1996; 16:4952-60.

Luo J., Li M., Tang Y., Laszkowska M., Roeder R. G., Gu W. Acetylation of p53 augments its site-specific DNA binding both in vitro and in vivo. Proc Natl Acad Sci U S A 2004.

Luo J., Nikolaev A. Y., Imai S., Chen D., Su F., Shiloh A., Guarente L., Gu W. Negative control of p53 by Sir2alpha promotes cell survival under stress. Cell 2001; 107:137-48.

Luo J., Su F., Chen D., Shiloh A., Gu W. Deacetylation of p53 modulates its effect on cell growth and apoptosis. Nature 2000; 408:377-81.

Manfredi J. J. p53 and apoptosis: it's not just in the nucleus anymore. Mol Cell 2003; 11:552-4.

Mazur S. J., Sakaguchi K., Appella E., Wang X. W., Harris C. C., Bohr V. A. Preferential binding of tumor suppressor p53 to positively or negatively supercoiled DNA involves the C-terminal domain. J Mol Biol 1999; 292:241-9.

McKinney K., Prives C. Efficient specific DNA binding by p53 requires both its central and C-terminal domains as revealed by studies with high-mobility group 1 protein. Mol Cell Biol 2002; 22:6797-808.

McLure K. G., Lee P. W. How p53 binds DNA as a tetramer. Embo J 1998; 17:3342-50.

Meplan C., Richard M. J., Hainaut P. Metalloregulation of the tumor suppressor protein p53: zinc mediates the renaturation of p53 after exposure to metal chelators in vitro and in intact cells. Oncogene 2000; 19:5227-36.

Michael D., Oren M. The p53-Mdm2 module and the ubiquitin system. Semin Cancer Biol 2003; 13:49-58.

Mihara M., Erster S., Zaika A., Petrenko O., Chittenden T., Pancoska P., Moll U. M. p53 has a direct apoptogenic role at the mitochondria. Mol Cell 2003; 11:577-90.

Miyashita T., Reed J. C. Tumor suppressor p53 is a direct transcriptional activator of the human bax gene. Cell 1995; 80:293-9.

Mujtaba S., He Y., Zeng L., Yan S., Plotnikova O., Sachchidanand, Sanchez R., Zeleznik-Le N. J., Ronai Z., Zhou M. M. Structural Mechanism of the Bromodomain of the Coactivator CBP in p53 Transcriptional Activation. Mol Cell 2004; 13:251-63.

Muller-Tiemann B. F., Halazonetis T. D., Elting J. J. Identification of an additional negative regulatory region for p53 sequence-specific DNA binding. Proc Natl Acad Sci U S A 1998; 95:6079-84.

Nagaich A. K., Appella E., Harrington R. E. DNA bending is essential for the site-specific recognition of DNA response elements by the DNA binding domain of the tumor suppressor protein p53. J Biol Chem 1997a; 272:14842-9.

Nagaich A. K., Zhurkin V. B., Durell S. R., Jernigan R. L., Appella E., Harrington R. E. p53-induced DNA bending and twisting: p53 tetramer binds on the outer side of a DNA loop and increases DNA twisting. Proc Natl Acad Sci U S A 1999; 96:1875-80.

Nagaich A. K., Zhurkin V. B., Sakamoto H., Gorin A. A., Clore G. M., Gronenborn A. M., Appella E., Harrington R. E. Architectural accommodation in the complex of four p53 DNA binding domain peptides with the p21/waf1/cip1 DNA response element. J Biol Chem 1997b; 272:14830-41.

Nie Y., Li H. H., Bula C. M., Liu X. Stimulation of p53 DNA binding by c-Abl requires the p53 C terminus and tetramerization. Mol Cell Biol 2000; 20:741-8.

Nikolova P. V., Henckel J., Lane D. P., Fersht A. R. Semirational design of active tumor suppressor p53 DNA binding domain with enhanced stability. Proc Natl Acad Sci U S A 1998; 95:14675-80.

Oda K., Arakawa H., Tanaka T., Matsuda K., Tanikawa C., Mori T., Nishimori H., Tamai K., Tokino T., Nakamura Y., Taya Y. p53AIP1, a potential mediator of p53-dependent apoptosis, and its regulation by Ser-46-phosphorylated p53. Cell 2000; 102:849-62.

Okorokov A. L., Rubbi C. P., Metcalfe S., Milner J. The interaction of p53 with the nuclear matrix is mediated by F-actin and modulated by DNA damage. Oncogene 2002; 21:356-67.

Olivier M., Eeles R., Hollstein M., Khan M. A., Harris C. C., Hainaut P. The IARC TP53 database: new online mutation analysis and recommendations to users. Hum Mutat 2002; 19:607-14.

Oren M. Decision making by p53: life, death and cancer. Cell Death Differ 2003; 10:431-42.

Ori A., Zauberman A., Doitsh G., Paran N., Oren M., Shaul Y. p53 binds and represses the HBV enhancer: an adjacent enhancer element can reverse the transcription effect of p53. Embo J 1998; 17:544-53.

Pagliaro L. C., Keyhani A., Williams D., Woods D., Liu B., Perrotte P., Slaton J. W., Merritt J. A., Grossman H. B., Dinney C. P. Repeated intravesical instillations of an adenoviral vector in patients with locally advanced bladder cancer: a phase I study of p53 gene therapy. J Clin Oncol 2003; 21:2247-53.

Palecek E., Brazda V., Jagelska E., Pecinka P., Karlovska L., Brazdova M. Enhancement of p53 sequence-specific binding by DNA supercoiling. Oncogene 2004.

Parks D., Bolinger R., Mann K. Redox state regulates binding of p53 to sequence-specific DNA, but not to non-specific or mismatched DNA. Nucleic Acids Res 1997; 25:1289-95.

Pavletich N. P., Chambers K. A., Pabo C. O. The DNA-binding domain of p53 contains the four conserved regions and the major mutation hot spots. Genes Dev 1993; 7:2556-64.

Rainwater R., Parks D., Anderson M. E., Tegtmeyer P., Mann K. Role of cysteine residues in regulation of p53 function. Mol Cell Biol 1995; 15:3892-903.

Reed M., Woelker B., Wang P., Wang Y., Anderson M. E., Tegtmeyer P. The C-terminal domain of p53 recognizes DNA damaged by ionizing radiation. Proc Natl Acad Sci U S A 1995; 92:9455-9.

Resnick M. A., Inga A. Functional mutants of the sequence-specific transcription factor p53 and implications for master genes of diversity. Proc Natl Acad Sci U S A 2003; 100:9934-9.

Resnick-Silverman L., St Clair S., Maurer M., Zhao K., Manfredi J. J. Identification of a novel class of genomic DNA-binding sites suggests a mechanism for selectivity in target gene activation by the tumor suppressor protein p53. Genes Dev 1998; 12:2102-7.

Rippin T. M., Freund S. M., Veprintsev D. B., Fersht A. R. Recognition of DNA by p53 core domain and location of intermolecular contacts of cooperative binding. J Mol Biol 2002; 319:351-8.

Rodriguez M. S., Desterro J. M., Lain S., Lane D. P., Hay R. T. Multiple C-terminal lysine residues target p53 for ubiquitin-proteasome-mediated degradation. Mol Cell Biol 2000; 20:8458-67.

Rodriguez M. S., Desterro J. M., Lain S., Midgley C. A., Lane D. P., Hay R. T. SUMO-1 modification activates the transcriptional response of p53. Embo J 1999; 18:6455-61.

Sakaguchi K., Herrera J. E., Saito S., Miki T., Bustin M., Vassilev A., Anderson C. W., Appella E. DNA damage activates p53 through a phosphorylation-acetylation cascade. Genes Dev 1998; 12:2831-41.

Sakamuro D., Sabbatini P., White E., Prendergast G. C. The polyproline region of p53 is required to activate apoptosis but not growth arrest. Oncogene 1997; 15:887-98.

Saller E., Tom E., Brunori M., Otter M., Estreicher A., Mack D. H., Iggo R. Increased apoptosis induction by 121F mutant p53. Embo J 1999; 18:4424-37.

Samuels-Lev Y., O'Connor D. J., Bergamaschi D., Trigiante G., Hsieh J. K., Zhong S., Campargue I., Naumovski L., Crook T., Lu X. ASPP proteins specifically stimulate the apoptotic function of p53. Mol Cell 2001; 8:781-94.

Scolnick D. M., Chehab N. H., Stavridi E. S., Lien M. C., Caruso L., Moran E., Berger S. L., Halazonetis T. D. CREB-binding protein and p300/CBP-associated factor are transcriptional coactivators of the p53 tumor suppressor protein. Cancer Res 1997; 57:3693-6.

Selivanova G., Iotsova V., Okan I., Fritsche M., Strom M., Groner B., Grafstrom R. C., Wiman K. G. Restoration of the growth suppression function of mutant p53 by a synthetic peptide derived from the p53 C-terminal domain. Nat Med 1997; 3:632-8.

Seo Y. R., Kelley M. R., Smith M. L. Selenomethionine regulation of p53 by a ref1-dependent redox mechanism. Proc Natl Acad Sci U S A 2002; 99:14548-53.

Shaulsky G., Goldfinger N., Ben-Ze'ev A., Rotter V. Nuclear accumulation of p53 protein is mediated by several nuclear localization signals and plays a role in tumorigenesis. Mol Cell Biol 1990; 10:6565-77.

Stenger J. E., Tegtmeyer P., Mayr G. A., Reed M., Wang Y., Wang P., Hough P. V., Mastrangelo I. A. p53 oligomerization and DNA looping are linked with transcriptional activation. Embo J 1994; 13:6011-20.

Stommel J. M., Marchenko N. D., Jimenez G. S., Moll U. M., Hope T. J., Wahl G. M. A leucine-rich nuclear export signal in the p53 tetramerization domain: regulation of subcellular localization and p53 activity by NES masking. Embo J 1999; 18:1660-72.

Strano S., Fontemaggi G., Costanzo A., Rizzo M. G., Monti O., Baccarini A., Del Sal G., Levrero M., Sacchi A., Oren M., Blandino G. Physical interaction with human tumor-derived p53 mutants inhibits p63 activities. J Biol Chem 2002; 277:18817-26.

Strano S., Munarriz E., Rossi M., Cristofanelli B., Shaul Y., Castagnoli L., Levine A. J., Sacchi A., Cesareni G., Oren M., Blandino G. Physical and functional interaction between p53 mutants and different isoforms of p73. J Biol Chem 2000; 275:29503-12.

Stros M., Muselikova-Polanska E., Pospisilova S., Strauss F. High-Affinity Binding of Tumor-Suppressor Protein p53 and HMGB1 to Hemicatenated DNA Loops. Biochemistry 2004; 43:7215-25.

Takenaka I., Morin F., Seizinger B. R., Kley N. Regulation of the sequence-specific DNA binding function of p53 by protein kinase C and protein phosphatases. J Biol Chem 1995; 270:5405-11.

Thornborrow E. C., Manfredi J. J. One mechanism for cell type-specific regulation of the bax promoter by the tumor suppressor p53 is dictated by the p53 response element. J Biol Chem 1999; 274:33747-56.

Ueno M., Masutani H., Arai R. J., Yamauchi A., Hirota K., Sakai T., Inamoto T., Yamaoka Y., Yodoi J., Nikaido T. Thioredoxin-dependent redox regulation of p53-mediated p21 activation. J Biol Chem 1999; 274:35809-15.

Unger T., Mietz J. A., Scheffner M., Yee C. L., Howley P. M. Functional domains of wild-type and mutant p53 proteins involved in transcriptional regulation, transdominant inhibition, and transformation suppression. Mol Cell Biol 1993; 13:5186-94.

Urist M. J., Di Como C. J., Lu M. L., Charytonowicz E., Verbel D., Crum C. P., Ince T. A., McKeon F. D., Cordon-Cardo C. Loss of p63 expression is associated with tumor progression in bladder cancer. Am J Pathol 2002; 161:1199-206.

Vaziri H., Dessain S. K., Ng Eaton E., Imai S. I., Frye R. A., Pandita T. K., Guarente L., Weinberg R. A. hSIR2(SIRT1) functions as an NAD-dependent p53 deacetylase. Cell 2001; 107:149-59.

Venot C., Maratrat M., Dureuil C., Conseiller E., Bracco L., Debussche L. The requirement for the p53 proline-rich functional domain for mediation of apoptosis is correlated with specific PIG3 gene transactivation and with transcriptional repression. Embo J 1998; 17:4668-79.

Venot C., Maratrat M., Sierra V., Conseiller E., Debussche L. Definition of a p53 transactivation function-deficient mutant and characterization of two independent p53 transactivation subdomains. Oncogene 1999; 18:2405-10.

Verhaegh G. W., Richard M. J., Hainaut P. Regulation of p53 by metal ions and by antioxidants: dithiocarbamate down-regulates p53 DNA-binding activity by increasing the intracellular level of copper. Mol Cell Biol 1997; 17:5699-706.

Vogelstein B., Kinzler K. W. p53 function and dysfunction. Cell 1992; 70:523-6.

Walker K. K., Levine A. J. Identification of a novel p53 functional domain that is necessary for efficient growth suppression. Proc Natl Acad Sci U S A 1996; 93:15335-40.

Wang W., Takimoto R., Rastinejad F., El-Deiry W. S. Stabilization of p53 by CP-31398 inhibits ubiquitination without altering phosphorylation at serine 15 or 20 or MDM2 binding. Mol Cell Biol 2003; 23:2171-81.

Wang Y., Reed M., Wang P., Stenger J. E., Mayr G., Anderson M. E., Schwedes J. F., Tegtmeyer P. p53 domains: identification and characterization of two autonomous DNA-binding regions. Genes Dev 1993; 7:2575-86.

Waterman M. J., Stavridi E. S., Waterman J. L., Halazonetis T. D. ATM-dependent activation of p53 involves dephosphorylation and association with 14-3-3 proteins. Nat Genet 1998; 19:175-8.

Waterman M. J., Waterman J. L., Halazonetis T. D. An engineered four-stranded coiled coil substitutes for the tetramerization domain of wild-type p53 and alleviates transdominant inhibition by tumor-derived p53 mutants. Cancer Res 1996; 56:158-63.

Wieczorek A. M., Waterman J. L., Waterman M. J., Halazonetis T. D. Structure-based rescue of common tumor-derived p53 mutants. Nat Med 1996; 2:1143-6.

Willis A., Jung E. J., Wakefield T., Chen X. Mutant p53 exerts a dominant negative effect by preventing wild-type p53 from binding to the promoter of its target genes. Oncogene 2004.

Wilson D. R. Viral-mediated gene transfer for cancer treatment. Curr Pharm Biotechnol 2002; 3:151-64.

Wong K. B., DeDecker B. S., Freund S. M., Proctor M. R., Bycroft M., Fersht A. R. Hot-spot mutants of p53 core domain evince characteristic local structural changes. Proc Natl Acad Sci U S A 1999; 96:8438-42.

Wu L., Bayle J. H., Elenbaas B., Pavletich N. P., Levine A. J. Alternatively spliced forms in the carboxy-terminal domain of the p53 protein regulate its ability to promote annealing of complementary single strands of nucleic acids. Mol Cell Biol 1995; 15:497-504.

Yakovleva T., Pramanik A., Kawasaki T., Tan-No K., Gileva I., Lindegren H., Langel U., Ekstrom T. J., Rigler R., Terenius L., Bakalkin G. p53 Latency. C-terminal domain prevents binding of p53 core to target but not to nonspecific DNA sequences. J Biol Chem 2001; 276:15650-8.

Yang A., Kaghad M., Caput D., McKeon F. On the shoulders of giants: p63, p73 and the rise of p53. Trends Genet 2002; 18:90-5.

Yang A., Kaghad M., Wang Y., Gillett E., Fleming M. D., Dotsch V., Andrews N. C., Caput D., McKeon F. p63, a p53 homolog at 3q27-29, encodes multiple products with transactivating, death-inducing, and dominant-negative activities. Mol Cell 1998; 2:305-16.

Yew P. R., Liu X., Berk A. J. Adenovirus E1B oncoprotein tethers a transcriptional repression domain to p53. Genes Dev 1994; 8:190-202.

Zhou J., Prives C. Replication of damaged DNA in vitro is blocked by p53. Nucleic Acids Res 2003; 31:3881-92.

Zhu J., Jiang J., Zhou W., Zhu K., Chen X. Differential regulation of cellular target genes by p53 devoid of the PXXP motifs with impaired apoptotic activity. Oncogene 1999; 18:2149-55.

Zhu J., Zhang S., Jiang J., Chen X. Definition of the p53 functional domains necessary for inducing apoptosis. J Biol Chem 2000; 275:39927-34.

Zhu J., Zhou W., Jiang J., Chen X. Identification of a novel p53 functional domain that is necessary for mediating apoptosis. J Biol Chem 1998; 273:13030-6.

Zilfou J. T., Hoffman W. H., Sank M., George D. L., Murphy M. The corepressor mSin3a interacts with the proline-rich domain of p53 and protects p53 from proteasome-mediated degradation. Mol Cell Biol 2001; 21:3974-85.

Zotchev S. B., Protopopova M., Selivanova G. p53 C-terminal interaction with DNA ends and gaps has opposing effect on specific DNA binding by the core. Nucleic Acids Res 2000; 28:4005-12.

Chapter 3

20 YEARS OF DNA DAMAGE SIGNALING TO P53

Kevin G. McLure and Michael B. Kastan
Department of Hematology-Oncology, St. Jude Children's Research Hospital, Memphis, TN, USA

INTRODUCTION

The short history of p53 contains an overwhelming number of facts and hypotheses, presenting the challenge of integrating diverse and sometimes mutually exclusive ideas into a coherent picture. It is important to make a distinction between p53 tumor suppressor activity, the mechanism of which remains speculative, and p53 responses to DNA damage, which are well characterized. Because critical steps in tumorigenesis involve genomic fixation of DNA damage-induced mutations, it seems reasonable to assume that DNA damage signaling to p53 would activate p53 tumor suppressor activity. However, this has not been demonstrated, and p53 tumor suppressor activity may not require the acute p53 response to DNA damage (Komarov et al., 1999). Nonetheless, the genotoxic chemicals and ionizing radiation that are clinically used to treat human cancer indisputably activate wild type p53.

DNA damage refers to alterations in the chemical bonds of constituent nucleotides, resulting in aberrant or mismatched base pairs, cross-linked bases, or single- and double-strand breaks in the phosphodiester DNA backbone. DNA damage can be induced by genotoxic chemicals, ultraviolet radiation, shortened telomeres, or reactive oxygen species generated by processes including mitochondrial respiration, ionizing radiation, or ischemia-reperfusion (Giaccia and Kastan, 1998). DNA damage can also be induced by oncogenic alterations in cancer cells, such as

P. Hainaut and K.G. Wiman (eds.), 25 Years of p53 Research, 53-71.

overexpressed/amplified c-myc, which has been reported to generate excess reactive oxygen species and damages DNA (Vafa et al., 2002).

There is an ongoing "background" of oxidative damage that is continuously repaired (Friedberg, 2003), and this does not detectably activate the p53 DNA damage response. Otherwise, since the p53 response can last for hours to days, p53 would be constantly activated. Thus, when we refer to DNA damage signaling to p53 we refer to a level of damage that is sufficiently above background to result in experimentally detectable changes in p53.

We will focus on how DNA damage signals to p53 and how DNA damage signaling regulates p53 function. Careful consideration of these events casts doubt on the pervasive assumption that post-translational modification of p53 is primarily responsible for DNA damage-induced p53 stabilization.

DNA DAMAGE DETECTION

Conceptually, p53 could be a direct sensor of DNA damage by binding directly to damaged DNA or to DNA damage repair products. *In vitro* p53 can directly bind to irradiated DNA, to DNA which has a short mismatch, or to DNA ends (Lee et al., 1995), (Reed et al., 1995), (Bakalkin et al., 1994). Binding to damaged DNA could be involved in the actual process of DNA repair (Offer et al., 1999) (Zhou et al., 2001) (Rubbi and Milner, 2003b). That lower organisms such as bacteria and yeast lack p53 but possess robust DNA repair systems indicates that p53 is dispensable for DNA repair. Nonetheless, a high level of DNA damage could directly signal to a relevant fraction of nuclear p53. As DNA damage does not induce a change in the intracellular distribution of p53, for example into subnuclear foci, p53 does not appear to relocalize to genomic sites of DNA damage.

The conceptual problem with p53 function being impacted by p53 directly recognizing damaged DNA or DNA repair products is that the signal does not directly result in persistent alteration of p53 functions. For example, to induce G1/S arrest p53 has to travel to the genomic p53 binding sites in the p21 gene in order to transcriptionally activate p21 (Dulic et al., 1994) (Szak et al., 2001). Although p53-dependent induction of apoptosis is mechanistically disputed, transcription-independent apoptosis might require p53 to translocate to mitochondria (Mihara et al., 2003). Another problem is that if p53 binds directly to damaged DNA or repair products, relatively few molecules of p53 would receive the DNA damage signal. In contrast, indirect signaling occurring via intermediate DNA damage sensors that transmit a signal to p53 conceptually allows for amplification of the DNA damage signal to a substantial fraction of the cellular p53.

Elucidating the interface between DNA damage detection and amplification systems is an ongoing endeavor with its roots in the yeast DNA damage response (Rouse and Jackson, 2002). These studies have guided the framework for the mammalian DNA damage detection systems, which have been closely linked to understanding how DNA damage signals to p53.

In the late 1970s it was generally assumed that viruses were the causative agent of human cancer, and tumor viruses were known to 'transform' mammalian cells into a cancer-like state, so the search was on for cellular proteins that bound to viral transforming antigens. Thus were discovered proteins that were bound by the SV40 tumor virus Large T Antigen, one of which migrated at ~53 kDa on an SDS-polyacrylamide gel and was descriptively named p53 (Linzer and Levine, 1979) (Lane and Crawford, 1979) (Deleo et al., 1979). The realization that the viral large TAg was inactivating a tumor suppressor gene rather than activating an oncogene would have to wait nearly 10 years, because mutated versions of p53 had been inadvertently cloned from cell lines and p53 was therefore assumed to be a proto-oncogene.

In this context, it is perhaps not surprising that the first report of DNA damage signaling to p53 in 1984 did not receive widespread attention (Maltzman and Czyzyk, 1984). The connection between DNA damage and p53 became much more significant when ionizing radiation was found to induce stabilization of the p53 protein, which by this time was known to be a tumor suppressor (Kastan et al., 1991). Such ionizing radiation-induced p53 stabilization was defective in radiosensitive cells derived from ataxia telangiectasia (AT) patients, and AT cells were also defective in an ionizing radiation-induced G1/S cell cycle checkpoint (Kastan et al., 1992). This was important because cell cycle checkpoints had been conceptualized as a mechanism by which yeast (and mammalian) cells could actively sense DNA damage and prevent fixation of genetic mutations before DNA repair could occur (Weinert and Hartwell, 1988) (Hartwell and Weinert, 1989). By this time a biochemical activity of p53 had been demonstrated, namely sequence-specific DNA binding (Kern et al., 1991) (El-Deiry et al., 1992) (Funk et al., 1992). Ionizing radiation induced both p53 protein stabilization and p53 sequence-specific DNA binding activity, both of which were deficient in AT cells (Kastan et al., 1992). Thus arose the concept that ionizing radiation could elevate p53 protein to bind DNA and induce G1/S cell cycle arrest, and that the gene defective in AT cells (later cloned and named ataxia telangiectasia mutated, ATM (Savitsky et al., 1995)) was required for such signaling.

ATM is not, however, required for ultraviolet radiation-induced p53 protein accumulation (Khanna and Lavin, 1993) (Canman et al., 1994). Instead, the ATM and Rad3-related protein ATR is responsible for ultraviolet radiation-induced p53 protein induction (Tibbetts et al., 1999). The mechanism by which ATM and ATR signal to p53 is dependent on their protein kinase activity (Canman et al., 1998) (Banin et al., 1998) (Tibbetts et al., 1999). As opposed to specific DNA lesions signaling directly to ATM or ATR, the current model is that a DNA damage-induced nuclear chromatin change, induced by ionizing radiation, activates ATM kinase activity (Bakkenist and Kastan, 2003). In contrast, processes that generate persistent single-stranded DNA, such as stalled DNA replication forks or nucleotide excision repair (e.g. to repair ultraviolet radiation-damaged DNA) may be the general signal to either recruit ATR or to activate ATR kinase activity via the ATR interacting protein ATRIP (Wang et al., 2003) (Unsal-Kacmaz and Sancar, 2004). In an interesting twist, DNA damage-induced disruption of nucleoli, rather than DNA damage *per se*, has been proposed to generate the signal for p53 protein stabilization (Rubbi and Milner, 2003a). The mechanism(s) by which DNA damage is detected by ATM and ATR, and then transmitted to p53 has received considerable attention.

P53 PHOSPHORYLATION AND PROTEIN ACCUMULATION

How does DNA damage cause p53 protein accumulation? DNA damage-induced accumulation of p53 is a rapid response that does not rely on changes in p53 mRNA expression. A process that has received little attention is the ionizing and ultraviolet radiation-induced increase in the rate of p53 mRNA translation (Fu and Benchimol, 1997) (Mazan-Mamczarz et al., 2003) (McLure, Takagi, and Kastan, unpublished observations). Much more effort has been directed at the rapid DNA damage-induced stabilization of the normally short-lived (~10 min) p53 protein (Maltzman and Czyzyk, 1984) (Kastan et al., 1991) (Tishler et al., 1993) (Maki and Howley, 1997). Clearly, it is important to unravel mechanisms by which DNA damage-induced increases in p53 translation or stabilization are achieved.

Most current hypotheses for p53 stabilization revolve around the p53 negative regulator hdm2 (the human homolog of murine mdm2). Hdm2 is critical for targeting p53 for degradation (Haupt et al., 1997) (Honda et al., 1997) (Kubbutat et al., 1997) (Midgley and Lane, 1997). Hdm2 is an E3 ubiquitin ligase that monoubiquitinates p53 (Ito et al., 2001), and upon subsequent polyubiquitination by a p300-dependent mechanism, p53 is targeted to the proteasome and degraded (Grossman et al., 2003). Although another regulator of p53 stability has been recently identified, Pirh2 (Cao et al., 2003), attention has focused on the role of the previously identified hdm2, which directly binds to an amino terminal domain of p53 (Momand et al., 1992) (Wu et al., 1993) (Oliner et al., 1993) (Kussie et al., 1996). Anything that disrupts the p53-hdm2 interaction is a candidate mechanism for DNA damage-induced p53 protein stabilization.

The first mechanism suggested for p53 stabilization involved the DNA-dependent protein kinase DNA-PK, which had been found to phosphorylate a p53 peptide on Ser15 (Lees-Miller et al., 1992). As DNA-PK is activated by double-stranded DNA ends, or by single to double-stranded transitions in DNA (Morozov et al., 1994), and ionizing radiation induces double-strand DNA breaks, it was assumed that ionizing radiation would induce DNA-PK to phosphorylate p53 *in vivo*. Indeed, ionizing radiation was shown to cause p53 Ser15 phosphorylation *in vivo* (Siliciano et al., 1997) (Shieh et al., 1997), and the Ser15 phosphorylation correlated with p53 protein induction. Because Ser15 phosphorylation weakly inhibited hdm2 binding, phosphorylation of p53 by DNA-PK on Ser15 was proposed to cause p53 stabilization (Shieh et al., 1997). However, although DNA-PK phosphorylates pre-existing p53 on Ser15 (Woo et al., 2002), DNA-PK is dispensable for DNA damage-induced p53 protein induction (Jimenez et al., 1999).

Another protein kinase that is related to DNA-PK by virtue of belonging to the PI3-kinase family is ATM (Savitsky et al., 1995). There was a correlation between ATM kinase activity being stimulated by ionizing radiation, ATM phosphorylating a p53 peptide on Ser15 (Canman et al., 1998) (Banin et al., 1998), and ATM being required for p53 protein accumulation following ionizing radiation (Kastan et al., 1992). Similarly, ATR is required for p53 Ser15 phosphorylation and for protein accumulation following ultraviolet irradiation (Tibbetts et al., 1999). However, germline mutation of the murine equivalent of Ser15 to alanine did not significantly impair DNA damage induction of p53 protein, finally burying the hypothesis that p53 protein stabilization occurred via phosphorylation of p53 on Ser15 (Chao et al., 2000) (Sluss et al., 2004). Nonetheless, the fact remains that ATM, ATR, and DNA-PK can all phosphorylate p53 on Ser15, suggesting a functional role for this modification.

There is evidence that Ser15 phosphorylation can enhance p53 transcriptional activity (Dumaz and Meek, 1999) (Turenne et al., 2001). This may be due to p53 phosphorylated on Ser15 binding better to p300, which enhances transcription by acetylating p53 to in turn enhance p53 DNA binding activity (Sakaguchi et al., 1998).

The next mechanism thought to be responsible for DNA damage-induced p53 protein stabilization was Ser20 phosphorylation, but with Chk1/Chk2 as the relevant p53 kinases. Ionizing radiation causes ATM-dependent activation of Chk2 kinase (Ahn et al., 2000) (Matsuoka et al., 2000) (Melchionna et al., 2000), and ultraviolet radiation causes ATR-dependent activation of Chk1 kinase (Guo et al., 2000) (Liu et al., 2000) (Zhao and Piwnica-Worms, 2001). Ionizing or ultraviolet radiation induce p53 Ser20 phosphorylation *in vivo*, and the Ser20 phosphorylation correlates with p53 protein induction (Shieh et al., 1999) (Chehab et al., 1999) (Hirao et al., 2000). Because Ser20 phosphorylation could inhibit hdm2 binding, phosphorylation of p53 by Chk1 or Chk2 on Ser20 was proposed to cause p53 stabilization (Chehab et al., 1999) (Shieh et al., 1999) (Unger et al., 1999). However, germline mutation of the murine equivalent of Ser20 to alanine did not significantly impair DNA damage induction of p53 protein, finally burying the hypothesis that p53 protein stabilization occurred via phosphorylation of p53 on Ser20 (Wu et al., 2002).

There is currently no clear function for Ser20 phosphorylation, but like Ser15 phosphorylation there may be a role in binding p300 (Dornan and Hupp, 2001). Ironically, if phosphorylation of any p53 residue in the hdm2 binding domain inhibits hdm2 binding, and thereby p53 degradation, it is Thr18, phosphorylation of which dramatically inhibits hdm2 binding *in vitro* (Sakaguchi et al., 2000). Although Thr18 phosphorylation is induced by DNA damage, the responsible *in vivo* kinase is not known. Interestingly, prior phosphorylation of Ser15 makes p53 Thr18 a substrate for phosphorylation by casein kinase I *in vitro* (Dumaz et al., 1999). Detailed analysis of the contributions of phosphorylation of one or more of Ser15, Thr18, and Ser20 to p53 protein induction or function is lacking because of the technical problems inherent with analyzing such interdependent effects.

Another reported target site of p53 itself is Ser45, which is phosphorylated in an ATM-dependent manner by homeodomain-interacting kinase-2 (D'Orazi et al., 2002) (Hofmann et al., 2002). Whether this ATM-dependent kinase that phosphorylates p53 is required for DNA damage-induced p53 stabilization via Ser45 phosphorylation *in vivo* remains to be determined.

An alternate mechanism by which the p53 protein level can be regulated is DNA damage-induced phosphorylation of hdm2 protein by ATM (Maya et al., 2001). Hdm2 that is phosphorylated on Ser395 loses its ability to bind p53, thereby stabilizing p53 independently from any signal to p53 itself. This could explain the observations that DNA damage-induced phosphorylations of p53 residues are not required for p53 protein stabilization.

Clearly, dephosphorylation should also occur on p53 residues that are phosphorylated, and ATM-dependent dephosphorylation of p53 on Ser376 creates a 14-3-3 binding site (Waterman et al., 1998). Based on the effect of 14-3-3 binding on other proteins, it might be predicted that 14-3-3 binding would sequester p53 in the cytoplasm, perhaps affecting p53 protein stability. However, 14-3-3 binding was proposed to activate nuclear p53 DNA binding activity (Waterman et al., 1998). Indeed, how phosphorylation/ dephosphorylation, and acetylation/ deacetylation, of any of the characterized p53 modification sites, contributes to the acute p53 response to DNA damage or to p53 tumor suppressor activity *in vivo* remains an open question.

P53 DNA BINDING ACTIVITY

A long-standing question is whether DNA damage signaling is required to activate p53 sequence-specific DNA binding. Although not commonly acknowledged, there are actually two models for how DNA damage functionally regulates p53. The first, often assumed, is that DNA damage signals directly to p53 to increase p53 sequence-specific DNA binding activity. This was initially demonstrated to occur by modifying the carboxy-terminal 30 amino acid regulatory domain of p53 by deletion, phosphorylation by casein kinase II, or protein binding (hsp70 or PAb421), all of which activate p53 sequence-specific DNA binding (Hupp et al., 1992) (Hupp et al., 1995). Activation of DNA binding is also conferred via modification of the carboxy terminal 30 amino acids by phosphorylation by protein kinase C (Takenaka et al., 1995), dephosphorylation of Ser376 and binding to 14-3-3 (Waterman et al., 1998), acetylation by p300/CBP (Gu and Roeder, 1997), binding to Ref-1 (Jayaraman et al., 1997), binding to c-abl (Nie et al., 2000), sumolation (Gostissa et al., 1999) (Rodriguez et al., 1999), binding to S100β (Lin.J. et al., 2001), or binding to short segments of single-stranded DNA (Jayaraman and Prives, 1995) (Okorokov et al., 1997). Although murine cells can alternatively splice p53 mRNA to produce a transcript lacking the carboxy terminal 30 amino acid regulatory domain, there is no comparable splice variant in human cells (Arai et al., 1986) (Will et al., 1995). A carboxy terminal phosphorylation event that is induced by ultraviolet but not ionizing radiation is Ser392 phosphorylation (Kapoor and Lozano, 1998) (Lu et al., 1998). Although this is not required for p53 protein accumulation, it does affect p53 sequence-specific DNA binding activity *in vitro*, although indirectly by regulating the affinity of p53 for non-specific DNA (Nichols and Matthews, 2002). Because Ser392 phosphorylation is not induced by ionizing radiation, this is either a function unique to ultraviolet radiation or there is a compensatory event induced by ionizing radiation.

A potential indirect signaling intermediate is the base excision repair endonuclease, Ref1, which can reduce and activate p53 DNA binding activity *in vitro* (Jayaraman et al., 1997). When damaged DNA is repaired by nucleotide excision repair, an ~29 base oligonucleotide is excised and released that can bind to the carboxy terminal regulatory domain of p53. This activates p53 sequence-specific DNA binding activity *in vitro*, and *in vivo* manipulation of the Ref1 protein level correlates with changes in transcriptional activation of some p53 targets (Jayaraman et al., 1997).

In vitro activation of p53 specific DNA binding can also occur in a manner dependent on DNA-PK plus an unidentified DNA damage-activated factor (Woo et al., 1998). However, no DNA damage signaling modification has been demonstrated to alter p53 tumor suppressor activity or DNA binding activity *in vivo*. In contrast, p53 present in nuclear extracts binds DNA even if accumulation was induced by inhibiting p53 degradation, in the absence of a DNA damaging signal. *In vivo*, DNA damage activates p53 DNA binding activity by virtue of an increased p53 protein level, but the actual specific DNA binding activity is not enhanced by DNA damage. This is shown by p53 binding similarly, after accounting for differences in total p53 protein, to some genomic binding sites *in vivo* in the absence of a DNA damage signal (Espinosa and Emerson, 2001) (Kaeser and Iggo, 2002). This apparent discrepancy between *in vitro* and *in vivo* results actually reflects regulation of p53 DNA binding activity by the conformation of the DNA. Even *in vitro*, p53 binding induces a severe bend and distortions in the DNA helix (Balagurumoorthy et al., 1995). Computer modeling predicted that pre-bent DNA, such as occurs on histones, would greatly increase the free energy available for p53 binding (Durell et al., 1998). This explains why *in vivo* p53 does not require activation of specific DNA binding in order to bind to some high-affinity binding sites.

Although p53 protein increases, and thus p53 DNA binding activity, it remains uncertain whether the specific DNA binding activity of p53 is regulated during a DNA damage response. It need not necessarily be regulated because the DNA damage-induced stabilization of p53 protein is sufficient to enhance binding to at least some genomic target sites that have a favorable chromatin conformation. Modifications of p53, such as phosphorylation, might affect DNA binding specificity directly, or indirectly by differentially recruiting other p53 binding molecules.

SIGNALING TO P53 VIA INTERACTING MOLECULES

This suggests a second model for regulating p53 activity, which is regulation of factors that cooperate with p53 to affect p53 DNA binding activity. Such a factor need not bind to p53 to affect p53 activity, but could be regulated by a DNA damage signal. For example, DNA damage-induced alteration of the chromatin structure around a p53 DNA binding site could affect the DNA torsion and bending, thereby regulating p53 affinity for the site.

As p53 is not only imported but also exported from the nucleus (Middeler et al., 1997), DNA damage could potentially affect either of these processes. As mentioned above, the p53-hdm2 interaction is inhibited by DNA damage-induced phosphorylation of hdm2 protein by ATM (Maya et al., 2001). This would inhibit the various negative regulatory effects of hdm2 on p53 as described above, and would additionally inhibit hdm2-dependent nuclear export of p53 (Tao and Levine, 1999). Interestingly, intra-molecular p53 interaction was also proposed to regulate nuclear export (Stommel et al., 1999). In that model, p53 dimers would be competent for export, whereas p53 tetramers would mask the nuclear export signal. This could potentially be regulated by a DNA damage signal at multiple levels, since p53 cotranslationally forms dimers, which subsequently associate post-translationally to form tetramers (Nicholls et al., 2002). It should be noted that although p53 was previously proposed to bind DNA as a dimer (Hupp et al., 1992), this was predicated upon the carboxy-terminal negative regulatory domain having been incorrectly identified as the tetramerization domain (Sturzbecher et al., 1992). In fact, p53 binds DNA as a tetramer, both when present in nuclear extract from DNA damaged cells (McLure and Lee, 1998) or when produced *in vitro* (Halazonetis and Kandil, 1993) (Cho et al., 1994) (Waterman et al., 1995) (Wang et al., 1995) (Balagurumoorthy et al., 1995).

The DNA binding activity of p53 can also potentially be affected by changes in extrinsic factors, including ASPP1/2, the full-length version of 53BP2 (Samuels-Lev et al., 2001) and the p53 family members p63/p73 (Flores et al., 2002). ASPP1/2 binds to p53 and enhances the affinity of p53 for the genomic binding site in pro-apoptotic promoters (Samuels-Lev et al., 2001). It is not clear how this is mechanistically achieved because ASPP1/2 binds to the DNA binding surface of p53, which appears to be incompatible with simultaneous DNA binding (Gorina and Pavletich, 1996). It is also unclear how p63/p73 enhance p53 binding to pro-apoptotic promoters, as p63/p73 do not heteroligomerize with p53 (Davison et al., 1999).

Another indirect regulator of p53 activity that can be activated by DNA damage is c-jun, which binds to the hdm2 promoter to coactivate transcription with p53 (Ries et al., 2000) (Phelps et al., 2003). Thus, DNA damage could signal in an ATM-dependent manner to the c-jun N-terminal kinase, JNK (Lee et al., 1998), or to the c-jun DNA binding activator and DNA repair endonuclease Ref1 (Xanthoudakis et al., 1992). Either of these could activate c-jun, which in turn can indirectly regulate p53 stability via hdm2 induction.

Another potential target for regulating p53 stability is p300. The region of p53 that binds to hdm2 also binds to p300, making it difficult to conclusively attribute p53 regulation to effects on hdm2 versus p300. The situation is further complicated by p300 binding to p53 via two different domains, which are differentially affected by phosphorylation of p53 on Ser15 and Ser20 (Dornan and Hupp, 2001). Additionally, hdm2 inhibits p300-mediated p53 acetylation, and acetylation of lysine residues in the carboxy terminus of p53 by p300 competes with ubiquitination of the same residues by hdm2, and p53 must be deacetylated in order to be degraded (Ito et al., 2001) (Ito et al., 2002). To top it off, p300 not only acetylates p53 to potentially regulate p53 DNA binding activity, but p300 targets p53 for poly-ubiquitination and degradation (Grossman et al., 2003). Clearly, effects of specific DNA damage signaling to p53, hdm2, and potentially p300, are very difficult to untangle. Nonetheless, this may be the most important regulatory network for DNA damage signaling to p53.

PHYSIOLOGIC EFFECTS

DNA damage signals to p53 to stabilize p53 protein, to induce post-translational modifications to p53, and possibly to regulate other factors that determine p53 activity. One activity of p53 is DNA binding, but the major question is what physiologic activities of p53 are activated by DNA damage signals? An important activity of p53 is preventing the occurance of, or replicative survival of cells that sustain, genetic abnormalities that may be tumorigenic, such as gene amplification (Livingstone et al., 1992) and rereplication or endoreduplication (Khan and Wahl, 1998) (Stewart et al., 1999a) (Vaziri et al., 2003). Regulation of these processes by p53 may be mediated indirectly by DNA damage-induced, p53-dependent cell cycle checkpoints. For example, DNA damage induces p53 to bind to a DNA binding site in the p21 promoter (El-Deiry et al., 1993), which induces p21-dependent G1/S arrest (Dulic et al., 1994).

An alternate DNA damage-induced cell fate that can be determined by p53 is apoptosis. Although it seems teleologically sound, there is no evidence to support the oft-stated hypothesis that cell cycle checkpoints afford a cell time to repair damaged DNA, but if damage cannot be repaired then p53 induces apoptosis.

For example, most ionizing radiation-induced single- and double-strand breaks are repaired in a timeframe of seconds to minutes. However, DNA damage-activated p53 does not induce p21 protein until several hours after ionizing radiation, at which time a p53-dependent G1/S arrest is initiated and persists for a timeframe of days (Kastan and Kuerbitz, 1993) (Dulic et al., 1994). In fact, in primary fibroblasts in tissue culture, a high dose of ionizing radiation induces permanent cell cycle arrest, or senescence (Di Leonardo et al., 1994). In contrast, relatively low doses of ionizing radiation induce apoptosis in primary murine thymocytes (Lowe et al., 1993) (Clarke et al., 1993). Thus, the p53 physiologic response to DNA damage is not determined by the amount of DNA damage. Neither do differential post-translational modifications of p53 cause the difference between arrest versus apoptosis. In fact, p53 itself does not appear to be the determinant of cell fate.

Instead, the intracellular environment that p53 feeds into is responsible for determining the response to DNA damage-induced p53 transcriptional and transcription-independent activities. For example, in one cell type growth factors can block ionizing radiation-induced apoptosis (Canman et al., 1995), and lethally irradiated mice can be rescued by inhibiting apoptosis downstream of p53 (Pestina et al., 2001). Another pervasive misconception has been that DNA damage-induced apoptosis occurs in a p53-dependent manner. In cell lines in tissue culture, the reverse is true. That is, isogenic cell lines that lack wild type p53 undergo greater DNA damage-induced apoptosis compared to cells that retain wild type p53 (Gupta et al., 1997) (Stewart et al., 1999b) (Han.Z. et al., 2002) (Magrini et al., 2002) (Galmarini et al., 2003) (Lee et al., 2003). This counterintuitive effect is due to p53-dependent activation of p21, and presumably the lack of p53-dependent apoptosis reflects prior selection against this process during tumorigenesis.

In contrast, cells in some organs *in vivo*, such as thymus, spleen, bone marrow, and small intestine, undergo ionizing radiation-induced cell death that is entirely dependent upon p53 (Lowe et al., 1993) (Clarke et al., 1993) (Komarov et al., 1999) (Fei et al., 2002). These are precisely the tissues that present dose-limiting toxicity for genotoxic cancer therapies in humans. Therefore, inhibition, rather than activation, of p53 might be expected to enhance the therapeutic index of genotoxic cancer therapeutics.

CONCLUSIONS

The clinical importance of p53 and the complexity of all aspects of the p53 DNA damage response have stimulated an overwhelming number of publications on the subject. From the initial discovery that DNA damage signals to p53, to the many identified covalently modified residues and the various proteins that interact with p53, there have been many hypotheses for how DNA damage signals to p53 and how such signaling regulates p53 activity. There have also been well-founded challenges to virtually every concept in the p53 field. We have been poignantly reminded that the simplest interpretation of experiments, while scientifically sound, may be incorrect. Moreover, apparently sound teleologic arguments have not borne fruit, highlighting our limited understanding of the complex cellular DNA damage response. How does DNA damage signaling stabilize p53 protein, which post-translational modifications are required for which p53 activities, and ultimately, what constitutes p53 tumor suppressor activity, how does DNA damage affect such activity, and how can we exploit such knowledge to improve cancer therapy? We can undoubtedly look forward to new approaches to sort out these old questions of how DNA damage signals to and regulates p53 function.

REFERENCES

Ahn,J.-Y., Schwarz,J.K., Piwnica-Worms,H., and Canman,C.E. (2000). Threonine 68 phosphorylation by ATM is required for efficient activation of Chk2 in response to ionizing radiation. Cancer Research *60*, 5934-5936.

Arai,N., Nomura,D., Yokota,K., Wolf,D., Brill,E., Shohat,O., and Rotter,V. (1986). Immunologically distinct p53 molecules generated by alternative splicing. Mol Cell Biol *6*, 3232-3239.

Bakalkin,G., Yakovleva,T., Selivanova,G., Magnusson,K.P., Szekely,L., Kiseleva,E., Klein,G., Terenius,L., and Wiman,K.G. (1994). p53 binds single-stranded DNA ends and catalyzes DNA renaturation and strand transfer. Proc Natl Acad Sci USA *91*, 413-417.

Bakkenist,C.J. and Kastan,M.B. (2003). DNA damage activates ATM through intermolecular autophosphorylation and dimer dissociation. Nature *421*, 499-506.

Balagurumoorthy,P., Sakamoto,H., Lewis,M.S., Zambrano,N., Clore,G.M., Gronenborn,A.M., Appella,E., and Harrington,R.E. (1995). Four p53 DNA-binding domain peptides bind natural p53-response elements and bend the DNA. Proc Natl Acad Sci USA *92*, 8591-8595.

Banin,S., Moyal,L., Shieh,S.-Y., Taya,Y., Anderson,C.W., Chessa,L., Smorodinsky,N.I., Prives,C., Reiss,Y., Shiloh,Y., and Ziv,Y. (1998). Enhanced phosphorylation of p53 by ATM in response to DNA damage. Science *281*, 1674-1677.

Canman,C.E., Gilmer,T., Coutts,S., and Kastan,M.B. (1995). Growth factor modulation of p53-mediated growth arrest vs. apoptosis. Genes Dev. *9*, 600-611.

Canman,C.E., Lim,D.-S., Cimprich,K.A., Taya,Y., Tamai,K., Sakaguchi,K., Appella,E., Kastan,M.B., and Siliciano,J.D. (1998). Activation of the ATM kinase by ionizing radiation and phosphorylation of p53. Science *281*, 1677-1679.

Canman,C.E., Wolff,A.C., Chen,C., Fornace,A.J., and Kastan,M.B. (1994). The p53-dependent G1 Cell Cycle Checkpoint Pathway and Ataxia-Telangiectasia. Cancer Res *54*, 5054-5058.

Cao,L., Li,W., Kim,S., Brodie,S.G., and Deng,C.X. (2003). Senescence, aging, and malignant transformation mediated by p53 in mice lacking the Brca1 full-length isoform. Genes Dev. *17*, 201-213.

Chao,C., Saito,S., Anderson,C.W., Appella,E., and Xu,Y. (2000). Phosphorylation of murine p53 at Ser-18 regulates the p53 responses to DNA damage. Proc Natl Acad Sci USA *97*, 11936-11941.

Chehab,N.H., Malikzay,A., Stavridi,E.S., and Halazonetis,T.D. (1999). Phosphorylation of Ser-20 mediates stabilization of human p53 in response to DNA damage. Proc Natl Acad Sci USA *96*, 13777-13782.

Cho,Y., Gorina,S., Jeffery,P.D., and Pavletich,N.P. (1994). Crystal structure of a p53 tumor suppressor-DNA complex: Understanding tumorigenic mutations. Science *265*, 346-355.

Clarke,A.R., Purdie,C.A., Harrison,D.J., Morris,R.G., Bird,C.C., Hooper,M.L., and Wyllie,A.H. (1993). Thymocyte apoptosis induced by p53-dependent and independent pathways. Nature *362*, 849-852.

D'Orazi,G., Cecchinelli,B., Bruno,T., Manni,I., Higashimoto,Y., Saito,S., Gostissa,M., Coen,S., Marchetti,A., Del Sal,G., Piaggio,G., Fanciulli,M., Appella,E., and Soddu,S. (2002). Homeodomain-interacting protein kinase-2 phosphorylates p53 at Ser 46 and mediates apoptosis. Nat Cell Biol *4*, 11-9.

Davison,T.S., Vagner,C., Kaghad,M., Ayed,A., Caput,D., and Arrowsmith,C.H. (1999). p73 and p63 are homotetramers capable of weak heterotypic interactions with each other but not with p53. J. Biol. Chem. *274*, 18709-18714.

Deleo,A.B., Jay,G., Appells,E., Dubois,G.C., Law,L.W., and Old,L.J. (1979). Detection of a transformation-related antigen in chemically induced sarcomas and other transfromed cells of the mouse. Proc Natl Acad Sci USA *76*, 2420-2424.

Di Leonardo,A., Linke,S.P., Clarkin,K., and Wahl,G.M. (1994). DNA damage triggers a prolonged p53-dependent G1 arrest and long-term induction of Cip1 in normal human fibroblasts. Genes Dev. *8*, 2540-2551.

Dornan,D. and Hupp,T.R. (2001). Inhibition of p53-dependent transcription by BOX-I phospho-peptide mimetics that bind to p300. EMBO Rep *2*, 139-144.

Dulic,V., Kaufmann,W.K., Wilson,S.J., Tlsty,T.D., Lees,E., Harper,W., Elledge,S.J., and Reed,S.I. (1994). p53-dependent inhibition of cyclin-dependent kinase activities in human fibroblasts during radiation-induced G1 arrest. Cell *76*, 1013-1023.

Dumaz,N. and Meek,D.W. (1999). Serine 15 phosphorylation stimulates p53 transactivation but does not directly influence interaction with HDM2. EMBO J *18*, 7002-7010.

Dumaz,N., Milne,D.M., and Meek,D.W. (1999). Protein kinase CK1 is a p53-threonine 18 kinase which requires prior phosphorylation of serine 15. FEBS Letters *463*, 312-316.

Durell,S.R., Appella,E., Nagaich,A.K., Harrington,R.E., Jernigan,R.L., and Zhurkin,V.B. (1998). DNA bending induced by tetrameric binding of the tumor-suppressive p53 protein: steric constraints on conformation. In Macromolecules, proc. structure, motion, interactions and expression of biological of the tenth conversation, R.H.Sarma and M.H.Sarma, eds. (Albany, NY: Adenine Press), pp. 277-295.

El-Deiry,W.S., Kern,S.E., Pietenpol,J.A., Kinzler,K.W., and Vogelstein,B. (1992). Definition of a consensus binding site for p53. Nature *356*, 215-221.

El-Deiry,W.S., Tokino,T., Velculescu,V.E., Levy,D.B., Parsons,R., Trent,J.M., Lin,D., Mercer,W.E., Kinzler,K.W., and Vogelstein,B. (1993). WAF1, a potential mediator of p53 tumor suppression. Cell *75*, 817-825.

Espinosa,J.M. and Emerson,B.M. (2001). Transcriptional regulation by p53 through intrinsic DNA/chromatin binding and site-directed cofactor recruitment. Mol Cell *8*, 57-69.

Fei,P., Bernhard,E.J., and El-Deiry,W.S. (2002). Tissue-specific induction of p53 targets in vivo. Cancer Res. *62*, 7316-7327.

Flores,E.R., Tsai,K.Y., Crowley,D., Sengupta,S., Yang,a., McKeon,F., and Jacks,T. (2002). p63 and p73 are required for p53-dependent apoptosis in response to DNA damage. Nature *416*, 560-564.

Friedberg,E.C. (2003). DNA damage and repair. Nature *421*, 436-440.

Fu,L. and Benchimol,S. (1997). Participation of the human p53 3'UTR in translational repression and activation following y-irradiation. EMBO J *16*, 4117-4125.

Funk,W.D., Pak,D.T., Karas,R.H., Wright,W.E., and Shay,J.W. (1992). A transcriptionally active DNA-binding site for human p53 protein complexes. Mol Cell Biol *12*, 2866-2871.

Galmarini,C.M., Voorzanger,N., Falette,N., Jordheim,L., Cros,E., Puisieux,A., and Dumontet,C. (2003). Influence of p53 and p21(WAF1) expression on sensitivity of cancer cells to cladribine. Biochem Pharmacol *65*, 121-129.

Giaccia,A.J. and Kastan,M.B. (1998). The complexity of p53 modulation: emerging patterns from divergent signals. Genes Dev. *12*, 2973-2983.

Gorina,S. and Pavletich,N.P. (1996). Structure of the p53 tumor suppressor bound to the ankyrin and SH3 domains of 53BP2. Science *274*, 1001-1005.

Gostissa,M., Hengstermann,A., Fogal,V., Sandy,P., Schwarz,S.E., Scheffner,M., and Del Sal,G. (1999). Activation of p53 by conjugation to the ubiquitin-like protein SUMO-1. EMBO J *18*, 6462-6471.

Grossman,S.R., Deato,M.E., Brignone,C., Chan,H.M., Kung,A.L., Tagami,H., Nakatani,Y., and Livingston,D.M. (2003). Polyubiquitination of p53 by a ubiquitin ligase activity of p300. Science *300*, 342-344.

Gu,W. and Roeder,R.G. (1997). Activation of p53 sequence-specific DNA binding by acetylation of the p53 C-terminal domain. Cell *90*, 595-606.

Guo,Z., Kumagai,A., Wang,S.X., and Dunphy,W.G. (2000). Requirement for Atr in phosphorylation of Chk1 and cell cycle regulation in response to DNA replication blocks and UV-damaged DNA in *Xenopus egg* extracts. Genes Dev. *14*, 2745-2756.

Gupta,M., Fan,S., Zhan,Q., Kohn,K.W., O'Connor,P.M., and Pommier,Y. (1997). Inactivation of p53 increases the cytotoxicity of camptothecin in human colon HCT116 and breast MCF-7 cancer cells. Clinical Cancer Research *3*, 1653-1660.

Halazonetis,T.D. and Kandil,A.N. (1993). Conformational shifts propagate from the oligomerization domain of p53 to its tetrameric DNA binding domain and restore DNA binding to select p53 mutants. EMBO J *12*, 5057-5064.

Han.Z., Wei,W., Dunaway,S., Darnowski,J.W., Calabresi,P., Sedivy,J., Hendrickson,E.A., Balan,K.V., Pantazis,P., and Wyche,J.H. (2002). Role of p21 in apoptosis and senescence of human colon cancer cells treated with camptothecin. J. Biol. Chem. *277*, 17154-17160.

Hartwell,L.H. and Weinert,T.A. (1989). Checkpoints: Controls that ensure the order of cell cycle enents. Science *246*, 629-634.

Haupt,Y., Maya,R., Kazaz,A., and Oren,M. (1997). Mdm2 promotes the rapid degradation of p53. Nature *387*, 296-299.

Hirao,A., Kong,Y.-Y., Matsuoka,S., Wakeham,A., Ruland,J., Yoshida,H., Liu,D., Elledge,S.J., and Mak,T.W. (2000). DNA damage-induced activation of p53 by the checkpoint kinase Chk2. Science *287*, 1824-1827.

Hofmann,T.G., Moller,A., Sirma,H., Zentgraf,H., Taya,Y., Droge,W., Will,H., and Schmitz,M.L. (2002). Regulation of p53 activity by its interaction with homeodomain-interacting protein kinase-2. Nat Cell Biol *4*, 1-10.

Honda,R., Tanaka,H., and Yasuda,H. (1997). Oncoprotein MDM2 is a ubiquitin ligase E3 for tumor suppressor p53. FEBS Letters *420*, 25-27.

Hupp,T.R., Meek,D.W., Midgley,C.A., and Lane,D.P. (1992). Regulation of the specific DNA binding function of p53. Cell *71*, 875-886.

Hupp,T.R., Sparks,A., and Lane,D.P. (1995). Small peptides activate the latent sequence-specific DNA binding function of p53. Cell *83*, 237-245.

Ito,A., Kawaguchi,Y., Lai,C.H., Kovacs,J.J., Higashimoto,Y., Appella,E., and Yao,T.P. (2002). MDM2-HDAC1-mediated deacetylation of p53 is required for its degradation. EMBO J *21*, 6236-6245.

Ito,A., Lai,C.H., Zhao,X., Saito,S., Hamilton,M.H., Appella,E., and Yao,T.P. (2001). p300/CBP-mediated p53 acetylation is commonly induced by p53-activating agents and inhibited by MDM2. EMBO J *20*, 1331-1340.

Jayaraman,L., Murthy,K.G.K., Zhu,C., Curran,T., Xanthoudakis,S., and Prives,C. (1997). Identification of redox-repair protein Ref-1 as a potent activator of p53. Genes Dev. *11*, 558-570.

Jayaraman,L. and Prives,C. (1995). Activation of p53 sequence-specific DNA binding by short single strands of DNA requires the p53 C-terminus. Cell *81*, 1021-1029.

Jimenez,G.S., Bryntesson,F., Torres-Arzayus,M.I., Priestley,A., Beeche,M., Saito,S., Sakaguchi,K., Appella,E., Jeggo,P.A., Tacciolo,G.E., Wahl,G.M., and Hubank,M. (1999). DNA-dependent protein kinase is not required for the p53-dependent response to DNA damage. Nature *400*, 81-83.

Kaeser,M.D. and Iggo,R.D. (2002). Chromatin immunoprecipition analysis fails to support the latency model for regulation of p53 DNA binding activity in vivo. Proc Natl Acad Sci USA *99*, 95-100.

Kapoor,M. and Lozano,G. (1998). Functional activation of p53 via phosphorylation following DNA damage by UV but not y radiation. Pro Natl Acad Sci USA *95*, 2834-2837.

Kastan,M.B. and Kuerbitz,S.J. (1993). Control of G1 arrest after DNA damage. Environ. Health Persp. *101 (suppl 5)*, 55-58.

Kastan,M.B., Onyekwere,O., Sidransky,D., Vogelstein,B., and Craig,R.W. (1991). Participation of p53 protein in the cellular response to DNA damage. Cancer Res *51*, 6304-6311.

Kastan,M.B., Zhan,Q., El-Deiry,W.S., Carrier,F., Jacks,T., Walsh,W.V., Plunkett,B.S., Vogelstein,B., and Fornace,A.J., Jr. (1992). A mammalian cell cycle checkpoint pathway utilizing p53 and GADD45 is defective in ataxia-telangiectasia. Cell *71*, 587-597.

Kern,S.E., Kinzler,K.W., Bruskin,A., Jarosz,D., Friedman,P., Prives,C., and Vogelstein,B. (1991). Identification of p53 as a sequence-specific DNA-binding protein. Science *252*, 1708-1711.

Khan,S.H. and Wahl,G.M. (1998). p53 and pRb prevent rereplication in response to microtubule inhibitors by mediating a reversible G_1 arrest. Cancer Res *58*, 396-401.

Khanna,K.K. and Lavin,M.F. (1993). Ionizing Radiation and UV induction of p53 protein by different pathways in ataxia-telangiectasia cells. Oncogene *8*, 3307-3312.

Komarov,P.G., Komarova,E.A., Kondratov,R.V., Christov-Tselkov,K. , Coon,J.S., Chernov,M.V., and Gudkov,A.V. (1999). A chemical inhibitor of p53 that protects mice from the side effects of cancer therapy. Science *285*, 1733-1737.

Kubbutat,M.H., Jones,S.N., and Vousden,K.H. (1997). Regulation of p53 stability by Mdm2. Nature *387*, 299-303.

Kussie,P.H., Gorina,S., Marechal,V., Elenbaas,B., Moreau,J., Levine,A.J., and Pavletich,N.P. (1996). Structure of the MDM2 oncoprotein bound to the p53 tumor suppressor transactivation domain. Science *274*, 948-953.

Lane,D.P. and Crawford,L.V. (1979). T antigen is bound to host protein in SV40 transformed cells. Nature *278*, 261-263.

Lee,E.J., Gerhold,M., Palmer,M.W., and Christen,R.D. (2003). p53 protein regulates the effects of amifostine on apoptosis, cell cycle progression, and cytoprotection. Br. J. Cancer *88*, 754-759.

Lee,S., Elenbaas,B., Levine,A., and Griffith,J. (1995). p53 and its 14kDa C-terminal domain recognize primary DNA damage in the form of insertion/deletion mismatches. Cell *81*, 1013-1020.

Lee,S.J., Dimtchev,A., Lavin,M., Dritschilo,A., and Jung,M. (1998). A novel ionizing radiation-induced signaling pathway that activates the transcription factor NF-*k*B. Oncogene *17*, 1821-1826.

Lees-Miller,S.P., Sakaguchi,K., Ullrich,S.J., Appella,E., and Anderson,C.W. (1992). Human DNA-activated protein kinase phosphorylates serines 15 and 37 in the amino-terminal transactivation domain of human p53. Mol Cell Biol *12*, 5041-5049.

Lin.J., Blake,M., Tang,C., Zimmer,D., Rustandi,R.R., Weber,D.J., and Carrier,F. (2001). Inhibition of p53 transcriptional activity by the S100B calcium-binding protein. J. Biol. Chem. *276*, 35037-35041.

Linzer,D.I. and Levine,A.J. (1979). Characterization of a 54K dalton cellular SV40 tumor antigen present in SV40-transformed cells and uninfected embryonal carcinoma cells. Cell *17*, 43-52.

Liu,Q., Guntuku,S., Cui,X.-S., Matsuoka,S., Cortez,D., Tamai,K., Luo,G., Carattini-Rivera,S., DeMayo,F., Bradley,A., Donehower,L.A., and Elledge,S.J. (2000). Chk1 is an essential kinase that is regulated by Atr and required for the G_2 /M DNA damage checkpoint. Genes Dev *14*, 1448-1459.

Livingstone,L.R., White,A., Sprouse,J., Livanos,E., Jacks,T., and Tlsty,T.D. (1992). Altered cell cycle arrest and gene amplification potential accompany loss of wild-type p53. Cell *70*, 923-935.

Lowe,S.W., Schmitt,S.W., Smith,S.W., Osborne,B.A., and Jacks,T. (1993). p53 is required for radiation-induced apoptosis in mouse thymocytes. Nature *362*, 847-849.

Lu,H., Taya,Y., Ikeda,M., and Levine,A.J. (1998). Ultraviolet radiation, but not gamma radiation or etoposide-induced DNA damage, results in the phosphorylation of the murine p53 protein at serine-389. Proc Natl Acad Sci USA *95*, 6399-6402.

Magrini,R., Bhonde,M.R., Hanski,M.L., Notter,M., Scherubl,H., Boland,C.R., Zeitz,M., and Hanski,C. (2002). Cellular effects of CPT-11 on colon carcinoma cells: dependence on p53 and hMLH1 status. Int J Cancer *101*, 23-31.

Maki,C.G. and Howley,P.M. (1997). Ubiquitination of p53 and p21 is differentially affected by ionizing and UV radiation. Mol Cell Biol *17*, 355-363.

Maltzman,W. and Czyzyk,L. (1984). UV irradiation stimulates levels of p53 cellular tumor antigen in nontransformed mouse cells. Mol Cell Biol *4(9)*, 1689-1694.

Matsuoka,S., Rotman,G., Ogawa,A., Shiloh,Y., Tamai,K., and Elledge,S.J. (2000). Ataxia telangiectasia-mutated phosphorylates Chk2 *in vivo* and *in vitro*. Proc Natl Acad Sci USA *97*, 10389-10394.

Maya,R., Balass,M., Kim,S.-T., Shkedy,D., Leal,J.-F.M., Shifman,O., Moas,M., Buschmann,T., Ronai,Z., Shiloh,Y., Kastan,M.B., Katzir,E., and Oren,M. (2001). ATM-dependent phosphorylation of Mdm2 on serine 395: role in p53 activation by DNA damage. Genes Dev. *15*, 1067-1077.

Mazan-Mamczarz,K., Galban,S., Lopez de Silanes,I., Martindale,J.L., Atasoy,U., Keene,J.D., and Gorospe,M. (2003). RNA-binding protein HuR enhances p53 translation in response to ultraviolet light irradiation. Proc Natl Acad Sci USA *100*, 8354-8359.

McLure,K.G. and Lee,P.W.K. (1998). How p53 binds DNA as a tetramer. EMBO J *17*, 3342-3350.

Melchionna,R., Chen,X.-B., Blasina,A., and McGowan,C.H. (2000). Threonine 68 is required for radiation-induced phosphorylation and activation of Cds1. Nature Cell Biol *2*, 762-765.

Middeler,G., Zerf,K., Jenovai,S., Thulig,A., Tschodrichrotter,M., Kubitscheck,U., and Peters,R. (1997). The tumor suppressor p53 is subject to both nuclear import and export, and both are fast, energy-dependent and lectin-inhibited. Oncogene *14*, 1407-1417.

Midgley,C.A. and Lane,D.P. (1997). p53 protein stability in tumour cells is not determined by mutation but is dependent on Mdm2 binding. Oncogene *15*, 1179-1189.

Mihara,M., Erster,S., Zaika,A., Petrenko,O., Chittenden,T., Pancoska,P., and Moll,U.M. (2003). p53 has a direct apoptogenic role at the mitochondria. Mol Cell *11*, 577-590.

Momand,J., Zambetti,G.P., Olson,D.C., George,D.L., and Levine,A.J. (1992). The mdm-2 oncogene product forms a complex with the p53 protein and inhibits p53-mediated transactivation. Cell *69*, 1237-1245.

Morozov,V.E., Falzon,M., Anderson,C.W., and Kuff,E.L. (1994). DNA-dependent protein kinase is activated by nicks and larger single-stranded gaps. J. Biol. Chem. *269*, 16684-16688.

Nicholls,C.D., McLure,K.G., Shields,M.A., and Lee,P.W. (2002). Biogenesis of p53 involves cotranslational dimerization of monomers and posttranslational dimerization of dimers. Implications on the dominant negative effect. J. Biol. Chem. *277*, 12937-12945.

Nichols,N.M. and Matthews,K.S. (2002). Human p53 phosphorylation mimic, S392E, increases nonspecific DNA affinity and thermal stability. Biochemistry *41*, 170-178.

Nie,Y., Li,H.-H., Bula,C.M., and Liu,X. (2000). Stimulation of p53 DNA binding by c-Abl requires the p53 C terminus and tetramerization. Mol Cell Biol *20*, 741-748.

Offer,H., Wolkowicz,R., Matas,D., Blumenstein,S., Livneh,A., and Rotter,V. (1999). Direct involvement of p53 in the base excision repair pathway of the DNA repair machinery. FEBS Letters *450*, 197-204.

Okorokov,A.L., Ponchel,F., and Milner,J. (1997). Induced N- and C-terminal cleavage of p53: a core fragment of p53, generated by interaction with damaged DNA, promotes cleavage of the N-terminus of full-length p53, whereas ssDNA induces C-terminal cleavage of p53. EMBO J *16*, 6008-6017.

Oliner,J.D., Pietenpol,J.A., Thiagalingam,S., Gyuris,J., Kinzler,K.W., and Vogelstein,B. (1993). Oncoprotein mdm2 conceals the activation domain of tumor suppressor p53. Nature *362*, 857-860.

Pestina,T.I., Cleveland,J.L., Yang,C., Zambetti,G.P., and Jackson,C.W. (2001). Mpl ligand prevents lethal myelosuppression by inhibiting p53-dependent apoptosis. Blood *98*, 2084-2090.

Phelps,M., Darley,M., Primrose,J.N., and Blaydes,J.P. (2003). p53-independent activation of the hdm2-P2 promoter through multiple transcription factor response elements results in elevated hdm2 expression in estrogen receptor alpha-positive breast cancer cells. Cancer Res. *63*, 2616-2623.

Reed,M., Woelker,B., Wang,P., Wang,Y., Anderson,M.E., and Tegrmeyer,P. (1995). The C-terminal domain of p53 recognizes DNA damaged by ionizing radiation. Proc Natl Acad Sci USA *92*, 9455-9459.

Ries,S., Biederer,C., Woods,D., Shifman,O., Shirasawa,S., Sasazuki,T., McMahon,M., Oren,M., and McCormick,F. (2000). Opposing effects of Ras on p53: transcriptional activation of mdm2 and induction of p19ARF. Cell *103*, 321-330.

Rodriguez, M. S., Desterro, J. M. P., Lain, S., Midgley, C. A., Lane, D. P., and Hay, R. T. (1999) SUMO-1 modification activates the transcriptional response of p53. EMBO J *18*, 6455-6461.

Rouse,J. and Jackson,S.P. (2002). Interfaces between the detection, signaling, and repair of DNA damage. Science *297*, 547-551.

Rubbi,C.P. and Milner,J. (2003a). Disruption of the nucleolus mediates stabilization of p53 in response to DNA damage and other stresses. EMBO J *22*, 6068-6077.

Rubbi,C.P. and Milner,J. (2003b). p53 is a chromatin accessibility factor for nucleotide excision repair of DNA damage. EMBO J *22*, 975-986.

Sakaguchi,K., Herrera,J.E., Saito,S., Miki,T., Bustin,M., Vassilev,A., Anderson,C.W., and Appella,E. (1998). DNA damage activates p53 through a phosphorylation-acetylation cascade. Genes Dev *12*, 2831-2841.

Sakaguchi,K., Saito,S., Higashimoto,Y., Roy,S., Anderson,C.W., and Appella,E. (2000). Damage-mediated phosphorylation of human p53 threonine 18 through a cascade mediated by a casein 1-like kinase. Effect on Mdm2 binding. J Biol Chem *275*, 9278-9283.

Samuels-Lev,Y., O'Connor,D.J., Bergamaschi,D., Trigiante,G., Hsieh,J.-K., Zhong,S., Campargue,I., Naumovski,L., Crook,T., and Lu,X. (2001). ASPP proteins specifically stimulate the apoptotic function of p53. Mol Cell *8*, 781-794.

Savitsky,K., Bar-Shira,A., Gilad,S., Rotman,G., Ziv,Y., Vanagaite,L., Tagle,D.A., Smith,S., Uziel,T., Sfez,S., Ashkenazi,M., Pecker,I., Frydman,M., Harnik,R., Patanjali,S.R., Simmons,A., Clines,G.A., Sartiel,A., Gatti,R.A., Chessa,L. , Sanal,O., Lavin,M.F., Jaspers,N.G.J., Taylor,A.M.R., Arlett,C.F., Miki,T., Weissman,S.M., Lovett,M., Collins,F.S., and Shiloh,Y. (1995). A single ataxia telangiectasia gene with a product similar to PI-3 kinase. Science *268*, 1749-1753.

Shieh,S.-Y., Ikeda,M., Taya,Y., and Prives,C. (1997). DNA damage-induced phosphorylation of p53 alleviates inhibition by MDM2. Cell *91*, 325-334.

Shieh,S.-Y., Taya,Y., and Prives,C. (1999). DNA damage-inducible phosphorylation of p53 at N-terminal sites including a novel site, Ser20, requires tetramerization. EMBO J *18*, 1815-1823.

Siliciano,J.D., Canman,C.E., Taya,Y., Sakaguchi,K., Appella,E., and Kastan,M.B. (1997). DNA damage induces phosphorylation of the amino terminus of p53 . Genes Dev *11*, 3471-3481.

Sluss,H.K., Armata,H., Gallant,J., and Jones,S.N. (2004). Phosphorylation of serine 18 regulates distinct p53 functions in mice. Mol Cell Biol *24*, 976-984.

Stewart,Z.A., Leach,D.L., and Pietenpol,J. (1999a). p21 [Waf1/Cip1] Inhibition of cyclin E/Cdk2 activity prevents endoreduplication after mitotic spindle disruption. Mol Cell Biol *19*, 205-215.

Stewart,Z.A., Mays,D., and Pietenpol,J.A. (1999b). Defective G $_1$-S cell cycle checkpoint function sensitizes cells to microtubule inhibitor-induced apoptosis. Cancer Research *59*, 3831-3837.

Stommel,J.M., Marchenko,N.D., Jimenez,G.S., Moll,U.M., Hope,T.J., and Wahl,G.M. (1999). A leucine-rich nuclear export signal in the p53 tetramerization domain: regulation of subcellular localization and p53 activity by NES masking. EMBO J *18*, 1660-1672.

Sturzbecher,H.W., Brain,R., Addison,C., Rudge,K., Remm,M., Grimaldi,M., Keenan,E., and Jenkins,J.R. (1992). A C-terminal alpha-helix plus basic region motif is the major structural determinant of p53 tetramerization. Oncogene *7*, 1513-1523.

Szak,S.T., Mays,D., and Pietenpol,J.A. (2001). Kinetics of p53 binding to promoter sites in vivo. Mol Cell Biol *21*, 3375-3386.

Takenaka,I., Morin,F., Seizinger,B.R., and Kley,N. (1995). Regulation of the sequence-specific DNA binding function of p53 by protein kinase C and protein phosphatases. J. Biol. Chem. *270*, 5405-5411.

Tao,W. and Levine,A.J. (1999). Nucleocytoplasmic shuttling of oncoprotein Hdm2 is required for Hdm2-mediated degradation of p53. Proc Natl Acad Sci USA *96*, 3077-3080.

Tibbetts,R.S., Brumbaugh,K.M., Williams,J.M., Sarkaria,J.N., Cliby,W.A., Shieh,S.Y., Taya,Y., Prives,C., and Abraham,R.T. (1999). A role for ATR in the DNA damage-induced phosphorylation of p53. Genes Dev. *13*, 152-157.

Tishler,R.B., Calderwood,S.K., Coleman,C.N., and Price,B.D. (1993). Increases in sequence specific DNA binding by p53 following treatment with chemotherapeutic and DNA damaging agents. Cancer Res *53*, 2212-2216.

Turenne,G.A., Paul,P., Laflair,L., and Price,B.D. (2001). Activation of p53 transcriptional activity requires ATM's kinase domain and multiple N-terminal serine residues of p53. Oncogene *20*, 5100-5110.

Unger,T., Juven-Gershon,T., Moallem,E., Berger,M., Sionov,R.V., Lozano,G., Oren,M., and Haupt,Y. (1999). Critical role for Ser20 of human p53 in the negative regulation of p53 by Mdm2 . EMBO J. *18*, 1805-1814.

Unsal-Kacmaz,K. and Sancar,A. (2004). Quaternary structure of ATR and effects of ATRIP and replication protein A on its DNA binding and kinase activities. Mol Cell Biol *24*, 1292-1300.

Vafa,O., Wade,M., Kern,S., Beeche,M., Pandita,T.K., Hampton,G.M., and Wahl,G.M. (2002). c-Myc can induce DNA damage, increase reactive oxygen species, and mitigate p53 function: a mechanism for oncogene-induced genetic instability. Mol Cell *9*, 1031-1044.

Vaziri,C., Saxena,S., Jeon,Y., Lee,C., Murata,K., Machida,Y., Wagle,N., Hwang,D.S., and Dutta,A. (2003). A p53-dependent checkpoint pathway prevents rereplication. Mol Cell *11*, 997-1008.

Wang,X., Zou,L., Zheng,H., Wei,Q., Elledge,S.J., and Li,L. (2003). Genomic instability and endoreduplication triggered by RAD17 deletion. Genes Dev. *17*, 965-970.

Wang,Y., Schwedes,J.F., Parks,D., Mann,K., and Tegtmeyer,P. (1995). Interaction of p53 with its consensus DNA-binding site. Mol Cell Biol *15*, 2157-2165.

Waterman,J.L., Shenk,J.L., and Halazonetis,T.D. (1995). The dihedral symmetry of the p53 tetramerization domain mandates a conformational switch upon DNA binding. EMBO J *14*, 512-519.

Waterman,M.J., Stavridi,E.S., Waterman,J.L.F., and Halazonetis,T.D. (1998). ATM-dependent activation of p53 involves dephosphorylation and association with 14-3-3 proteins. Nature Genetics *19*, 175.

Weinert,T.A. and Hartwell,L.H. (1988). The RAD9 gene controls the cell cycle response to DNA damage in saccharomyces cerevisiae. Science *241*, 317-322.

Will,K., Warnecke,G., Bergmann,S., and Deppert,W. (1995). Species- and tissue-specific expression of the C-terminal alternatively spliced form of the tumor suppressor p53. Nucleic Acids Res *23*, 4023-4028.

Woo,R.A., Jack,M.T., Xu,Y., Burma,S., Chen,D.J., and Lee,P.W. (2002). DNA damage-induced apoptosis requires the DNA-dependent protein kinase, and is mediated by the latent population of p53. EMBO J *21*, 3000-3008.

Woo,R.A., McLure,K.G., Lees-Miller,S.P., Rancourt,D.E., and Lee,P.W.K. (1998). DNA-dependent protein kinase acts upstream of p53 in response to DNA damage. Nature *394*, 700-705.

Wu,X., Bayle,H., Olson,D., and Levine,A.J. (1993). The p53-mdm-2 autoregulatory feedback loop. Genes Dev. *7*, 1126-1132.

Wu,Z., Earle,J., Saito,S., Anderson,C.W., Appella,E., and Xu,Y. (2002). Mutation of mouse p53 Ser23 and the response to DNA damage. Mol Cell Biol. *22*, 2441-2449.

Xanthoudakis,S., Miao,G., Wang,F., Pan,Y.-C.E., and Curran,T. (1992). Redox activation of Fos-Jun DNA binding activity is mediated by a DNA repair enzyme. EMBO J. *11*, 3323-3335.

Zhao,H. and Piwnica-Worms,H. (2001). ATR-mediated checkpoint pathways regulate phosphorylation and activation of human Chk1. Mol Cell Biol *21*, 4129-4139.

Zhou,J., Ahn,J., Wilson,S.H., and Prives,C. (2001). A role for p53 in base excision repair. EMBO J *20*, 914-923.

Chapter 4

GATEKEEPERS OF THE GUARDIAN: P53 REGULATION BY POST-TRANSLATIONAL MODIFICATION, MDM2 AND MDMX

Geoffrey M. Wahl, Jayne M. Stommel, Kurt Krummel and Mark Wade
The Salk Institute for Biological Studies, La Jolla, CA, USA

INTRODUCTION

Happy 25th Anniversary p53! Since this is such a special occasion, I (GW) thought of explaining how fate brought p53 and me together.

It was a snowy day in Utah when Arnie Levine came to the University of Utah in 1976 to present a lecture on genetic approaches to differentiation of teratocarcinoma cells. My graduate work with Mario Capecchi (starting at Harvard and continuing at the University of Utah) led me to appreciate the potential power of genetics in cancer research. I therefore arranged to visit Arnie's lab to learn more about his research program. We discussed many topics, but not about how the large transforming protein (T antigen) of SV40 (SV40TAg) interacted with a putative ~54kDα cellular protein (Linzer and Levine, 1979).

I next went to Stanford to visit George Stark, whose lab had isolated mutant cancer cell lines resistant to PALA, an inhibitor of *de novo* uridine synthesis. Unlike most mutants described to that time, PALA resistance developed incrementally, and was associated with progressive increases in the levels of CAD, the enzyme targeted by the drug. Molecular cloning was just starting at Stanford, and I could see a clear route to solving the genetic mechanism(s) of this unusual form of drug resistance in cancer cells. I joined George's group in January 1977, and soon met a sabbatical visitor named Lionel Crawford. Lionel told me about work his post-doc, David Lane, was doing with SV40, and how SV40TAγ associated with an ~54kDα protein,

P. Hainaut and K.G. Wiman (eds.), 25 Years of p53 Research, 73-113.
© 2005 *Springer. Printed in the Netherlands.*

presumably of cellular origin (Lane and Crawford, 1979). I thought this was an interesting curiosity, but I didn't see the links to cellular transformation at that time. I certainly couldn't imagine at that early date how SV40TAγ interactions with cellular proteins might relate to understanding the mechanisms of PALA resistance.

My studies in George's lab with a graduate student, Richard Padgett, showed that gene amplification was the sole mechanism accounting for PALA resistance in the cell lines we investigated (Wahl et al., 1979). When I left Stanford, I recall having a discussion with George in which I asked him whether normal cells treated with the same drug also acquired PALA resistance. He answered that they had just done one experiment to address this question, and he told me something that would affect my research program for the next decade: he said that normal cells appeared to stop dividing, while cancer cells died when treated with equivalent PALA concentrations. My interpretation of his comment was that normal cells have controls that prevent them from cycling in response to this drug, while cancer cells may have lost such controls. Thus, my objective for the future was to explore the validity of this hypothesis, and to try to elucidate genes involved in the control circuitry.

Fast-forward a decade. Work from my lab and others showed that DNA breakage initiates gene amplification, and that breakage is induced when cells enter and proceed through S-phase under nucleotide-limiting conditions (Morgan et al., 1986; Windle et al., 1991). I therefore started to look for genes that prevent cells from entering S-phase under conditions that induce chromosome breakage. An exciting candidate emerged after I read some papers from Mike Tainsky's group in which *in vitro* passage of cells derived from Li-Fraumeni patients led to chromosome abnormalities similar to those in cells undergoing gene amplification (compare Bischoff et al., 1990; Morgan et al., 1986; Windle et al., 1991). As Li-Fraumeni patients have germ line p53 mutations, we began to investigate whether there was a link between p53 loss and gene amplification. More specifically, we predicted that normal cells treated with PALA would arrest prior to S-phase, while p53-deficient cells would enter S-phase, undergo chromosome breakage, and generate rare survivors with amplification of the CAD gene (which gives rise to PALA-resistance by enabling over-expression of the CAD protein (Yin et al., 1992). In retrospect, the seminal work of Michael Kastan and colleagues linking p53 to a G1 damage checkpoint (Kastan et al., 1991) makes this a logical expectation.

It was a memorable day when a post-doc in my lab, Yuxin Yin, excitedly showed me the results that corroborated our hypothesis linking loss of p53 to failed cell cycle control and gain of amplification competence (Yin et al., 1992). Interestingly, in contrast to the extensive death PALA induced in cancer cell lines, we noticed that a significant fraction of a normal cell culture treated with PALA re-entered the cell cycle when the drug was removed. We later showed this was due to the ability of ribonucleoside depletion to cause cells in G1 to activate p53, which prevents them from entering S-phase and undergoing breakage (Linke et al., 1996). These studies demonstrated p53 could serve as a "Guardian of the Genome" (term attributed to David Lane; Lane, 1992) by its ability to halt cell cycle progression in response to conditions that could induce genetic instability, such as DNA damage or ribonucleoside depletion; conversely, our work also demonstrated that loss of p53 enabled tumor cells to proliferate under DNA damaging conditions to generate genetically unstable variants (Yin et al., 1992). Similar conclusions were reached by work performed independently in the labs of Thea Tlsty and George Stark (Livingstone et al., 1992; Perry et al., 1992).

Twenty-five years later, more than 30,000 articles have been published on the small cellular ~54kDa protein bound by SV40T-antigen and the increasing number of proteins that regulate it. It is clear that this protein, now referred to as p53, is a tumor suppressor gene (Malkin et al., 1990; Srivastava et al., 1990) that is inactivated by mutation in about half of all human cancers (Hollstein et al., 1991). A substantial fraction of the remaining cancers have functionally compromised p53 due to alterations in its regulators such as MDM2 and the related protein MDMX (e.g., Momand et al., 1998; Riemenschneider et al., 1999). Below, we will refer to the mouse and human homologs (Migliorini et al., 2002a) as MDM2 and MDMX for simplicity.

MECHANISMS OF P53 MEDIATED TUMOR SUPPRESSION

Why is p53 so frequently inactivated in cancer? The answer likely relates to the ability of p53 to eliminate cells that encounter conditions that could induce genetic instability or promote unscheduled cell division. We now know that p53 is activated by small amounts of various types of DNA damage (e.g., see Huang et al., 1996; Kastan et al., 1991; Wahl and Carr, 2001), short or abnormally structured telomeres (Chin et al., 1999; Karlseder et al., 1999), metabolic and other consequences of high level oncogene signaling (Denko et al., 1994; Felsher and Bishop, 1999; Mai et al., 1996; Sherr, 2001; Vafa et al., 2002), microtubule dysfunction (Di Leonardo et al., 1997; Khan and Wahl, 1998; Lanni and Jacks, 1998; Minn et al., 1996), loss of nucleolar integrity (Rubbi and Milner, 2003), hypoxia (Alarcon et al., 1999) and perturbation of the endoplasmic reticulum (ER) (Qu et al., 2004).

The list has grown continuously over the years, so it wouldn't be surprising if more p53 activating conditions were identified in the future.

The mammalian p53 pathway generates responses as varied as reversible cell cycle arrest and apoptosis based on the nature of the activating signal and cell type. Since p53 output can kill cells, stringent regulatory mechanisms must have evolved to prevent its errant activation, as well as to allow it to rapidly initiate a response when appropriate. We will review studies that are starting to provide insight into how the p53 regulatory circuit evolved to control genetic stability, the cell cycle, and apoptosis to limit tumor formation. We will analyze *in vitro* and *in vivo* data that raise questions about the contributions of highly conserved phosphorylation sites to p53 control, and how mouse models are indicating that these sites may only be important in specific tissues. Finally, we will discuss recent studies describing an important new contribution to p53 control: the requirement for DNA damage to induce the degradation of MDM2 to activate p53.

p53 suppresses tumor formation largely by transcriptional regulation of a diverse set of target genes, but the importance of transcription-independent mechanisms is still being investigated (see below). p53 binds degenerate consensus sequences consisting of two inverted repeats in each half-site (el-Deiry et al., 1992; Funk et al., 1992). It binds most efficiently to its response elements as a tetramer, but the binding efficiencies and kinetics are likely affected by factors such as the precise sequence of the response elements, the type of other regulatory elements in the control region of the target gene, and chromatin context (Espinosa et al., 2003; Friedman et al., 1993; Inga et al., 2002; McLure and Lee, 1998; Szak et al., 2001). Most p53 mutations in human cancers affect the structure of its large DNA binding domain, or the residues used to contact the DNA backbone (see Cho et al., 1994; Gorina and Pavletich, 1996). These data imply that effective tumor suppression requires that p53 contact its response elements in chromatin.

Control of cell cycle arrest

Many mechanisms have been suggested to account for p53-mediated tumor suppression, but its abilities to induce cell cycle arrest or apoptosis and as a consequence to prevent unscheduled proliferation and to limit genetic instability appear paramount. Each of these functions can be largely accounted for by transcriptional activation of appropriate target genes. For example, p53's ability to induce a G1 arrest in response to DNA damage mainly depends on induction of the cyclin-cdk inhibitor p21/waf1/cip1/sdi1 (el-Deiry et al., 1993; Noda et al., 1994). One piece of evidence supporting this conclusion is that p21 deletion in mice and in human cell lines almost entirely prevents DNA damage from inducing a G1 arrest (Deng et al., 1995;

Waldman et al., 1995). Consistent with the data in mammals, the p53 ortholog in Drosophila is activated by DNA damage, but it does not induce the p21-like gene Dacapo, and consequently does not induce a cell cycle arrest (Ollmann et al., 2000). p53 also participates in G2 arrest through induction of 14-3-3 sigma and GADD 45 (for examples, see (Hermeking et al., 1997; Jin et al., 2000; Yang et al., 2000). GADD45 function is required for efficient G2 arrest induced by base-alteration mutagens but not ionizing radiation (Hollander et al., 1999), while p21 helps to sustain G2 arrest triggered by DNA damage (Bunz et al., 1998).

Control of apoptosis

The mechanisms by which p53 regulates apoptosis continue to be debated. A strong case can be made for transcription-dependent mechanisms as p53 regulates many pro-apoptotic genes including BAX, PUMA, PERP, NOXA, AIP1, FAS1/APO1, and IGF-BP3 in mammals and, hid, sickle, EIGER and reaper in Drosophila (see Wahl and Carr for mammalian references; representative fly references are Brodsky et al., 2000; Lee et al., 2003; Peters et al., 2002). Target gene activation by fly p53 is required for DNA damage-induced apoptosis since combined deletion of hid, sickle and reaper abrogates the apoptotic response in flies with wild type p53 (Brodsky et al., 2004). It is likely that apoptosis regulation in mammals is more complex, with different genes or gene sets being determined by the cell type and activating stimulus. For example, BAX appears to be a key gene for inducing apoptosis by p53 in an oncogene (Eμ-Myc) model of lymphomagenesis (Eischen et al., 2001), whereas BAX loss only partially reduces DNA damage-induced apoptosis in E1A expressing MEFs (McCurrach et al., 1997). PUMA has recently emerged as a critical p53 pro-apoptotic BH3-only target gene in several tissues since PUMA knockout mice are completely deficient in damage-induced apoptosis in the CNS and thymus (Jeffers et al., 2003; Villunger et al., 2003). The tissue and gene-specific requirements complicates the problem of defining the transcriptional targets and transcriptional-dependence of p53-activated apoptotic programs.

On the other hand, it has long been debated whether the sole mechanism by which p53 induces apoptosis involves transcriptional regulation (Caelles et al., 1994; Chipuk et al., 2004; Mihara et al., 2003; Moll and Zaika, 2001). Recent papers suggest that p53 can interact with apoptotic regulators in the cytoplasm to induce an apoptotic program without gene activation, but the specific interactions do not seem consistent in different studies (Chipuk et al., 2004; Mihara et al., 2003). For example, one study showed that p53 may interact with anti-apoptotic proteins such as BCL2 to liberate BAX and BAK (Mihara et al., 2003), while another reported that direct interaction between p53 and BAX enabled BAX to associate with mitochondria to induce cytochrome C-release (Chipuk et al., 2004). The latter study also showed that mouse embryo fibroblasts encoding a transcriptionally inactive and nuclear restricted endogenous p53^{QS} allele (Jimenez et al., 2000) could be made to undergo apoptosis by using wheat germ agglutinin to accumulate p53^{QS} in the cytoplasm (Chipuk et al., 2004). However, the use of wheat germ agglutinin to force cytoplasmic accumulation could sensitize the cells to apoptotic signals since it blocks nuclear import of proteins, and nuclear export of proteins and RNA (Middeler et al., 1997; Watanabe et al., 1999; Yoneda et al., 1987). The resulting macromolecular mislocalization might enable cytoplasmic p53 to tip the balance towards apoptosis. For example, treatment of cells with the nuclear export inhibitor leptomycin B (LMB) creates a stress that activates p53 and can induce apoptosis (Smart et al., 1999). The biological significance of a cytoplasmic component for p53-induced apoptosis is also uncertain as the cytoplasmic abundance of p53 is highest in unstressed, exponential cells and p53 is almost exclusively nuclear when cells are exposed to apoptotic stresses (see Shirangi et al., 2002; Stommel and Wahl, 2004 for recent analyses). Quantifying the contributions of transcriptional and non-transcriptional mechanisms for p53-induced apoptosis will require analysis of p53 mutants that are transcriptionally inactive and cytoplasmically sequestered that are expressed at normal levels.

Control of genetic stability

p53 has been reported to limit genetic instability in two broad ways. First, p53 can prevent cells with irreparable lesions from proliferating by inducing a permanent arrest resembling senescence or apoptosis (see Wahl et al., 1997) for a review). Second, depending on the type of damage induced, the cell cycle phase in which p53 is activated, and the cell type, p53 increases repair efficiency by inducing cell cycle delays, activating repair genes, or participating directly in some forms of repair. For example, p53-induced expression of the cyclin-dependent kinase inhibitor p21 can modulate the G2/M interval in response to ionizing radiation (Bunz et al., 1998). This may limit instability by allowing additional time for the cell to repair double strand breaks. Conversely, p21 deficiency increases the chance that cells with unrepaired chromosomes will enter G1 to generate descendants with chromosome anomalies (Bunz et al., 1998; Wouters et al., 1997). Consistent with this, irradiated cells deficient in p53 or p21 progressed more rapidly through G2/M-phase. This resulted in increased chromosome anomalies, cell death, and sensitization to radiation (Bunz et al., 1998; Wouters et al., 1997). The G2 delay may increase double strand break repair due to the availability of the sister chromatid as a template for homologous recombination and error-free repair.

Repair of double strand breaks by homologous recombination is not likely in G1 as the sister chromatid is not present (see Wahl and Carr, 2001) for review). Therefore, double strand breaks are repaired in G1 by an error-prone process such as non-homologous end joining (Lees-Miller and Meek, 2003). p53 limits the probability of cells with unrepaired DNA in G1 from generating mutant offspring by inducing a permanent arrest or apoptosis to remove such cells from the proliferating pool (Wahl and Carr, 2001). The importance of p53 for policing the repair process is vividly illustrated in analyses of cells deficient in enzymes that participate in non-homologous end joining or the histone γH2AX that either protects broken ends or organizes the chromatin to optimize repair efficiency. Cells deficient in these proteins are hyper-sensitive to breakage, which can lead to lethality in mice with wild type p53 (Bassing and Alt, 2004; Bassing et al., 2003; Gao et al., 2000). Loss of p53 rescues the lethality, but makes the animals tumor prone due to an increased rate of accumulation of chromosome abnormalities such as oncogene amplification that can drive tumor progression (Bassing et al., 2003; Gao et al., 2000).

p53 has also been reported to modulate the DNA repair process by transcription-dependent and transcription-independent mechanisms. For example, p53 modulates the repair efficiency of base DNA damage induced by UV and ionizing radiation through induction of genes such as DDB2 (p48) (Fitch et al., 2003; Hwang et al., 1999), GADD45a (Smith et al., 2000) and XPC (Amundson et al., 2002). The proteins encoded by these genes participate in the global genomic repair subpathway of nucleotide excision repair. p53 may also participate directly in base excision repair to correct damage induced by alkylating agents such as MMS (Offer et al., 1999; Seo et al., 2002; Zhou et al., 2001b). In this case, interactions between p53, apurinic endonuclease, and DNA polymerase beta appear to be important (Zhou et al., 2001b). p53 has been proposed to be able to facilitate repair through a putative strand annealing function and an intrinsic 3'-5' exonucleolytic activity in the DNA binding domain (Janus et al., 1999). The relative importance of p53-mediated transcriptional and non-transcriptional mechanisms to DNA repair and the control of genomic stability remain to be determined.

REGULATING P53

p53 must be tightly regulated as it has the potential to either kill a cell or to prevent it from dividing again. It is likely that most of this control is through post-translational regulation that activates or suppresses p53 as a transcription factor. Strong evidence that p53 mediated tumor suppression requires a functional transactivation domain was obtained using homologous recombination to generate cell lines or mice encoding a transcriptionally inert p53 protein (p5325Q26S=p53QS) (Chao et al., 2000b; Jimenez et al., 2000). This p53 mutant binds to its consensus sequences in EMSA (*ibid.*) and ChIP assays (M. Tang and G. Wahl, unpublished), but fails to induce or repress known target genes as the two amino acid changes it contains prevents interaction with the basal (Lu and Levine, 1995) transcription machinery (Thut et al., 1995; Xiao et al., 1994) . It does not elicit apoptosis or cell cycle arrest *in vitro* or *in vivo* (Chao et al., 2000b; Jimenez et al., 2000; M. Nister, M. Tang, M. Beeche, T. van Dyke, G..M. Wahl, manuscript in preparation). Importantly, mice with this mutation exhibit the same tumor spectrum and latency as animals completely lacking p53 protein (M. Nister, M. Tang, M. Beeche, T. van Dyke, G. M. Wahl, manuscript in preparation). These data, along with others summarized above, indicate that nuclear, presumably transcription-dependent, functions of p53 are critical for it to suppress tumor formation. Therefore, the remainder of this chapter will focus on the factors and mechanisms that regulate the nuclear functions of p53.

A small digression concerning MDM2 is necessary to enable a discussion of p53 control mechanisms (see below for more thorough discussion of MDM2). p53 levels are kept low mainly through the combined actions of two related RING finger proteins, MDM2 and MDMX (MDM4), that can associate as homo- or heterodimers through their RING domains (Ashcroft and Vousden, 1999; Gu et al., 2002; Haupt et al., 1997; Kubbutat et al., 1997; Michael and Oren, 2003; Migliorini et al., 2002a; Sharp et al., 1999; Tanimura et al., 1999). MDM2 was first identified as the 90kDa protein encoded by a gene amplified on mouse "double minute (DM) chromosomes" and has since been observed to be amplified in a subset of human tumors expressing wild type p53 (Oliner et al., 1992). Overexpression of MDM2 can prevent p53 induced cell cycle arrest and apoptosis (Chen et al., 1994; Oliner et al., 1993). Similarly, MDMX is the likely target gene in the 1q32 amplicon detected in a subset of gliomas with wild type p53 (Riemenschneider et al., 1999). Thus, MDM2 and MDMX appear to be oncogenes in human cancers. Both genes are also essential since deletion of either leads to early embryonic lethality in mice (Jones et al., 1995; Migliorini et al., 2002c; Montes de Oca Luna et al., 1995; Parant et al., 2001). Importantly, deleting p53 eliminates the lethality of MDM2 or MDMX deficiency (*ibid.*). These data establish p53 as the key downstream target of MDM2 and MDMX, and MDM2 and MDMX as essential negative regulators of p53.

MDM2 is a ring finger E3 ubiquitin ligase that mediates the ubiquitination and degradation of p53 (Fang et al., 2000; Fuchs et al., 1998; Haupt et al., 1997; Honda et al., 1997; Honda and Yasuda, 2000; Kubbutat et al., 1997; Lai et al., 2001). MDM2 mediated ubiquitination mainly occurs on C-terminal lysines, and transfection analyses show that p53 mutants in which all these lysines were changed to arginine are stable, active and nuclear (Nakamura et al., 2000; Rodriguez et al., 2000). Other studies show that p53 mutations that prevent MDM2 association also generate stable, nuclear p53 (Jimenez et al., 2000; Lin et al., 1994). These studies establish links between MDM2 interactions with p53, p53 protein abundance, and p53 subcellular localization.

Subcellular localization

p53 is a very unstable protein that is typically nuclear and present at very low levels. p53 appears to shuttle between nucleus and cytoplasm during the cell cycle (David-Pfeuty et al., 1996; Moll et al., 1996; Ostermeyer et al., 1996; Shaulsky et al., 1991); its nuclear entry and exit are mediated by specific import and export machinery as it exceeds the 40-50kDa limit for passive nuclear shuttling (Gorlich and Kutay, 1999). p53 contains nuclear localization and nuclear export signals, and its subcellular localization reflects a balance between the rates of import and export (Henderson and Eleftheriou, 2000; Shaulsky et al., 1990; Stommel et al., 1999; Zhang and Xiong, 2001). Given the importance of nuclear functions of p53 in tumor suppression, it is not surprising that some tumors have evolved mechanisms to accumulate p53 in the cytoplasm to inactivate it (e.g., see Moll et al., 1995; Moll et al., 1992; Sun et al., 1992).

p53 has two reported nuclear export signals (NES), a C-terminal one within the tetramerization domain (Stommel et al., 1999), and a second that overlaps the N-terminal transactivation domain (Zhang and Xiong, 2001). Because treatment of cells with leptomycin B (LMB), which inhibits the nuclear export receptor CRM1 (Kudo et al., 1998; Wolff et al., 1997), results in p53 nuclear localization, either or both are potential CRM1 targets (Stommel et al., 1999).

The C-terminal NES has the potential to link p53 structure with subcellular localization and nuclear functions. The crystal structure of the tetramerization domain indicates that the NES it contains should be concealed in the tetramer, but exposed in monomers or dimers. The importance of the C-terminal NES for controlling p53 subcellular localization is indicated by the p53 nuclear restriction caused by C-terminal NES mutations (Stommel et al., 1999). Thus, the positioning of an NES in the tetramerization domain allows for factors that affect p53 tetramerization and dissociation to be linked to subcellular localization and binding of p53 to its response elements. As an example, some studies indicate that phosphorylation of serine 392 (human p53) in the C-terminus might stabilize p53 tetramers, while phosphorylation of serines 315 and 392 might destabilize tetramers (Sakaguchi et al., 1997; see Jimenez et al., 1999; Liang and Clarke, 2001 for reviews). Other studies have been interpreted to indicate that MDM2-mediated ubiquitination of p53 may expose the p53 C-terminal NES to enable p53 export to the cytoplasm (Boyd et al., 2000; Geyer et al., 2000; Lohrum et al., 2001).

The N-terminal NES has also been proposed to induce p53 nuclear export. This NES is proposed to be active in unstressed cells, but is inactivated by DNA damage to allow for rapid nuclear accumulation (Zhang and Xiong, 2001). However, this putative NES lies within the transactivation domain, and overlaps the sequences known to bind MDM2 (see Michael and Oren, 2003) for a recent MDM2 review and references). While DNA damage induced N-terminal phosphorylations were proposed to inactivate the N-terminal NES (Zhang and Xiong, 2001), these modifications occur in regions that could affect MDM2-p53 association (e.g., see Dumaz et al., 2001; Shieh et al., 1999; Siliciano et al., 1997; Unger et al., 1999a). DNA damage also induces modifications on MDM2 (Maya et al., 2001) that can reduce MDM2 stability (Stommel and Wahl, 2004), which also impedes MDM2-p53 interaction. Furthermore, mutations in the proposed N-terminal NES designed to limit interaction with the export receptor were made in residues that prevent association with MDM2 (Kussie et al., 1996; Lin et al., 1994). Consequently, these mutations stabilize p53, leading to its tetramerization, and constitutive nuclear localization (Jimenez et al., 2000; Stommel et al., 1999). Thus, the data used to support the existence of an N-terminal NES can also be explained by the fact that each treatment or condition antagonizes MDM2 binding, leading to increased p53 abundance, tetramerization and masking of the C-terminal NES. Alternatively, conditions that reduce MDM2-p53 association should also reduce p53 C-terminal ubiquitination, which could also reduce p53 nuclear export by preventing unmasking of the C-terminal NES (see above).

As mentioned above, a number of cancer cell lines have cytoplasmic p53. Several mechanisms could account for this. First, these cells could have mutations that lead to an excess of p53 nuclear export over import. In support of this idea, treatment of neuroblastoma cells exhibiting cytoplasmic p53 with either p53 C-terminal peptides that bind the C-terminal NES or with the export inhibitor LMB (Ostermeyer et al., 1996; Smart et al., 1999; Stommel et al., 1999) leads to nuclear accumulation of p53. Another mechanism for cytoplasmic accumulation of p53 involves association with a cytoplasmic anchor protein. One candidate for such a molecule is Parc (p53-

associated parkin-like cytoplasmic protein; Nikolaev et al., 2003). This large protein is overproduced in neuroblastomas with cytoplasmic p53, and reducing Parc in these cells by siRNA induced p53 nuclear localization, apoptosis and sensitization to chemotherapy (Nikolaev et al., 2003). Interestingly, the same neuroblastoma cells in which elevated Parc was proposed to bind to p53 and sequester it in the cytoplasm were shown previously to exhibit hyperactive p53 export (Ostermeyer et al., 1996; Smart et al., 1999; Stommel et al., 1999). The basis for these different results remains to be determined. It is also uncertain how much of a role Parc plays in controlling subcellular localization of p53 in normal, unstressed cells as they contain very low levels of p53 that is predominantly nuclear.

MDM2 and MDMX as inhibitors of p53 transactivation

MDM2 inhibits p53 function in at least two ways, though it apparently needs MDMX to do so with optimal efficiency (Gu et al., 2002; Migliorini et al., 2002a). First, similar N-terminal regions of MDM2 and MDMX interact with hydrophobic side chains of an amphipathic alpha-helix in the p53 N-terminal transactivation domain (Bottger et al., 1999; Chen et al., 1993; Kussie et al., 1996). Consequently, MDM2 and MDMX, by binding to the transactivation domain, could inhibit transactivation by preventing the basal transcription machinery from binding and/or by preventing p53 acetylation by histone acetyl transferases such as p300 and CBP (Gu et al., 1997; Lu and Levine, 1995; Momand et al., 1992; Oliner et al., 1993; Shvarts et al., 1996; Thut et al., 1995; Xiao et al., 1994). MDM2 may also inhibit p53 transactivation by recruiting co-repressors such as CtBP2 or by titrating basic transcription factors (Mirnezami et al., 2003; Thut et al., 1997).

Regulating p53 stability by MDM2 and MDMX

The second mechanism by which MDM2 inhibits p53 is by acting as a co-factor for p53 degradation. p53 was initially shown to be targeted for degradation by the oncogenic papilloma virus E6 protein, which binds to p53 and recruits a cellular ubiquitin ligase (E6-AP) to mediate p53 ubiquitination (Scheffner et al., 1993). Support for a completely host-encoded mechanism for p53 proteasomal turnover was indicated by the substantial increase in ubiquitinated p53 caused by the proteasome inhibitor MG132 in cells that were not virally infected (Maki et al., 1996).

The MDM2 ubiquitin ligase that mediates p53 degradation (see above) also mediates its own poly-ubiquitination (Lai et al., 2001). However, there is debate about whether MDM2 mediates the mono- (Lai et al., 2001), or poly-ubiquitination of p53 (Li et al., 2003). A recent report showing that low levels of MDM2 mediate p53 mono-ubiquitination, while higher levels induce poly-ubiquitination (Li et al., 2003) raises the possibility that MDM2 level regulates a ubiquitination switch. The question of which type of ubiquitination occurs could be important since mono-ubiquitination can mediate changes in subcellular distribution but not degradation, while chains containing at least four ubiquitins are required for proteasomal degradation (Thrower et al., 2000; see Hicke, 2001 for a review). Indeed, p53 mono-ubiquitination was reported to lead to nuclear export of p53, while poly-ubiquitination led to p53 degradation (Li et al., 2003). Either consequence of MDM2-mediated ubiquitination would diminish p53's capacity to regulate gene expression.

The notion that MDM2 could induce mono-ubiquitination for p53 export or polyubiquitination for degradation raises the question of whether p53 must be exported to be degraded in the cytoplasm (Boyd et al., 2000; Freedman and Levine, 1998; Geyer et al., 2000; Inoue et al., 2001; O'Keefe et al., 2003; Roth et al., 1998; Tao and Levine, 1999). The model summarized above predicts that in unstressed cells, MDM2 should be at low levels, leading to p53 mono-ubiquitination, nuclear export, and cytoplasmic accumulation. However, in unstressed cells, p53 is present at low abundance but is predominantly nuclear (e.g., see Stommel and Wahl, 2004). It is also important to consider that p53 half-life is about 30 minutes in unstressed cells (e.g., see Oren et al., 1981; Stommel and Wahl, 2004), while its export to the cytoplasm takes hours as the C-terminal p53 NES is very weak (Henderson and Eleftheriou, 2000; Stommel et al., 1999). The slow export rate and short half-life are incompatible with models requiring that p53 export is required for its degradation in the cytoplasm. As proteasome inhibitors lead to p53 accumulation in the nucleus and the cytoplasm, it appears to be unstable in both locations (see Stommel and Wahl, 2004 for a recent example). Consistent with this interpretation, nuclear and cytoplasmic p53 can be ubiquitinated and degraded, implying that proteasomes in both compartments accept it as a substrate (Geyer et al., 2000; Joseph et al., 2003; Lohrum et al., 2001; Shirangi et al., 2002; Stommel and Wahl, 2004; Xirodimas et al., 2001; Yu et al., 2000). However, the slow export kinetics of p53, and its co-localization with MDM2 in the nucleus, suggest the nucleus as a preferred site for p53 turnover.

The precise mechanisms by which MDM2 leads to p53 degradation remain to be defined. p53 polyubiquitination might be achieved by MDM2 alone in cells expressing high levels of MDM2, such as in cancers with MDM2 amplification or overexpression. This begs the question of how poly-ubiquitination is achieved in normal cells with unstable p53 and low levels of MDM2 if MDM2 only induces mono-ubiquitination under such conditions (Fang et al., 2000; Li et al., 2003). One solution is that MDM2 could associate with another ubiquitin ligase, an "E4", that adds poly-ubiquitin chains to the lysines previously mono-ubiquitinated by MDM2. This interaction is likely to involve the MDM2 RING domain, since replacing it with the RING domain of another protein, Praja1, enables MDM2 to poly-ubiquitinate itself but prevents p53 ubiquitination (Fang et al., 2000). However, this situation may be more complex as the MDM2 RING domain has also been implicated in other processes such as ATP binding and acetylation that may also affect MDM2 function (Poyurovsky et al., 2003; Wang et al., 2004). The central region of MDM2 containing an acidic domain is also needed for p53 degradation (Argentini et al., 2001; Kawai et al., 2003b; Meulmeester et al., 2003). Interestingly, hHR23A a human homologue of a yeast DNA repair protein, binds to the 26S proteasome and to the acidic domain of MDM2 (Hiyama et al., 1999; Zhu et al., 2001a; Brignone et al., 2004).

One potential candidate for a p53 E4 polyubiquitin ligase is the histone acetyl transferase p300 (Grossman et al., 2003). At first glance, this seems to be a surprising finding, since p300 binds to the same N-terminal region of p53 as MDM2, it acetylates the same C-terminal lysines that MDM2 mono-ubiquitinates, and it has been reported to be a p53 co-activator (Barlev et al., 2001; Gu and Roeder, 1997). However, consistent with a role for p300 in p53 degradation, MDM2, p300 and p53 form ternary complexes, and MDM2 mutants that cannot bind p300 can mediate p53 ubiquitination but not degradation (Grossman et al., 1998; Kobet et al., 2000; Zhu et al., 2001b). It is also possible that p300 serves as a bridge to either the proteasome, to other ligases that mediate polyubiquitination, or to proteins that associate with the proteasome, such as hHR23A (Zhu et al., 2001a).

The precise mechanisms by which MDM2 and MDMX collaborate to regulate p53 are important to define, as both proteins are clearly required for optimal inactivation of p53. MDM2 mediates the ubiquitination and degradation of MDMX (Kawai et al., 2003a; Pan and Chen, 2003; Tanimura et al., 1999). On the other hand, MDMX cannot induce MDM2 degradation as it is not an E3 ubiquitin ligase (Stad et al., 2001). Several studies employing transfection and overexpression indicated that MDMX can inhibit p53 degradation (Jackson and Berberich, 2000; Sharp et al., 1999; Stad et al., 2001). This is contrary to genetic analyses in mice and more recent siRNA studies showing that MDMX depletion activates and stabilizes p53 (Kawai et al., 2003a; Migliorini et al., 2002b; Parant et al., 2001). These disparate observations now seem to be resolved by a study that changed the ratio of MDMX relative to MDM2 and then determined the effects on p53 degradation. These studies showed that MDMX can stabilize MDM2, and at an appropriate ratio, increases significantly the ability of MDM2 to degrade p53 (Gu et al., 2002). These data explain how both MDMX and MDM2 assist each other to maximize p53 inhibition, and how deletion of either elicits embryonic lethality. If MDM2 is deleted, MDMX may bind but cannot degrade p53, and MDMX is apparently not present at a high enough concentration to effectively inhibit p53 activated death or arrest programs. If MDMX is deleted, MDM2 may either be too unstable to inactivate p53, or it may be less efficient at mediating p53 ubiquitination. As MDM2 and MDMX interact with each other, an implication of these studies is that a heterodimer of MDMX and MDM2 may be the most potent p53 inhibitor. An extension of these data is that conditions that interfere with MDMX binding to MDM2, that reduce the levels of either protein, or that affect MDM2 E3 ubiquitin ligase function could have profound effects on p53 regulation.

The impact of other ubiquitin ligases, de-ubiquitinating enzymes, and ARF on p53 and MDM2 stability and function

Pirh2 and COP1

Recent evidence suggests the existence of other ubiquitin ligases and de-ubiquitinating enzymes capable of modulating p53 and MDM2 levels and activities. Pirh2 is a RING domain protein that binds to the p53 DNA binding domain, and appears to induce p53 ubiquitination and degradation (Leng et al., 2003). Like MDM2, the Pirh2 gene is induced by p53. Pirh2 functions independently of MDM2, and is expressed in many tissues.

The human homolog of the Arabadopsis gene COP1 (constitutively photomorphogenic 1) has recently been identified as a p53 interacting RING finger protein able to ubiquitinate p53 (Dornan et al., 2004). Like Pirh2, it appears to be a p53-inducible gene, and when overexpressed, can reduce p53-dependent cell cycle arrest or apoptosis in cancer cell lines. Reducing COP1 levels by siRNA increases p53 levels, and sensitizes cells to damage induced activation of p53. COP1 overexpression reduced p53 levels in a co-transfection experiment using p53-/mdm2-null MEFs, suggesting that COP1 functions independently of MDM2.

The physiologic significance of both COP1 and Pirh2 remains to be determined in light of the early embryonic lethality caused by MDM2 deletion. The failure of either COP1 or Pirh2 to rescue MDM2 deficiency is puzzling since cells lacking MDM2 should have activated p53 to induce high levels of these other putative p53 E3 ubiquitin ligases. It will be important to determine whether COP1 and Pirh2 regulate p53 in specific tissues, perhaps where MDM2/MDMX are limiting (Mendrysa et al., 2003).

HAUSP

Nearly 100 de-ubiquitinating proteins (DUBs) have been identified in the human genome (see Lima, 2003 for a review). A DUB that targets p53, MDM2 or MDMX, should affect p53 pathway regulation, but it is not easy to predict the effects. For example, a DUB directed against p53 should stabilize it, while one directed against MDM2 might have complex consequences depending on the ability of the de-ubiquitinated and presumably stabilized MDM2 to interact with and ubiquitinate p53. Work over the past several years has shown that there is at least one DUB, HAUSP (herpes associated ubiquitin-specific protease, also known as USP7), that targets p53 and MDM2, and that the consequences for p53 activation are indeed complex (Li et al., 2002; Lim et al., 2004; Wood, 2002).

HAUSP was identified as a p53-interacting protein (Li et al., 2002). It was initially proposed that HAUSP stabilized and activated p53 by removing ubiquitins from the p53 C-terminus. Importantly, although HAUSP may have many substrates, its overexpression only causes growth arrest in cells expressing wild type p53, implying a p53-dependence to its growth inhibitory effects (Li et al., 2002).

Recent studies challenge the notion that HAUSP directly activates p53 via p53 de-ubiquitination. Two groups, one using siRNA (Li et al., 2004) and the other using homologous recombination in a tumor cell line to knock out HAUSP function (Cummings et al., 2004), showed that complete elimination of HAUSP caused p53 stabilization and growth arrest. This is contrary to expectation if the key HAUSP target is p53, as HAUSP elimination should leave p53 ubiquitinated, leading to its *destabilization.* However, the observed p53 activation by HAUSP knockdown can be explained if HAUSP's main target is MDM2. In this case, eliminating HAUSP should increase MDM2 ubiquitination, leading to its rapid degradation, and consequent activation of p53. In support of this, eliminating HAUSP caused accelerated MDM2 degradation and p53 stabilization (Li et al., 2004). Importantly, partial reduction of HAUSP produced the opposite result, in that p53 was partly destabilized (Li et al., 2004). It remains to be determined whether HAUSP activity or abundance can be regulated by growth conditions or stress and whether this affects MDM2 activity and p53 regulation *in vivo.*

ARF

Factors that affect the ability of MDM2 to ubiquitinate p53, or to control access of MDM2 to p53, should also contribute to p53 regulation. ARF, an alternative reading frame product of the INK4A locus (Kamijo et al., 1997;

Sherr, 2001), binds to MDM2, p53, or both (Kamijo et al., 1998). Several mechanisms have been proposed for ARF-mediated regulation of p53. ARF levels increase significantly in response to high level persistent signaling by Myc or oncogenically mutated Ras, or as MEFs and human fibroblasts become senescent (Dimri et al., 2000; Kamijo et al., 1997; Sherr, 2001). ARF is a nucleolar protein, and when induced to high levels in cells expressing oncogenes or nearing senescence, co-localizes with MDM2 in nucleoli (Weber et al., 1999). These observations led to the proposal that ARF may sequester MDM2 in the nucleolus, leading to MDM2 depletion from the nucleoplasm, and consequent activation of p53 (Weber et al., 1999). However, other studies show that ARF can activate p53 in the nucleoplasm (Llanos et al., 2001). This may be explained by the ability of ARF to bind to and inhibit the E3 ubiquitin ligase function of MDM2 (Honda and Yasuda, 1999), or by increasing the MDM2-mediated degradation of MDMX (Pan and Chen, 2003). In the latter case, decreasing MDMX levels should in turn destabilize MDM2, depleting the cell of both p53 negative regulators (Gu et al., 2002).

ARF contributes to efficient p53-dependent induction of apoptosis or cell cycle arrest in response to a subset of the signals that activate p53 in a subset of tissues. Interestingly, one tumor in an ARF null animal contained mutated or inactivated p53, implying that both genes can collaborate in tumor progression and that p53 and ARF deficiencies are not functionally equivalent. For example, ARF contributes significantly to p53 activation induced by over-expressed Myc in B-cells, in p53-dependent senescence in MEFs growing *in vitro*, and in damage-induced responses of MEFs but not of other cell types such as intestinal epithelial cells (e.g., see Eischen et al., 1999; Kamijo et al., 1999b; Khan et al., 2000; Zindy et al., 1998). Also, ARF-null mice exhibit a different tumor spectrum and develop tumors with different latencies than p53 null mice (Kamijo et al., 1999a). Importantly, ARF does not appear to play a role in p53 dependent-apoptosis or tumor suppression in the mouse choroid plexus in which tumor progression is initiated by inactivating the retinoblastoma (Rb) tumor suppressor (Tolbert et al., 2002). This is noteworthy since Rb inactivation increases E2F1 activity in this system, and E2F1 has been shown to activate ARF in several *in vitro* systems (Bates et al., 1998; Dimri et al., 2000). Furthermore, in human fibroblasts, recent data show that decreasing ARF expression by siRNA enhances growth but does little to stimulate transformation induced by oncogenic ras, and that oncogenic ras still activates p53 when little if any ARF is present (Voorhoeve and Agami, 2004). Together, these data imply that ARF is an important, albeit species, cell and developmental stage specific modulator of p53 function, and that backup systems exist for activating p53 in tissues that do not express ARF.

P53 ACTIVATION BY POST-TRANSLATIONAL MODIFICATION

The current model

The data summarized above show that p53 function can be regulated by inhibitors including MDM2 and MDMX, by proteins that modulate the functions of MDM2/MDMX (e.g., ARF, HAUSP, etc.), and by proteins with both activating and inactivating capacities such as p300. However, p53 and MDM2 are also subject to rapid post-translational phosphorylation on highly conserved serine and threonine residues by numerous protein kinases, and the functional impacts of these modifications are still uncertain (for detailed reviews and references, see (Appella and Anderson, 2001; Hay and Meek, 2000; Meek, 2002; Meek and Knippschild, 2003; Stewart and Pietenpol, 2001; Wahl and Carr, 2001). The clearest example for an essential role for phosphorylation in p53 activation comes from studies in *Drosophila*. DNA damage activates MNK, the *Drosophila* homolog of mammalian Chk2, to phosphorylate serine 4 in the p53 N-terminus (Brodsky et al., 2004; Peters et al., 2002). Mutation of MNK (Chk2), or of p53 serine 4 to alanine, prevented ionizing radiation from activating p53 or eliciting an apoptotic response (Brodsky et al., 2004; Peters et al., 2002). Importantly, p53 activation in flies occurred with a phosphorylation associated mobility shift, but was not accompanied by an increase in p53 abundance. These data indicate that phosphorylation does not activate fly p53 by changing its stability. This observation is consistent with the absence of a recognizable fly ortholog of MDM2. These data also provide compelling evidence that in *Drosophila*, an off-on switch for p53 activation is created by N-terminal phosphorylation by MNK (Chk2). However, the situation is not nearly so clear in mammalian cells.

Many studies in mammalian cells demonstrate that phosphorylation of multiple N-terminal serines in p53 is induced by DNA damage, as occurs for serine 4 in *Drosophila*. But in mammals, these phosphorylations seem to have more subtle effects on p53 function than was observed in flies. The first kinase identified for this role is one mutated in patients with ataxia telangiectasia (ATM), a disease associated with cancer predisposition, radiation sensitivity, chromosome abnormalities, and a failure to efficiently activate p53 in response to ionizing radiation (Banin et al., 1998; Canman et al., 1998; Kastan et al., 1992; Siliciano et al., 1997). The ATM kinase phosphorylates serine 15 in the p53 N-terminus, which is adjacent to the MDM2-binding domain (Shieh et al., 1997). An ATM-Rad3 related kinase (ATR) has also been reported to target serine 15 (Tibbetts et al., 1999).

While serine 15 phosphorylation was initially proposed to prevent or reduce association of p53 with MDM2 (Shieh et al., 1997), other studies showed that serine 15 phosphorylation does not markedly affect MDM2 binding (Dumaz and Meek, 1999; Kane et al., 2000; Schon et al., 2002). Rather, serine 15 phosphorylation may enhance binding of the CBP co-activator (Dumaz and Meek, 1999; Lambert et al., 1998). Studies using phosphorylated peptides, *in vitro* binding, and transfection of relevant mutants suggested that phosphorylation of serine 20, probably in combination with threonine 18 and an N-terminal proline-rich region, mediate structural changes resulting in reduced affinity for MDM2 (Craig et al., 1999; Dumaz et al., 2001; Jabbur et al., 2002; Sakaguchi et al., 1998; Sakaguchi et al., 2000; Schon et al., 2002). However, other analyses of p53 mutants with one or more of the phosphorylation sites mutated have generated inconsistent results. For example, mutation of serine 20 alone, or in combination with five other N-terminal serines including serine 15 produced only a 50% reduction in the ability to induce apoptosis after transfection into H1299 cells (Unger et al., 1999a; Unger et al., 1999b). Two other studies showed that, in contrast to the work summarized above, mutation of N-terminal serine or threonine phosphorylation sites as well as others in the C-terminus, alone or in combination, had little effect on p53 stability or activation in cell culture models (Ashcroft et al., 1999; Blattner et al., 1999). As the magnitude of the effects of p53 substitution mutations depends on the amount of transfected p53 relative to MDM2 expressed in the cells (Dumaz et al., 2001), the relevance of such analyses to control at normal physiologic levels remains uncertain.

A model has emerged emphasizing the importance of N-terminal phosphorylation in p53 activation (Sakaguchi et al., 1998). Phosphorylation of serines 15 and 20, and threonine 18 is proposed to induce a conformational change that prevents MDM2 from interacting with p53. This results in increased binding of p300/CBP, and presumably, the basal transcription machinery. As p300/CBP and the basal transcription machinery bind p53 in a region that partially overlaps that bound by MDM2, co-activator recruitment would compete for MDM2 binding (De Guzman et al., 2000; Lu and Levine, 1995; Thut et al., 1995; Xiao et al., 1994). Preventing MDM2 binding would increase p53 transcriptional output by increasing p53 abundance. p300/CBP binding to p53 should also lead to acetylation of p53 C-terminal lysines; this could stabilize p53 by preventing MDM2-mediated ubiquitination of the same residues (Gu and Roeder, 1997; Nakamura et al., 2000; Rodriguez et al., 2000). p53 C-terminal acetylation has also been proposed to increase its ability to associate with chromatin, and to enable recruitment of another histone acetyl transferase, PCAF, that induces histone acetylation beyond that induced by p300/CBP (Barlev et al., 2001; Gu and Roeder, 1997; Liu et al., 1999).

While the model nicely integrates p53 N-terminal structure with the potential impact of phosphorylation on MDM2, p300/CBP and basal machinery binding, it has not been validated by analyses of the behavior of p53 phosphorylation site mutants. This could be explained in many ways, including the inability of the methods used to achieve physiologic levels of p53 or to generate the proper stoichiometric relationships between p53, MDM2 and MDMX. Therefore, a more rigorous test of the model is to use animal models, as described below.

Reconsidering p53 N-terminal phosphorylation

One prediction of the phoshporylation-acetylation cascade model for p53 activation is that binding of p300/CBP to the N-terminal transactivation domain is required for p53 to bind to chromatin. This prediction was tested using a mouse mutant in which residues leucine 25 and tryptophan 26 were changed to glutamine 25 and serine 26 (i.e., p53^{QS}). These residues are in the amphipathic alpha-helix that binds MDM2, p300/CBP, and the basal transcription machinery. The indicated substitutions prevent p53 acetylation and transcriptional function (Chao et al., 2000b; Jimenez et al., 2000). However, p53^{QS}still binds as well as wild type p53 to p53 response elements in electrophoretic mobility shift experiments *in vitro* and to chromatin in MEFs (Chao et al., 2000b; Jimenez et al., 2000; M. Tang and G. Wahl, unpublished observations). This indicates that p300/CBP mediated p53 or chromatin acetylation is not required for p53 to bind the response elements of its target genes *in vivo*. These data are consistent with studies showing that stresses including leptomycin B treatment can activate p53 without inducing detectable C-terminal acetylation (Smart et al., 1999; Stommel and Wahl, 2004). Another implication is that C-terminal acetylation is not required for p53-mediated transcriptional regulation.

The regulatory importance of p53 N-terminal phosphorylation is being studied by making mutations of the conserved serine residues suggested by transfection experiments to be key contributors to p53 activation and stability. Mouse serines 18 and 23 (equivalent to human serines 15 and 20) have individually been mutated to alanine (S18A or S23A, respectively). The effects of the S18A mutation are important to determine as the phosphorylation-acetylation model predicts that C-terminal acetylation is dependent on prior phosphorylation of serine 18. Importantly, threonine 21 was not phosphorylated after DNA damage in the S18A mutant, which is consistent with the phosphorylation cascade initiating at serine 18. However, the *in vivo* data generated thus far do not point to an essential role for serine 18 phosphorylation in p53 activation in a majority of mouse tissues. p53 S18A in mouse embryonic stem cells, differentiated ES cells, or MEFs was present at nearly normal levels in unstressed cells, and was induced almost as well as wild type p53 in response to UV or ionizing radiation (Chao et al., 2003; Chao et al., 2000a; Sluss et al., 2004). While C-terminal acetylation was unaffected in differentiated ES cells, it appeared to be significantly reduced in MEFs (Chao et al., 2003; Chao et al., 2000a). S18A p53 exhibited equivalent binding to p53 response elements in chromatin using ChIP analysis, but some p53 target genes may be expressed at reduced levels (Chao et al., 2003). S18A MEFs arrested like wild type cells after ionizing radiation, but apoptosis in the thymus and spleen was reduced by 50% (Chao et al., 2003; Sluss et al., 2004). By contrast, S18A retinal cells exhibited only 20% of the wild type apoptotic response at 2Gy, but at 14 Gy appeared to undergo apoptosis at nearly wild type level (Borges et al., 2004). Of note, S18A was as effective at tumor suppression as wild type p53 (Sluss et al., 2004). Together, these *in vivo* analyses indicate that serine 18 phosporylation in mice is not essential for p53 activation in all tissues, and that inability to phosphorylate this residue does not compromise tumor suppression. Therefore, the phosphorylation-acetylation cascade may not be essential for p53 activation in mice. Alternatively, it may be important for p53 activation in only certain tissues, or there may be additional mechanisms that contribute to p53 control that are independent of serine 18 and threonine 21 phosphorylation.

S23A mice have also been generated, and exhibited nearly wild type patterns of p53 activation and induction of apoptosis in ES cells or in thymocytes derived by RAG reconstitution (Wu et al., 2002). Therefore, S23A mutation did not compromise p53 stabilization or function in the tissues analyzed. This is not the result expected if phosphorylation at this position is required to reduce MDM2 binding after DNA damage (Craig et al., 1999; Shieh et al., 1999; Unger et al., 1999a). However, the only study performed thus far did not generate mice in which every cell contained the S23A mutation. Consequently, firm conclusions regarding the impact of S23A mutation on p53 function and tumor suppression will await analysis of mice expressing the mutation constitutionally so that tissue specific effects can be evaluated.

The observations made thus far in mouse models indicate that phosphorylation of serines 18 and 23 and threonine 21 is not essential for p53 activation. They also suggest that p53 can be activated by mechanisms that are independent of N-terminal phosphorylation. This interpretation is consistent with studies showing that merely disrupting p53-MDM2 interaction is sufficient to activate p53. For example, diffusible peptides that prevent MDM2 from binding to p53 induce cell cycle arrest or apoptosis without N-terminal phosphorylation (Bottger et al., 1997; Bottger et al., 1996; Garcia-Echeverria et al., 2000). In addition, a recent study described cis-imidazoline compounds (Nutlins) that mimic the p53 N-terminal alpha helical region that binds MDM2. Nanomolar concentrations of Nutlins activate p53 to induce either cell cycle arrest or apoptosis without measurable N-terminal modifications (Vassilev et al., 2004). The Nutlin effect is specific for MDM2 binding, as only Nutlin enantiomers that matched the p53 side chain conformation in the MDM2 binding site were active. These compounds effectively block MDM2-p53 interaction (Stommel and Wahl, 2004; Vassilev et al., 2004). These data demonstrate that preventing MDM2 from associating with p53 enables full p53 activation without stress-induced modifications.

Regulated MDM2 degradation is important for p53 activation

As a consequence of MDM2 being a p53 target gene, MDM2 transcripts and protein accumulate after p53 is activated (Barak et al., 1993; Michael and Oren, 2003; Perry et al., 1993; Wu et al., 1993). The activation of MDM2 by p53 establishes an auto-regulatory negative feedback loop to allow for finer tuning of the p53 response, and to reduce the chance of errant p53 activation (Lev Bar-Or et al., 2000; Wu et al., 1993). However, the activation of MDM2 by p53 also creates a problem in that p53 needs to be kept in an active form to initiate and maintain a stress response at the same time as MDM2 levels are increasing. N-terminal p53 phosphorylations were supposed to allow for this by blocking MDM2-p53 interactions, but as discussed above, such modifications do not appear to be sufficient. Conversely, as the damage response wanes, mechanisms to turn off p53 will need to be restored, and it is currently unclear what these might entail. Below, we discuss data showing that another important response to DNA damage involves the accelerated degradation of MDM2. MDM2 destabilization is required for p53 activation to occur as MDM2 levels rise during a damage response. This process is triggered by damage-activated kinases, and requires the MDM2 RING domain (Stommel and Wahl, 2004). We therefore refer to this step in p53 activation as damage-activated MDM2 auto-degradation.

Efficient p53 activation requires the activity of damage-activated kinases such as ATM (Banin et al., 1998; Canman et al., 1998; Kastan et al., 1992; Siliciano et al., 1997). It has largely been assumed that p53 is the critical target for these modifications, but the nearly full activation of p53 N-terminal phosphorylation mutants suggests that this interpretation is incorrect. Rather, the data are more consistent with at least one additional substrate that is targeted by these kinases being involved in p53 activation. Importantly, MDM2 is phosphorylated by ATM with kinetics that are compatible with p53 activation (Khosravi et al., 1999; Maya et al., 2001). MDM2 phosphorylation was initially proposed to impede MDM2's ability to promote p53 export to the cytoplasm for degradation. However, as discussed above, p53 can be degraded efficiently in the nucleus, suggesting that MDM2 phosphorylation is required for a different step needed for p53 activation.

Our recent observations reveal that a critical step in p53 activation by DNA damage involves accelerated degradation of MDM2 (Stommel and Wahl, 2004). We first noticed that DNA damage decreased the stability of transfected MDM2, and confirmed this in normal human fibroblasts. We then observed a tight temporal correlation between the timing of accelerated MDM2 degradation and p53 activation. Within minutes after induction of DNA damage, ATM was activated, and p53 became phosphorylated on serine 15; but early after damage induction, serine 15 phosphorylated p53 was unstable, and was transcriptionally inactive. The instability of p53, along with its transcriptional inactivity, are consistent with p53 being able to interact with MDM2 at this time. Co-immune precipitation studies confirmed this idea, but we could only detect serine 15 phosphorylated p53 associated with MDM2 when we used proteasome inhibitors (for reasons discussed below). In contrast to the behavior of p53 soon after damage, between 1-2 hrs after damage induction, p53 became stabilized and p53 target genes were activated. Importantly, MDM2 was relatively stable at early times, and was significantly destabilized at 1-2hrs after damage induction, which correlates nicely with p53 becoming stable and active. Later, as the damage response waned, p53 became unstable, transcriptionally inactive, and this correlated with re-stabilization of MDM2 (Stommel and Wahl, 2004).

We next asked whether MDM2 degradation is required for p53 activation. We reasoned that if accelerated MDM2 degradation is required for p53 activation, then preventing its degradation with proteasome inhibitors should prevent p53 from being activated. On the contrary, if phosphorylation of p53 and/or MDM2 could prevent their association, then proteasome inhibitors should not prevent p53 activation. Consistent with the model invoking MDM2 degradation in p53 activation, we observed that proteasome inhibitors prevented p53 mediated activation of p21 and MDM2. Importantly, under the conditions used, proteasome inhibitors did not prevent p53 from being phosphorylated on serine 15. One explanation of these results is that stabilizing MDM2 allows it to interact with p53 following DNA damage, resulting in p53 inactivation. This interpretation is consistent with the co-immune precipitation of serine 15 phosphorylated p53 with MDM2 in the presence of proteasome inhibitors. Importantly, adding an active Nutlin to prevent MDM2-p53 association prior to proteasome inhibition restored the ability of DNA damage to activate p53. This control shows that proteasome inhibitors do not block p53 activation by a non-specific mechanism. Rather, it is the MDM2 stabilization produced by the proteasome inhibitors, and the ability of the stabilized MDM2 to interact with serine 15 phosphorylated p53, that prevents p53 activation. These data imply that MDM2 destabilization during DNA damage contributes significantly to the inability of MDM2 to block p53 activation during a damage response.

Accelerated MDM2 degradation following DNA damage requires phosphorylation by damage activated kinases and is dependent on a functional MDM2 RING domain. Studies with ATM deficient cells, and phosphorylation site mutants of MDM2 suggest that more than one kinase, or more than one kinase target site in MDM2, may be involved in MDM2 destabilization (Stommel and Wahl, 2004). On the other hand, mutation of cysteine 464 in the MDM2 RING domain stabilized MDM2 and made it resistant to damage induced degradation. The mechanism by which damage-kinase mediated phosphorylation destabilizes MDM2 remains to be defined. One potential mechanism is that damage induced phosphorylation(s) enable recruitment of the E2 ubiquitin transferase via the RING domain. As MDMX prevents MDM2 auto-ubiquitination, and MDMX and MDM2 associate via their RING domains, it is also possible that destabilization results from phosphorylation induced dissociation of MDMX from MDM2. Interestingly, Yuan and colleagues recently found that DNA damage destabilized both MDM2 and MDMX (Kawai et al., 2003a). Therefore, it is also possible that DNA damage could enhance MDM2 mediated ubiquitination of MDMX, which would accelerate both MDMX and MDM2 degradation.

Small changes in MDM2 abundance can affect p53 activation

The summary above makes it reasonable to expect that other factors that affect the abundance, stability, functionality or interaction of MDM2 and MDMX could set the threshold for p53 activation. Mitogenic levels of signaling from Raf and activation of NF-kB increase MDM2 levels sufficiently to make it more difficult to activate p53 by DNA damage (Ries et al., 2000; Tergaonkar et al., 2002). Similarly, activated AKT induces phosphorylation of MDM2 on at least two sites (S166, 186) resulting in nuclear accumulation (Gottlieb et al., 2002; Mayo and Donner, 2001; Ogawara et al., 2002; Zhou et al., 2001a). While this has been reported to result from increased nuclear localization of MDM2 (Mayo and Donner,

2001), MDM2 is typically a predominantly nuclear protein, so the observed nuclear accumulation may instead result from increased stabilization of otherwise unstable nuclear MDM2. Stabilizing MDM2 should enable it to interact with and inhibit p53 more effectively.

Just as small increases in MDM2 levels can blunt p53 activation, small decreases in MDM2 levels make it easier to activate p53. For example, decreasing MDM2 levels by an average of 50% led to decreased body weight and reduced the size of multiple organs in p53+ but not p53- animals (Mendrysa et al., 2003). Decreasing MDM2 levels by at least 50% led to significantly increased sensitivity to ionizing radiation (Mendrysa et al., 2003). In another study, MDM2+/- mice exhibited greater resistance to Eu-Myc induced lymphomagenesis and to have greatly increased life spans due to drastic reduction of peripheral B cells by p53 (Alt et al., 2003). Peripheral and primary B cells from Eµ-Myc-MDM2+/- or MDM2+/- mice were far more sensitive to spontaneous apoptosis than those of wild type littermates, and loss of p53 rescued this sensitivity. Similarly, knockdown of MDM2 in zebra fish induced apoptosis, and arrested development at a very early stage (Langheinrich et al., 2002). These data provide compelling examples of how modest alterations in MDM2 abundance or stability produce profound effects on "spontaneous", as well as oncogene and damage induced activation of p53.

A REVISED MODEL FOR P53 ACTIVATION

We propose the following revised model to account for p53 activation by DNA damage (see Figure 1). We envision two coordinated processes being involved. First, N-terminal phosphorylations in p53 may be important for it to cooperate with individual or preassembled components of the transcriptional regulatory machinery to send signals to RNA polymerase to convert it from an inactive to an active state. Although p53 N-terminal phosphorylations were initially proposed to disrupt p53-MDM2 interactions, the data emerging from studies of mouse mutations, the inconsistencies in transfection analyses, and our immune-precipitation analyses, do not strongly support this idea. Also, in *Drosophila*, while N-terminal phosphorylation is critical for activation, flies appear to lack an MDM2 homolog.

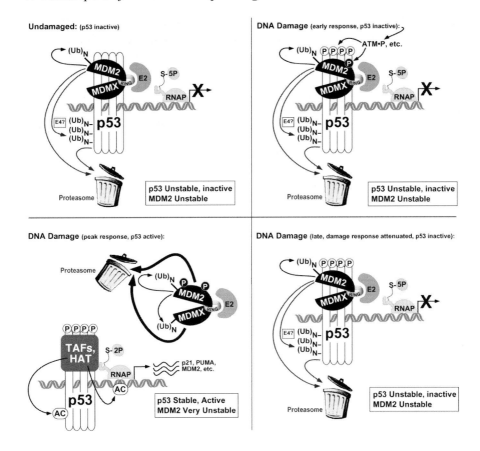

Figure 1. **Revised model for p53 activation.** The figure summarizes data discussed in the text concerning the time course of events occurring during p53 activation. **A. Unstressed cells**. We suggest that in unstressed cells, p53 tetramers are inactive, but able to bind their response elements in chromatin. As MDM2 and MDMX are both required to inactivate p53, we show them heterodimerized via their respective RING domains and bound to the N-terminal p53 transactivation domain. The MDM2-MDMX heterodimer may engage an E2 to enable poly-ubiquitination of MDM2 and MDMX and the mono-ubiquitination of p53. According to current information, p53 degradation would require poly-ubiquitination, which would imply the activity of at least one other protein (or protein complex). In the unstressed state, RNA polymerase is bound in an inactive state to the promoter, indicated by the phosphorylation of serine 5 in its C-terminal tail. **B. Early after DNA damage.** Soon after DNA damage, ATM and other DNA damage-activated kinases phosphorylate both p53 and MDM2. However, at the early time, even though p53 is phosphorylated on serine 15 (and presumably other sites), MDM2 (and MDMX) are still bound, and all are unstable. p53 is inactive at this early time after damage induction. **C. Peak DNA damage response.** The peak of the DNA damage response correlates with the stabilization of p53 and increased p53 abundance (due to its increased stability). p53 is stabilized because of the accelerated auto-degradation of MDM2, and the accelerated MDM2-mediated degradation of MDMX, and perhaps reduced affinity for the MDM2-MDMX complex due to p53 N-terminal phosphorylation at residues ser15, thr18 and ser20. We propose that the accelerated

degradation of MDM2/MDMX, along with p53 N-terminal phosphorylation, allows sufficient time for p53 to interact with the transcriptional co-activators that are required to convert RNA polymerase into its active form, which is phosphorylated at serine 2 in its C-terminal domain. Due to increased transcription of the MDM2 target gene, MDM2 increases in abundance, though the newly synthesized protein is modified by ongoing damage kinase activity, rendering it very unstable and incapable of inactivating p53. **D. Late in the DNA damage response.** If the DNA damage is successfully repaired, ATM (and, we presume, other damage kinases) become inactivated. The MDM2 mRNA that accumulated due to p53-mediated activation is translated into a more stable form of MDM2 that can hetero-dimerize with MDMX, leading to effective inactivation of p53 and attenuation of the DNA damage response. At this time, p53, MDM2, and MDMX once again appear to exhibit similar short half-lives. We did not indicate specific timing of these events, as they may be affected by cell type, genetic background, and mitogenic and survival signaling.

The second step involves the ability of DNA damage to convert MDM2 into a more active E3 to accelerate the degradation of itself and MDMX. It is possible that accelerated degradation of MDM2 is sufficient since MDMX alone cannot efficiently inhibit p53-mediated biological responses. However, recent data indicate that MDMX degradation should further destabilize MDM2, which could contribute to reinforcing the activation process. A tempting speculation is that the MDM2 accumulation that results from p53 mediated transactivation of the MDM2 gene actually creates a feed-forward loop for p53 activation. This positive regulation could result when the MDM2:MDMX ratio becomes sufficient for MDM2 to induce MDMX degradation, which would in turn destabilize MDM2, resulting in more p53 activation. We infer that accelerated MDM2 degradation is a critical step for p53 activation as inhibiting it prevents p53 activation, even when the p53 contains damage-associated modifications.

It is possible that p53, MDM2 and p300 exist in chromatin-bound complexes since all of these are (mainly) nuclear proteins, MDM2 binds to p53 best when p53 is tetrameric, and tetrameric p53 binds best to chromatin (Hainaut et al., 1994; Marston et al., 1995; McLure and Lee, 1998). Elimination of MDM2 from such complexes may enable rapid activation of p53 in response to stresses. This view is consistent with recent kinetic and ChIP analyses of p53 target gene activation showing that p53 and the basal transcription machinery are poised to function (Espinosa et al., 2003). Activating signals then induce phosphorylation of the C-terminal tail of the poised RNA polymerase to enable it to translocate along the DNA (Espinosa et al., 2003).

It is noteworthy that, like p53, other short-lived transcription factors including Myc, Hif1a, etc., have transactivation domains that overlap with the residues that mediate their destruction (Muratani and Tansey, 2003; Salghetti et al., 2000). Perhaps this organization evolved to allow for competition between the factors required for transcription factor activation with those needed for removal of the transcription factor from chromatin. We speculate that such an organization may have evolved to enable rapid proteolysis of chromatin bound transcription factors to prevent their errant activation; on the other hand, regulated degradation of an inhibitory E3 ubiquitin ligase enables its rapid removal from the complex to expedite transcriptional activation in response to the appropriate signal.

Advances in bioinformatics, molecular biology, biochemistry, and genetics are providing the bases for a detailed understanding of the circuitry that regulates p53 and tunes its output. This will provide a model for other pathways that process numerous signals to generate diverse responses. We are hopeful that advances in structural biology and chemistry will propel efforts to identify additional drugs capable of activating mutant p53 or wild type p53 in MDM2/MDMX over-expressing cells to enable selective activation of the p53 pathway. This class of targeted therapeutics should add significantly to the existing armamentarium to increase the success of cancer treatment in a broad range of neoplasms. This would be a great present to all who are still around to celebrate the 50th Anniversary of the discovery of p53!

ACKNOWLEDGEMENTS

The authorship indicates the relative level of contributions to the writing of this Chapter, with KK and MW contributing equally. We thank Dr. Franck Toledo and Ms. Ee-tsin Wong for their helpful comments on the manuscript that led to this Chapter. Work from the Wahl lab referenced in this Chapter was supported by grants from the National Institutes of Health and National Cancer Institute to GMW, fellowships from the Pioneer Fund for KK and National Science Foundation for JS.

REFERENCES

The authors apologize for our inability to include all of the meritorious work of our colleagues due to the enormous volume of literature on the topics reviewed and the limited amount of space to cover it.

Alarcon, R., Koumenis, C., Geyer, R. K., Maki, C. G., and Giaccia, A. J. (1999). Hypoxia induces p53 accumulation through MDM2 down-regulation and inhibition of E6-mediated degradation. Cancer Res *59*, 6046-6051.

Alt, J. R., Greiner, T. C., Cleveland, J. L., and Eischen, C. M. (2003). Mdm2 haplo-insufficiency profoundly inhibits Myc-induced lymphomagenesis. EMBO J *22*, 1442-1450.

Amundson, S. A., Patterson, A., Do, K. T., and Fornace, A. J., Jr. (2002). A nucleotide excision repair master-switch: p53 regulated coordinate induction of global genomic repair genes. Cancer Biol Ther *1*, 145-149.

Appella, E., and Anderson, C. W. (2001). Post-translational modifications and activation of p53 by genotoxic stresses. Eur J Biochem *268*, 2764-2772.

Argentini, M., Barboule, N., and Wasylyk, B. (2001). The contribution of the acidic domain of MDM2 to p53 and MDM2 stability. Oncogene *20*, 1267-1275.

Ashcroft, M., Kubbutat, M. H., and Vousden, K. H. (1999). Regulation of p53 function and stability by phosphorylation. Mol Cell Biol *19*, 1751-1758.

Ashcroft, M., and Vousden, K. H. (1999). Regulation of p53 stability. Oncogene *18*, 7637-7643.

Banin, S., Moyal, L., Shieh, S., Taya, Y., Anderson, C. W., Chessa, L., Smorodinsky, N. I., Prives, C., Reiss, Y., Shiloh, Y., and Ziv, Y. (1998). Enhanced phosphorylation of p53 by ATM in response to DNA damage. Science *281*, 1674-1677.

Barak, Y., Juven, T., Haffner, R., and Oren, M. (1993). mdm2 expression is induced by wild type p53 activity. Embo J *12*, 461-468.

Barlev, N. A., Liu, L., Chehab, N. H., Mansfield, K., Harris, K. G., Halazonetis, T. D., and Berger, S. L. (2001). Acetylation of p53 activates transcription through recruitment of coactivators/histone acetyltransferases. Mol Cell *8*, 1243-1254.

Bassing, C. H., and Alt, F. W. (2004). H2AX may function as an anchor to hold broken chromosomal DNA ends in close proximity. Cell Cycle *3*, 149-153.

Bassing, C. H., Suh, H., Ferguson, D. O., Chua, K. F., Manis, J., Eckersdorff, M., Gleason, M., Bronson, R., Lee, C., and Alt, F. W. (2003). Histone H2AX: a dosage-dependent suppressor of oncogenic translocations and tumors. Cell *114*, 359-370.

Bates, S., Phillips, A. C., Clark, P. A., Stott, F., Peters, G., Ludwig, R. L., and Vousden, K. H. (1998). p14ARF links the tumour suppressors RB and p53. Nature *395*, 124-125.

Bischoff, F. Z., Yim, S. O., Pathak, S., Grant, G., Siciliano, M. J., Giovanella, B. C., Strong, L. C., and Tainsky, M. A. (1990). Spontaneous abnormalities in normal fibroblasts from patients with Li-Fraumeni cancer syndrome: aneuploidy and immortalization. Cancer Res *50*, 7979-7984.

Blattner, C., Tobiasch, E., Litfen, M., Rahmsdorf, H. J., and Herrlich, P. (1999). DNA damage induced p53 stabilization: no indication for an involvement of p53 phosphorylation. Oncogene *18*, 1723-1732.

Borges, H. L., Chao, C., Xu, Y., Linden, R., and Wang, J. Y. (2004). Radiation-induced apoptosis in developing mouse retina exhibits dose-dependent requirement for ATM phosphorylation of p53. Cell Death Differ *11*, 494-502.

Bottger, A., Bottger, V., Sparks, A., Liu, W. L., Howard, S. F., and Lane, D. P. (1997). Design of a synthetic Mdm2-binding mini protein that activates the p53 response in vivo. Curr Biol *7*, 860-869.

Bottger, V., Bottger, A., Garcia-Echeverria, C., Ramos, Y. F., van der Eb, A. J., Jochemsen, A. G., and Lane, D. P. (1999). Comparative study of the p53-mdm2 and p53-MDMX interfaces. Oncogene *18*, 189-199.

Bottger, V., Bottger, A., Howard, S. F., Picksley, S. M., Chene, P., Garcia-Echeverria, C., Hochkeppel, H. K., and Lane, D. P. (1996). Identification of novel mdm2 binding peptides by phage display. Oncogene *13*, 2141-2147.

Boyd, S. D., Tsai, K. Y., and Jacks, T. (2000). An intact HDM2 RING-finger domain is required for nuclear exclusion of p53. Nat Cell Biol *2*, 563-568.

Brignone, C., Bradley, K. E., Kisselev, A. F., and Grossman, S. R. (2004). A post-ubiquitination role for MDM2 and hHR23A in the p53 degradation pathway. Oncogene.

Brodsky, M. H., Nordstrom, W., Tsang, G., Kwan, E., Rubin, G. M., and Abrams, J. M. (2000). Drosophila p53 binds a damage response element at the reaper locus. Cell *101*, 103-113.

Brodsky, M. H., Weinert, B. T., Tsang, G., Rong, Y. S., McGinnis, N. M., Golic, K. G., Rio, D. C., and Rubin, G. M. (2004). Drosophila melanogaster MNK/Chk2 and p53 regulate multiple DNA repair and apoptotic pathways following DNA damage. Mol Cell Biol *24*, 1219-1231.

Bunz, F., Dutriaux, A., Lengauer, C., Waldman, T., Zhou, S., Brown, J. P., Sedivy, J. M., Kinzler, K. W., and Vogelstein, B. (1998). Requirement for p53 and p21 to sustain G2 arrest after DNA damage. Science *282*, 1497-1501.

Caelles, C., Helmberg, A., and Karin, M. (1994). p53-dependent apoptosis in the absence of transcriptional activation of p53-target genes. Nature *370*, 220-223.

Canman, C. E., Lim, D. S., Cimprich, K. A., Taya, Y., Tamai, K., Sakaguchi, K., Appella, E., Kastan, M. B., and Siliciano, J. D. (1998). Activation of the ATM kinase by ionizing radiation and phosphorylation of p53. Science *281*, 1677-1679.

Chao, C., Hergenhahn, M., Kaeser, M. D., Wu, Z., Saito, S., Iggo, R., Hollstein, M., Appella, E., and Xu, Y. (2003). Cell type- and promoter-specific roles of Ser18 phosphorylation in regulating p53 responses. J Biol Chem *278*, 41028-41033.

Chao, C., Saito, S., Anderson, C. W., Appella, E., and Xu, Y. (2000a). Phosphorylation of murine p53 at ser-18 regulates the p53 responses to DNA damage. Proc Natl Acad Sci U S A *97*, 11936-11941.

Chao, C., Saito, S., Kang, J., Anderson, C. W., Appella, E., and Xu, Y. (2000b). p53 transcriptional activity is essential for p53-dependent apoptosis following DNA damage. Embo J *19*, 4967-4975.

Chen, C. Y., Oliner, J. D., Zhan, Q., Fornace, A. J., Jr., Vogelstein, B., and Kastan, M. B. (1994). Interactions between p53 and MDM2 in a mammalian cell cycle checkpoint pathway. Proc Natl Acad Sci U S A *91*, 2684-2688.

Chen, J., Marechal, V., and Levine, A. J. (1993). Mapping of the p53 and mdm-2 interaction domains. Mol Cell Biol *13*, 4107-4114.

Chin, L., Artandi, S. E., Shen, Q., Tam, A., Lee, S. L., Gottlieb, G. J., Greider, C. W., and DePinho, R. A. (1999). p53 deficiency rescues the adverse effects of telomere loss and cooperates with telomere dysfunction to accelerate carcinogenesis. Cell *97*, 527-538.

Chipuk, J. E., Kuwana, T., Bouchier-Hayes, L., Droin, N. M., Newmeyer, D. D., Schuler, M., and Green, D. R. (2004). Direct activation of Bax by p53 mediates mitochondrial membrane permeabilization and apoptosis. Science *303*, 1010-1014.

Cho, Y., Gorina, S., Jeffrey, P. D., and Pavletich, N. P. (1994). Crystal structure of a p53 tumor suppressor-DNA complex: understanding tumorigenic mutations [see comments]. Science *265*, 346-355.

Craig, A. L., Burch, L., Vojtesek, B., Mikutowska, J., Thompson, A., and Hupp, T. R. (1999). Novel phosphorylation sites of human tumour suppressor protein p53 at Ser20 and Thr18 that disrupt the binding of mdm2 (mouse double minute 2) protein are modified in human cancers. Biochem J *342*, 133-141.

Cummings, J. M., Rago, C., Kohli, M., Kinzler, K. W., Lengauer, C., and Vogelstein, B. (2004). Tumour suppression: disruption of HAUSP gene stabilizes p53. Nature *428*, 1 p following 486.

David-Pfeuty, T., Chakrani, F., Ory, K., and Nouvian-Dooghe, Y. (1996). Cell cycle-dependent regulation of nuclear p53 traffic occurs in one subclass of human tumor cells and in untransformed cells. Cell Growth Differ *7*, 1211-1225.

De Guzman, R. N., Liu, H. Y., Martinez-Yamout, M., Dyson, H. J., and Wright, P. E. (2000). Solution structure of the TAZ2 (CH3) domain of the transcriptional adaptor protein CBP. J Mol Biol *303*, 243-253.

Deng, C., Zhang, P., Harper, J. W., Elledge, S. J., and Leder, P. (1995). Mice lacking p21CIP1/WAF1 undergo normal development, but are defective in G1 checkpoint control. Cell *82*, 675-684.

Denko, N. C., Giaccia, A. J., Stringer, J. R., and Stambrook, P. J. (1994). The human Ha-ras oncogene induces genomic instability in murine fibroblasts within one cell cycle. Proc Natl Acad Sci U S A *91*, 5124-5128.

Di Leonardo, A., Khan, S. H., Linke, S. P., Greco, V., Seidita, G., and Wahl, G. M. (1997). DNA rereplication in the presence of mitotic spindle inhibitors in human and mouse fibroblasts lacking either p53 or pRb function. Cancer Res *57*, 1013-1019.

Dimri, G. P., Itahana, K., Acosta, M., and Campisi, J. (2000). Regulation of a senescence checkpoint response by the E2F1 transcription factor and p14(ARF) tumor suppressor. Mol Cell Biol *20*, 273-285.

Dornan, D., Wertz, I., Shimizu, H., Arnott, D., Frantz, G. D., Dowd, P., K, O. R., Koeppen, H., and Dixit, V. M. (2004). The ubiquitin ligase COP1 is a critical negative regulator of p53. Nature.

Dumaz, N., and Meek, D. W. (1999). Serine15 phosphorylation stimulates p53 transactivation but does not directly influence interaction with HDM2. Embo J *18*, 7002-7010.

Dumaz, N., Milne, D. M., Jardine, L. J., and Meek, D. W. (2001). Critical roles for the serine 20, but not the serine 15, phosphorylation site and for the polyproline domain in regulating p53 turnover. Biochem J *359*, 459-464.

Eischen, C. M., Roussel, M. F., Korsmeyer, S. J., and Cleveland, J. L. (2001). Bax loss impairs Myc-induced apoptosis and circumvents the selection of p53 mutations during Myc-mediated lymphomagenesis. Mol Cell Biol *21*, 7653-7662.

Eischen, C. M., Weber, J. D., Roussel, M. F., Sherr, C. J., and Cleveland, J. L. (1999). Disruption of the ARF-Mdm2-p53 tumor suppressor pathway in Myc-induced lymphomagenesis. Genes Dev *13*, 2658-2669.

el-Deiry, W. S., Kern, S. E., Pietenpol, J. A., Kinzler, K. W., and Vogelstein, B. (1992). Definition of a consensus binding site for p53. Nat Genet *1*, 45-49.

el-Deiry, W. S., Tokino, T., Velculescu, V. E., Levy, D. B., Parsons, R., Trent, J. M., Lin, D., Mercer, W. E., Kinzler, K. W., and Vogelstein, B. (1993). WAF1, a potential mediator of p53 tumor suppression. Cell *75*, 817-825.

Espinosa, J. M., Verdun, R. E., and Emerson, B. M. (2003). p53 functions through stress- and promoter-specific recruitment of transcription initiation components before and after DNA damage. Mol Cell *12*, 1015-1027.

Fang, S., Jensen, J. P., Ludwig, R. L., Vousden, K. H., and Weissman, A. M. (2000). Mdm2 is a RING finger-dependent ubiquitin protein ligase for itself and p53. J Biol Chem *275*, 8945-8951.

Felsher, D. W., and Bishop, J. M. (1999). Transient excess of MYC activity can elicit genomic instability and tumorigenesis. Proc Natl Acad Sci U S A *96*, 3940-3944.

Fitch, M. E., Cross, I. V., Turner, S. J., Adimoolam, S., Lin, C. X., Williams, K. G., and Ford, J. M. (2003). The DDB2 nucleotide excision repair gene product p48 enhances global genomic repair in p53 deficient human fibroblasts. DNA Repair (Amst) *2*, 819-826.

Freedman, D. A., and Levine, A. J. (1998). Nuclear export is required for degradation of endogenous p53 by MDM2 and human papillomavirus E6. Mol Cell Biol *18*, 7288-7293.

Friedman, P. N., Chen, X., Bargonetti, J., and Prives, C. (1993). The p53 protein is an unusually shaped tetramer that binds directly to DNA [published erratum appears in Proc Natl Acad Sci U S A 1993 Jun 15;90(12):5878]. Proc Natl Acad Sci U S A *90*, 3319-3323.

Fuchs, S. Y., Adler, V., Buschmann, T., Wu, X., and Ronai, Z. (1998). Mdm2 association with p53 targets its ubiquitination. Oncogene *17*, 2543-2547.

Funk, W. D., Pak, D. T., Karas, R. H., Wright, W. E., and Shay, J. W. (1992). A transcriptionally active DNA-binding site for human p53 protein complexes. Mol Cell Biol *12*, 2866-2871.

Gao, Y., Ferguson, D. O., Xie, W., Manis, J. P., Sekiguchi, J., Frank, K. M., Chaudhuri, J., Horner, J., DePinho, R. A., and Alt, F. W. (2000). Interplay of p53 and DNA-repair protein XRCC4 in tumorigenesis, genomic stability and development. Nature *404*, 897-900.

Garcia-Echeverria, C., Chene, P., Blommers, M. J., and Furet, P. (2000). Discovery of potent antagonists of the interaction between human double minute 2 and tumor suppressor p53. J Med Chem *43*, 3205-3208.

Geyer, R. K., Yu, Z. K., and Maki, C. G. (2000). The MDM2 RING-finger domain is required to promote p53 nuclear export. Nat Cell Biol *2*, 569-573.

Gorina, S., and Pavletich, N. P. (1996). Structure of the p53 tumor suppressor bound to the ankyrin and SH3 domains of 53BP2 [see comments]. Science *274*, 1001-1005.

Gorlich, D., and Kutay, U. (1999). Transport between the cell nucleus and the cytoplasm. Annu Rev Cell Dev Biol *15*, 607-660.

Gottlieb, T. M., Leal, J. F., Seger, R., Taya, Y., and Oren, M. (2002). Cross-talk between Akt, p53 and Mdm2: possible implications for the regulation of apoptosis. Oncogene *21*, 1299-1303.

Grossman, S. R., Deato, M. E., Brignone, C., Chan, H. M., Kung, A. L., Tagami, H., Nakatani, Y., and Livingston, D. M. (2003). Polyubiquitination of p53 by a ubiquitin ligase activity of p300. Science *300*, 342-344.

Grossman, S. R., Perez, M., Kung, A. L., Joseph, M., Mansur, C., Xiao, Z. X., Kumar, S., Howley, P. M., and Livingston, D. M. (1998). p300/MDM2 complexes participate in MDM2-mediated p53 degradation. Mol Cell *2*, 405-415.

Gu, J., Kawai, H., Nie, L., Kitao, H., Wiederschain, D., Jochemsen, A. G., Parant, J., Lozano, G., and Yuan, Z. M. (2002). Mutual dependence of MDM2 and MDMX in their functional inactivation of p53. J Biol Chem *277*, 19251-19254.

Gu, W., and Roeder, R. G. (1997). Activation of p53 sequence-specific DNA binding by acetylation of the p53 C-terminal domain. Cell *90*, 595-606.

Gu, W., Shi, X. L., and Roeder, R. G. (1997). Synergistic activation of transcription by CBP and p53. Nature *387*, 819-823.

Hainaut, P., Hall, A., and Milner, J. (1994). Analysis of p53 quaternary structure in relation to sequence-specific DNA binding. Oncogene *9*, 299-303.

Haupt, Y., Maya, R., Kazaz, A., and Oren, M. (1997). Mdm2 promotes the rapid degradation of p53. Nature *387*, 296-299.

Hay, T. J., and Meek, D. W. (2000). Multiple sites of in vivo phosphorylation in the MDM2 oncoprotein cluster within two important functional domains. FEBS Lett *478*, 183-186.

Henderson, B. R., and Eleftheriou, A. (2000). A comparison of the activity, sequence specificity, and CRM1-dependence of different nuclear export signals. Exp Cell Res *256*, 213-224.

Hermeking, H., Lengauer, C., Polyak, K., He, T. C., Zhang, L., Thiagalingam, S., Kinzler, K. W., and Vogelstein, B. (1997). 14-3-3 sigma is a p53-regulated inhibitor of G2/M progression. Mol Cell *1*, 3-11.

Hicke, L. (2001). Protein regulation by monoubiquitin. Nat Rev Mol Cell Biol *2*, 195-201.

Hiyama, H., Yokoi, M., Masutani, C., Sugasawa, K., Maekawa, T., Tanaka, K., Hoeijmakers, J. H., and Hanaoka, F. (1999). Interaction of hHR23 with S5a. The ubiquitin-like domain of hHR23 mediates interaction with S5a subunit of 26 S proteasome. J Biol Chem *274*, 28019-28025.

Hollander, M. C., Sheikh, M. S., Bulavin, D. V., Lundgren, K., Augeri-Henmueller, L., Shehee, R., Molinaro, T. A., Kim, K. E., Tolosa, E., Ashwell, J. D., *et al.* (1999). Genomic instability in Gadd45a-deficient mice. Nat Genet *23*, 176-184.

Hollstein, M., Sidransky, D., Vogelstein, B., and Harris, C. C. (1991). p53 mutations in human cancers. Science *253*, 49-53.

Honda, R., Tanaka, H., and Yasuda, H. (1997). Oncoprotein MDM2 is a ubiquitin ligase E3 for tumor suppressor p53. FEBS Lett *420*, 25-27.

Honda, R., and Yasuda, H. (1999). Association of p19(ARF) with Mdm2 inhibits ubiquitin ligase activity of Mdm2 for tumor suppressor p53. Embo J *18*, 22-27.

Honda, R., and Yasuda, H. (2000). Activity of MDM2, a ubiquitin ligase, toward p53 or itself is dependent on the RING finger domain of the ligase. Oncogene *19*, 1473-1476.

Huang, L. C., Clarkin, K. C., and Wahl, G. M. (1996). Sensitivity and selectivity of the DNA damage sensor responsible for activating p53-dependent G1 arrest. Proc Natl Acad Sci U S A *93*, 4827-4832.

Hwang, B. J., Ford, J. M., Hanawalt, P. C., and Chu, G. (1999). Expression of the p48 xeroderma pigmentosum gene is p53-dependent and is involved in global genomic repair. Proc Natl Acad Sci U S A *96*, 424-428.

Inga, A., Storici, F., Darden, T. A., and Resnick, M. A. (2002). Differential transactivation by the p53 transcription factor is highly dependent on p53 level and promoter target sequence. Mol Cell Biol *22*, 8612-8625.

Inoue, T., Geyer, R. K., Howard, D., Yu, Z. K., and Maki, C. G. (2001). MDM2 can promote the ubiquitination, nuclear export, and degradation of p53 in the absence of direct binding. J Biol Chem *276*, 45255-45260.

Jabbur, J. R., Tabor, A. D., Cheng, X., Wang, H., Uesugi, M., Lozano, G., and Zhang, W. (2002). Mdm-2 binding and TAF(II)31 recruitment is regulated by hydrogen bond disruption between the p53 residues Thr18 and Asp21. Oncogene *21*, 7100-7113.

Jackson, M. W., and Berberich, S. J. (2000). MdmX protects p53 from Mdm2-mediated degradation. Mol Cell Biol *20*, 1001-1007.

Janus, F., Albrechtsen, N., Dornreiter, I., Wiesmuller, L., Grosse, F., and Deppert, W. (1999). The dual role model for p53 in maintaining genomic integrity. Cell Mol Life Sci *55*, 12-27.

Jeffers, J. R., Parganas, E., Lee, Y., Yang, C., Wang, J., Brennan, J., MacLean, K. H., Han, J., Chittenden, T., Ihle, J. N., *et al.* (2003). Puma is an essential mediator of p53-dependent and -independent apoptotic pathways. Cancer Cell *4*, 321-328.

Jimenez, G. S., Khan, S. H., Stommel, J. M., and Wahl, G. M. (1999). p53 regulation by post-translational modification and nuclear retention in response to diverse stresses. Oncogene *18*, 7656-7665.

Jimenez, G. S., Nister, M., Stommel, J. M., Beeche, M., Barcarse, E. A., Zhang, X. Q., O'Gorman, S., and Wahl, G. M. (2000). A transactivation-deficient mouse model provides insights into Trp53 regulation and function. Nat Genet *26*, 37-43.

Jin, S., Antinore, M. J., Lung, F. D., Dong, X., Zhao, H., Fan, F., Colchagie, A. B., Blanck, P., Roller, P. P., Fornace, A. J., Jr., and Zhan, Q. (2000). The GADD45 inhibition of Cdc2 kinase correlates with GADD45-mediated growth suppression. J Biol Chem *275*, 16602-16608.

Jones, S. N., Roe, A. E., Donehower, L. A., and Bradley, A. (1995). Rescue of embryonic lethality in Mdm2-deficient mice by absence of p53. Nature *378*, 206-208.

Joseph, T. W., Zaika, A., and Moll, U. M. (2003). Nuclear and cytoplasmic degradation of endogenous p53 and HDM2 occurs during down-regulation of the p53 response after multiple types of DNA damage. Faseb J *17*, 1622-1630.

Kamijo, T., Bodner, S., van de Kamp, E., Randle, D. H., and Sherr, C. J. (1999a). Tumor spectrum in ARF-deficient mice. Cancer Res *59*, 2217-2222.

Kamijo, T., van de Kamp, E., Chong, M. J., Zindy, F., Diehl, J. A., Sherr, C. J., and McKinnon, P. J. (1999b). Loss of the ARF tumor suppressor reverses premature replicative arrest but not radiation hypersensitivity arising from disabled atm function. Cancer Res *59*, 2464-2469.

Kamijo, T., Weber, J. D., Zambetti, G., Zindy, F., Roussel, M. F., and Sherr, C. J. (1998). Functional and physical interactions of the ARF tumor suppressor with p53 and Mdm2. Proc Natl Acad Sci U S A *95*, 8292-8297.

Kamijo, T., Zindy, F., Roussel, M. F., Quelle, D. E., Downing, J. R., Ashmun, R. A., Grosveld, G., and Sherr, C. J. (1997). Tumor suppression at the mouse INK4a locus mediated by the alternative reading frame product p19ARF. Cell *91*, 649-659.

Kane, S. A., Fleener, C. A., Zhang, Y. S., Davis, L. J., Musselman, A. L., and Huang, P. S. (2000). Development of a binding assay for p53/HDM2 by using homogeneous time-resolved fluorescence. Anal Biochem *278*, 29-38.

Karlseder, J., Broccoli, D., Dai, Y., Hardy, S., and de Lange, T. (1999). p53- and ATM-dependent apoptosis induced by telomeres lacking TRF2. Science *283*, 1321-1325.

Kastan, M. B., Onyekwere, O., Sidransky, D., Vogelstein, B., and Craig, R. W. (1991). Participation of p53 protein in the cellular response to DNA damage. Cancer Res *51*, 6304-6311.

Kastan, M. B., Zhan, Q., el-Deiry, W. S., Carrier, F., Jacks, T., Walsh, W. V., Plunkett, B. S., Vogelstein, B., and Fornace, A. J., Jr. (1992). A mammalian cell cycle checkpoint pathway utilizing p53 and GADD45 is defective in ataxia-telangiectasia. Cell *71*, 587-597.

Kawai, H., Wiederschain, D., Kitao, H., Stuart, J., Tsai, K. K., and Yuan, Z. M. (2003a). DNA damage-induced MDMX degradation is mediated by MDM2. J Biol Chem *278*, 45946-45953.

Kawai, H., Wiederschain, D., and Yuan, Z. M. (2003b). Critical contribution of the MDM2 acidic domain to p53 ubiquitination. Mol Cell Biol *23*, 4939-4947.

Khan, S. H., Moritsugu, J., and Wahl, G. M. (2000). Differential requirement for p19ARF in the p53-dependent arrest induced by DNA damage, microtubule disruption, and ribonucleotide depletion. Proc Natl Acad Sci U S A *97*, 3266-3271.

Khan, S. H., and Wahl, G. M. (1998). p53 and pRb prevent rereplication in response to microtubule inhibitors by mediating a reversible G1 arrest. Cancer Res *58*, 396-401.

Khosravi, R., Maya, R., Gottlieb, T., Oren, M., Shiloh, Y., and Shkedy, D. (1999). Rapid ATM-dependent phosphorylation of MDM2 precedes p53 accumulation in response to DNA damage. Proc Natl Acad Sci U S A *96*, 14973-14977.

Kobet, E., Zeng, X., Zhu, Y., Keller, D., and Lu, H. (2000). MDM2 inhibits p300-mediated p53 acetylation and activation by forming a ternary complex with the two proteins. Proc Natl Acad Sci U S A *97*, 12547-12552.

Kubbutat, M. H., Jones, S. N., and Vousden, K. H. (1997). Regulation of p53 stability by Mdm2. Nature *387*, 299-303.

Kudo, N., Wolff, B., Sekimoto, T., Schreiner, E. P., Yoneda, Y., Yanagida, M., Horinouchi, S., and Yoshida, M. (1998). Leptomycin B inhibition of signal-mediated nuclear export by direct binding to CRM1. Exp Cell Res *242*, 540-547.

Kussie, P. H., Gorina, S., Marechal, V., Elenbaas, B., Moreau, J., Levine, A. J., and Pavletich, N. P. (1996). Structure of the MDM2 oncoprotein bound to the p53 tumor suppressor transactivation domain. Science *274*, 948-953.

Lai, Z., Ferry, K. V., Diamond, M. A., Wee, K. E., Kim, Y. B., Ma, J., Yang, T., Benfield, P. A., Copeland, R. A., and Auger, K. R. (2001). Human mdm2 mediates multiple mono-ubiquitination of p53 by a mechanism requiring enzyme isomerization. J Biol Chem *276*, 31357-31367.

Lambert, P. F., Kashanchi, F., Radonovich, M. F., Shiekhattar, R., and Brady, J. N. (1998). Phosphorylation of p53 serine 15 increases interaction with CBP. J Biol Chem *273*, 33048-33053.

Lane, D. P. (1992). Cancer. p53, guardian of the genome. Nature *358*, 15-16.

Lane, D. P., and Crawford, L. V. (1979). T antigen is bound to a host protein in SV40-transformed cells. Nature *278*, 261-263.

Langheinrich, U., Hennen, E., Stott, G., and Vacun, G. (2002). Zebrafish as a model organism for the identification and characterization of drugs and genes affecting p53 signaling. Curr Biol *12*, 2023-2028.

Lanni, J. S., and Jacks, T. (1998). Characterization of the p53-dependent postmitotic checkpoint following spindle disruption. Mol Cell Biol *18*, 1055-1064.

Lee, J. H., Lee, E., Park, J., Kim, E., Kim, J., and Chung, J. (2003). In vivo p53 function is indispensable for DNA damage-induced apoptotic signaling in Drosophila. FEBS Lett *550*, 5-10.

Lees-Miller, S. P., and Meek, K. (2003). Repair of DNA double strand breaks by non-homologous end joining. Biochimie *85*, 1161-1173.

Leng, R. P., Lin, Y., Ma, W., Wu, H., Lemmers, B., Chung, S., Parant, J. M., Lozano, G., Hakem, R., and Benchimol, S. (2003). Pirh2, a p53-induced ubiquitin-protein ligase, promotes p53 degradation. Cell *112*, 779-791.

Lev Bar-Or, R., Maya, R., Segel, L. A., Alon, U., Levine, A. J., and Oren, M. (2000). Generation of oscillations by the p53-Mdm2 feedback loop: a theoretical and experimental study. Proc Natl Acad Sci U S A *97*, 11250-11255.

Li, M., Brooks, C. L., Kon, N., and Gu, W. (2004). A dynamic role of HAUSP in the p53-Mdm2 pathway. Mol Cell *13*, 879-886.

Li, M., Brooks, C. L., Wu-Baer, F., Chen, D., Baer, R., and Gu, W. (2003). Mono- versus polyubiquitination: differential control of p53 fate by Mdm2. Science *302*, 1972-1975.

Li, M., Chen, D., Shiloh, A., Luo, J., Nikolaev, A. Y., Qin, J., and Gu, W. (2002). Deubiquitination of p53 by HAUSP is an important pathway for p53 stabilization. Nature *416*, 648-653.

Liang, S. H., and Clarke, M. F. (2001). Regulation of p53 localization. Eur J Biochem *268*, 2779-2783.

Lim, S. K., Shin, J. M., Kim, Y. S., and Baek, K. H. (2004). Identification and characterization of murine mHAUSP encoding a deubiquitinating enzyme that regulates the status of p53 ubiquitination. Int J Oncol *24*, 357-364.

Lima, C. D. (2003). Regulating UBP-mediated ubiquitin deconjugation. Structure (Camb) *11*, 3-4.

Lin, J., Chen, J., Elenbaas, B., and Levine, A. J. (1994). Several hydrophobic amino acids in the p53 amino-terminal domain are required for transcriptional activation, binding to mdm-2 and the adenovirus 5 E1B 55-kD protein. Genes Dev *8*, 1235-1246.

Linke, S. P., Clarkin, K. C., Di Leonardo, A., Tsou, A., and Wahl, G. M. (1996). A reversible, p53-dependent G0/G1 cell cycle arrest induced by ribonucleotide depletion in the absence of detectable DNA damage. Genes Dev *10*, 934-947.

Linzer, D. I., and Levine, A. J. (1979). Characterization of a 54K dalton cellular SV40 tumor antigen present in SV40-transformed cells and uninfected embryonal carcinoma cells. Cell *17*, 43-52.

Liu, L., Scolnick, D. M., Trievel, R. C., Zhang, H. B., Marmorstein, R., Halazonetis, T. D., and Berger, S. L. (1999). p53 sites acetylated in vitro by PCAF and p300 are acetylated in vivo in response to DNA damage. Mol Cell Biol *19*, 1202-1209.

Livingstone, L. R., White, A., Sprouse, J., Livanos, E., Jacks, T., and Tlsty, T. D. (1992). Altered cell cycle arrest and gene amplification potential accompany loss of wild-type p53. Cell *70*, 923-935.

Llanos, S., Clark, P. A., Rowe, J., and Peters, G. (2001). Stabilization of p53 by p14ARF without relocation of MDM2 to the nucleolus. Nat Cell Biol *3*, 445-452.

Lohrum, M. A., Woods, D. B., Ludwig, R. L., Balint, E., and Vousden, K. H. (2001). C-terminal ubiquitination of p53 contributes to nuclear export. Mol Cell Biol *21*, 8521-8532.

Lu, H., and Levine, A. J. (1995). Human TAFII31 protein is a transcriptional coactivator of the p53 protein. Proc Natl Acad Sci U S A *92*, 5154-5158.

Mai, S., Fluri, M., Siwarski, D., and Huppi, K. (1996). Genomic instability in MycER-activated Rat1A-MycER cells. Chromosome Res *4*, 365-371.

Maki, C. G., Huibregtse, J. M., and Howley, P. M. (1996). In vivo ubiquitination and proteasome-mediated degradation of p53(1). Cancer Res *56*, 2649-2654.

Malkin, D., Li, F. P., Strong, L. C., Fraumeni, J. F., Jr., Nelson, C. E., Kim, D. H., Kassel, J., Gryka, M. A., Bischoff, F. Z., Tainsky, M. A., and et al. (1990). Germ line p53 mutations in a familial syndrome of breast cancer, sarcomas, and other neoplasms. Science *250*, 1233-1238.

Marston, N. J., Jenkins, J. R., and Vousden, K. H. (1995). Oligomerisation of full length p53 contributes to the interaction with mdm2 but not HPV E6. Oncogene *10*, 1709-1715.

Maya, R., Balass, M., Kim, S. T., Shkedy, D., Leal, J. F., Shifman, O., Moas, M., Buschmann, T., Ronai, Z., Shiloh, Y., et al. (2001). ATM-dependent phosphorylation of Mdm2 on serine 395: role in p53 activation by DNA damage. Genes Dev *15*, 1067-1077.

Mayo, L. D., and Donner, D. B. (2001). A phosphatidylinositol 3-kinase/Akt pathway promotes translocation of Mdm2 from the cytoplasm to the nucleus. Proc Natl Acad Sci U S A *98*, 11598-11603.

McCurrach, M. E., Connor, T. M., Knudson, C. M., Korsmeyer, S. J., and Lowe, S. W. (1997). bax-deficiency promotes drug resistance and oncogenic transformation by attenuating p53-dependent apoptosis. Proc Natl Acad Sci U S A *94*, 2345-2349.

McLure, K. G., and Lee, P. W. (1998). How p53 binds DNA as a tetramer. Embo J *17*, 3342-3350.

Meek, D. W. (2002). p53 Induction: phosphorylation sites cooperate in regulating. Cancer Biol Ther *1*, 284-286.

Meek, D. W., and Knippschild, U. (2003). Posttranslational modification of MDM2. Mol Cancer Res *1*, 1017-1026.

Mendrysa, S. M., McElwee, M. K., Michalowski, J., O'Leary, K. A., Young, K. M., and Perry, M. E. (2003). mdm2 Is critical for inhibition of p53 during lymphopoiesis and the response to ionizing irradiation. Mol Cell Biol *23*, 462-472.

Meulmeester, E., Frenk, R., Stad, R., de Graaf, P., Marine, J. C., Vousden, K. H., and Jochemsen, A. G. (2003). Critical role for a central part of Mdm2 in the ubiquitylation of p53. Mol Cell Biol *23*, 4929-4938.

Michael, D., and Oren, M. (2003). The p53-Mdm2 module and the ubiquitin system. Semin Cancer Biol *13*, 49-58.

Middeler, G., Zerf, K., Jenovai, S., Thulig, A., Tschodrich-Rotter, M., Kubitscheck, U., and Peters, R. (1997). The tumor suppressor p53 is subject to both nuclear import and export, and both are fast, energy-dependent and lectin-inhibited. Oncogene *14*, 1407-1417.

Migliorini, D., Danovi, D., Colombo, E., Carbone, R., Pelicci, P. G., and Marine, J. C. (2002a). Hdmx recruitment into the nucleus by Hdm2 is essential for its ability to regulate p53 stability and transactivation. J Biol Chem *277*, 7318-7323.

Migliorini, D., Denchi, E. L., Danovi, D., Jochemsen, A., Capillo, M., Gobbi, A., Helin, K., Pelicci, P. G., and Marine, J. C. (2002b). Mdm4 (Mdmx) regulates p53-induced growth arrest and neuronal cell death during early embryonic mouse development. Mol Cell Biol *22*, 5527-5538.

Migliorini, D., Denchi, E. L., Danovi, D., Jochemsen, A., Capillo, M., Gobbi, A., Helin, K., Pelicci, P. G., and Marine, J. C. (2002c). Mdm4 (Mdmx) regulates p53-induced growth arrest and neuronal cell death during early embryonic mouse development. Mol Cell Biol *22*, 5527-5538.

Mihara, M., Erster, S., Zaika, A., Petrenko, O., Chittenden, T., Pancoska, P., and Moll, U. M. (2003). p53 has a direct apoptogenic role at the mitochondria. Mol Cell *11*, 577-590.

Minn, A. J., Boise, L. H., and Thompson, C. B. (1996). Expression of Bcl-xL and loss of p53 can cooperate to overcome a cell cycle checkpoint induced by mitotic spindle damage. Genes Dev *10*, 2621-2631.

Mirnezami, A. H., Campbell, S. J., Darley, M., Primrose, J. N., Johnson, P. W., and Blaydes, J. P. (2003). Hdm2 recruits a hypoxia-sensitive corepressor to negatively regulate p53-dependent transcription. Curr Biol *13*, 1234-1239.

Moll, U. M., LaQuaglia, M., Benard, J., and Riou, G. (1995). Wild-type p53 protein undergoes cytoplasmic sequestration in undifferentiated neuroblastomas but not in differentiated tumors. Proc Natl Acad Sci U S A *92*, 4407-4411.

Moll, U. M., Ostermeyer, A. G., Haladay, R., Winkfield, B., Frazier, M., and Zambetti, G. (1996). Cytoplasmic sequestration of wild-type p53 protein impairs the G1 checkpoint after DNA damage. Mol Cell Biol *16*, 1126-1137.

Moll, U. M., Riou, G., and Levine, A. J. (1992). Two distinct mechanisms alter p53 in breast cancer: mutation and nuclear exclusion. Proc Natl Acad Sci U S A *89*, 7262-7266.

Moll, U. M., and Zaika, A. (2001). Nuclear and mitochondrial apoptotic pathways of p53. FEBS Lett *493*, 65-69.

Momand, J., Jung, D., Wilczynski, S., and Niland, J. (1998). The MDM2 gene amplification database. Nucleic Acids Res *26*, 3453-3459.

Momand, J., Zambetti, G. P., Olson, D. C., George, D., and Levine, A. J. (1992). The mdm-2 oncogene product forms a complex with the p53 protein and inhibits p53-mediated transactivation. Cell *69*, 1237-1245.

Montes de Oca Luna, R., Wagner, D. S., and Lozano, G. (1995). Rescue of early embryonic lethality in mdm2-deficient mice by deletion of p53. Nature *378*, 203-206.

Morgan, W. F., Bodycote, J., Fero, M. L., Hahn, P. J., Kapp, L. N., Pantelias, G. E., and Painter, R. B. (1986). A cytogenetic investigation of DNA rereplication after hydroxyurea treatment: implications for gene amplification. Chromosoma *93*, 191-196.

Muratani, M., and Tansey, W. P. (2003). How the ubiquitin-proteasome system controls transcription. Nat Rev Mol Cell Biol *4*, 192-201.

Nakamura, S., Roth, J. A., and Mukhopadhyay, T. (2000). Multiple lysine mutations in the C-terminal domain of p53 interfere with MDM2-dependent protein degradation and ubiquitination. Mol Cell Biol *20*, 9391-9398.

Nikolaev, A. Y., Li, M., Puskas, N., Qin, J., and Gu, W. (2003). Parc: a cytoplasmic anchor for p53. Cell *112*, 29-40.

Noda, A., Ning, Y., Venable, S. F., Pereira-Smith, O. M., and Smith, J. R. (1994). Cloning of senescent cell-derived inhibitors of DNA synthesis using an expression screen. Exp Cell Res *211*, 90-98.

O'Keefe, K., Li, H., and Zhang, Y. (2003). Nucleocytoplasmic shuttling of p53 is essential for MDM2-mediated cytoplasmic degradation but not ubiquitination. Mol Cell Biol *23*, 6396-6405.

Offer, H., Wolkowicz, R., Matas, D., Blumenstein, S., Livneh, Z., and Rotter, V. (1999). Direct involvement of p53 in the base excision repair pathway of the DNA repair machinery. FEBS Lett *450*, 197-204.

Ogawara, Y., Kishishita, S., Obata, T., Isazawa, Y., Suzuki, T., Tanaka, K., Masuyama, N., and Gotoh, Y. (2002). Akt enhances Mdm2-mediated ubiquitination and degradation of p53. J Biol Chem *277*, 21843-21850.

Oliner, J. D., Kinzler, K. W., Meltzer, P. S., George, D. L., and Vogelstein, B. (1992). Amplification of a gene encoding a p53-associated protein in human sarcomas [see comments]. Nature *358*, 80-83.

Oliner, J. D., Pietenpol, J. A., Thiagalingam, S., Gyuris, J., Kinzler, K. W., and Vogelstein, B. (1993). Oncoprotein MDM2 conceals the activation domain of tumour suppressor p53. Nature *362*, 857-860.

Ollmann, M., Young, L. M., Di Como, C. J., Karim, F., Belvin, M., Robertson, S., Whittaker, K., Demsky, M., Fisher, W. W., Buchman, A., *et al.* (2000). Drosophila p53 is a structural and functional homolog of the tumor suppressor p53. Cell *101*, 91-101.

Oren, M., Maltzman, W., and Levine, A. J. (1981). Post-translational regulation of the 54K cellular tumor antigen in normal and transformed cells. Mol Cell Biol *1*, 101-110.

Ostermeyer, A. G., Runko, E., Winkfield, B., Ahn, B., and Moll, U. M. (1996). Cytoplasmically sequestered wild-type p53 protein in neuroblastoma is relocated to the nucleus by a C-terminal peptide. Proc Natl Acad Sci U S A *93*, 15190-15194.

Pan, Y., and Chen, J. (2003). MDM2 Promotes Ubiquitination and Degradation of MDMX. Mol Cell Biol *23*, 5113-5121.

Parant, J., Chavez-Reyes, A., Little, N. A., Yan, W., Reinke, V., Jochemsen, A. G., and Lozano, G. (2001). Rescue of embryonic lethality in Mdm4-null mice by loss of Trp53 suggests a nonoverlapping pathway with MDM2 to regulate p53. Nat Genet *29*, 92-95.

Perry, M. E., Commane, M., and Stark, G. R. (1992). Simian virus 40 large tumor antigen alone or two cooperating oncogenes convert REF52 cells to a state permissive for gene amplification. Proc Natl Acad Sci U S A *89*, 8112-8116.

Perry, M. E., Piette, J., Zawadzki, J. A., Harvey, D., and Levine, A. J. (1993). The mdm-2 gene is induced in response to UV light in a p53-dependent manner. Proc Natl Acad Sci U S A *90*, 11623-11627.

Peters, M., DeLuca, C., Hirao, A., Stambolic, V., Potter, J., Zhou, L., Liepa, J., Snow, B., Arya, S., Wong, J., *et al.* (2002). Chk2 regulates irradiation-induced, p53-mediated apoptosis in Drosophila. Proc Natl Acad Sci U S A *99*, 11305-11310.

Poyurovsky, M. V., Jacq, X., Ma, C., Karni-Schmidt, O., Parker, P. J., Chalfie, M., Manley, J. L., and Prives, C. (2003). Nucleotide binding by the Mdm2 RING domain facilitates Arf-independent Mdm2 nucleolar localization. Mol Cell *12*, 875-887.

Qu, L., Huang, S., Baltzis, D., Rivas-Estilla, A. M., Pluquet, O., Hatzoglou, M., Koumenis, C., Taya, Y., Yoshimura, A., and Koromilas, A. E. (2004). Endoplasmic reticulum stress induces p53 cytoplasmic localization and prevents p53-dependent apoptosis by a pathway involving glycogen synthase kinase-3beta. Genes Dev *18*, 261-277.

Riemenschneider, M. J., Buschges, R., Wolter, M., Reifenberger, J., Bostrom, J., Kraus, J. A., Schlegel, U., and Reifenberger, G. (1999). Amplification and overexpression of the MDM4 (MDMX) gene from 1q32 in a subset of malignant gliomas without TP53 mutation or MDM2 amplification. Cancer Res *59*, 6091-6096.

Ries, S., Biederer, C., Woods, D., Shifman, O., Shirasawa, S., Sasazuki, T., McMahon, M., Oren, M., and McCormick, F. (2000). Opposing effects of Ras on p53: transcriptional activation of mdm2 and induction of p19ARF. Cell *103*, 321-330.

Rodriguez, M. S., Desterro, J. M., Lain, S., Lane, D. P., and Hay, R. T. (2000). Multiple C-terminal lysine residues target p53 for ubiquitin-proteasome- mediated degradation. Mol Cell Biol *20*, 8458-8467.

Roth, J., Dobbelstein, M., Freedman, D. A., Shenk, T., and Levine, A. J. (1998). Nucleo-cytoplasmic shuttling of the hdm2 oncoprotein regulates the levels of the p53 protein via a pathway used by the human immunodeficiency virus rev protein. Embo J *17*, 554-564.

Rubbi, C. P., and Milner, J. (2003). Disruption of the nucleolus mediates stabilization of p53 in response to DNA damage and other stresses. Embo J *22*, 6068-6077.

Sakaguchi, K., Herrera, J. E., Saito, S., Miki, T., Bustin, M., Vassilev, A., Anderson, C. W., and Appella, E. (1998). DNA damage activates p53 through a phosphorylation-acetylation cascade. Genes Dev *12*, 2831-2841.

Sakaguchi, K., Saito, S., Higashimoto, Y., Roy, S., Anderson, C. W., and Appella, E. (2000). Damage-mediated phosphorylation of human p53 threonine 18 through a cascade mediated by a casein 1-like kinase. Effect on Mdm2 binding. J Biol Chem *275*, 9278-9283.

Sakaguchi, K., Sakamoto, H., Lewis, M. S., Anderson, C. W., Erickson, J. W., Appella, E., and Xie, D. (1997). Phosphorylation of serine 392 stabilizes the tetramer formation of tumor suppressor protein p53. Biochemistry *36*, 10117-10124.

Salghetti, S. E., Muratani, M., Wijnen, H., Futcher, B., and Tansey, W. P. (2000). Functional overlap of sequences that activate transcription and signal ubiquitin-mediated proteolysis. Proc Natl Acad Sci U S A *97*, 3118-3123.

Scheffner, M., Huibregtse, J. M., Vierstra, R. D., and Howley, P. M. (1993). The HPV-16 E6 and E6-AP complex functions as a ubiquitin-protein ligase in the ubiquitination of p53. Cell *75*, 495-505.

Schon, O., Friedler, A., Bycroft, M., Freund, S. M., and Fersht, A. R. (2002). Molecular mechanism of the interaction between MDM2 and p53. J Mol Biol *323*, 491-501.

Seo, Y. R., Fishel, M. L., Amundson, S., Kelley, M. R., and Smith, M. L. (2002). Implication of p53 in base excision DNA repair: in vivo evidence. Oncogene *21*, 731-737.

Sharp, D. A., Kratowicz, S. A., Sank, M. J., and George, D. L. (1999). Stabilization of the MDM2 oncoprotein by interaction with the structurally related MDMX protein. J Biol Chem *274*, 38189-38196.

Shaulsky, G., Goldfinger, N., Ben-Ze'ev, A., and Rotter, V. (1990). Nuclear accumulation of p53 protein is mediated by several nuclear localization signals and plays a role in tumorigenesis. Mol Cell Biol *10*, 6565-6577.

Shaulsky, G., Goldfinger, N., Tosky, M. S., Levine, A. J., and Rotter, V. (1991). Nuclear localization is essential for the activity of p53 protein. Oncogene *6*, 2055-2065.

Sherr, C. J. (2001). The INK4a/ARF network in tumour suppression. Nat Rev Mol Cell Biol *2*, 731-737.

Shieh, S. Y., Ikeda, M., Taya, Y., and Prives, C. (1997). DNA damage-induced phosphorylation of p53 alleviates inhibition by MDM2. Cell *91*, 325-334.

Shieh, S. Y., Taya, Y., and Prives, C. (1999). DNA damage-inducible phosphorylation of p53 at N-terminal sites including a novel site, Ser20, requires tetramerization. EMBO J *18*, 1815-1823.

Shirangi, T. R., Zaika, A., and Moll, U. M. (2002). Nuclear degradation of p53 occurs during down-regulation of the p53 response after DNA damage. FASEB J *16*, 420-422.

Shvarts, A., Steegenga, W. T., Riteco, N., van Laar, T., Dekker, P., Bazuine, M., van Ham, R. C., van der Houven van Oordt, W., Hateboer, G., van der Eb, A. J., and Jochemsen, A. G.

(1996). MDMX: a novel p53-binding protein with some functional properties of MDM2. Embo J *15*, 5349-5357.

Siliciano, J. D., Canman, C. E., Taya, Y., Sakaguchi, K., Appella, E., and Kastan, M. B. (1997). DNA damage induces phosphorylation of the amino terminus of p53. Genes Dev *11*, 3471-3481.

Sluss, H. K., Armata, H., Gallant, J., and Jones, S. N. (2004). Phosphorylation of serine 18 regulates distinct p53 functions in mice. Mol Cell Biol *24*, 976-984.

Smart, P., Lane, E. B., Lane, D. P., Midgley, C., Vojtesek, B., and Lain, S. (1999). Effects on normal fibroblasts and neuroblastoma cells of the activation of the p53 response by the nuclear export inhibitor leptomycin B. Oncogene *18*, 7378-7386.

Smith, M. L., Ford, J. M., Hollander, M. C., Bortnick, R. A., Amundson, S. A., Seo, Y. R., Deng, C. X., Hanawalt, P. C., and Fornace, A. J., Jr. (2000). p53-mediated DNA repair responses to UV radiation: studies of mouse cells lacking p53, p21, and/or gadd45 genes. Mol Cell Biol *20*, 3705-3714.

Srivastava, S., Zou, Z. Q., Pirollo, K., Blattner, W., and Chang, E. H. (1990). Germ-line transmission of a mutated p53 gene in a cancer-prone family with Li-Fraumeni syndrome. Nature *348*, 747-749.

Stad, R., Little, N. A., Xirodimas, D. P., Frenk, R., van der Eb, A. J., Lane, D. P., Saville, M. K., and Jochemsen, A. G. (2001). Mdmx stabilizes p53 and Mdm2 via two distinct mechanisms. EMBO Rep *2*, 1029-1034.

Stewart, Z. A., and Pietenpol, J. A. (2001). p53 Signaling and cell cycle checkpoints. Chem Res Toxicol *14*, 243-263.

Stommel, J. M., Marchenko, N. D., Jimenez, G. S., Moll, U. M., Hope, T. J., and Wahl, G. M. (1999). A leucine-rich nuclear export signal in the p53 tetramerization domain: regulation of subcellular localization and p53 activity by NES masking. EMBO J *18*, 1660-1672.

Stommel, J. M., and Wahl, G. M. (2004). Accelerated MDM2 auto-degradation induced by DNA-damage kinases is required for p53 activation. Embo J *23*, 1547-1556.

Sun, X. F., Carstensen, J. M., Zhang, H., Stal, O., Wingren, S., Hatschek, T., and Nordenskjold, B. (1992). Prognostic significance of cytoplasmic p53 oncoprotein in colorectal adenocarcinoma. Lancet *340*, 1369-1373.

Szak, S. T., Mays, D., and Pietenpol, J. A. (2001). Kinetics of p53 binding to promoter sites in vivo. Mol Cell Biol *21*, 3375-3386.

Tanimura, S., Ohtsuka, S., Mitsui, K., Shirouzu, K., Yoshimura, A., and Ohtsubo, M. (1999). MDM2 interacts with MDMX through their RING finger domains. FEBS Lett *447*, 5-9.

Tao, W., and Levine, A. J. (1999). Nucleocytoplasmic shuttling of oncoprotein Hdm2 is required for Hdm2- mediated degradation of p53. Proc Natl Acad Sci U S A *96*, 3077-3080.

Tergaonkar, V., Pando, M., Vafa, O., Wahl, G., and Verma, I. (2002). p53 stabilization is decreased upon NFkappaB activation: a role for NFkappaB in acquisition of resistance to chemotherapy. Cancer Cell *1*, 493-503.

Thrower, J. S., Hoffman, L., Rechsteiner, M., and Pickart, C. M. (2000). Recognition of the polyubiquitin proteolytic signal. Embo J *19*, 94-102.

Thut, C. J., Chen, J. L., Klemm, R., and Tjian, R. (1995). p53 transcriptional activation mediated by coactivators TAFII40 and TAFII60. Science *267*, 100-104.

Thut, C. J., Goodrich, J. A., and Tjian, R. (1997). Repression of p53-mediated transcription by MDM2: a dual mechanism. Genes Dev *11*, 1974-1986.

Tibbetts, R. S., Brumbaugh, K. M., Williams, J. M., Sarkaria, J. N., Cliby, W. A., Shieh, S. Y., Taya, Y., Prives, C., and Abraham, R. T. (1999). A role for ATR in the DNA damage-induced phosphorylation of p53. Genes Dev *13*, 152-157.

Tolbert, D., Lu, X., Yin, C., Tantama, M., and Van Dyke, T. (2002). p19(ARF) is dispensable for oncogenic stress-induced p53-mediated apoptosis and tumor suppression in vivo. Mol Cell Biol *22*, 370-377.

Unger, T., Juven-Gershon, T., Moallem, E., Berger, M., Vogt Sionov, R., Lozano, G., Oren, M., and Haupt, Y. (1999a). Critical role for Ser20 of human p53 in the negative regulation of p53 by Mdm2. Embo J *18*, 1805-1814.

Unger, T., Sionov, R. V., Moallem, E., Yee, C. L., Howley, P. M., Oren, M., and Haupt, Y. (1999b). Mutations in serines 15 and 20 of human p53 impair its apoptotic activity. Oncogene *18*, 3205-3212.

Vafa, O., Wade, M., Kern, S., Beeche, M., Pandita, T. K., Hampton, G. M., and Wahl, G. M. (2002). c-Myc can induce DNA damage, increase reactive oxygen species, and mitigate p53 function: a mechanism for oncogene-induced genetic instability. Mol Cell *9*, 1031-1044.

Vassilev, L. T., Vu, B. T., Graves, B., Carvajal, D., Podlaski, F., Filipovic, Z., Kong, N., Kammlott, U., Lukacs, C., Klein, C., *et al.* (2004). In Vivo Activation of the p53 Pathway by Small-Molecule Antagonists of MDM2. Science *303*, 844-848.

Villunger, A., Michalak, E. M., Coultas, L., Mullauer, F., Bock, G., Ausserlechner, M. J., Adams, J. M., and Strasser, A. (2003). p53- and drug-induced apoptotic responses mediated by BH3-only proteins puma and noxa. Science *302*, 1036-1038.

Voorhoeve, P. M., and Agami, R. (2004). Unraveling Human Tumor Suppressor Pathways: A Tale of the INK4A Locus. Cell Cycle *3*.

Wahl, G. M., and Carr, A. M. (2001). The evolution of diverse biological responses to DNA damage: insights from yeast and p53. Nat Cell Biol *3*, E277-286.

Wahl, G. M., Linke, S. P., Paulson, T. G., and Huang, L. C. (1997). Maintaining genetic stability through TP53 mediated checkpoint control. Cancer Surv *29*, 183-219.

Wahl, G. M., Padgett, R. A., and Stark, G. R. (1979). Gene amplification causes overproduction of the first three enzymes of UMP synthesis in N-(phosphonacetyl)-L-aspartate-resistant hamster cells. J Biol Chem *254*, 8679-8689.

Waldman, T., Kinzler, K. W., and Vogelstein, B. (1995). p21 is necessary for the p53-mediated G1 arrest in human cancer cells. Cancer Res *55*, 5187-5190.

Wang, X., Taplick, J., Geva, N., and Oren, M. (2004). Inhibition of p53 degradation by Mdm2 acetylation. FEBS Lett *561*, 195-201.

Watanabe, M., Fukuda, M., Yoshida, M., Yanagida, M., and Nishida, E. (1999). Involvement of CRM1, a nuclear export receptor, in mRNA export in mammalian cells and fission yeast. Genes Cells *4*, 291-297.

Weber, J. D., Taylor, L. J., Roussel, M. F., Sherr, C. J., and Bar-Sagi, D. (1999). Nucleolar Arf sequesters Mdm2 and activates p53. Nat Cell Biol *1*, 20-26.

Windle, B., Draper, B. W., Yin, Y. X., O'Gorman, S., and Wahl, G. M. (1991). A central role for chromosome breakage in gene amplification, deletion formation, and amplicon integration. Genes Dev *5*, 160-174.

Wolff, B., Sanglier, J. J., and Wang, Y. (1997). Leptomycin B is an inhibitor of nuclear export: inhibition of nucleo- cytoplasmic translocation of the human immunodeficiency virus type 1 (HIV-1) Rev protein and Rev-dependent mRNA. Chem Biol *4*, 139-147.

Wood, S. A. (2002). Dubble or nothing? Is HAUSP deubiquitylating enzyme the final arbiter of p53 levels? Sci STKE *2002*, PE34.

Wouters, B. G., Giaccia, A. J., Denko, N. C., and Brown, J. M. (1997). Loss of p21Wafl/Cipl sensitizes tumors to radiation by an apoptosis-independent mechanism. Cancer Res *57*, 4703-4706.

Wu, X., Bayle, J. H., Olson, D., and Levine, A. J. (1993). The p53-mdm-2 autoregulatory feedback loop. Genes Dev *7*, 1126-1132.

Wu, Z., Earle, J., Saito, S., Anderson, C. W., Appella, E., and Xu, Y. (2002). Mutation of mouse p53 Ser23 and the response to DNA damage. Mol Cell Biol *22*, 2441-2449.

Xiao, H., Pearson, A., Coulombe, B., Truant, R., Zhang, S., Regier, J. L., Triezenberg, S. J., Reinberg, D., Flores, O., Ingles, C. J., and et al. (1994). Binding of basal transcription factor TFIIH to the acidic activation domains of VP16 and p53. Mol Cell Biol *14*, 7013-7024.

Xirodimas, D. P., Stephen, C. W., and Lane, D. P. (2001). Cocompartmentalization of p53 and Mdm2 is a major determinant for Mdm2- mediated degradation of p53. Exp Cell Res *270*, 66-77.

Yang, Q., Manicone, A., Coursen, J. D., Linke, S. P., Nagashima, M., Forgues, M., and Wang, X. W. (2000). Identification of a functional domain in a GADD45-mediated G2/M checkpoint. J Biol Chem *275*, 36892-36898.

Yin, Y., Tainsky, M. A., Bischoff, F. Z., Strong, L. C., and Wahl, G. M. (1992). Wild-type p53 restores cell cycle control and inhibits gene amplification in cells with mutant p53 alleles. Cell *70*, 937-948.

Yoneda, Y., Imamoto-Sonobe, N., Yamaizumi, M., and Uchida, T. (1987). Reversible inhibition of protein import into the nucleus by wheat germ agglutinin injected into cultured cells. Exp Cell Res *173*, 586-595.

Yu, Z. K., Geyer, R. K., and Maki, C. G. (2000). MDM2-dependent ubiquitination of nuclear and cytoplasmic p53. Oncogene *19*, 5892-5897.

Zhang, Y., and Xiong, Y. (2001). A p53 amino-terminal nuclear export signal inhibited by DNA damage- induced phosphorylation. Science *292*, 1910-1915.

Zhou, B. P., Liao, Y., Xia, W., Zou, Y., Spohn, B., and Hung, M. C. (2001a). HER-2/neu induces p53 ubiquitination via Akt-mediated MDM2 phosphorylation. Nat Cell Biol *3*, 973-982.

Zhou, J., Ahn, J., Wilson, S. H., and Prives, C. (2001b). A role for p53 in base excision repair. Embo J *20*, 914-923.

Zhu, Q., Wani, G., Wani, M. A., and Wani, A. A. (2001a). Human homologue of yeast Rad23 protein A interacts with p300/cyclic AMP-responsive element binding (CREB)-binding protein to down-regulate transcriptional activity of p53. Cancer Res *61*, 64-70.

Zhu, Q., Yao, J., Wani, G., Wani, M. A., and Wani, A. A. (2001b). Mdm2 mutant defective in binding p300 promotes ubiquitination but not degradation of p53: Evidence for the role of p300 in integrating ubiquitination and proteolysis. J Biol Chem *4*, 4.

Zindy, F., Eischen, C. M., Randle, D. H., Kamijo, T., Cleveland, J. L., Sherr, C. J., and Roussel, M. F. (1998). Myc signaling via the ARF tumor suppressor regulates p53-dependent apoptosis and immortalization. Genes Dev *12*, 2424-2433.

Chapter 5

REGULATION OF THE P53 RESPONSE BY CELLULAR GROWTH AND SURVIVAL FACTORS

Lauren Brown and Samuel Benchimol
Ontario Cancer Institute / Princess Margaret Hospital and Department of Medical Biophysics, University of Toronto, Toronto, Ontario, Canada

INTRODUCTION

In response to abnormal proliferative signals and many forms of cellular stress including DNA damage and ribonucleotide depletion, p53 induces cells to undergo a transient arrest in G1 that is believed to allow time for repair of damaged DNA before the initiation of S phase. Failure to arrest in G1 can lead to chromosome aberrations and genomic instability. Activated p53 can also eliminate cells from the proliferative population through mechanisms that involve prolonged arrest in G1 (as seen during telomere-initiated replicative senescence and stress/DNA damage-induced premature senescence) and apoptosis (Levine, 1997; Oren, 2003; Vogelstein et al., 2000). The elimination of damaged, stressed or abnormally proliferating cells by p53 is considered to be the principal means by which p53 mediates tumour suppression (Symonds et al., 1994; Schmitt et al., 2002). Inappropriate or prolonged activation of p53 in normal tissues, however, can lead to tissue damage and has been associated with multiple sclerosis (Wosik et al., 2003), neurodegenerative disorders and exacerbation of ischemic damage from stroke or cardiac arrest (Mattson et al., 2001; Komarova and Gudkov, 2001). Accordingly, the regulation of p53 function is important for the maintenance of tissue homeostasis.

P. Hainaut and K.G. Wiman (eds.), 25 Years of p53 Research, 115-140.

p53-mediated apoptosis is dependent on the Apaf-1/caspase-9 pathway (Soengas et al., 1999) and involves mitochondrial cytochrome c release (Schuler et al., 2000). How p53 elicits the release of cytochrome c to promote caspase activation remains elusive. p53-mediated apoptosis involves transcriptional regulation of target genes (Chao et al., 2000; Jimenez et al., 2000) as well as transcription-independent functions of p53, possibly reflecting distinct mechanisms of p53 action in different cell types (Oren, 2003; Vousden, 2000; Benchimol, 2001). A number of p53-regulated genes have been identified and some of these promote apoptosis when overexpressed. A subset has been shown, additionally, to attenuate apoptosis when disrupted through antisense RNA, siRNA or gene deletion methods including: *Bax* (Miyashita and Reed, 1995), *Noxa* (Oda E. et al., 2000; Shibue et al., 2003; Villunger et al., 2003), *Puma* (Villunger et al., 2003; Nakano and Vousden, 2001; Yu et al., 2001, 2003; Jeffers et al., 2003), *PERP* (Ihrie et al., 2003), *p53AIP1* (Oda K. et al., 2000), *Pidd/Lrdd* (Lin et al., 2000), *p53DINP1* (Okamura et al., 2001), *PAC1* (Yin et al., 2003), *UNC5H2* (Tanikawa et al., 2003), and *TSAP6* (Passer et al., 2003). Bax, Noxa, Puma and p53AIP1 proteins are localized at the mitochondria and each has been shown to associate with Bcl-2. So far, however, no single molecule can be considered to be the principal mediator of p53-dependent apoptosis.

It remains unclear why certain cells undergo apoptosis in response to p53 activation while other cells undergo p53-dependent cell cycle arrest. Differences in the cellular response to p53 activation have been attributed to extracellular survival factors and to intrinsic factors that might reflect differences in DNA repair, p53 expression and activation, intracellular death/survival pathways, oncogene activation, or selective transactivation/ repression of p53-target genes in different cell types. For example, normal fibroblasts undergo p53-dependent G1 arrest in response to DNA damage whereas hyperproliferative fibroblasts such as those expressing ectopic E1A, c-myc or E2F-1 undergo p53-dependent apoptosis (Levine, 1997); cells expressing ectopic Bcl-2 or Bcl-X$_L$ are protected from p53-dependent apoptosis (Chiou et al., 1994; Schott et al., 1995; Wang et al., 1993) and constitutively active PI3K and PKB delay the onset of p53-mediated apoptosis (Lin et al., 2002; Sabbatini and McCormick., 1999). Promoter selectivity by p53 may also contribute to cellular outcome (Oren, 2003). This could reflect differences in the affinity of various promoters for p53, such that some are responsive only to high levels of p53 or to certain modified forms of p53 (Resnick-Silverman et al., 1998). Beside covalent modification of p53, promoter selectivity leading to cell cycle arrest or apoptosis can be regulated by the interaction of p53 with other proteins including ASPP, JMY, WT1, BRCA1, p63 and p73 (Oren, 2003; Flores et al., 2002; Vousden and Lu, 2002). Here we describe how anti-apoptotic Bcl-2 family members and the MAPK and PI3K/PKB signalling pathways regulate the cellular response to p53 activation.

ANTI-APOPTOTIC BCL-2 FAMILY MEMBERS

The cellular decision to undergo apoptosis is governed by the integration of death and survival signals. The mitochondrial death pathway is triggered by a variety of stress-induced signals, including genotoxic agents, metabolic inhibitors and inadequate growth factor stimulation. These signals act initially on proapoptotic members of the BH3-only subset of the Bcl-2 family of proteins (e.g. Bid, Bim, Bmf, Bik, Noxa, Puma), which associate with anti-apoptotic Bcl-2 family members (e.g. Bcl-2, Bcl-X_L, Mcl-1) residing in the outer mitochondrial membrane and neutralize their ability to maintain membrane integrity. This, combined with the oligomerization of other pro-apoptotic family members (e.g. Bax and Bak), results in mitochondrial damage and release of mitochondrial proteins including cytochrome c and other apoptogenic factors that lead to caspase activation and apoptosis (Cory and Adams, 2002) . The ratio of anti- to pro-apoptotic Bcl-2 family members is thought to determine the susceptibility of a cell to undergo apoptosis. Survival and death signals influence the concentration and activity of anti- and pro-apoptotic Bcl-2 family members, tipping the balance in favour of cell survival or cell death. Overexpression of Bcl-2 and other anti-apoptotic family members in cancer attests to the importance of this family of oncoproteins in suppressing apoptosis and prolonging malignant cell survival (Cory et al., 2003). The expression of anti-apoptotic Bcl-2 proteins correlates with the survival of numerous hematopoietic cell lines in the presence of their lineage-specific cytokines (Lotem and Sachs, 1999).

Cytokine suppression of p53 apoptosis by up-regulation of anti-apoptotic Bcl-2 proteins

Cytokines have a well-documented role in apoptosis suppression, illustrated by the requirement of colony stimulating factors (G-CSF, M-CSF and GM-CSF), interleukin-3 (IL-3) and erythropoietin (EPO) to maintain the viability of hematopoietic cells in culture (Lotem et al., 1991; Williams et al., 1990; Koury and Bondurant, 1990). In addition to apoptosis induced by growth factor withdrawal, hematopoietic cells undergo apoptosis upon exposure to γ-irradiation, treatment with chemotherapeutic agents as well as forced expression of wild-type p53 (Yonish-Rouash et al., 1991; Canman et al., 1995; Abrahamson et al., 1995; Quelle et al., 1998; Lin and Benchimol, 1995). In some cases, apoptosis that is dependent upon p53 can be suppressed when cells are cultured in the presence of their lineage-specific cytokines. Cells that are rescued from apoptosis remain in a viable, growth arrested state. The common ability of certain cytokines to suppress p53-induced apoptosis is striking and may reflect a mechanism by which tumours that retain wild-type p53 gain resistance to apoptosis-inducing anti-cancer agents (Lotem and Sachs, 1999).

EPO and IL-3 bind to type I cytokine receptors, causing receptor dimerization. Lacking intrinsic kinase activity, type I cytokine receptors recruit members of the Janus kinase (JAK) tyrosine kinase family to mediate phosphorylation of tyrosine residues located within the intracellular portion of the receptor dimer (Wojchowski et al., 1999). An immediate downstream target of JAK2 after EPO activation is Signal Transducer and Activator of Transcription 5 (STAT5) and it has been proposed that STAT5-dependent transcriptional up-regulation of Bcl-X_L mediates survival downstream of EPO (Socolovsky et al., 1999, 2001). In contrast, EPO has been shown to up-regulate Bcl-2 and Bcl-X_L transcripts in cells expressing EPO-R mutants incapable of activating STAT5 (Quelle et al., 1998). Using an erythroleukemia cell line expressing a temperature sensitive p53 mutant (p53ts) that can be induced to undergo p53 dependent apoptosis at 32°C, we have shown that EPO promotes survival and suppresses p53-dependent apoptosis through a mechanism that is dependent on JAK2 but independent of STAT5. Moreover, we observed that EPO stimulation resulted in an increase in Bcl-X_L expression that was regulated primarily through a posttranscriptional mechanism involving Bcl-X_L protein modification (Lin et al., 2002). Although the mechanism regulating Bcl-X_L expression in response to EPO is controversial (Socolovsky et al., 1999; Teglund et al., 1998), the importance of Bcl-X_L as a mediator of EPO-dependent erythroid survival is well established by animal studies. Bcl-X_L deficient mice have severe hematopoietic defects resulting from massive cell death of erythroid progenitors and JAK2 deficient mice die in utero from a block in definitive eythropoiesis, a maturation program during embryogenesis when red blood cell production switches from the yolk sac to the fetal liver (Motoyama et al., 1995; Parganas et al., 1998). The phenotype of JAK2 deficient mice bears a striking resemblance to that of EPO and EPO-R deficient mice (Wu et al., 1995). Ectopic Bcl-X_L expression alone has been shown to substitute for EPO during differentiation of primary mouse erythroblasts in culture. Hence, the primary role of EPO during erythropoiesis appears to be apoptosis protection through the up regulation of Bcl-X_L protein expression, and terminal erythroid differentiation of the surviving cells is thought to depend on an intrinsic default differentiation program (Dolznig et al., 2002).

How do cytokines rescue cells from p53-dependent apoptosis and regulate Bcl-X$_L$ and/or Bcl-2 expression? The dependency of this survival signal upon JAK2 is established, however, the signalling components that connect JAK2 activation and the activation of anti-apoptotic Bcl-2 proteins is not fully understood (Lin et al., 2002; Quelle et al., 1998). Pro-survival cytokines activate STAT5, MAPK and PI3K signalling pathways, and the relative importance of these pathways in providing protection against p53-induced apoptosis, is an area of intense investigation. We have observed that EPO-suppression of p53-dependent apoptosis is independent of PI3K (Lin et al., 2002) and the three MAPK pathways (unpublished data). These experiments also revealed that chemical inhibition of PI3K markedly increased p53-dependent apoptosis suggesting that intrinsic levels of activated PI3K/PKB, commonly present in transformed cells, limit the ability of p53 to induce cell death (Lin et al., 2002). This could be problematic for gene therapy approaches that attempt to reconstitute p53 expression in p53 null tumours with the expectation of inducing apoptosis. The observation that survival pathways impinge on p53-dependent cell death is widespread across many cell types. The following sections discuss mechanisms by which the MAPK and PI3K/PKB pathways interact with p53 and regulate the cellular response to p53 activation.

MAPK PATHWAYS

Ras/RAF/MEK/ERK

The Ras/Raf/MEK/ERK mitogenic activated protein kinase signalling pathway (Ras/ERK) has a well-documented role in suppressing apoptosis downstream of survival-promoting growth factors in cell types ranging from cultured murine fibroblasts and rat neurons to the developing Drosophila eye and nervous system (Bergmann et al., 1998, 2002; Xia et al., 1995; Gardner and Johnson, 1996; Parrizas et al., 1997; Kurada and White, 1998) . Upon growth factor binding, Receptor Tyrosine Kinases (RTKs) dimerize and activate Ras through the interaction of adaptor proteins that recognize phosphorylated tyrosines residues within the cytoplasmic domain and recruit GDP-bound Ras to the membrane. SOS, a guanine nucleotide exchange factor, then catalyzes the exchange of GDP for GTP, generating activated GTP-bound Ras, which in turn activates downstream kinases in the signalling cascade (Figure 1). The Ras/ERK signalling pathway promotes survival through transcriptional and post-transcriptional processes.

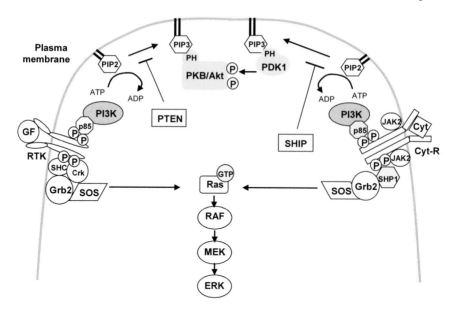

Figure 1. Activation of the Ras/MAPK and PI3K signalling pathways by growth factor (GF) binding to growth factor receptor tyrosine kinases (RTK) and pro-survival cytokine (Cyt) binding to cytokine receptors (Cyt-R).

ERK1/2 activate pp90 ribosomal S6 kinase (pp90rsk) which in turn phosphorylates and inactivates the pro-apoptotic Bcl-2 family member BAD on Serine residue 112 (Shimamura et al., 2000; Bonni et al., 1999). Phosphorylated BAD is bound by 14-3-3 proteins and sequestered in the cytoplasm, rendering it incapable of inhibiting the action of anti-apoptotic Bcl-2 family members at the mitochondrial membrane (Zha et al., 1996). In neurons, BDNGF-mediated survival is dependent on Ras/ERK-mediated phosphorylation and activation of the CREB transcription factor (Bonni et al., 1999). In hematopoietic cells treated with GM-CSF and thrombopoietin, cell survival involves pp90rsk-mediated phosphorylation of CREB on Ser 133 (Kwon et al., 2000; Zauli et al., 1998). Transcriptional targets of CREB1 that may play a direct role in apoptosis suppression downstream of the Ras/ERK pathway include the anti-apoptotic genes *bcl-2*, and *bag-l* (Riccio et al., 1999; Wilson et al., 1996; Perkins et al., 2003). pp90rsk activation requires phosphorylation by both ERK and phospho-inositide-dependent kinase 1 (PDK1), activated by phospholipid second messengers generated by PI3K (Richards et al., 1999; Jensen et al., 1999). Thus, Ras/ERK signalling represents one of two pathways that contribute to cell survival through pp90rsk .

Ras/ERK signalling and p53

A number of studies have investigated the connection between the Ras/ERK signalling pathway and p53 activation (Figure 2). A complex and incomplete picture has emerged in which the Ras/ERK pathway converges upon p53 and has opposing effects on p53 function. The outcome of these opposing effects is likely determined by cell type or growth conditions (Ries et al., 2000). Some studies place p53 and Ras/ERK signalling components within the same linear pathway with p53 acting upstream or downstream of Ras/ERK. Other studies propose that Ras/ERK signalling operates in a parallel pathway to facilitate/oppose p53 functions.

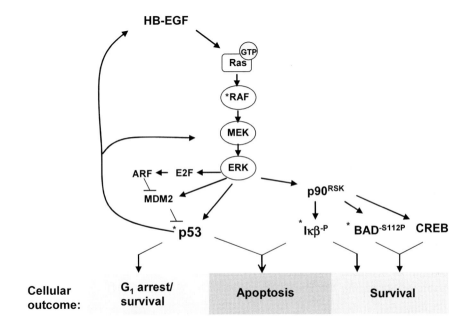

Figure 2. Apoptotic and Survival pathways induced downstream of the Ras/MAPK signalling pathway. * Denotes proteins regulated by multiple signaling pathways.

Over the past decade, a number of groups have shown that primary human cells exposed to DNA damage or oncogenic stimulation undergo a prolonged p53-dependent and Rb-dependent arrest in G1, and exhibit a senescence-like state that is commonly referred to as "premature senescence" (Di Leonardo et al., 1994; Serrano et al., 1997; Wright and Shay, 2002). Ras-induced growth arrest is dependent upon Raf-1 and MEK1 kinases, and is associated with an increase in ERK kinase activity (Serrano et al., 1997; Lin et al., 1998; Zhu et al., 1998). Oncogenic Ras, as well as constitutive activation of the Ras/ERK signalling cascade is associated with increased expression of p53, p16INK4a and p19ARF (Ries et al., 2000; Serrano et al., 1997; Agarwal et al., 2001). Ras/ERK signalling is essential for activation of cyclin D transcription, resulting in the generation of cyclin D/cdk4 activity that leads to Rb phosphorylation and E2F1 activation. E2F1 induces p19ARF expression, likely through direct transcriptional activation via E2F sites in the ARF promoter. p19ARF binds to Mdm2 and blocks its interaction with p53 resulting in p53 stabilization (Pomerantz et al., 1998; Kamijo et al., 1998). Mdm2 acts as a negative regulator of p53 through a direct interaction that targets p53 for ubiquitin-mediated degradation. Activation of the Ras/ERK pathway also results in elevated levels of Mdm2 (Ries et al., 2000). Thus, p53 protein levels are determined by a balance between these opposing effects of the Ras/ERK pathway.

ERK activation was also shown to increase the level of p53 mRNA and this effect could be blocked by treatment with the MEK inhibitor U0126 (Agarwal et al., 2001). Two studies reported that ERK could phosphorylate p53 on Ser15, a modification that disrupts the MDM2-p53 interaction resulting in p53 protein accumulation (Persons et al., 2000; Wang et al., 2001). Two recent studies indicate that p53 can activate the Ras/ERK pathway. Using a p53-inducible cell model, Ryan et al. (Ryan et al., 2000) reported that p53 expression resulted in NF-kB activation involving the Ras/ERK pathway and activation of pp90rsk. NF-kB activation and apoptosis in response to inducible p53 expression were blocked by treatment with a MEK1 inhibitor (Ryan et al., 2000). This study provides a rare instance in which NF-kB is associated with pro-apoptotic activity rather than survival. Aaronson and colleagues have identified HB-EGF (heparin-binding EGF-like growth factor) as the product of a p53-responsive gene. HB-EGF is secreted and through its interaction with the EGF receptor is capable of activating the Ras/ERK pathway. p53-induced HB-EGF protects cells from death in response to oxidative stress and DNA damage through ERK activation and might facilitate cell cycle rentry after DNA repair is complete (Fang et al., 2001; Lee et al., 2000).

It is pertinent to consider potential differences between oncogenic mutant Ras and normal Ras proteins in initiating the Ras/ERK signalling cascade and how this might impact on cell survival or cell death. Physiological activation of this pathway by normal Ras proteins might produce a transient and less intense signal compared with oncogenic mutant Ras proteins that produce an intense and prolonged signal (Sewing et al., 1997; Woods et al., 1997). The cellular response to these two types of signals may be profoundly different; the former leading to proliferation and survival and the latter leading to p53 activation, and cell cycle arrest or apoptosis in an effort to suppress neoplasia and eliminate oncogene-expressing cells. Sustained

ERK activation in response to oncogenic Ras may lead to inappropriate accumulation of phosphorylated substrates and activation of transcription factors that would otherwise not occur in response to a transient signal from normal Ras (Marshall, 1995).

MEKK1/MKK(4 and 7)/JNK

Of the three known JNK family members, JNK1 and 2 are ubiquitously expressed whereas JNK 3 is expressed primarily in the brain, heart and testis (Gupta et al., 1996; Ip et al., 1998). Each is able to activate the c-jun transcription factor by phosphorylating Ser residues 63 and 73, located within the N-terminal transactivation domain (Hibi et al., 1993; Pulverer et al., 1991; Adler et al., 1992). As with ERK, JNKs are activated by sequential phosphorylation of protein kinases involved in an archetypical MAPK cascade. Based on their initial identification as stress-activated kinases, early research focused on the role of JNKs in apoptosis. Indeed, when activated by stress stimuli such as UV irradiation and growth factor withdrawal JNK has an apoptotic role (Xia et al., 1995; Tournier et al., 2000); emerging evidence, however, suggests that JNK additionally functions to promote cell survival.

In neurons, JNK1/2 play a critical role in stress-induced apoptosis in response to nerve-growth factor (NGF) withdrawal. PC12 neuronal cells deprived of NGF undergo rapid cell death, blocked by the expression of a dominant-interfering JNK mutant. Conversely, PC12 cells expressing constitutively activated MEKK1, the upstream kinase activator of JNK, undergo apoptosis (Xia et al., 1995). Overexpression of c-jun in cultured sympathetic neurons induces apoptosis, and expression of a dominant-interfering c-jun mutant protects against apoptosis due to NGF-withdrawal, implicating it as one of the downstream targets of JNK in this type of neuronal cell death (Ham et al., 1995). In PC12 cells, death from NGF withdrawal is associated with an increase in Fas ligand and cognate death receptor activation (Le-Niculescu et al., 1999).

Mice deficient for either JNK1 or 2 show no obvious phenotype, with the exception of immunodeficiency due to a defect in T-cell function (Constant et al., 2000; Sabapathy et al., 1999a). In response to UV irradiation, only Jnk1-/- single knockout MEFs display impaired apoptosis compared to their wild-type or Jnk2-/- counterparts, yet still undergo some cell death (Tournier et al., 2000). The lack of resistance to UV stress in the single knockout studies is believed to result from the ability of JNK1 and 2 to function in a compensatory manner, supported by the fact that JNK1/2 double knockout (Jnk1/2-/-) MEFs are completely resistant to death from UV irradiation. Jnk1/2-/- mice are embryonic lethal and show exencephaly of the hindbrain at E9.25 due to a reduction in hindbrain apoptosis. Also evident is an increase in apoptosis in the forebrain and hindbrain post neural tube closure at approximately E10.5 (Sabapathy et al., 1999b; Kuan et al., 1999). This points to a role for JNK1/2 in both apoptosis and survival at different times during fetal mouse brain development. Evidence from tumour cell models suggests that JNK acts as a potent survival factor. Several transformed cell lines express constitutive activated JNK, and expression of a c-jun S63/73A mutant, lacking JNK phosphorylation sites, suppresses the transforming ability of several oncogenes (Ip et al., 1998; Behrens et al., 2000). In addition, JNK suppresses apoptosis via inhibitory phosphorylation of the proapoptotic Bcl-2 family protein BAD on Thr201 (Yu et al., 2004).

JNK signalling and p53

In response to stress stimuli, p53 undergoes a complex series of post-translation modifications including phosphorylation and acetylation that lead to protein stabilization, accumulation and transcriptional activation (Prives and Hall, 1999). JNK along with other kinases can phosphorylate and activate p53 (Milne et al., 1995; Hu et al., 1997; Fuchs et al., 1998b; She et al., 2002), however, the role of p53 in JNK-induced apoptosis/survival and the specific phosphorylation events that mediate these responses have yet to be determined. In addition, JNK can bind p53 and target it for ubiquitin-mediated proteosomal degradation (Fuchs et al., 1998a). These opposing effects of JNK on p53 depend in part on cell type, the stimulus used to activate JNK signalling, and cellular growth conditions.

In addition to p53 and c-jun, JNK also activates JunB and ATF-2 by phosphorylation (Davis 2000; Lin, 2003) and targets these transcriptions factors for ubiquitin-mediated degradation, but only when they are in their unphosphorylated state (Fuchs et al., 1996, 1997; Musti et al., 1997). In non-stressed, proliferating cells an estimated 30 % of p53 is found in complex with JNK. Binding is associated with p53 ubiquitination and decreased p53 protein levels suggesting that JNK and/or associated factors target p53 for ubiquitin-mediated proteosomal degradation (Fuchs et al., 1998a), (Figure 3). In cells exposed to UV-irradiation (a known activator of JNK), or expressing constitutively activated MEKK1, p53 is phosphorylated, no longer ubiquitinated, accumulates and becomes transcriptionally active (Fuchs et al., 1998b). The current view is that in unstressed cells, JNK binds p53 and other targets to promote ubiquitin-dependent degradation. In response to certain cellular stresses, in particular UV-irradiation, activated JNK phosphorylates bound targets resulting in their dissociation from JNK and associated factors that mediate degradation (Fuchs et al., 1996, 1997, 1998b; Musti et al., 1997). Thus, in UV-irradiated cells, JNK switches from an ubiquitin-targeting enzyme to a pro-apoptotic kinase that phosphorylates p53 and protects it from degradation. This model is consistent with other observations including our own that suggest that basal JNK activity in proliferating cells under non-stressed conditions plays a critical role in cell survival (see below).

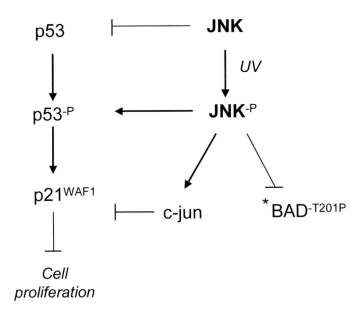

Figure 3. Mechanisms of cell survival and proliferation mediated by JNK under stressed and non-stressed conditions.

JNK-mediated degradation of p53 occurs independently of Mdm2. This is supported by the observation that mutant p53, unable to bind Mdm2, is still degraded by JNK and mutant p53 unable to bind JNK is degraded by Mdm2. In synchronously growing cells JNK/p53 complexes are observed as cells enter G1, whereas Mdm2/p53 complexes are observed as cells enter the G2/M phase of the cell cycle (Fuchs et al., 1998a). These studies suggest that p53 stability is affected by JNK independently of Mdm2 in a cell cycle-dependent manner. One intriguing possibility suggested by these findings is that JNK may normally be involved in regulating the level of latent p53 protein in unstressed cells whereas Mdm2, which is induced by stress in a p53-dependent manner, may serve to down-regulate activated p53 and to terminate the p53-dependent stress response.

JNK activation results in the induction of c-jun following UV-irradiation. c-jun has been shown to inhibit the association of p53 with p21 promoter DNA in UV-irradiated cells thereby suppressing p53-mediated activation of p21WAF1 expression (Shaulian et al., 2000). As a result, c-jun has been implicated in promoting cell cycle re-entry following p53-dependent G1 arrest, presumably once damaged DNA has been repaired. In the absence of c-jun, UV-activated p53 results in a prolonged growth arrest that is associated with protection from apoptosis. In cells that express c-jun constitutively, p21WAF1 induction is blocked and the predominant cellular response to activated p53 is apoptosis (Shaulian et al., 2000). Potapova et al. (2000) reported that inhibition of JNK in p53-null cells caused growth suppression due to apoptosis. In p53 intact cells, JNK inhibition resulted in p53-dependent increase in p21WAF1 expression and survival of growth arrested cells (Potapova et al., 2000). This agrees with the model in which basal JNK in nonstressed cells suppresses p53 by targeting it for ubiquitin-mediated degradation. Therefore, in nonstressed cells JNK promotes p53 degradation, whereas in stressed cells JNK activates p53 and c-jun by phosphorylation and c-jun attenuates p53-dependent activation of p21WAF1 mRNA expression (Figure 3). We have observed that basal levels of JNK protect cells from p53-dependent cell death. Murine erythroleukemia cells expressing a p53ts allele show enhanced p53-dependent apoptosis upon treatment with the chemical inhibitor SP600125 or following expression of a dominant interfering JNK mutant. Neither treatment alone induces apoptosis of parental cells or p53ts-expressing cells grown at the non-permissive temperature (unpublished observations).

MKK(3 and 6)/p38

The p38 MAPKs exists in 4 isoforms, α, β, γ and δ, with the α and β isoforms having the widest range of mammalian tissue expression (Martin-Blanco, 2000). Like JNK, initial identification of p38 as a kinase activated by cellular stress and inflammatory cytokines linked it with an apoptotic cellular response. p38 is now also implicated in cell proliferation and survival (Kyriakis and Avruch, 2001). p38 has been shown to play a role in apoptosis in response to stress due to growth factor withdrawal. PC12 neuronal cells undergo apoptosis upon NGF withdrawal and this can be blocked with a p38 chemical inhibitor (PD169316) or with a dominant-interfering p38 mutant kinase. Rat-1 cells showed a similar p38-dependent apoptotic response upon serum-depletion, and in both cell lines, factor withdrawal was associated with an increase in p38 kinase activity (Xia et al., 1995; Kummer et al., 1997). Notably, dominant-interfering kinases of both p38 and JNK were able to block apoptosis induced by NGF withdrawal, suggesting that they may act in concert in mediating this type of neuronal cell death (Xia et al., 1995). Treatment of normal human diploid fibroblasts with the non-steroidal anti-inflammatory drug Sodium Salicylate (NSAID) activates p38 and leads to apoptosis that can be blocked with the p38 chemical inhibitor, SB203580. p38 activation might represent a mechanism by which NSAIDs exert their anti-neoplastic effect (Schwenger et al., 1997). Excitatory amino acids such as glutamate induce p38-dependent apoptosis of rat cerebellar granule neurons (Kawasaki, et al., 1997). *Jnk3-/-* mice are resistant to apoptosis induced in hippocampal neurons with kainite, an acid agonist of glutamate, suggesting that JNK3 also plays a role in excitatory-induced neuron apoptosis (Yang et al., 1997).

In hematopoietic cells, treatment with EPO and IL-3 have been shown to activate p38 MAPK activity and promote survival and differentiation (Nagata et al., 1997, 1998). Blocking expression of either JNK1/2 or p38 with antisense oligonucleotides inhibited erythroid differentiation (Nagata, et al., 1998). The phenotype of p38 deficient mice further illustrates that p38 has a critical role in EPO-mediated survival, at least during embryogenesis when red blood cell production switches from the yolk sac to the fetal liver (Klingmuller, 1997). Viable *p38-/-* mice are severely anemic due to a defect in definitive erythropoiesis, however, this failure of erythropoiesis is attributed to diminished EPO gene expression, placing EPO downstream of p38 in this process (Tamura et al., 2000). We have observed that EPO-mediated rescue of p53-dependent apoptosis in erythroid cells occurs independently of p38 (unpublished data). Intriguingly, the p53ts erythroid cell line used in these investigations has basal p38 kinase activity that effectively limits p53-dependent death (unpublished observations). This suggests that, like JNK, basal p38 plays a role in cell survival.

p38 signalling and p53

In response to UV-irradiation, p38 MAPK phosphorylates p53 on Ser residues 33, 46 and 389 (Huang et al., 1999; Keller et al., 1999, Bulavin et al., 1999; Takekawa et al., 2000). Although the physiological relevance of p53 Ser389 phosphorylation is controversial, p38-mediated phosphorylation of p53 on Ser33 and 46 is important for transcriptional activation and for the ability of p53 to induce arrest and/or apoptosis in response to UV (Bulavin et al., 1999). Takekawa et al. (2000) identified Wip1/PPM1D as a serine/threonine protein phosphatase that dephosphorylates and inactivates p38 thereby attenuating the cellular response to UV-irradiation. Overexpression of Wip1/PPM1D reduced p38-dependent p53 phosphorylation at Ser 33 and 46. Wip1/PPM1D, was originally identified as a p53 regulated gene (Fiscella et al., 1997). Moreover, Wip1/PPM1D expression was shown to be dependent upon p38, as treatment with SB203580 prevented its induction with UV. p38, p53 and Wip1/PPM1D, therefore, function in a negative regulatory loop in response to UV-irradiation (Figure 4); p38, activated in response to UV, phosphorylates p53 on Ser33 and 46, and activated p53 induces transcription of Wip1/PPM1D which terminates the UV-response by dephosphorylating p38 (Takekawa et al., 2000). The proposed function of this loop is to downregulate the p38-p53 response to UV-irradiation, allowing cells to re-enter the cell cycle once genetic lesions are repaired. In the event of irreparable DNA damage, sustained p38 MAPK activity overcomes the action of Wip1/PPM1D and p53-dependent apoptosis ensues. According to this model, Wip1/PPM1D acts as a key regulator of the p53 decision to induce cellular arrest or apoptosis in response to UV-irradiation.

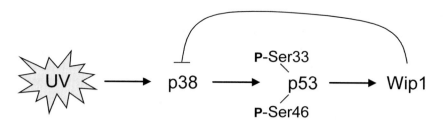

Figure 4. UV-irradiation induces p53-dependent Wip1 expression which functions in a negative regulatory loop to suppress p38 activity.

Nitric oxide (NO)-induced death of cultured chondrocytes has also been linked to p38 and p53 (Kim et al., 2002a). In this model, p38 activates p53 by at least two mechanisms: p38 activation of NF-kB which regulates p53 transcription, and direct p38 phosphorylation of p53 on Ser15, which disrupts the p53-Mdm2 interaction and leads to p53 stabilization (Kim et al., 2002b). Chemotherapeutic agents, such as cisplatin and doxorubicin, induce p38-dependent phosphorylation of p53 on Ser 33, illustrating a putative mechanism utilized by these agents to induce apoptosis during cancer therapy (Sanchez-Prieto et al, 2000). Overall, p38 regulation of p53 occurs through multiple mechanisms that are stimulus-dependent.

PI3K/PKB

A major pathway of cell survival upon activation of RTKs and cytokine receptors is through the activation of PI3K/PKB. Phospho-tyrosine residues within the cytoplasmic domains of these receptors are recognized by the p85 regulatory subunit of PI3K, which recruits the p110 catalytic subunit to the plasma membrane where it catalyzes the addition of a phosphate group to the D3 position of membrane-bound phosphatidylinositol-4,5-bisphosphate, generating phosphatidylinositol-3,4,5-triphosphate (PIP3) (Klingmuller,

1997; Cantley, 2002). PI3K-generated PIP3 acts as a second messenger to activate a number of downstream pathways involved in cell growth, migration and survival (Cantley, 2002). Key to the survival response is the recognition of PIP3 by PKB/Akt and PDK1 through their lipid binding Pleckstrin Homology (PH) domains (Scheid and Woodgett, 2003). Once localized to the membrane, PDK1 phosphorylates PKB within its catalytic domain activation loop (Thr308) to allow substrate binding (Alessi et al., 1996). To become fully active, PKB also requires phosphorylation within a hydrophobic carboxy-proximal region (Ser473), thought to occur through auto-phosphorylation or phosphorylation by an as yet unidentified kinase (Scheid and Woodgett, 2001). Disruption of PDK1 by gene targeting or anti-sense inhibition renders cells unresponsive to PKB activation in response to growth factor stimulation, evidence that PDK1 is the major kinase responsible for PKB phosphorylation and activation (Flynn et al., 2000; Williams et al., 2000).

In addition to being activated by a number of growth factors to prevent apoptosis of factor dependent cells, PKB activation is known to protect cells from apoptosis in response to a number of death-inducing stimuli, such as UV irradiation, treatment with sorbitol, cyclohexamide and TNF-α (Sabattini and McCormick, 1999; Kulik et al., 1997; Ulrich et al., 1998; Ahmed et al., 1997; Stambolic et al., 1998). Activated PKB phosphorylates a number of downstream targets involved in cell survival such as glycogen synthase kinase (GSK), Forkhead transcription factors FKHR1 and AFX, pro-apoptotic BAD and IkB kinase; in all cases, phosphorylation inhibits the function of these proteins (Datta et al., 1997; del Peso et al., 1997; Liang et al., 2003; Brunet et al., 1999; Ozes et al., 1999; Romashkova et al., 1999). Src-homology 2 (SH2)-containing phosphatase (SHIP) and Phosphatase and tensin homologue deleted from chromosome ten (PTEN), serve as negative regulators of PI3K/PKB signalling through their ability to dephosphorylate PIP3 to phosphatidylinosotol-3,4-bisphosphate and phosphatidylinositol-4,5-biphosphate, respectively (Maehama and Dixon, 1999; Liu et al., 1999; Aman et al., 1998).

PI3K/PKB signalling and p53

The expression of PKB alone has been demonstrated to overcome p53-dependent apoptosis, an effect associated with a decrease in p53 DNA-binding and transcriptional activation of pro-apoptotic targets like Bax (Sabbatini and McCormick, 1999; Yamaguchi et al., 2001). These observations lead to the idea that some opposing regulation between p53 and PKB exists (Oren et al., 2002). One link between these two pathways involves Mdm2. PKB, whether activated by IL-3, IGF-1 or an oncogenic

RTK, binds and phosphorylates Mdm2 at two serine residues (Ser166 and Ser186). PKB-mediated phosphorylation of Mdm2 results in its translocation to the nucleus where it binds p53 and targets it for ubiquitin-mediated proteosomal degradation (Zhou et al., 2001; Mayo and Donner, 2001; Gottlieb et al., 2002). Earlier work suggested that the binding of nuclear Mdm2 to p53 is facilitated by p300, which participates in the formation of a ternary complex that stabilizes the Mdm2-p53 interaction (Grossman et al., 1998). This leads to a decrease in p53 protein and transcriptional activity and is consistent with the view that the E3-ligase activity and nuclear import and export signals of Mdm2, encompassing Ser166 and 186, are important for Mdm2-dependent p53 degradation (Zhou et al., 2001; Mayo and Donner, 2001; Woods and Vousden, 2001). p19ARF also binds Mdm2 and inhibits its ability to promote p53 degradation (Pomerantz et al., 1998; Kamijo et al., 1998). Zhou et al. (2001) have proposed that in the presence of survival factors, PKB-dependent phosphorylation of Mdm2 leads to ternary complex formation with p300 and p53 in the nucleus and p53 degradation; unphosphorylated Mdm2 (e.g. in the absence of activated of PKB) is bound by p19ARF and is incapable of targeting p53 for degradation (Figure 5).

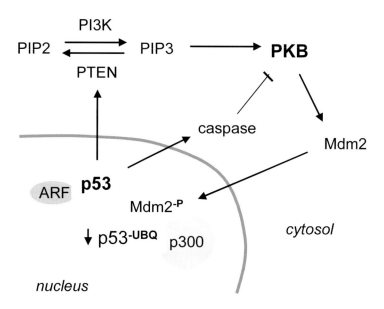

Figure 5. Opposing regulation of PI3K/PKB and p53.

The finding that PTEN is a transcriptional target of p53 adds an intriguing link between the p53 apoptotic program and PKB survival pathway. There are 2 half sites within the PTEN promoter identical to the p53 consensus binding site, with the exception that the PTEN spacer region does not conform to the typical spacer region being 14 bp as opposed to 10-13 bp. Nevertheless, p53 binds this region in a sequence-specific manner to activate PTEN transcription; both promoter binding and transcriptional activation are inhibited by mutation within the p53-consensus binding site of PTEN (Stambolic et al., 2001). PTEN-/- cells are impaired in their apoptotic response to death-inducing stimuli such as UV-irradiation and TNF-α treatment. In addition, PTEN-/- MEFs are resistant to apoptosis induced by forced expression of p53 (Stambolic et al., 1998). These observations suggest that p53-dependent regulation of PTEN expression is important for p53 induced cell death (Stambolic et al., 1998, 2001). This is consistent with our own observations that chemical inhibition of PI3K markedly potentiates p53-dependent apoptosis in cells with a constitutively activated PI3K/PKB pathway (Lin et al., 2002). Thus, in order to effect maximal killing, p53 must not only induce effectors of apoptosis such as Bax, Noxa, Puma and PIDD (Benchimol, 2001), it must also down-regulate intrinsic survival pathways such as PI3K/PKB. Restoring PTEN function in tumour cells that lack PTEN or that overexpress Mdm2 restores their sensitivity to apoptosis-inducing chemotherapeutic agents such as etoposide and doxorubicin, respectively, further supporting a role for PKB in suppressing p53-dependent apoptosis (Zhou et al., 2003; Mayo et al., 2002). The opposing effects of p53 and PKB on death and survival are depicted in the model shown in Figure 5. In cells primed to undergo apoptosis (e.g. from growth-factor deprivation), p53 signals prevail and PKB activation is decreased either through caspase-mediated degradation of PKB protein (Gottlieb et al., 2002), or through PTEN-mediated dephosphorylation of PIP3 (Stambolic et al., 1998, 2001). Under conditions that favour survival, PKB phosphorylates and activates Mdm2 leading to p53 degradation. PKB has many other targets that promote survival independently of any direct effect upon p53.

SUMMARY

Death and life decisions within a cell are regulated through a complex and integrated network that we are still trying to understand. Protooncogenes like c-myc and tumour suppressor genes like p53 encode proteins that can promote survival under certain conditions and death under other conditions. How these decisions are determined remains elusive and under investigations in numerous laboratories. In a similar vein, the three arms of the MAPK pathway, once thought to regulate proliferation/survival (ERK) or apoptosis (JNK, p38) are now known to act in a far more complex fashion promoting death or survival in a context-dependent manner. We have focussed on intrinsic and extrinsic factors that govern death/survival pathways (Bcl-2 family, MAPK pathways, and PI3K/PKB pathway) that ultimately converge on p53 either directly or indirectly to determine the final cellular outcome. .

REFERENCES

Abrahamson J.L., Lee J.M., Bernstein A. Regulation of p53-mediated apoptosis and cell cycle arrest by Steel factor. Mol Cell Biol. 1995; 15:6953-6960.

Adler V., Franklin C.C., Kraft A.S. Phorbol esters stimulate the phosphorylation of c-Jun but not v-Jun: regulation by the N-terminal delta domain. Proc Natl Acad Sci USA. 1992; 89:5341-5345.

Agarwal M.L., Ramana C.V., Hamilton M., Taylor W.R, DePrimo S.E., Bean L.J., Agarwal A., Agarwal M.K., Wolfman A., Stark G.R. Regulation of p53 expression by the RAS-MAP kinase pathway. Oncogene. 2001; 20:2527-2536.

Ahmed N.N., Grimes H.L., Bellacosa A., Chan T.O., Tsichlis P.N. Transduction of interleukin-2 antiapoptotic and proliferative signals via Akt protein kinase. Proc Natl Acad Sci USA. 1997; 94:3627-3632.

Alessi D.R., Andjelkovic M., Caudwell B., Cron P., Morrice N., Cohen P., Hemmings B.A. Mechanism of activation of protein kinase B by insulin and IGF-1. EMBO J. 1996; 15:6541-6551.

Aman M.J., Lamkin T.D., Okada H., Kurosaki T., Ravichandran K.S. The inositol phosphatase SHIP inhibits Akt/PKB activation in B cells. J Biol Chem. 1998; 273:33922-33928.

Behrens A., Jochum W., Sibilia M., Wagner E.F. Oncogenic transformation by ras and fos is mediated by c-Jun N-terminal phosphorylation. Oncogene. 2000; 19:2657-2663.

Benchimol S. p53-dependent pathways of apoptosis. Cell Death Differ. 2001; 8:1049-1051

Bergmann A., Agapite J., McCall K., Steller H. The Drosophila gene hid is a direct molecular target of Ras-dependent survival signaling. Cell. 1998; 95:331-341.

Bergmann A., Tugentman M., Shilo B.Z., Steller H. Regulation of cell number by MAPK-dependent control of apoptosis: a mechanism for trophic survival signaling. Dev Cell. 2002; 2:159-170.

Bonni A., Brunet A., West A.E., Datta S.R., Takasu M.A., Greenberg M.E. Cell survival promoted by the Ras-MAPK signaling pathway by transcription-dependent and -independent mechanisms. Science. 1999; 286:1358-1362

Brunet A., Bonni A., Zigmond M.J, Lin M.Z., Juo P., Hu L.S., Anderson M.J., Arden K.C., Blenis J., Greenberg M.E. Akt promotes cell survival by phosphorylating and inhibiting a Forkhead transcription factor. Cell. 1999; 96:857-868.

Bulavin D.V., Saito S., Hollander M.C., Sakaguchi K., Anderson C.W., Appella E., Fornace A.J., Jr. Phosphorylation of human p53 by p38 kinase coordinates N-terminal phosphorylation and apoptosis in response to UV radiation. EMBO J. 1999; 18:6845-6854.

Canman C.E., Gilmer T.M., Coutts S.B., Kastan M.B. Growth factor modulation of p53-mediated growth arrest versus apoptosis. Genes Dev. 1995; 9:600-611.

Cantley L.C. The phosphoinositide 3-kinase pathway. Science. 2002;296:1655-1657.

Chao C., Saito S., Kang J., Anderson C.W., Appella E. Xu Y. p53 transcriptional activity is essential for p53-dependent apoptosis following DNA damage. EMBO J. 2000; 19:4967-4975

Chiou S.K., Rao L., White E. Bcl-2 blocks p53-dependent apoptosis. Mol Cell Biol. 1994; 14:2556-2563

Constant S.L., Dong C., Yang D.D., Wysk M., Davis R.J., Flavell R.A. JNK1 is required for T cell-mediated immunity against Leishmania major infection. J Immunol. 2000; 165:2671-2676.

Cory S., Adams J.M. The Bcl2 family: regulators of the cellular life-or-death switch. Nat Rev Cancer. 2002; 2:647-656.

Cory S., Huang D.C., Adams J.M. The Bcl-2 family: roles in cell survival and oncogenesis. Oncogene. 2003; 22:8590-8607.

Datta S.R,. Dudek H., Tao X., Masters S., Fu H., Gotoh Y., Greenberg M.E. Akt phosphorylation of BAD couples survival signals to the cell-intrinsic death machinery. Cell. 1997; 91:231-241.

Davis R.J. Signal transduction by the JNK group of MAP kinases. Cell. 2000; 103:239-252.

del Peso L., Gonzalez-Garcia M., Page C., Herrera R., Nunez G. Interleukin-3-induced phosphorylation of BAD through the protein kinase Akt. Science. 1997; 278:687-689.

Di Leonardo A., Linke S.P., Clarkin K., Wahl G.M. DNA damage triggers a prolonged p53-dependent G1 arrest and long-term induction of Cip1 in normal human fibroblasts. Genes Dev. 1994; 8:2540-2551.

Dolznig H., Habermann B., Stangl K., Deiner E.M., Moriggl R., Beug H., Mullner E.W. Apoptosis protection by the Epo target Bcl-X(L) allows factor-independent differentiation of primary erythroblasts. Curr Biol. 2002; 12:1076-1085.

Fang L., Li G., Liu G., Lee S.W., Aaronson S.A. p53 induction of heparin-binding EGF-like growth factor counteracts p53 growth suppression through activation of MAPK and PI3K/Akt signaling cascades. EMBO J. 2001; 20:1931-1939.

Fiscella M., Zhang H., Fan S., Sakaguchi K., Shen S., Mercer W.E., Vande Woude G.F., O'Connor P.M., Appella E. Wip1, a novel human protein phosphatase that is induced in response to ionizing radiation in a p53-dependent manner. Proc Natl Acad Sci USA. 1997; 94:6048-6053.

Flores E.R., Tsai K.Y., Crowley D., Sengupta S., Yang A., McKeon F., Jacks T. p63 and p73 are required for p53-dependent apoptosis in response to DNA damage. Nature. 2002; 416:560-564

Flynn P., Wongdagger M., Zavar M., Dean N.M., Stokoe D. Inhibition of PDK-1 activity causes a reduction in cell proliferation and survival. Curr Biol. 2000; 10:1439-1442.

Fuchs SY, Dolan L, Davis RJ, Ronai Z. Phosphorylation-dependent targeting of c-Jun ubiquitination by Jun N-kinase. Oncogene. 1996; 13:1531-1535.

Fuchs S.Y., Xie B., Adler V., Fried V.A., Davis R.J. Ronai Z. c-Jun NH2-terminal kinases target the ubiquitination of their associated transcription factors. J Biol Chem. 1997; 272:32163-32168.

Fuchs S.Y., Adler V., Buschmann T., Yin Z., Wu X., Jones S.N., Ronai Z. JNK targets p53 ubiquitination and degradation in nonstressed cells. Genes Dev. 1998a; 12:2658-2663.

Fuchs S.Y., Adler V., Pincus M.R., Ronai Z. MEKK1/JNK signaling stabilizes and activates p53. Proc Natl Acad Sci USA. 1998b; 95:10541-10546.

Gardner A.M., Johnson G.L. Fibroblast growth factor-2 suppression of tumor necrosis factor alpha-mediated apoptosis requires Ras and the activation of mitogen-activated protein kinase. J Biol Chem. 1996; 271:14560-14566.

Gottlieb T.M., Leal J.F., Seger R., Taya Y., Oren M. Cross-talk between Akt, p53 and Mdm2: possible implications for the regulation of apoptosis. Oncogene. 2002; 21:1299-1303.

Grossman S.R., Perez M., Kung A.L., Joseph M., Mansur C., Xiao Z.X., Kumar S., Howley P.M., Livingston D.M. p300/MDM2 complexes participate in MDM2-mediated p53 degradation. Mol Cell. 1998; 2:405-415.

Gupta S., Barrett T., Whitmarsh A.J., Cavanagh J., Sluss H.K., Derijard B., Davis R.J. Selective interaction of JNK protein kinase isoforms with transcription factors. EMBO J. 1996; 15:2760-2770.

Ham J., Babij C., Whitfield J., Pfarr C.M., Lallemand D., Yaniv M., Rubin L.L. A c-Jun dominant negative mutant protects sympathetic neurons against programmed cell death. Neuron. 1995; 14:927-939.

Hibi M., Lin A., Smeal T., Minden A., Karin M. Identification of an oncoprotein- and UV-responsive protein kinase that binds and potentiates the c-Jun activation domain. Genes Dev. 1993; 7:2135-2148.

Hu M.C., Qiu W.R., Wang Y.P. JNK1, JNK2 and JNK3 are p53 N-terminal serine 34 kinases. Oncogene. 1997; 15:2277-2287.

Huang C., Ma W.Y., Maxiner A., Sun Y., Dong Z. p38 kinase mediates UV-induced phosphorylation of p53 protein at serine 389. J Biol Chem. 1999; 274:12229-12235.

Ihrie R.A., Reczek E., Horner J.S., Khachatrian L., Sage J., Jacks T., Attardi L.D. PERP is a mediator of p53-dependent apoptosis in diverse cell types. Curr Biol. 2003; 13:1985-1990.

Ip Y.T., Davis RJ. Signal transduction by the c-Jun N-terminal kinase (JNK)--from inflammation to development. Curr Opin Cell Biol. 1998; 10:205-219.

Jeffers J.R., Parganas E., Lee Y., Yang C., Wang J., Brennan J., MacLean K.H., Han J., Chittenden T., Ihle J.N., McKinnon P.J., Cleveland J.L., Zambetti G.P. Puma is an essential mediator of p53-dependent and -independent apoptotic pathways. Cancer Cell. 2003; 4:321-328.

Jensen CJ, Buch MB, Krag TO, Hemmings BA, Gammeltoft S, Frodin M. 90-kDa ribosomal S6 kinase is phosphorylated and activated by 3-phosphoinositide-dependent protein kinase-1. J Biol Chem. 1999; 274:27168-27176.

Jimenez G.S., Nister M,, Stommel J.M., Beeche M., Barcarse E.A., Zhang X.Q., O'Gorman S., Wahl G.M. A transactivation-deficient mouse model provides insights into Trp53 regulation and function. Nat Genet. 2000; 26:37-43

Kamijo T., Weber J.D., Zambetti G., Zindy F., Roussel M.F., Sherr C.J. Functional and physical interactions of the ARF tumor suppressor with p53 and Mdm2. Proc Natl Acad Sci USA. 1998; 95:8292-8297.

Kawasaki H., Morooka T., Shimohama S., Kimura J., Hirano T., Gotoh Y., Nishida E. Activation and involvement of p38 mitogen-activated protein kinase in glutamate-induced apoptosis in rat cerebellar granule cells. J Biol Chem. 1997; 272:18518-18521.

Keller D., Zeng X., Li X., Kapoor M., Iordanov M.S., Taya Y., Lozano G., Magun B., Lu H. The p38MAPK inhibitor SB203580 alleviates ultraviolet-induced phosphorylation at serine 389 but not serine 15 and activation of p53. Biochem Biophys Res Commun. 1999; 261:464-471.

Kim S.J., Hwang S.G., Shin D.Y., Kang S.S., Chun J.S. p38 kinase regulates nitric oxide-induced apoptosis of articular chondrocytes by accumulating p53 via NFkappa B-dependent transcription and stabilization by serine 15 phosphorylation. J Biol Chem. 2002b; 277:33501-33508.

Kim S.J., Ju J.W., Oh C.D., Yoon Y.M., Song W.K., Kim J.H., Yoo Y.J., Bang O.S., Kang S.S., Chun J.S. ERK-1/2 and p38 kinase oppositely regulate nitric oxide-induced apoptosis of chondrocytes in association with p53, caspase-3, and differentiation status. J Biol Chem. 2002a; 277:1332-1339.

Klingmuller U. The role of tyrosine phosphorylation in proliferation and maturation of erythroid progenitor cells--signals emanating from the erythropoietin receptor. Eur J Biochem. 1997; 249:637-647.

Komarova E.A. Gudkov A.V. Chemoprotection from p53-dependent apoptosis: potential clinical applications of the p53 inhibitors. Biochem Pharmacol. 2001; 62:657-667.

Koury M.J., Bondurant M.C. Erythropoietin retards DNA breakdown and prevents programmed death in erythroid progenitor cells. Science. 1990; 248:378-381.

Kuan C.Y., Yang D.D., Samanta Roy D.R., Davis R.J., Rakic P., Flavell R.A. The Jnk1 and Jnk2 protein kinases are required for regional specific apoptosis during early brain development. Neuron. 1999; 22:667-676.

Kulik G., Klippel A., Weber M.J. Antiapoptotic signalling by the insulin-like growth factor I receptor, phosphatidylinositol 3-kinase, and Akt. Mol Cell Biol. 1997; 17:1595-1606.

Kummer J.L., Rao P.K., Heidenreich K.A. Apoptosis induced by withdrawal of trophic factors is mediated by p38 mitogen-activated protein kinase. J Biol Chem. 1997; 272:20490-20494.

Kurada P., White K. Ras promotes cell survival in Drosophila by downregulating hid expression. Cell. 1998; 95:319-329.

Kwon E.M., Raines M.A., Blenis J., Sakamoto K.M. Granulocyte-macrophage colony-stimulating factor stimulation results in phosphorylation of cAMP response element-binding protein through activation of pp90RSK. Blood. 2000; 95:2552-2558.

Kyriakis J.M., Avruch J. Mammalian mitogen-activated protein kinase signal transduction pathways activated by stress and inflammation. Physiol Rev. 2001; 81:807-869.

Lee S.W., Fang L., Igarashi M., Ouchi T., Lu K.P., Aaronson S.A. Sustained activation of Ras/Raf/mitogen-activated protein kinase cascade by the tumor suppressor p53. Proc Natl Acad Sci USA. 2000; 97:8302-8305.

Le-Niculescu H., Bonfoco E., Kasuya Y., Claret F.X, Green D.R., Karin M. Withdrawal of survival factors results in activation of the JNK pathway in neuronal cells leading to Fas ligand induction and cell death. Mol Cell Biol. 1999; 19:751-763.

Levine A.J. p53, the cellular gatekeeper for growth and division. Cell. 1997; 88:323-331

Liang J., Slingerland J.M. Multiple roles of the PI3K/PKB (Akt) pathway in cell cycle progression. Cell Cycle. 2003; 2:339-345.

Lin A. Activation of the JNK signaling pathway: breaking the brake on apoptosis. Bioessays. 2003; 25:17-24.

Lin A.W., Barradas M., Stone J.C., van Aelst L., Serrano M., Lowe S.W. Premature senescence involving p53 and p16 is activated in response to constitutive MEK/MAPK mitogenic signaling. Genes Dev. 1998; 12:3008-3019.

Lin Y., Benchimol S. Cytokines inhibit p53-mediated apoptosis but not p53-mediated G1 arrest. Mol Cell Biol. 1995; 15:6045-6054.

Lin Y., Ma W., Benchimol S. Pidd, a new death-domain-containing protein, is induced by p53 and promotes apoptosis. Nat Genet. 2000; 26:122-127

Lin Y., Brown L., Hedley D.W., Barber D.L., Benchimol S. The death-promoting activity of p53 can be inhibited by distinct signaling pathways. Blood. 2002; 100:3990-4000.

Liu Q., Sasaki T., Kozieradzki I., Wakeham A., Itie A., Dumont D.J., Penninger J.M. SHIP is a negative regulator of growth factor receptor-mediated PKB/Akt activation and myeloid cell survival. Genes Dev. 1999; 13:786-791.

Lotem J., Cragoe E.J., Jr., Sachs L. Rescue from programmed cell death in leukemic and normal myeloid cells. Blood. 1991; 78:953-960.

Lotem J., Sachs L. Cytokines as suppressors of apoptosis. Apoptosis. 1999; 4:187-196

Maehama T., Dixon J.E. PTEN: a tumour suppressor that functions as a phospholipid phosphatase. Trends Cell Biol. 1999; 9:125-128.

Marshall C.J. Specificity of receptor tyrosine kinase signaling: transient versus sustained extracellular signal-regulated kinase activation. Cell. 1995; 80:179-185.

Martin-Blanco E. p38 MAPK signalling cascades: ancient roles and new functions. Bioessays. 2000; 22:637-645.

Mattson M.P., Duan W., Pedersen W.A., Culmsee C. Neurodegenerative disorders and ischemic brain diseases. Apoptosis. 2001; 6:69-81.

Mayo L.D., Dixon J.E., Durden D.L., Tonks N.K., Donner D.B. PTEN protects p53 from Mdm2 and sensitizes cancer cells to chemotherapy. J Biol Chem. 2002; 277:5484-5489.

Mayo L.D., Donner D.B. A phosphatidylinositol 3-kinase/Akt pathway promotes translocation of Mdm2 from the cytoplasm to the nucleus. Proc Natl Acad Sci USA. 2001; 98:11598-11603.

Milne D.M., Campbell L.E., Campbell D.G., Meek D.W. p53 is phosphorylated in vitro and in vivo by an ultraviolet radiation-induced protein kinase characteristic of the c-Jun kinase, JNK1. J Biol Chem. 1995; 270:5511-5518.

Miyashita T., Reed J.C. Tumor suppressor p53 is a direct transcriptional activator of the human bax gene. Cell. 1995; 80:293-299

Motoyama N., Wang F., Roth K.A., Sawa H., Nakayama K., Negishi I., Senju S., Zhang Q., Fujii S., et al. Massive cell death of immature hematopoietic cells and neurons in Bcl-x-deficient mice. Science. 1995; 267:1506-1510.

Musti A.M., Treier M., Bohmann D. Reduced ubiquitin-dependent degradation of c-Jun after phosphorylation by MAP kinases. Science. 1997; 275:400-402.

Nagata Y., Moriguchi T., Nishida E., Todokoro K. Activation of p38 MAP kinase pathway by erythropoietin and interleukin-3. Blood. 1997; 90:929-934.

Nagata Y., Takahashi N., Davis R.J., Todokoro K. Activation of p38 MAP kinase and JNK but not ERK is required for erythropoietin-induced erythroid differentiation. Blood. 1998; 92:1859-1869.

Nakano K., Vousden K.H. PUMA, a novel proapoptotic gene, is induced by p53. Mol Cell. 2001; 7:683-694

Oda E., Ohki R., Murasawa H., Nemoto J., Shibue T., Yamashita T., Tokino T., Taniguchi T., Tanaka N. Noxa, a BH3-only member of the Bcl-2 family and candidate mediator of p53-induced apoptosis. Science. 2000; 288:1053-1058

Oda K., Arakawa H., Tanaka T., Matsuda K., Tanikawa C., Mori T., Nishimori H., Tamai K., Tokino T., Nakamura Y., Taya Y. p53AIP1, a potential mediator of p53-dependent apoptosis, and its regulation by Ser-46-phosphorylated p53. Cell. 2000; 102:849-862

Okamura S., Arakawa H., Tanaka T., Nakanishi H., Ng C.C., Taya Y., Monden M., Nakamura Y. p53DINP1, a p53-inducible gene, regulates p53-dependent apoptosis. Mol Cell. 2001; 8:85-94

Oren M. Decision making by p53: life, death and cancer. Cell Death Differ. 2003; 10:431-442

Oren M., Damalas A., Gottlieb T., Michael D., Taplick J., Leal J.F., Maya R., Moas M., Seger R., Taya Y., Ben-Ze'Ev A. Regulation of p53: intricate loops and delicate balances. Ann N Y Acad Sci. 2002; 973:374-383.

Ozes O.N., Mayo L.D., Gustin J.A., Pfeffer S.R., Pfeffer L.M., Donner D.B. NF-kappaB activation by tumour necrosis factor requires the Akt serine-threonine kinase. Nature. 1999; 401:82-85.

Parganas E., Wang D., Stravopodis D., Topham D.J., Marine J.C., Teglund S., Vanin E.F., Bodner S., Colamonici O.R., van Deursen J.M., Grosveld G., Ihle J.N. Jak2 is essential for signaling through a variety of cytokine receptors. Cell. 1998; 93:385-395.

Parrizas M., Saltiel AR, LeRoith D. Insulin-like growth factor 1 inhibits apoptosis using the phosphatidylinositol 3'-kinase and mitogen-activated protein kinase pathways. J Biol Chem. 1997; 272:154-161.

Passer B.J., Nancy-Portebois V., Amzallag N., Prieur S., Cans C., Roborel de Climens A., Fiucci G., Bouvard V., Tuynder M., Susini L., Morchoisne S., Crible V., Lespagnol A., Dausset J., Oren M., Amson R., Telerman A. The p53-inducible TSAP6 gene product regulates apoptosis and the cell cycle and interacts with Nix and the Myt1 kinase. Proc Natl Acad Sci U S A. 2003; 100:2284-2289

Perkins D., Pereira E.F.R., Aurelian L. The Herpes Simplex Virus Type 2 R1 Protein Kinase (ICP10 PK) Functions as a Dominant Regulator of Apoptosis in Hippocampal Neurons Involving Activation of the ERK Survival Pathway and Upregulation of the Antiapoptotic Protein Bag-1. J. Virol. 2003; 77:1292-1305

Persons D.L., Yazlovitskaya E.M., Pelling J.C. Effect of extracellular signal-regulated kinase on p53 accumulation in response to cisplatin. J Biol Chem. 2000; 275:35778-35785.

Pomerantz J., Schreiber-Agus N., Liegeois N.J., Silverman A., Alland L., Chin L., Potes J., Chen K., Orlow I., Lee H.W., Cordon-Cardo C., DePinho R.A. The Ink4a tumor suppressor gene product, p19Arf, interacts with MDM2 and neutralizes MDM2's inhibition of p53. Cell. 1998; 92:713-723.

Potapova O., Gorospe M., Dougherty R.H., Dean N.M., Gaarde W.A., Holbrook N.J. Inhibition of c-Jun N-terminal kinase 2 expression suppresses growth and induces apoptosis of human tumor cells in a p53-dependent manner. Mol Cell Biol. 2000; 20:1713-1722.

Prives C., Hall P.A. The p53 pathway. J Pathol. 1999; 187:112-126.

Pulverer B.J., Kyriakis J.M., Avruch J., Nikolakaki E., Woodgett J.R. Phosphorylation of c-jun mediated by MAP kinases. Nature. 1991; 353:670-674.

Quelle F.W., Wang J., Feng J., Wang D., Cleveland J.L., Ihle J.N., Zambetti G.P. Cytokine rescue of p53-dependent apoptosis and cell cycle arrest is mediated by distinct Jak kinase signaling pathways. Genes Dev. 1998; 12:1099-1107.

Resnick-Silverman L., St Clair S., Maurer M., Zhao K., Manfredi J.J. Identification of a novel class of genomic DNA-binding sites suggests a mechanism for selectivity in target gene activation by the tumor suppressor protein p53. Genes Dev. 1998; 12:2102-2107

Riccio A., Ahn S., Davenport C.M., Blendy J.A., Ginty D.D. Mediation by a CREB family transcription factor of NGF-dependent survival of sympathetic neurons. Science. 1999; 286:2358-2361

Richards S.A., Fu J., Romanelli A., Shimamura A., Blenis J. Ribosomal S6 kinase 1 (RSK1) activation requires signals dependent on and independent of the MAP kinase ERK. Curr Biol. 1999; 9:810-820.

Ries S., Biederer C., Woods D., Shifman O., Shirasawa S., Sasazuki T., McMahon M., Oren M., McCormick F. Opposing effects of Ras on p53: transcriptional activation of mdm2 and induction of p19ARF. Cell. 2000; 103:321-330.

Romashkova J.A. Makarov S.S. NF-kappaB is a target of AKT in anti-apoptotic PDGF signalling. Nature. 1999; 401:86-90.

Ryan K.M., Ernst M.K., Rice N.R, Vousden K.H. Role of NF-kappaB in p53-mediated programmed cell death. Nature. 2000; 404:892-897.

Sabapathy K., Hu Y., Kallunki T., Schreiber M., David J.P., Jochum W., Wagner E.F., Karin M. JNK2 is required for efficient T-cell activation and apoptosis but not for normal lymphocyte development. Curr Biol. 1999a; 9:116-125.

Sabapathy K., Jochum W., Hochedlinger K., Chang L., Karin M., Wagner E.F. Defective neural tube morphogenesis and altered apoptosis in the absence of both JNK1 and JNK2. Mech Dev. 1999b; 89:115-124.

Sabbatini P., McCormick F.. Phosphoinositide 3-OH kinase (PI3K) and PKB/Akt delay the onset of p53-mediated, transcriptionally dependent apoptosis. J Biol Chem. 1999; 274:24263-24269.

Sanchez-Prieto R., Rojas J.M., Taya Y., Gutkind J.S. A role for the p38 mitogen-acitvated protein kinase pathway in the transcriptional activation of p53 on genotoxic stress by chemotherapeutic agents. Cancer Res. 2000; 60:2464-2472.

Scheid M.P., Woodgett J.R. PKB/AKT: functional insights from genetic models. Nat Rev Mol Cell Biol. 2001; 2:760-768.

Scheid M.P., Woodgett J.R. Unravelling the activation mechanisms of protein kinase B/Akt. FEBS Lett. 2003; 546:108-112.

Schmitt C.A., Fridman J.S., Yang M., Baranov E., Hoffman R.M., Lowe S.W. Dissecting p53 tumor suppressor functions in vivo. Cancer Cell. 2002;1:289-298

Schott A.F., Apel I.J., Nunez G., Clarke M.F. Bcl-X$_L$ protects cancer cells from p53-mediated apoptosis. Oncogene. 1995; 11:1389-1394

Schuler M., Bossy-Wetzel E., Goldstein J.C., Fitzgerald P., Green D.R. p53 induces apoptosis by caspase activation through mitochondrial cytochrome c release. J Biol Chem. 2000; 275:7337-7342

Schwenger P., Bellosta P., Vietor I., Basilico C., Skolnik E.Y., Vilcek J. Sodium salicylate induces apoptosis via p38 mitogen-activated protein kinase but inhibits tumor necrosis factor-induced c-Jun N-terminal kinase/stress-activated protein kinase activation. Proc Natl Acad Sci USA. 1997; 94:2869-2873.

Serrano M., Lin A.W., McCurrach M.E., Beach D., Lowe S.W. Oncogenic ras provokes premature cell senescence associated with accumulation of p53 and p16INK4a. Cell. 1997; 88:593-602.

Sewing A., Wiseman B., Lloyd A., Land H. High-intensity Raf signal causes cell cycle arrest mediated by p21Cip1. Mol. Cell. Biol. 1997; 17:5588-5597

Shaulian E., Schreiber M., Piu F., Beeche M., Wagner E.F., Karin M. The mammalian UV response: c-Jun induction is required for exit from p53-imposed growth arrest. Cell. 2000; 103:897-907.

She Q.B., Ma W.Y., Dong Z. Role of MAP kinases in UVB-induced phosphorylation of p53 at serine 20. Oncogene. 2002; 21:1580-1589.

Shibue T., Takeda K., Oda E., Tanaka H., Murasawa H., Takaoka A., Morishita Y., Akira S., Taniguchi T., Tanaka N. Integral role of Noxa in p53-mediated apoptotic response. Genes Dev. 2003; 17:2233-2238.

Shimamura A., Ballif B.A., Richards S.A., Blenis J. Rsk1 mediates a MEK-MAP kinase cell survival signal. Curr Biol. 2000; 10:127-135.

Socolovsky M., Fallon A.E., Wang S., Brugnara C., Lodish H.F. Fetal anemia and apoptosis of red cell progenitors in Stat5a-/-5b-/- mice: a direct role for Stat5 in Bcl-X(L) induction. Cell. 1999; 98:181-191.

Socolovsky M., Nam H., Fleming M.D., Haase V.H., Brugnara C., Lodish H.F. Ineffective erythropoiesis in Stat5a(-/-)5b(-/-) mice due to decreased survival of early erythroblasts. Blood. 2001; 98:3261-3273

Soengas M.S., Alarcon R.M., Yoshida H., Giaccia A.J., Hakem R., Mak T.W., Lowe S.W. Apaf-1 and caspase-9 in p53-dependent apoptosis and tumor inhibition. Science. 1999; 284:156-159

Stambolic V., MacPherson D., Sas D., Lin Y., Snow B., Jang Y., Benchimol S., Mak TW. Regulation of PTEN transcription by p53. Mol Cell. 2001; 8:317-325.

Stambolic V., Suzuki A., de la Pompa J.L., Brothers G.M., Mirtsos C., Sasaki T., Ruland J., Penninger J.M., Siderovski D.P., Mak T.W. Negative regulation of PKB/Akt-dependent cell survival by the tumor suppressor PTEN. Cell. 1998; 95:29-39.

Symonds H., Krall L., Remington L., Saenz-Robles M., Lowe S., Jacks T., Van Dyke T. p53-dependent apoptosis suppresses tumor growth and progression in vivo. Cell. 1994; 78:703-711

Takekawa M., Adachi M., Nakahata A., Nakayama I., Itoh F., Tsukuda H., Taya Y., Imai K. p53-inducible wip1 phosphatase mediates a negative feedback regulation of p38 MAPK-p53 signaling in response to UV radiation. EMBO J. 2000; 19:6517-6526.

Tamura K., Sudo T., Senftleben U., Dadak A.M., Johnson R., Karin M. Requirement for p38alpha in erythropoietin expression: a role for stress kinases in erythropoiesis. Cell. 2000; 102:221-231.

Tanikawa C., Matsuda K., Fukuda S., Nakamura Y., Arakawa H. p53RDL1 regulates p53-dependent apoptosis. Nat Cell Biol. 2003; 5:216-223

Teglund S., McKay C., Schuetz E., van Deursen J.M., Stravopodis D., Wang D., Brown M., Bodner S., Grosveld G., Ihle J.N. Stat5a and Stat5b proteins have essential and nonessential, or redundant, roles in cytokine responses. Cell. 1998; 93:841-850.

Tournier C., Hess P., Yang D.D., Xu J., Turner T.K., Nimnual A., Bar-Sagi D., Jones S.N., Flavell R.A., Davis R.J. Requirement of JNK for stress-induced activation of the cytochrome c-mediated death pathway. Science. 2000; 288:870-874.

Ulrich E., Duwel A., Kauffmann-Zeh A., Gilbert C., Lyon D., Rudkin B., Evan G., Martin-Zanca D. Specific TrkA survival signals interfere with different apoptotic pathways. Oncogene. 1998; 16:825-832.

Villunger A., Michalak E.M., Coultas L., Mullauer F., Bock G., Ausserlechner M.J., Adams J.M., Strasser A. p53- and drug-induced apoptotic responses mediated by BH3-only proteins puma and noxa. Science. 2003; 302:1036-1038.

Vogelstein B., Lane D., Levine A.J. Surfing the p53 network. Nature. 2000; 408:307-310

Vousden K.H. p53: death star. Cell. 2000; 103:691-694

Vousden K.H., Lu X.. Live or let die: the cell's response to p53. Nat Rev Cancer. 2002; 2:594-604

Wang S., Shi X. Mechanisms of Cr(VI)-induced p53 activation: the role of phosphorylation, mdm2 and ERK. Carcinogenesis. 2001; 22:757-762.

Wang Y., Szekely L., Okan I., Klein G., Wiman K.G. Wild-type p53-triggered apoptosis is inhibited by bcl-2 in a v-myc-induced T-cell lymphoma line. Oncogene. 1993; 8:3427-3431

Williams G.T., Smith C.A., Spooncer E., Dexter T.M., Taylor D.R. Haemopoietic colony stimulating factors promote cell survival by suppressing apoptosis. Nature. 1990; 343:76-79.

Williams M.R., Arthur J.S., Balendran A., van der Kaay J., Poli V., Cohen P., Alessi D.R. The role of 3-phosphoinositide-dependent protein kinase 1 in activating AGC kinases defined in embryonic stem cells. Curr Biol. 2000; 10:439-448.

Wilson B., Mochon E., Boxer L. Induction of bcl-2 expression by phosphorylated CREB proteins during B- cell activation and rescue from apoptosis. Mol. Cell. Biol. 1996; 16:5546-5556

Wojchowski D.M., Gregory R.C., Miller C.P., Pandit A.K., Pircher T.J. Signal transduction in the erythropoietin receptor system. Exp Cell Res. 1999; 253:143-156.

Woods D., Parry D., Cherwinski H., Bosch E., Lees E., McMahon M. Raf-induced proliferation or cell cycle arrest is determined by the level of Raf activity with arrest mediated by p21Cip1. Mol. Cell. Biol. 1997; 17:5598-5611

Woods D.B., Vousden K.H. Regulation of p53 function. Exp Cell Res. 2001; 264:56-66.

Wosik K., Antel J., Kuhlmann T. Bruck W., Massie B., Nalbantoglu J. Oligodendrocyte injury in multiple sclerosis: a role for p53. J Neurochem. 2003; 85:635-644.

Wright W.E., Shay J.W. Historical claims and current interpretations of replicative aging. Nat Biotechnol. 2002; 20:682-688.

Wu H., Liu X., Jaenisch R., Lodish H.F. Generation of committed erythroid BFU-E and CFU-E progenitors does not require erythropoietin or the erythropoietin receptor. Cell. 1995; 83:59-67.

Xia Z., Dickens M., Raingeaud J., Davis R.J., Greenberg M.E. Opposing effects of ERK and JNK-p38 MAP kinases on apoptosis. Science. 1995; 270:1326-1331.

Yamaguchi A., Tamatani M., Matsuzaki H., Namikawa K., Kiyama H., Vitek M.P., Mitsuda N., Tohyama M. Akt activation protects hippocampal neurons from apoptosis by inhibiting transcriptional activity of p53. J Biol Chem. 2001; 276:5256-5264.

Yang D.D., Kuan C.Y., Whitmarsh A.J., Rincon M., Zheng T.S., Davis R.J., Rakic P., Flavell R.A. Absence of excitotoxicity-induced apoptosis in the hippocampus of mice lacking the Jnk3 gene. Nature. 1997; 389:865-870.

Yin Y., Liu Y.X., Jin Y.J., Hall E.J., Barrett J.C. PAC1 phosphatase is a transcription target of p53 in signalling apoptosis and growth suppression. Nature. 2003; 422:527-531

Yonish-Rouach E., Resnitzky D., Lotem J., Sachs L., Kimchi A., Oren M. Wild-type p53 induces apoptosis of myeloid leukaemic cells that is inhibited by interleukin-6. Nature. 1991; 352:345-347.

Yu C., Minemoto Y., Zhang J., Liu J., Tang F., Bui T.N., Xiang J., Lin A. JNK suppresses apoptosis via phosphorylation of the proapoptotic Bcl-2 family protein BAD. Mol Cell. 2004; 13:329-340.

Yu J., Zhang L., Hwang P.M., Kinzler K.W., Vogelstein B. PUMA induces the rapid apoptosis of colorectal cancer cells. Mol Cell. 2001; 7:673-682

Yu J., Wang Z., Kinzler K.W., Vogelstein B., Zhang L. PUMA mediates the apoptotic response to p53 in colorectal cancer cells. Proc Natl Acad Sci USA. 2003; 100:1931-1936

Zauli G., Gibellini D., Vitale M., Secchiero P., Celeghini C. Bassini A., Pierpaoli S., Marchisio M., Guidotti L., Capitani S. The induction of megakaryocyte differentiation is accompanied by selective Ser133 phosphorylation of the transcription factor CREB in both HEL cell line and primary CD34+ cells. Blood. 1998; 92:472-480.

Zha J., Harada H., Yang E., Jockel J., Korsmeyer S.J. Serine phosphorylation of death agonist BAD in response to survival factor results in binding to 14-3-3 not BCL-X(L). Cell. 1996; 87:619-628.

Zhou B.P., Liao Y., Xia W., Zou Y., Spohn B., Hung M.C. HER-2/neu induces p53 ubiquitination via Akt-mediated MDM2 phosphorylation. Nat Cell Biol. 2001; 3:973-982.

Zhou M., Gu L., Findley H.W., Jiang R., Woods W.G. PTEN reverses MDM2-mediated chemotherapy resistance by interacting with p53 in acute lymphoblastic leukemia cells. Cancer Res. 2003; 63:6357-6362.

Zhu J., Woods D., McMahon M., Bishop J.M. Senescence of human fibroblasts induced by oncogenic Raf. Genes Dev. 1998; 12:2997-3007

Chapter 6

P53, CELL CYCLE ARREST AND APOPTOSIS

Shulin Wang and Wafik S. El-Deiry
Departments of Medicine, Genetics, Pharmacology, and Abramson Cancer Center, University of Pennsylvania School of Medicine, Philadelphia, PA, USA

INTRODUCTION

The p53 gene, first described in 1979, was the first tumor suppressor gene to be identified (Lane and Crawford, 1979; Linzer and Levine, 1979). It was originally identified as an oncogene- a cell cycle accelerator, but subsequent studies ten years after its discovery confirmed it to be a tumor suppressor gene that is highly mutated in a wide variety of tumors (Baker et al., 1990; Finlay et al., 1989). In about half of the tumors, p53 is inactivated directly as a result of mutations in the p53 gene. In many others, it is inactivated indirectly through binding to viral proteins, or as a result of alterations in the genes whose products interact with p53 or transmit information to or from p53. The tumor suppressor protein p53 acts as a major node in a complex signaling pathway that evolved to sense a broad range of cellular stresses such as DNA damage, oncogene activation, nucleotide depletion, and hypoxia (Figure 1). In the absence of cellular stress, the p53 protein is expressed at low steady-state levels and exerts little, if any, effect on cell fate. However, in response to various types of stress, p53 becomes activated and this is reflected in elevated protein levels as well as augmented biochemical functions. As a consequence of p53 activation, cells can undergo marked phenotype changes, ranging from cell cycle arrest, senescence, or apoptosis (Bourdon et al., 2003; Dumont et al., 2003; El-Deiry, 2003; Oren, 2003) (Figure 1). The practical implication of these facts is that when a cell undergoes alteration that predisposes it to become cancerous, p53 is activated to trigger a response that either takes care of the damage (by augmented DNA repair) or eliminates the affected cells from the

P. Hainaut and K.G. Wiman (eds.), 25 Years of p53 Research, 141-163.
© 2005 *Springer. Printed in the Netherlands.*

replicative pool through induction of apoptosis, thereby preventing its expansion into a large population of malignant progeny (Figure 1). The protective p53 response is how p53 signals tumor suppression, and explains why its inactivation is so frequently selected for in almost all types of human cancer (Velculescu and El-Deiry, 1996; Vogelstein and Kinzler, 1992; Vogelstein et al., 2000).

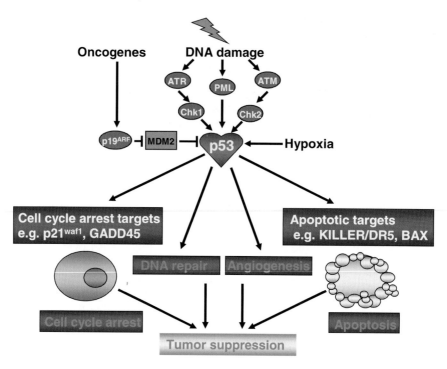

Figure 1. The p53 pathways and p53-mediated tumor suppression. p53 mediates the responses to various stress signals such as DNA damage, oncogenes activation, or hypoxia. In general, these signals induce p53 by stabilizing p53 protein, which leads to an increase in cellular p53 levels. Several cellular responses to p53 activation have been described, and the choice of responses depends on factors such as cell type, cellular environment or oncogenic alterations. The effect of p53 activation is to inhibit cell growth through cell cycle arrest (senescence) or induction of apoptosis, thereby suppressing tumor formation.

The mechanism(s) by which p53 accomplishes its biological functions are still not completely understood. However, in the last decade, it has become clear that p53 is a transcription factor with potential to bind to several hundred different promoter elements in the genome, broadly altering the patterns of specific gene expression. The regions of p53 responsible for binding to specific DNA sequences which in turn activate transcription have been defined. Virtually all naturally occurring mutations in the p53 gene reduce or eliminate the ability of the encoded p53 protein to activate transcription, supporting the idea that this activity is critical to p53's role as a tumor suppressor. Several dozen critical genes that are regulated by p53 have been identified. In this review, we will focus on the functions of p53 and its downstream target genes including the mechanisms of p53-mediated cell cycle arrest and apoptosis.

P53 FUNCTION

p53 is a sequence-specific DNA-binding transcription factor that binds DNA as a tetramer and activates or represses transcription from a large and increasing number of target genes (El-Deiry, 1998). Many of the genes induced by p53 can be divided into categories that reflect the responses to p53, such as cell cycle arrest genes, DNA repair genes, and apoptosis-inducing genes (El-Deiry, 2003; Sax and El-Deiry, 2003) (Figure 1). The coordinated expression of these genes, depending on cell type, environment, and stimulus may determine the outcome in response to cellular stress. Although it is generally believed that p53 effects are exerted through its activation of transcription, it is becoming evident that p53 is also capable of repressing transcription (Ho and Benchimol, 2003). Other activities of p53 independent of transcription have been described, including the ability of p53 to re-localize death receptors from the Golgi to the cell surface (Bennett et al., 1998) and a possible direct role in the mitochondria, where p53 specifically interacts with Bcl-2 family members (Chipuk et al., 2004; Mihara et al., 2003).

p53 is an efficient inhibitor of cell growth and causes cell cycle arrest and apoptosis (Figure 1). Regulation of p53 activity is therefore critical to allow both normal cell growth and tumor suppression. The basal activity of p53 remains low in normal or un-stressed cells due to its rapid turnover. Multiple mechanisms exist to regulate p53 activity, underscoring the importance of restraining p53 activity in un-stressed conditions. Regulation of p53 expression by transcriptional factors such as NFκB (Webster and Perkins, 1999) or HOXA5 (Raman et al., 2000) and the mechanisms that control p53 translation (Fu et al., 1996) likely contribute to the overall activity of p53. However, the principal mechanisms that govern p53 activity appear to be exerted at the protein level. These include regulation of p53 protein stability, post-translational modification, protein-protein interaction, and sub-cellular localization. One of the key components regulating p53 is MDM2, a protein that functions as an E3 ubiquitin ligase for p53, mediating ubiquitination of p53 and allowing it to be recognized and degraded by the proteasome (Kubbutat et al., 1997). MDM2 is a p53 transcriptional target gene and establishes a feedback loop in which p53 drives expression of its negative regulator (Wu et al., 1993). More recent reports showed that the p53-induced ubiquitin ligase Pirh2 (Leng et al., 2003) and COP1 (Dornan et al., 2004), like Mdm2, also participate in an autoregulatory feedback loop that controls p53 function. Yin Yang1 was identified as a negative regulator of p53 and induce Hdm2-mediated polyubiquitination (Sui et al., 2004). Other post-translational modifications of p53 including phosphorylation, acetylation, sumoylation and MDM2-mediated NEDD8 conjugation of p53, affect p53 stability and function (Brooks and Gu, 2003; Grossman et al., 2003; Leng et al., 2003; Vousden and Lu, 2002; Woods and Vousden, 2001; Xirodimas et al., 2004). These mechanisms keep a strong check on p53 in normal circumstances, but allow rapid activation of p53 response to cellular stress that might be caused by, or contribute to, oncogenic progression. MDM2 has been shown to monoubiquitinate p53 and to direct it out of the nucleus whereas other E3 ligase like p300 polyubiquitinate p53 (Grossman et al., 2003; Li et al., 2003).

P53 AND CELL CYCLE ARREST

Transient alterations in cell cycle progression after exposure to various different DNA damaging agents have been observed in many cell types (Hartwell and Weinert, 1989; Weinert and Hartwell, 1990). These alterations presumably permit optimal repair of damage before the cell reinitiates replicative DNA synthesis (G1 arrest) and/or begins mitosis (G2 arrest) (Figure 2). Failure to repair DNA damage prior to replicative synthesis or mitosis could result in propagation of mutagenic lesions and could contribute to the progressive accumulation of genomic changes necessary for neoplastic transformation to occur. The first evidence that p53 controls cell cycle progression came from the work from Kastan and co-workers who identified three participants (AT gene(s), p53, and GADD45) in a signal transduction pathway that controls cell cycle arrest following DNA damage; abnormalities in this pathway probably contribute to tumor development (Kastan et al., 1992). Later on, p53 was shown to be required for DNA-damage-induced G1 arrest primarily through transactivation of the best characterized p53 downstream target p21, a cyclin-dependent kinase inhibitor (El-Deiry et al., 1993; Harper et al., 1993). The elevated p21 binds and inactivates cyclin E/Cdk2 or cyclinD/Cdk4 complexes resulting in pRB hypophosphorylation and cell cycle arrest (Harper et al., 1993; Stewart and Pietenpol, 2001). RB is a negative regulator of the transcription factor E2F, which is required for expression of S-phase-specific genes.

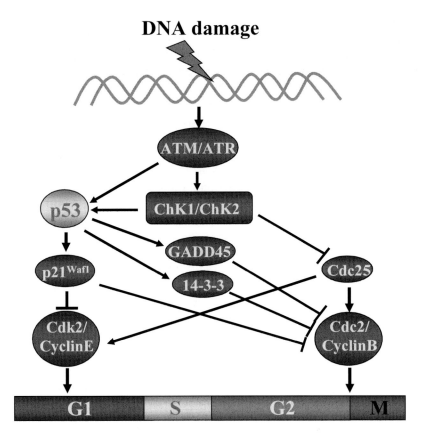

Figure 2. Participation of p53 in cell cycle regulation. ATM and ATR are required for checkpoint-dependent phsophorylation of multiple checkpoint components such as p53, ChK1 and ChK2. Checkpoint activation results in phosphorylation and activation of ChK1 and ChK2. ChK1 or ChK2 phosphorylate p53 in response to UV or irradiation, leading to p53 stabilization. The p53 protein induces transcription of p21Wafl, resulting in cell-cycle-arrest in G1. p53 also contributes to G2 arrest by inducing the transcription of GADD45, p21 and 14-3-3 and by repressing the transcription of Cyclin B. ChK1 and ChK2 phosphorylate Cdc25, resulting in 14-3-3-dependent checkpoint activation.

That RB acts downstream of p53 is suggested by experiments showing that loss of RB function can bypass p53-induced G1 arrest (Slebos et al., 1994). Studies on p21 knockout mice showed that p21$^{-/-}$ embryonic fibroblasts are significantly deficient in their ability to arrest in G1 in response to DNA damage and nucleotide pool perturbation (Deng et al., 1995). p21$^{-/-}$ cells also exhibit a significant growth alteration in vitro, achieving a saturation density as high as that observed in p53$^{-/-}$ cells. These results establish the role of p21 in the G1 checkpoint. However, other aspects of p53 function, such as thymocyte apoptosis and the mitotic spindle checkpoint, appear normal, suggesting that p21 is not the only p53 target gene required for p53-mediated tumor suppression (Deng et al., 1995). Kastan et al. observed that ataxia-telangiectasia cells failed to induce p53 in response to ionizing irradiation and that a mammalian cell cycle checkpoint pathway utilizing p53 and GADD45 is defective in ataxia-telangiectasia, suggesting that the ATM protein acts upstream of p53 (Kastan et al., 1992). Consistent with this, activity of the CDK inhibitor p21 was not detected in X-irradiated ataxia-telangiectasia cells, while it was detected in wild-type controls (Dulic et al., 1994). However, Lu and Lane observed induction of p53 protein in a large number of X-irradiated ataxia-telangiectasia cells, although with delayed kinetics (Lu and Lane, 1993). Thus, ATM is not required for p53 induction, although it can influence the timing of this process. In addition to its role in G1 arrest, wild-type p53 positively modulates the exit from the gamma-ray-induced G2 checkpoint (Guillouf et al., 1995; Taylor and Stark, 2001). The biochemical pathways involved in the DNA damage-induced G2 arrest are thought to involve signaling cascades that converge to inhibit the activation of Cdc2 (Herzinger et al., 1995; Nurse, 1990; Taylor and Stark, 2001) (Figure 2). Cdc2 is inhibited simultaneously by three transcriptional targets of p53, GADD45 (Zhan et al., 1998), p21 and 14-3-3 (Hermeking et al., 1997). Binding of Cdc2 to Cyclin B 1 is required for its activity, and repression of the Cyclin B1 gene by p53 also contributes to blocking entry into mitosis. p53 also represses Cdc2 gene, to help ensure that cells do not escape from the initial block (Taylor and Stark, 2001). DNA damage also activates p53-independent pathways that inhibit Cdc2 activity. In response to genotoxic stress, members of the PI-3K family become activated and initiate signal transduction pathways that regulate DNA repair and cell cycle progression. Several members of the PI-3K family can directly phosphorylate p53, including DNA-PK, ATM, ATR (Canman and Lim, 1998; Canman et al., 1998; Matsuoka et al., 1998). ATM-dependent signaling induced by DNA damage also results in activation of the ChK1 and ChK2 kinases (Caspari, 2000). ChK1 and ChK2 can phosphorylate Cdc25C, which in turn generates a consensus binding site for 14-3-3 proteins. Binding of 14-3-3 proteins to Cdc25C results in the nuclear export of Cdc25C, sequestration of the phosphatase in the cytoplasm, and thus inhibition of Cdc2 activity (Lopez-Girona et al., 1999). CAK is composed of Cdk7, Cyclin H and Mat 1, and activates Cdc2 by phosphorylation. Purified wild-type p53 binds and inhibits CAK in vitro but the tumor-derived R175H mutant of p53 apparently does not (Schneider et al., 1998). Exposure of $p21^{-/-}$ MEFs to UV-radiation inhibits endogenous CAK activity, an effect not observed with $p53^{-/-}$ MEFs (Schneider et al., 1998). These results suggest that inhibition of CAK by p53 may contribute to G2 arrest. p21 also participates in the G2 checkpoint and several mechanisms have been postulated for how p21 inhibits Cdc2 activity to cause G2 arrest (Taylor and Stark, 2001). First, p21 inhibits CDK activity by binding directly to CDK/Cyclin complexes (Boulaire et al., 2000). A second mechanism for inhibiting Cdc2 is suggested by experiments in Xenopus showing that active Cdk2 is involved in generating active Cdc2. p21, by inhibiting Cdk2, causes loss of Cdc2 activity (Guadagno and Newport, 1996). Thirdly, binding of p21 to Cdk2 can block access to CAK, and a similar effect may underlie the inhibition of Cdc2 by p21 (Hitomi et al., 1998). A fourth mechanism depends on the binding between p21 and proliferating-cell nuclear antigen (PCNA), the principal replicative DNA polymerase accessory subunit, required for DNA synthesis and DNA repair (Waga et al., 1994). Since PCNA is required for DNA synthesis processivity, p21 might inactivate it, causing DNA damage during S phase and thus leading to inhibition of Cdc2 and G2 arrest by p53-independent mechanisms.

P53 AND APOPTOSIS

One of the most extensively studied areas in the p53 field is its ability to induce apoptosis. The landmark finding that p53 can control apoptosis came from the work by Oren and colleagues, and they found that re-introduction of p53 into p53-deficient myeloid leukemia cells can induce apoptosis in a manner that could be countered by a pro-survival cytokine (Yonish-Rouach et al., 1991). Earlier studies by Scott Lowe and colleagues, using thymocytes from p53 knockout mice (Donehower et al., 1992) showed that p53 is required for radiation-induced apoptosis in the thymus but is not necessary for all forms of apoptosis (Lowe et al., 1993b). The Van Dyke laboratory demonstrated that p53-dependent apoptosis contributes to suppressing tumor growth and progression in vivo (Symonds et al., 1994). In addition to its role in suppressing tumorigenesis, p53-dependent apoptosis contributes to chemotherapy-induced cell death and inactivation of p53 can produce treatment-resistant tumors, therefore suggesting that p53 status may be a determinant for tumor response to therapy (Lowe et al., 1994; Lowe et al., 1993a). p53 induces expression of a wide array of death effectors, and the p53-inducible genes that might contribute to the induction of apoptosis through multiple pathways have been described (El-Deiry, 2003; Sax and El-Deiry, 2003). The p53-inducible pro-apoptotic genes are involved in several death pathways, such as the death receptor pathway, the mitochondrial pathway, and the recently identified endoplasmic reticulum (ER) pathway (Bourdon et al., 2002) (Figure 3).

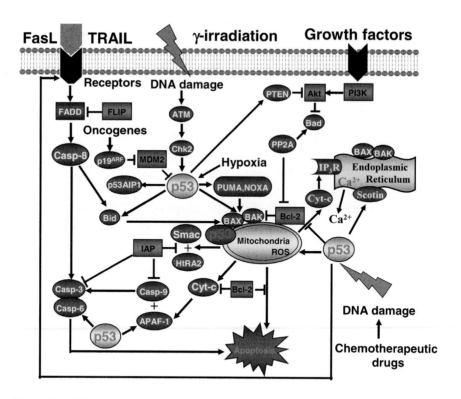

Figure 3. Multiple apoptotic pathways activated by p53. In response to cellular stresses such as DNA damage, oncogene activation or hypoxia, p53 transcriptionally represses (e.g. survivin or Bcl2) or transactivates its down-stream targets to induce apoptosis through (1) mitochondrial pathway (Bax, Bak, NUMA, NOXA); (2) death receptor pathway (Fas, KILLER/DR5, Bid); (3) endoplasmic recticulum pathway (Scotin). These pathways cross-communicate with each other and converge to a common downstream pathway that subsequently leads to cell death. Downstream of the mitochondria, caspase 6 and APAF1 can also be transactivated by p53, thus modulating the sensitivity of the cells to undergo apoptosis. p53 can translocate to mitochondria, where it directly interacts with Bcl-2 family members and leads to cell death in a p53-transcription-independent manner.

DNA damage can induce transcriptional up-regulation of some death receptors such as FAS, KILLER/DR5 (Takimoto and El-Deiry, 2000; Wu et al., 1997) and the recently described p53RDL1 (Tanikawa et al., 2003), through p53-dependent as well as p53-independent mechanisms (El-Deiry, 2001). This up-regulation increases the cellular sensitivity to death-receptor ligands. In Type I cells, death-receptor engagement of the cell-extrinsic pathway suffices for commitment to apoptotic death. In Type II cells, commitment to apoptosis requires amplification of the death receptor signaling by the cell-intrinsic pathway (Wang and El-Deiry, 2003a; Wang and El-Deiry, 2003b). Death receptors can activate the cell intrinsic pathway by caspase-8 mediated cleavage of the apical-proapoptotic Bcl-2 family member Bid (Li et al., 1998; Luo et al., 1998). Bid interacts with the proapoptotic Bcl-2 family members, which cause the release of mitochondrial cytochrome c and Smac/DIABLO, activating caspase-9 and − 3. Tumor-necrosis-factor-related apoptosis inducing ligand (TRAIL) which is a member of the TNF family and has been shown to induce apoptosis in a wide variety of tumor cells but not most normal cells, appears to be a promising agent for cancer therapy. TRAIL induces apoptosis through engagement of its receptors, KILLER/DR5 and DR4. Recently, the pro-apoptotic Bcl-2 family member Bax was found to be required for TRAIL-induced apoptosis in human colon carcinomas and pretreatment of Bax$^{-/-}$ human colon cancer cells to chemotherapeutic agents restored the TRAIL sensitivity (Burns and El-Deiry, 2001; LeBlanc et al., 2002; Wang and El-Deiry, 2003a). p53 was found to be involved in this process and p53-dependent up-regulation of KILLER/DR5 mainly contributes to the restoration of TRAIL sensitivity to human colon cancer cells with the mitochondrial apoptotic defect (Wang and El-Deiry, 2003a). In addition to its role in TRAIL-induced apoptosis, KILLER/DR5 was also found to be highly induced in vivo in response to γ-irradiation in a tissue-specific manner suggesting a potential function of this p53 target gene in apoptosis in vivo (Burns et al., 2001; Fei et al., 2002). Silencing of KILLER/DR5 in vivo significantly promotes tumor progression, thereby providing evidence that KILLER/DR5 might be a tumor suppressor (Wang and El-Deiry, 2004). Most chemotherapeutics and irradiation trigger tumor-cell apoptosis through the cell-intrinsic pathway, as an indirect consequence of causing cellular damage. Engagement of this pathway usually requires p53 function. TRAIL treatment in combination with chemo- or radiotherapy can enhance TRAIL sensitivity or reverse TRAIL resistance by p53-dependent up-regulation of its downstream targets such as KILLER/DR5 (Wang and El-Deiry, 2003a).

The mechanistic link between p53-mediated transactivation and apoptosis came from its ability to regulate the transcription of proapoptotic Bcl2 family members. These include the multidomain Bcl-2 family member Bax (Miyashita et al., 1994b) as well as BH3 only proteins such as Bid (Sax et al., 2002), Puma (Nakano and Vousden, 2001), and Noxa (Oda et al., 2000a). p53 binds to the consensus p53 response elements in the promoters of these genes and induces their transcription. Gene knockout studies demonstrated that the Bcl-2 family members act downstream of p53 during apoptosis. Bax-deficient MEFs are resistant to oncogene-induced apoptosis, leading to increased transformation in vitro and tumorigenesis in vivo (McCurrach et al., 1997; Yin et al., 1997). PUMA was recently identified as an essential mediator of p53-dependent and in-dependent apoptosis in vivo and knockout of PUMA recapitulates some of the apoptotic deficiency observed in p53 knockout mice (Jeffers et al., 2003; Villunger et al., 2003; Yu and Zhang, 2003).

In addition to the transactivation function, p53 also has transrepression capabilities that may contribute to apoptosis. The first antiapoptotic protein whose expression was reported to be transcriptionally blocked by p53 is Bcl-2 (Miyashita et al., 1994a). Subsequently, p53 was found to repress the promoter of Bcl-xl and survivin (Hoffman et al., 2002; Sugars et al., 2001). p53 thus appears to be a highly sophisticated executioner, simultaneously upregulating death-promoting genes and turning off protective genes.

p53 also transcriptionally up-regulates a number of cytoplasmic proteins including the death domain containing protein PIDD, the PIG genes (Polyak et al., 1997) involved in the generation of reactive oxygen species, as well as Bid which serves as a link between death receptor signaling and mitochondrial cytochrome c release (Sax et al., 2002). p53 also transactivates several components downstream of mitochondria in the apoptotic machinery. One of these components is APAF-1 (Kannan et al., 2001; Moroni et al., 2001), which binds to cytochrome c and form apoptosome initiating the cleavage of caspase 9. In addition, p53 can transactivate caspases such as caspase 6 or caspase 10 leading to enhanced chemosensitivity of some types of cells (MacLachlan and El-Deiry, 2002; Rikhof et al., 2003) (Figure 3).

Previous work suggests that the promelocytic leukemia gene (*PML*) can act up-stream of p53 to enhance transcription of p53 targets by recruiting p53 to nuclear bodies (NBs) (Bischof et al., 2002). PML binds and recruits the negative p53 regulator Mdm2 into NBs, therefore protecting p53 from Mdm2-mediated degradation (Kurki et al., 2003; Louria-Hayon et al., 2003; Wei et al., 2003). Collectively, these studies suggest that PML can act as an up-stream regulator of p53. However, a recent study showed that PML is itself a p53 target gene that also acts downstream of p53 to participate in additional p53-mediated programs, including cell cycle arrest and apoptosis (de Stanchina et al., 2004).

The ER is the main intracellular storage compartment for Ca^{2+}, which is an important secondary messenger that is required for numerous cellular functions. Apoptosis occurs upon the perturbation of cellular Ca^{2+} homeostasis, such as cytosolic Ca^{2+} overload, ER Ca^{2+} depletion, and mitochondrial Ca^{2+} increase (Boehning et al., 2003) (Figure 3). The close physical contact of mitochondria and the ER results in a higher exposure to Ca^{2+} of mitochondria than the rest of the cytosol, when Ca^{2+} is released from the ER. Bcl-2 family proteins have been previously implicated in controlling apoptosis by affecting cellular Ca^{2+} homeostasis (Bassik et al., 2004; Scorrano et al., 2003; Zong et al., 2003). One of the p53 targets, Scotin, is localized to the ER and nuclear membrane. Inhibition of endogenous Scotin expression increases resistance to p53-dependent apoptosis induced by DNA damage, suggesting that Scotin plays a role in p53-dependent apoptosis (Bourdon et al., 2002).

p53 can also induce apoptosis through short-circuiting cell survival pathways. The direct evidence came from the observation that p53 regulates PTEN, a negative regulator of the PI3K pathway. p53 transactivates the promoter of PTEN and therefore increases PTEN expression. PTEN serves to prevent the activation of Akt, thereby facilitating apoptosis (Stambolic et al., 2001). p53 also represses the expression of the catalytic subunit of PI3K (Singh et al., 2002). Since PI3K is a critical upstream activator of Akt, this inhibitory effect of p53 will also lead to Akt inactivation, which may cooperate with the induction of PTEN and the degradation of Akt to achieve effective p53-mediated attenuation of Akt function. Activated p53 can also cause a rapid decrease steady-state level of Akt, through a mechanism involving caspase-mediated Akt degradation (Gottlieb et al., 2002). Akt can phosphorylate MDM2 to promote its negative regulatory effect towards p53 (Zhou et al., 2001).

p53 functions indisputably in the nucleus to regulate transcription of genes involved in processes including cell cycle arrest and apoptosis. Tumor-derived mutant p53 proteins exhibit defects in the ability to bind DNA and therefore to affect gene expression, arguing for an important role for the DNA binding and transcriptional functions in the tumor suppressor activity of p53. It has been shown previously that in certain cell types, p53 can induce apoptosis independent of its effect on transcription (Chipuk et al., 2003; Haupt et al., 1995). Ectopic expression of p53 mutants including p53[1-102] lacking the DNA binding domain, p53[Q22/S23] mutated in the transactivation and Mdm2 interaction regions or p53 ΔNLS without a nuclear localization signal could still induce apoptosis. Moreover, the transcription-independent p53-mediated apoptosis can be inhibited by overexpressing of the anti-apoptotic Bcl2 family member, Bcl-XL, thus demonstrating the crucial role of the Bcl-2 family of proteins in the regulation of this process (Chipuk et al., 2003). Mihara et al. provided mechanistic insight by showing that p53 localizes to mitochondria and directly interacts with the anti-apoptotic proteins Bcl2 and Bcl-XL, and subsequently induces the release of cytochrome c. The binding region for Bcl2 and Bcl-XL on p53 is localized to the core region involved in sequence-specific DNA binding, the same region that harbors the vast majority of "hot spot" mutations found in human cancers (Mihara et al., 2003). Dumont et al. reported that a common p53 polymorphic variant R72 increases nuclear export, mitochondrial localization, and apoptosis (Dumont et al., 2003). It was recently reported by Doug Green's laboratory that when p53 accumulates in the cytosol, it can release both proapoptotic multidomain proteins as well as BH3-only proteins that were sequestered by Bcl-XL and function analogously to the BH3-only subset of proapoptotic Bcl-2 proteins to activate Bax and trigger apoptosis (Chipuk et al., 2004). Leu et al. showed that p53 interacts with the proapoptotic mitochondrial protein Bak and the interaction of p53 with Bak causes oligomerization of Bak and the release of cytochrome c from the mitochondria. They also showed that the formation of the p53-Bak complex coincides with loss of an interaction between Bak and the anti-apoptotic Bcl-2 family member Mcl1 (Leu et al., 2004). Taken together, these groups have made a number of observations that provide insights into the transcription-independent functions of p53 in promoting apoptosis (Figure 4).

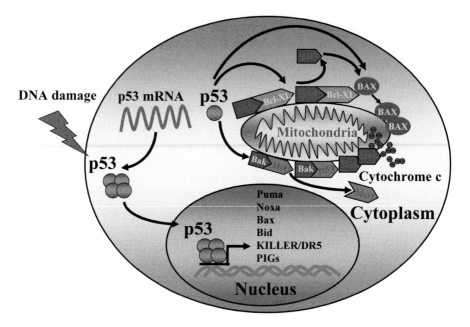

Figure 4. Nuclear and cytoplasmic roles of p53 in apoptosis. p53 induces apoptosis in a transcription-dependent and –independent manner. Stress signals stabilize nuclear p53 protein, which in turn transactivates or transcriptionally represses p53 targets, subsequently leading to cell death. Cytoplasmic p53 can translocate to mitochondria, where it interacts directly with Bax or Bak and causes the oligomerization of these two proapoptotic Bcl-2 family members, or liberates proapoptotic BH3-only proteins bound to Bcl-XL/Bcl2 at the mitochondria. The released BH-3 proteins can then activate Bax and cause cytochrome c release resulting in apoptosis. The interaction of p53 with Bak coincides with loss of an interaction between Bak and anti-apoptotic Bcl-2 family member Mcl1.

LIFE AND DEATH DECISIONS MADE BY P53

When cells respond adequately to a p53-activating signal, the biological outcomes may vary greatly. In particular, normal cells seem to be more refractory to the effects of p53 than their tumor-derived counterparts. Moreover, rapidly proliferating cells appear to be more sensitive to p53 activation than resting or slowly proliferating cells. One of the major questions and areas of intense investigation, in part owing to its paramount relevance to the successful application of cancer therapy, is how a cell makes decision to either undergo apoptosis versus induction of a viable growth arrest or senescence upon p53 activation (El-Deiry, 2003; Vousden and Lu, 2002; Vousden and Woude, 2000). Much of the choice is not in the hand of p53 itself, but rather is determined by cooperating or ameliorating intracellular and extracellular signals, which dictate whether p53 activation will spare the cell or lead to its apoptotic demise (Figure 5).

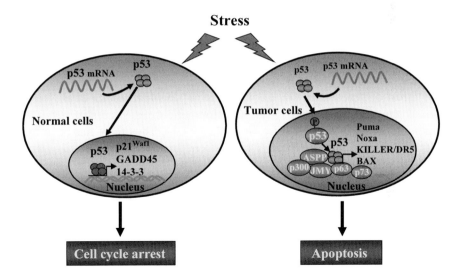

Figure 5. Modulation of the choice of the responses to p53. There are several factors that might contribute to the life and death decisions of normal or tumor cells in response to stress stimuli. Activation of p53 in normal cells usually leads to selective activation of cell-cycle-arrest target genes such as p21Waf1, GADD45, or 14-3-3, resulting in inhibition of cell proliferation. In tumor cells, the tissue-specific p53-interacting and p53-modulating factors target p53 towards specific subsets of promoters. These factors can bind p53 (directly or indirectly) as shown for ASPP and JMY or as shown for p63 and p73, and enhance p53 DNA binding affinity to the promoters of proapoptotic genes. Phosphorylation alters p53 conformation to either enhance the interaction with the apoptotic cofactors or allow binding to the promoters of apoptotic target genes.

The initial suggestion that p53 can independently regulate cell-cycle-arrest and apoptosis came from observations of tumor-derived p53 point mutants, which retain the ability to activate G1 arrest but fail to suppress the transformation of cells by oncogenes, such as the human Papillomavirus (HPV) E7 protein (Crook et al., 1994; Rowan et al., 1996). It became clear later that suppression of transformation in these assays was a reflection of the ability of p53 to induce apoptosis, and that the tumor-derived mutants being studied show a selective loss of apoptotic function that correlates with a selective defect in the ability to activate expression of some of the apoptotic p53 target genes (Friedlander et al., 1996; Ludwig et al., 1996). This may be due to a variable affinity of p53 for the binding sites in the different promoters it regulates. p53 mutants with slight conformation modification still bind to high affinity binding sites in the promoter region of cell-cycle-arrest genes, but are incapable of interacting with low binding affinity sites in the promoters of apoptotic target genes. The degree to which p53 accumulates in a given cells can also directly influence the outcome of arrest versus death (Blagosklonny and El-Deiry, 1996; Blagosklonny and El-Deiry, 1998; Wu and El-Deiry, 1996). Cells expressing low levels of p53 generally arrest in G1, while those that express high levels undergo apoptosis.

The induction of p53 in primary fibroblasts is usually associated with cell cycle arrest or senescence (Di Leonardo et al., 1994; Kuerbitz et al., 1992), whereas the activation of p53 in hematopoietic cells generally results in apoptosis (Clarke et al., 1993; Lowe et al., 1993b). Thus there has been long standing evidence that cell type is likely to influence whether a cell is more likely to arrest or die. In case of cells of different origins such as hematopoietic versus epithelial cells, one could invoke tissue specific factors or modifications in p53 that might affect its ability to cause cell cycle arrest or apoptosis. No such tissue-specific factors have been identified yet, although recent studies have provided some clues that p53-dependent gene activation profiles may contribute to making the decisions whether a cell might undergo apoptosis or cell cycle arrest (Fei et al., 2002) and the outcome is predetermined by the spectrum of p53-responsive genes that are available for modulation: different cell lineages might keep different sets of p53-responsive genes in active chromatin, thereby predetermining a specific pattern of transactivation and transrepression that dictates the final outcome. So, cells expressing more apoptotic genes would be more likely to undergo apoptosis than growth arrest.

In addition, several studies have implicated that covalent modification on p53 may play a critical role in its target gene preference. Phosphorylation is the most widely studied type of p53 modification (Lohrum and Scheidtmann, 1996) and a more recent striking example involves phosphorylation of serine

46, which is specifically required for the efficient transactivation of the proapoptotic p53AIP1 gene (Oda et al., 2000b). The covalent modification on p53 may lead to conformation change, which subsequently might alter directly its DNA-binding specificity.

Besides covalent modification of p53 itself, p53 binding proteins appear to be implicated in modulating the selection of target genes. In order to function efficiently as an activator of gene expression, p53 forms complexes with other transcriptional regulators, including acetyltransferases such as p300/CBP. These interactions allow for the acetylation of histones that surround the p53-binding sites, opening up the surrounding chromatin and allow access of the basal transcriptional machinery (Espinosa and Emerson, 2001). JMY, a transcriptional cofactor, cooperates with p300 to enhance the ability of p53 to selectively activate the expression of genes such as Bax, but without a significant effect on the activation of other genes such as p21 (Shikama et al., 1999). Although the purified recombinant p53 protein binds directly to DNA in a sequence-specific manner, it does so within the cell in complex with other proteins. ASPP1 and ASPP2 have recently been identified and appear to stimulate the expression of endogenous Bax in cells containing wild-type p53. This response correlates well with the ability of ASPP proteins to enhance the selectively p53 binding to the Bax promoter in vivo and to stimulate the promoters of the proapoptotic responsive genes (Bax and PIG13), but not other targets such as p21 (Samuels-Lev et al., 2001). As discussed above, proapoptotic p53 target genes usually have low-affinity binding sites, and ASPP proteins may stimulate p53 DNA binding activity sufficiently to trigger the expression of this subset of genes. Conversely, inhibition of ASPP expression was shown to selectively block the apoptotic response to p53 (Bergamaschi et al., 2003).

An un-expected insight into the p53 promoter selectivity was provided by experiments that induction of cell death by p53 requires the presence of at least one of the other p53 family members, p63 or p73 (Flores et al., 2002). These experiments revealed that both p63 and p73 enable activation of proapoptotic genes such as Bax and PERP by p53. Apparently, p63 or p73 are constitutively associated with these genes within the chromatin, and are required for the recruitment of p53 to those sites once p53 is induced in response to appropriate stress. However, recent studies have challenged the importance of p63 or p73 in p53-dependent apoptosis or tumor suppression (Senoo et al., 2004). Recently another tumor suppressor protein, BRCA1 was found to selectively direct p53 to transactivate target genes involved in cell cycle arrest and DNA repair, but not apoptosis (MacLachlan et al., 2002).

The choice between cell death and survival is also dependent on the activity of survival signals that can be mediated by soluble ligand binding to cell-surface receptors, or direct interactions with neighboring cells. Rescue of p53-induced apoptosis by survival factors has been associated with the activation of the Akt kinase (Sabbatini and McCormick, 1999). The inhibition of p53 by Akt is counteracted by the ability of p53 to induce the expression of PTEN (Stambolic et al., 2001). The induction of PTEN has been shown to be essential for p53-mediated apoptosis in mouse cells, underscoring the importance of survival signaling in determining the final outcome of the p53 responses.

CONCLUSIONS AND PERSPECTIVES

p53 is a key component of pathways regulating cellular responses to stress and controls numerous downstream targets that can result in variable cellular phenotypic outcomes, including apoptosis, transient growth arrest, sustained growth arrest or senescence. This network is tightly regulated and complex with multiple feedback loops, posttranslational modifications, protein-protein interactions, and tissue specificity. It is becoming apparent that upstream signals may drive cell fate responses. As these signals are relayed, p53 levels and post-translational modifications are changed to accommodate the setting, and different subsets of p53 effectors are transcribed to yield the appropriate responses. These responses maintain genomic integrity and stability, and are essential to protect against tumorigenesis. As the cell cycle arrest and apoptosis activities of p53 can be independently regulated, a more subtle variation on the theme of reactivating p53 for cancer therapy becomes how specifically to activate the death-inducing functions of p53 particularly in tumor cells. The increasing number of p53 cofactors that are required for p53 to induce apoptosis is providing some attractive targets for the development of therapies to restore the apoptotic functions of p53. The main challenge in future therapeutic development is to develop global approaches to analyze the complexity of the interactions between p53 and other cellular factors and extend the picture of the role of p53 to consider not only single pathways but also entire networks of interactions.

ACKNOWLEDGEMENTS

W. S. El-Deiry is an Assistant Investigator of Howard Hughes Medical Institute.

REFERENCES

Baker, S. J., Markowitz, S., Fearon, E. R., Willson, J. K., and Vogelstein, B. (1990). Suppression of human colorectal carcinoma cell growth by wild-type p53. Science *249*, 912-915.

Bassik, M. C., Scorrano, L., Oakes, S. A., Pozzan, T., and Korsmeyer, S. J. (2004). Phosphorylation of BCL-2 regulates ER Ca(2+) homeostasis and apoptosis. Embo J *23*, 1207-1216.

Bennett, M., Macdonald, K., Chan, S. W., Luzio, J. P., Simari, R., and Weissberg, P. (1998). Cell surface trafficking of Fas: a rapid mechanism of p53-mediated apoptosis. Science *282*, 290-293.

Bergamaschi, D., Samuels, Y., O'Neil, N. J., Trigiante, G., Crook, T., Hsieh, J. K., O'Connor, D. J., Zhong, S., Campargue, I., Tomlinson, M. L., *et al.* (2003). iASPP oncoprotein is a key inhibitor of p53 conserved from worm to human. Nat Genet *33*, 162-167.

Bischof, O., Kirsh, O., Pearson, M., Itahana, K., Pelicci, P. G., and Dejean, A. (2002). Deconstructing PML-induced premature senescence. Embo J *21*, 3358-3369.

Blagosklonny, M. V., and El-Deiry, W. S. (1996). In vitro evaluation of a p53-expressing adenovirus as an anti-cancer drug. Int J Cancer *67*, 386-392.

Blagosklonny, M. V., and El-Deiry, W. S. (1998). Acute overexpression of wt p53 facilitates anticancer drug-induced death of cancer and normal cells. Int J Cancer *75*, 933-940.

Boehning, D., Patterson, R. L., Sedaghat, L., Glebova, N. O., Kurosaki, T., and Snyder, S. H. (2003). Cytochrome c binds to inositol (1,4,5) trisphosphate receptors, amplifying calcium-dependent apoptosis. Nat Cell Biol *5*, 1051-1061.

Boulaire, J., Fotedar, A., and Fotedar, R. (2000). The functions of the cdk-cyclin kinase inhibitor p21WAF1. Pathol Biol (Paris) *48*, 190-202.

Bourdon, J. C., Laurenzi, V. D., Melino, G., and Lane, D. (2003). p53: 25 years of research and more questions to answer. Cell Death Differ *10*, 397-399.

Bourdon, J. C., Renzing, J., Robertson, P. L., Fernandes, K. N., and Lane, D. P. (2002). Scotin, a novel p53-inducible proapoptotic protein located in the ER and the nuclear membrane. J Cell Biol *158*, 235-246.

Brooks, C. L., and Gu, W. (2003). Ubiquitination, phosphorylation and acetylation: the molecular basis for p53 regulation. Curr Opin Cell Biol *15*, 164-171.

Burns, T. F., Bernhard, E. J., and El-Deiry, W. S. (2001). Tissue specific expression of p53 target genes suggests a key role for KILLER/DR5 in p53-dependent apoptosis in vivo. Oncogene *20*, 4601-4612.

Burns, T. F., and El-Deiry, W. S. (2001). Identification of inhibitors of TRAIL-induced death (ITIDs) in the TRAIL-sensitive colon carcinoma cell line SW480 using a genetic approach. J Biol Chem *276*, 37879-37886.

Canman, C. E., and Lim, D. S. (1998). The role of ATM in DNA damage responses and cancer. Oncogene *17*, 3301-3308.

Canman, C. E., Lim, D. S., Cimprich, K. A., Taya, Y., Tamai, K., Sakaguchi, K., Appella, E., Kastan, M. B., and Siliciano, J. D. (1998). Activation of the ATM kinase by ionizing radiation and phosphorylation of p53. Science *281*, 1677-1679.

Caspari, T. (2000). How to activate p53. Curr Biol *10*, R315-317.

Chipuk, J. E., Kuwana, T., Bouchier-Hayes, L., Droin, N. M., Newmeyer, D. D., Schuler, M., and Green, D. R. (2004). Direct activation of Bax by p53 mediates mitochondrial membrane permeabilization and apoptosis. Science *303*, 1010-1014.

Chipuk, J. E., Maurer, U., Green, D. R., and Schuler, M. (2003). Pharmacologic activation of p53 elicits Bax-dependent apoptosis in the absence of transcription. Cancer Cell *4*, 371-381.

Clarke, A. R., Purdie, C. A., Harrison, D. J., Morris, R. G., Bird, C. C., Hooper, M. L., and Wyllie, A. H. (1993). Thymocyte apoptosis induced by p53-dependent and independent pathways. Nature *362*, 849-852.

Crook, T., Marston, N. J., Sara, E. A., and Vousden, K. H. (1994). Transcriptional activation by p53 correlates with suppression of growth but not transformation. Cell *79*, 817-827.

de Stanchina, E., Querido, E., Narita, M., Davuluri, R. V., Pandolfi, P. P., Ferbeyre, G., and Lowe, S. W. (2004). PML is a direct p53 target that modulates p53 effector functions. Mol Cell *13*, 523-535.

Deng, C., Zhang, P., Harper, J. W., Elledge, S. J., and Leder, P. (1995). Mice lacking p21CIP1/WAF1 undergo normal development, but are defective in G1 checkpoint control. Cell *82*, 675-684.

Di Leonardo, A., Linke, S. P., Clarkin, K., and Wahl, G. M. (1994). DNA damage triggers a prolonged p53-dependent G1 arrest and long-term induction of Cip1 in normal human fibroblasts. Genes Dev *8*, 2540-2551.

Donehower, L. A., Harvey, M., Slagle, B. L., McArthur, M. J., Montgomery, C. A., Jr., Butel, J. S., and Bradley, A. (1992). Mice deficient for p53 are developmentally normal but susceptible to spontaneous tumours. Nature *356*, 215-221.

Dornan, D., Wertz, I., Shimizu, H., Arnott, D., Frantz, G. D., Dowd, P., K, O. R., Koeppen, H., and Dixit, V. M. (2004). The ubiquitin ligase COP1 is a critical negative regulator of p53. Nature *429*, 86-92.

Dulic, V., Kaufmann, W. K., Wilson, S. J., Tlsty, T. D., Lees, E., Harper, J. W., Elledge, S. J., and Reed, S. I. (1994). p53-dependent inhibition of cyclin-dependent kinase activities in human fibroblasts during radiation-induced G1 arrest. Cell *76*, 1013-1023.

Dumont, P., Leu, J. I., Della Pietra, A. C., 3rd, George, D. L., and Murphy, M. (2003). The codon 72 polymorphic variants of p53 have markedly different apoptotic potential. Nat Genet *33*, 357-365.

El-Deiry, W. S. (1998). Regulation of p53 downstream genes. Semin Cancer Biol *8*, 345-357.

El-Deiry, W. S. (2001). Insights into cancer therapeutic design based on p53 and TRAIL receptor signaling. Cell Death Differ *8*, 1066-1075.

El-Deiry, W. S. (2003). The role of p53 in chemosensitivity and radiosensitivity. Oncogene *22*, 7486-7495.

El-Deiry, W. S., Tokino, T., Velculescu, V. E., Levy, D. B., Parsons, R., Trent, J. M., Lin, D., Mercer, W. E., Kinzler, K. W., and Vogelstein, B. (1993). WAF1, a potential mediator of p53 tumor suppression. Cell *75*, 817-825.

Espinosa, J. M., and Emerson, B. M. (2001). Transcriptional regulation by p53 through intrinsic DNA/chromatin binding and site-directed cofactor recruitment. Mol Cell *8*, 57-69.

Fei, P., Bernhard, E. J., and El-Deiry, W. S. (2002). Tissue-specific induction of p53 targets in vivo. Cancer Res *62*, 7316-7327.

Finlay, C. A., Hinds, P. W., and Levine, A. J. (1989). The p53 proto-oncogene can act as a suppressor of transformation. Cell *57*, 1083-1093.

Flores, E. R., Tsai, K. Y., Crowley, D., Sengupta, S., Yang, A., McKeon, F., and Jacks, T. (2002). p63 and p73 are required for p53-dependent apoptosis in response to DNA damage. Nature *416*, 560-564.

Friedlander, P., Haupt, Y., Prives, C., and Oren, M. (1996). A mutant p53 that discriminates between p53-responsive genes cannot induce apoptosis. Mol Cell Biol *16*, 4961-4971.

Fu, L., Minden, M. D., and Benchimol, S. (1996). Translational regulation of human p53 gene expression. Embo J *15*, 4392-4401.

Gottlieb, T. M., Leal, J. F., Seger, R., Taya, Y., and Oren, M. (2002). Cross-talk between Akt, p53 and Mdm2: possible implications for the regulation of apoptosis. Oncogene *21*, 1299-1303.

Grossman, S. R., Deato, M. E., Brignone, C., Chan, H. M., Kung, A. L., Tagami, H., Nakatani, Y., and Livingston, D. M. (2003). Polyubiquitination of p53 by a ubiquitin ligase activity of p300. Science *300*, 342-344.

Guadagno, T. M., and Newport, J. W. (1996). Cdk2 kinase is required for entry into mitosis as a positive regulator of Cdc2-cyclin B kinase activity. Cell *84*, 73-82.

Guillouf, C., Rosselli, F., Krishnaraju, K., Moustacchi, E., Hoffman, B., and Liebermann, D. A. (1995). p53 involvement in control of G2 exit of the cell cycle: role in DNA damage-induced apoptosis. Oncogene *10*, 2263-2270.

Harper, J. W., Adami, G. R., Wei, N., Keyomarsi, K., and Elledge, S. J. (1993). The p21 Cdk-interacting protein Cip1 is a potent inhibitor of G1 cyclin-dependent kinases. Cell *75*, 805-816.

Hartwell, L. H., and Weinert, T. A. (1989). Checkpoints: controls that ensure the order of cell cycle events. Science *246*, 629-634.

Haupt, Y., Rowan, S., Shaulian, E., Vousden, K. H., and Oren, M. (1995). Induction of apoptosis in HeLa cells by trans-activation-deficient p53. Genes Dev *9*, 2170-2183.

Hermeking, H., Lengauer, C., Polyak, K., He, T. C., Zhang, L., Thiagalingam, S., Kinzler, K. W., and Vogelstein, B. (1997). 14-3-3 sigma is a p53-regulated inhibitor of G2/M progression. Mol Cell *1*, 3-11.

Herzinger, T., Funk, J. O., Hillmer, K., Eick, D., Wolf, D. A., and Kind, P. (1995). Ultraviolet B irradiation-induced G2 cell cycle arrest in human keratinocytes by inhibitory phosphorylation of the cdc2 cell cycle kinase. Oncogene *11*, 2151-2156.

Hitomi, M., Shu, J., Agarwal, M., Agarwal, A., and Stacey, D. W. (1998). p21Waf1 inhibits the activity of cyclin dependent kinase 2 by preventing its activating phosphorylation. Oncogene *17*, 959-969.

Ho, J., and Benchimol, S. (2003). Transcriptional repression mediated by the p53 tumour suppressor. Cell Death Differ *10*, 404-408.

Hoffman, W. H., Biade, S., Zilfou, J. T., Chen, J., and Murphy, M. (2002). Transcriptional repression of the anti-apoptotic survivin gene by wild type p53. J Biol Chem *277*, 3247-3257.

Jeffers, J. R., Parganas, E., Lee, Y., Yang, C., Wang, J., Brennan, J., MacLean, K. H., Han, J., Chittenden, T., Ihle, J. N., *et al.* (2003). Puma is an essential mediator of p53-dependent and -independent apoptotic pathways. Cancer Cell *4*, 321-328.

Kannan, K., Kaminski, N., Rechavi, G., Jakob-Hirsch, J., Amariglio, N., and Givol, D. (2001). DNA microarray analysis of genes involved in p53 mediated apoptosis: activation of Apaf-1. Oncogene *20*, 3449-3455.

Kastan, M. B., Zhan, Q., el-Deiry, W. S., Carrier, F., Jacks, T., Walsh, W. V., Plunkett, B. S., Vogelstein, B., and Fornace, A. J., Jr. (1992). A mammalian cell cycle checkpoint pathway utilizing p53 and GADD45 is defective in ataxia-telangiectasia. Cell *71*, 587-597.

Kubbutat, M. H., Jones, S. N., and Vousden, K. H. (1997). Regulation of p53 stability by Mdm2. Nature *387*, 299-303.

Kuerbitz, S. J., Plunkett, B. S., Walsh, W. V., and Kastan, M. B. (1992). Wild-type p53 is a cell cycle checkpoint determinant following irradiation. Proc Natl Acad Sci U S A *89*, 7491-7495.

Kurki, S., Latonen, L., and Laiho, M. (2003). Cellular stress and DNA damage invoke temporally distinct Mdm2, p53 and PML complexes and damage-specific nuclear relocalization. J Cell Sci *116*, 3917-3925.

Lane, D. P., and Crawford, L. V. (1979). T antigen is bound to a host protein in SV40-transformed cells. Nature *278*, 261-263.

LeBlanc, H., Lawrence, D., Varfolomeev, E., Totpal, K., Morlan, J., Schow, P., Fong, S., Schwall, R., Sinicropi, D., and Ashkenazi, A. (2002). Tumor-cell resistance to death receptor--induced apoptosis through mutational inactivation of the proapoptotic Bcl-2 homolog Bax. Nat Med *8*, 274-281.

Leng, R. P., Lin, Y., Ma, W., Wu, H., Lemmers, B., Chung, S., Parant, J. M., Lozano, G., Hakem, R., and Benchimol, S. (2003). Pirh2, a p53-induced ubiquitin-protein ligase, promotes p53 degradation. Cell *112*, 779-791.

Leu, J. I., Dumont, P., Hafey, M., Murphy, M. E., and George, D. L. (2004). Mitochondrial p53 activates Bak and causes disruption of a Bak-Mcl1 complex. Nat Cell Biol.

Li, H., Zhu, H., Xu, C. J., and Yuan, J. (1998). Cleavage of BID by caspase 8 mediates the mitochondrial damage in the Fas pathway of apoptosis. Cell *94*, 491-501.

Li, M., Brooks, C. L., Wu-Baer, F., Chen, D., Baer, R., and Gu, W. (2003). Mono- versus polyubiquitination: differential control of p53 fate by Mdm2. Science *302*, 1972-1975.

Linzer, D. I., and Levine, A. J. (1979). Characterization of a 54K dalton cellular SV40 tumor antigen present in SV40-transformed cells and uninfected embryonal carcinoma cells. Cell *17*, 43-52.

Lohrum, M., and Scheidtmann, K. H. (1996). Differential effects of phosphorylation of rat p53 on transactivation of promoters derived from different p53 responsive genes. Oncogene *13*, 2527-2539.

Lopez-Girona, A., Furnari, B., Mondesert, O., and Russell, P. (1999). Nuclear localization of Cdc25 is regulated by DNA damage and a 14-3-3 protein. Nature *397*, 172-175.

Louria-Hayon, I., Grossman, T., Sionov, R. V., Alsheich, O., Pandolfi, P. P., and Haupt, Y. (2003). The promyelocytic leukemia protein protects p53 from Mdm2-mediated inhibition and degradation. J Biol Chem *278*, 33134-33141.

Lowe, S. W., Bodis, S., McClatchey, A., Remington, L., Ruley, H. E., Fisher, D. E., Housman, D. E., and Jacks, T. (1994). p53 status and the efficacy of cancer therapy in vivo. Science *266*, 807-810.

Lowe, S. W., Ruley, H. E., Jacks, T., and Housman, D. E. (1993a). p53-dependent apoptosis modulates the cytotoxicity of anticancer agents. Cell *74*, 957-967.

Lowe, S. W., Schmitt, E. M., Smith, S. W., Osborne, B. A., and Jacks, T. (1993b). p53 is required for radiation-induced apoptosis in mouse thymocytes. Nature *362*, 847-849.

Lu, X., and Lane, D. P. (1993). Differential induction of transcriptionally active p53 following UV or ionizing radiation: defects in chromosome instability syndromes? Cell *75*, 765-778.

Ludwig, R. L., Bates, S., and Vousden, K. H. (1996). Differential activation of target cellular promoters by p53 mutants with impaired apoptotic function. Mol Cell Biol *16*, 4952-4960.

Luo, X., Budihardjo, I., Zou, H., Slaughter, C., and Wang, X. (1998). Bid, a Bcl2 interacting protein, mediates cytochrome c release from mitochondria in response to activation of cell surface death receptors. Cell *94*, 481-490.

MacLachlan, T. K., and El-Deiry, W. S. (2002). Apoptotic threshold is lowered by p53 transactivation of caspase-6. Proc Natl Acad Sci U S A *99*, 9492-9497.

MacLachlan, T. K., Takimoto, R., and El-Deiry, W. S. (2002). BRCA1 directs a selective p53-dependent transcriptional response towards growth arrest and DNA repair targets. Mol Cell Biol *22*, 4280-4292.

Matsuoka, S., Huang, M., and Elledge, S. J. (1998). Linkage of ATM to cell cycle regulation by the Chk2 protein kinase. Science *282*, 1893-1897.

McCurrach, M. E., Connor, T. M., Knudson, C. M., Korsmeyer, S. J., and Lowe, S. W. (1997). bax-deficiency promotes drug resistance and oncogenic transformation by attenuating p53-dependent apoptosis. Proc Natl Acad Sci U S A *94*, 2345-2349.

Mihara, M., Erster, S., Zaika, A., Petrenko, O., Chittenden, T., Pancoska, P., and Moll, U. M. (2003). p53 has a direct apoptogenic role at the mitochondria. Mol Cell *11*, 577-590.

Miyashita, T., Harigai, M., Hanada, M., and Reed, J. C. (1994a). Identification of a p53-dependent negative response element in the bcl-2 gene. Cancer Res *54*, 3131-3135.

Miyashita, T., Krajewski, S., Krajewska, M., Wang, H. G., Lin, H. K., Liebermann, D. A., Hoffman, B., and Reed, J. C. (1994b). Tumor suppressor p53 is a regulator of bcl-2 and bax gene expression in vitro and in vivo. Oncogene *9*, 1799-1805.

Moroni, M. C., Hickman, E. S., Denchi, E. L., Caprara, G., Colli, E., Cecconi, F., Muller, H., and Helin, K. (2001). Apaf-1 is a transcriptional target for E2F and p53. Nat Cell Biol *3*, 552-558.

Nakano, K., and Vousden, K. H. (2001). PUMA, a novel proapoptotic gene, is induced by p53. Mol Cell *7*, 683-694.

Nurse, P. (1990). Universal control mechanism regulating onset of M-phase. Nature *344*, 503-508.

Oda, E., Ohki, R., Murasawa, H., Nemoto, J., Shibue, T., Yamashita, T., Tokino, T., Taniguchi, T., and Tanaka, N. (2000a). Noxa, a BH3-only member of the Bcl-2 family and candidate mediator of p53-induced apoptosis. Science *288*, 1053-1058.

Oda, K., Arakawa, H., Tanaka, T., Matsuda, K., Tanikawa, C., Mori, T., Nishimori, H., Tamai, K., Tokino, T., Nakamura, Y., and Taya, Y. (2000b). p53AIP1, a potential mediator of p53-dependent apoptosis, and its regulation by Ser-46-phosphorylated p53. Cell *102*, 849-862.

Oren, M. (2003). Decision making by p53: life, death and cancer. Cell Death Differ *10*, 431-442.

Polyak, K., Xia, Y., Zweier, J. L., Kinzler, K. W., and Vogelstein, B. (1997). A model for p53-induced apoptosis. Nature *389*, 300-305.

Raman, V., Martensen, S. A., Reisman, D., Evron, E., Odenwald, W. F., Jaffee, E., Marks, J., and Sukumar, S. (2000). Compromised HOXA5 function can limit p53 expression in human breast tumours. Nature *405*, 974-978.

Rikhof, B., Corn, P. G., and El-Deiry, W. S. (2003). Caspase 10 levels are increased following DNA damage in a p53-dependent manner. Cancer Biol Ther *2*, 707-712.

Rowan, S., Ludwig, R. L., Haupt, Y., Bates, S., Lu, X., Oren, M., and Vousden, K. H. (1996). Specific loss of apoptotic but not cell-cycle arrest function in a human tumor derived p53 mutant. Embo J *15*, 827-838.

Sabbatini, P., and McCormick, F. (1999). Phosphoinositide 3-OH kinase (PI3K) and PKB/Akt delay the onset of p53-mediated, transcriptionally dependent apoptosis. J Biol Chem *274*, 24263-24269.

Samuels-Lev, Y., O'Connor, D. J., Bergamaschi, D., Trigiante, G., Hsieh, J. K., Zhong, S., Campargue, I., Naumovski, L., Crook, T., and Lu, X. (2001). ASPP proteins specifically stimulate the apoptotic function of p53. Mol Cell *8*, 781-794.

Sax, J. K., and El-Deiry, W. S. (2003). p53 downstream targets and chemosensitivity. Cell Death Differ *10*, 413-417.

Sax, J. K., Fei, P., Murphy, M. E., Bernhard, E., Korsmeyer, S. J., and El-Deiry, W. S. (2002). BID regulation by p53 contributes to chemosensitivity. Nat Cell Biol *4*, 842-849.

Schneider, E., Montenarh, M., and Wagner, P. (1998). Regulation of CAK kinase activity by p53. Oncogene *17*, 2733-2741.

Scorrano, L., Oakes, S. A., Opferman, J. T., Cheng, E. H., Sorcinelli, M. D., Pozzan, T., and Korsmeyer, S. J. (2003). BAX and BAK regulation of endoplasmic reticulum Ca2+: a control point for apoptosis. Science *300*, 135-139.

Senoo, M., Manis, J. P., Alt, F. W., and McKeon, F. (2004). p63 and p73 are not required for the development and p53-dependent apoptosis of T cells. Cancer Cell *6*, 85-89.

Shikama, N., Lee, C. W., France, S., Delavaine, L., Lyon, J., Krstic-Demonacos, M., and La Thangue, N. B. (1999). A novel cofactor for p300 that regulates the p53 response. Mol Cell *4*, 365-376.

Singh, B., Reddy, P. G., Goberdhan, A., Walsh, C., Dao, S., Ngai, I., Chou, T. C., P, O. C., Levine, A. J., Rao, P. H., and Stoffel, A. (2002). p53 regulates cell survival by inhibiting PIK3CA in squamous cell carcinomas. Genes Dev *16*, 984-993.

Slebos, R. J., Lee, M. H., Plunkett, B. S., Kessis, T. D., Williams, B. O., Jacks, T., Hedrick, L., Kastan, M. B., and Cho, K. R. (1994). p53-dependent G1 arrest involves pRB-related proteins and is disrupted by the human papillomavirus 16 E7 oncoprotein. Proc Natl Acad Sci U S A *91*, 5320-5324.

Stambolic, V., MacPherson, D., Sas, D., Lin, Y., Snow, B., Jang, Y., Benchimol, S., and Mak, T. W. (2001). Regulation of PTEN transcription by p53. Mol Cell *8*, 317-325.

Stewart, Z. A., and Pietenpol, J. A. (2001). p53 Signaling and cell cycle checkpoints. Chem Res Toxicol *14*, 243-263.

Sugars, K. L., Budhram-Mahadeo, V., Packham, G., and Latchman, D. S. (2001). A minimal Bcl-x promoter is activated by Brn-3a and repressed by p53. Nucleic Acids Res *29*, 4530-4540.

Sui, G., Affar el, B., Shi, Y., Brignone, C., Wall, N. R., Yin, P., Donohoe, M., Luke, M. P., Calvo, D., and Grossman, S. R. (2004). Yin Yang 1 is a negative regulator of p53. Cell *117*, 859-872.

Symonds, H., Krall, L., Remington, L., Saenz-Robles, M., Lowe, S., Jacks, T., and Van Dyke, T. (1994). p53-dependent apoptosis suppresses tumor growth and progression in vivo. Cell *78*, 703-711.

Takimoto, R., and El-Deiry, W. S. (2000). Wild-type p53 transactivates the KILLER/DR5 gene through an intronic sequence-specific DNA-binding site. Oncogene *19*, 1735-1743.

Tanikawa, C., Matsuda, K., Fukuda, S., Nakamura, Y., and Arakawa, H. (2003). p53RDL1 regulates p53-dependent apoptosis. Nat Cell Biol *5*, 216-223.

Taylor, W. R., and Stark, G. R. (2001). Regulation of the G2/M transition by p53. Oncogene *20*, 1803-1815.

Velculescu, V. E., and El-Deiry, W. S. (1996). Biological and clinical importance of the p53 tumor suppressor gene. Clin Chem *42*, 858-868.

Villunger, A., Michalak, E. M., Coultas, L., Mullauer, F., Bock, G., Ausserlechner, M. J., Adams, J. M., and Strasser, A. (2003). p53- and drug-induced apoptotic responses mediated by BH3-only proteins puma and noxa. Science *302*, 1036-1038.

Vogelstein, B., and Kinzler, K. W. (1992). p53 function and dysfunction. Cell *70*, 523-526.

Vogelstein, B., Lane, D., and Levine, A. J. (2000). Surfing the p53 network. Nature *408*, 307-310.

Vousden, K. H., and Lu, X. (2002). Live or let die: the cell's response to p53. Nat Rev Cancer *2*, 594-604.

Vousden, K. H., and Woude, G. F. (2000). The ins and outs of p53. Nat Cell Biol *2*, E178-180.

Waga, S., Hannon, G. J., Beach, D., and Stillman, B. (1994). The p21 inhibitor of cyclin-dependent kinases controls DNA replication by interaction with PCNA. Nature *369*, 574-578.

Wang, S., and El-Deiry, W. S. (2003a). Requirement of p53 targets in chemosensitization of colonic carcinoma to death ligand therapy. Proc Natl Acad Sci U S A *100*, 15095-15100.

Wang, S., and El-Deiry, W. S. (2003b). TRAIL and apoptosis induction by TNF-family death receptors. Oncogene *22*, 8628-8633.

Wang, S., and El-Deiry, W. S. (2004). Inducible silencing of KILLER/DR5 in vivo promotes bioluminescent colon tumor xenograft growth and confers resistance to chemotherapeutic agent 5-Fluorouracil. Cancer Res 64, 6666-6672.

Webster, G. A., and Perkins, N. D. (1999). Transcriptional cross talk between NF-kappaB and p53. Mol Cell Biol *19*, 3485-3495.

Wei, X., Yu, Z. K., Ramalingam, A., Grossman, S. R., Yu, J. H., Bloch, D. B., and Maki, C. G. (2003). Physical and functional interactions between PML and MDM2. J Biol Chem *278*, 29288-29297.

Weinert, T. A., and Hartwell, L. H. (1990). Characterization of RAD9 of Saccharomyces cerevisiae and evidence that its function acts posttranslationally in cell cycle arrest after DNA damage. Mol Cell Biol *10*, 6554-6564.

Woods, D. B., and Vousden, K. H. (2001). Regulation of p53 function. Exp Cell Res *264*, 56-66.

Wu, G. S., Burns, T. F., McDonald, E. R., 3rd, Jiang, W., Meng, R., Krantz, I. D., Kao, G., Gan, D. D., Zhou, J. Y., Muschel, R., *et al.* (1997). KILLER/DR5 is a DNA damage-inducible p53-regulated death receptor gene. Nat Genet *17*, 141-143.

Wu, G. S., and El-Deiry, W. S. (1996). Apoptotic death of tumor cells correlates with chemosensitivity, independent of p53 or bcl-2. Clin Cancer Res *2*, 623-633.

Wu, X., Bayle, J. H., Olson, D., and Levine, A. J. (1993). The p53-mdm-2 autoregulatory feedback loop. Genes Dev *7*, 1126-1132.

Xirodimas, D. P., Saville, M. K., Bourdon, J. C., Hay, R. T., and Lane, D. P. (2004). Mdm2-mediated NEDD8 conjugation of p53 inhibits its transcriptional activity. Cell *118*, 83-97.

Yin, C., Knudson, C. M., Korsmeyer, S. J., and Van Dyke, T. (1997). Bax suppresses tumorigenesis and stimulates apoptosis in vivo. Nature *385*, 637-640.

Yonish-Rouach, E., Resnitzky, D., Lotem, J., Sachs, L., Kimchi, A., and Oren, M. (1991). Wild-type p53 induces apoptosis of myeloid leukaemic cells that is inhibited by interleukin-6. Nature *352*, 345-347.

Yu, J., and Zhang, L. (2003). No PUMA, no death: implications for p53-dependent apoptosis. Cancer Cell *4*, 248-249.

Zhan, Q., Chen, I. T., Antinore, M. J., and Fornace, A. J., Jr. (1998). Tumor suppressor p53 can participate in transcriptional induction of the GADD45 promoter in the absence of direct DNA binding. Mol Cell Biol *18*, 2768-2778.

Zhou, B. P., Liao, Y., Xia, W., Zou, Y., Spohn, B., and Hung, M. C. (2001). HER-2/neu induces p53 ubiquitination via Akt-mediated MDM2 phosphorylation. Nat Cell Biol *3*, 973-982.

Zong, W. X., Li, C., Hatzivassiliou, G., Lindsten, T., Yu, Q. C., Yuan, J., and Thompson, C. B. (2003). Bax and Bak can localize to the endoplasmic reticulum to initiate apoptosis. J Cell Biol *162*, 59-69.

Chapter 7

P53 HAS A DIRECT PRO-APOPTOTIC ACTION AT THE MITOCHONDRIA

Ute M. Moll
Department of Pathology Stony Brook Univeristy, Stony Brook, NY, USA

P53 TRANSCRIPTION FUNCTION AND APOPTOTIC TARGET GENES

The basis for p53's striking apoptotic and tumor suppressive potency lies in its pleiotropism that includes transcription-dependent and –independent functions. p53 kills cells predominantly via the mitochondrial death pathway rather than the death receptor pathway (Schuler et al. 2001). p53 can mediate apoptosis by transcriptional activation of pro-apoptotic genes like the BH3-only proteins Noxa and Puma, Bax, p53 AIP1, Apaf-1, DRAL and PERP, and by transcriptional repression of Bcl2 and IAPs. For Noxa, Puma and PIDD, downregulation decreases - but does not abolish – the extent of death after stress. Of note, induction of these target gene products show variable kinetics, with some being delayed in their response (over 24 h), e.g. Bax and p53AIP1 (Attardi et al. 2000; Nakano et al. 2001). Analysis of p53-regulated global gene expression shows that the type, strength and kinetics of the target gene profiles depends on p53 levels, stress type and cell type (Zhao et al. 2000). This indicates that only individual genes will be chosen from the complex spectrum of potentially inducible genes to mediate a specific p53 response in a given physiological situation.

Often, these genes were found in screens that compared a particular tumor cell line lacking p53 with its counterpart overexpressing ectopic p53, and some form of subtractive methodology (e.g. subtractive hybridization, differential display) was used. This was the case, e.g. for Bax, PIDD and the PIG group. PERP and Noxa were also found using a subtractive approach

P. Hainaut and K.G. Wiman (eds.), 25 Years of p53 Research,
© 2005 *Springer. Printed in the Netherlands.*

but relied on endogenous p53 by comparing p53+/+ MEFs with their p53-/-
counterparts.

The candidacy of these genes for being p53 apoptosis genes is based on
the following criteria: (i) demonstrating p53-dependency of their induction
(usually accompanied by the identification of a putative p53-binding site in
their 5' UTR-regulatory region), and (ii) demonstrating that ectopic
overexpression of the candidate gene is sufficient to induce apoptosis in p53-
deficient tissue culture cells. In some cases (e.g. Noxa and PIDD),
downregulation of the endogenous gene by antisense methods decreases (but
never abolishes) cell death rate after stress.

However, most known p53 target genes are induced to similar levels
during p53-mediated G1 arrest and apoptosis (Nakano et al. 2001). This
strongly suggests that they function more generally in transducing p53 stress
signals, but that they are not the decisive death determinant in the cell's
decision fork whether to arrest or to undergo cell death. This situation holds
true for the p53 target genes Bax, IGF-BP3, Killer/DR5, the PIGs, PIDD,
and Noxa. Each of these genes is similarly induced in doxorubicin-stressed
p53+/+ MEFs, which only arrest, and in doxorubicin-stressed E1A p53+/+
MEFs, which only apoptose (Nakano et al. 2001). Notable exceptions in
apoptosis specificity is PERP and possibly, p53AIP1. PERP, a member of
the tetraspan transmembrane protein family PMP22/gas3, was cloned after
subtracting against G1 arrest-associated p53-induced messages. PERP is
specifically induced during p53-mediated apoptosis in E1A-harboring MEF
cells but not during G1 arrest (Nakano et al. 2001) Also, stress-mediated
serine 46 phosphorylation on p53 is specifically associated with induction of
p53AIP1 but not with induction of p21Wafl, Noxa and PIG3, suggesting
that p53AIP1 might be an apoptosis gene induced by this particular
modification of p53 (Komarova et al. 1997) The induction of candidate
apoptosis genes exhibit variable kinetics with some being rather slow in their
inductive response (24 h or longer), e.g. PIG3 and p53AIP1. While many of
the above-mentioned candidates are widely expressed, DRAL is an example
of a tissuerestricted response gene which is exclusively expressed in
cardiomyocytes (Lowe et al. 1993).

The apoptotic response to DNA damage in the mouse has been well
characterized. It has been clearly established that p53 activity is the prime
determinant of radiation and drug sensitivity in vivo. In radiosensitive organs
such as thymus, spleen and small intestine, activation of caspase 3 and
subsequent apoptosis after γIR and DNA-damaging drugs is largely or
completely p53-dependent (Clarke et al. 1993; Fei et al. 2002; Komarova et
al. 1997; Lowe et al. 1993; MacCallum et al. 2001). In contrast,
radioresistant organs such as liver, kidney, lung and skeletal muscle do not
activate caspase 3 and apoptosis after 5-10 Gy γIR (Clarke et al. 1993; Fei et

al. 2002; Komarova et al. 1997; Lowe et al. 1993). However, the mechanism of how p53 kills cells in vivo is clearly pleiotropic and tissue-specific. The pattern of p53-induced apoptotic target gene expression is highly selective and complex. For example, the key effectors of apoptosis in colon and the red pulp of spleen is Bid, while in thymus and the white pulp of spleen it is PUMA (Fei et al. 2002; Sax JK et al. 2002). Bid and PUMA have been shown to trigger the mitochondrial apoptotic pathway. They cause permeabilization of the outer membrane and release of activators of caspases and nucleases such as Cytochrome C, SMAC/Diablo, Omi/HtrA2, AIF and Endonuclease G.

An important question in γIR-induced cell death is the relative contribution made by individual apoptogenic p53 target gene products. Although about a dozen different p53 target genes have been characterized that can mediate apoptosis when forcibly overexpressed, it remains unclear whether any single transcriptional target is critical. Genes whose products act directly at the mitochondria like the BH3-only proteins PUMA, Noxa and Bid or the BH123 protein Bax stand out. However, with knock-out mice for PUMA, Noxa and Bax now available, it became clear that none of them has the power to phenocopy p53-/- mice (Jeffers et al. 2003; Nakano et al. 2001; Shibue et al. 2003; Villunger et al. 2003). Instead, each gene only captures an aspect of the pleiotropic p53 action at best. Moreover, each of them exhibits considerable tissue-specificity in their action. Surprisingly, thymocytes from Bax -/- knock-out mice and Noxa -/- knock-out mice, as well as thymocytes from Bax-/-, Noxa -/- double knock-out mice are capable of undergoing apoptosis after DNA damage as efficiently as wild type thymocytes (Shibue et al. 2003; Villunger et al. 2003), clearly indicating that in thymocytes Bax and Noxa are fully dispensable for DNA damage-induced death. Bax-deficient mice only develop benign B- and T-cell hyperplasias but no tumors (Nakano et al. 2001). On the other hand, Noxa does make an important contribution in gut stem cells. Noxa-/- mice are tumor-free when unchallenged, but show resistance to X-ray induced gastrointestinal death due to impaired apoptosis of the epithelial cells in the crypts of the small intestine (Shibue et al. 2003). Moreover, Noxa-/- mouse embryo fibroblasts exhibit partial resistance to E1A-induced apoptosis in response to DNA damage by adriamycin (Shibue et al. 2003)) or etoposide (Villunger et al. 2003), which is further enhanced by Bax deficiency (Shibue et al. 2003). Thus, Noxa contributes to p53-mediated death in fibroblasts and crypt intestinal cells, but plays no role in thymocytes.

PUMA-/- thymocytes fail to phenocopy p53-/- thymocytes and PUMA knock-out mice remain completely tumor-free, including the predominant Tcell lymphomas that are so typical of p53 knock-out mice (Jeffers et al. 2003; Villunger et al. 2003). Yet, PUMA is an important mediator of p53-

induced normal thymocyte death after challenge (Fei et al. 2002; Jeffers et al. 2003; Villunger et al. 2003)(our unpublished work). Ex vivo cultured PUMA -/- thymocytes exhibit a partial protection from γ-irradiation after 1.25 - 2.5 Gy, showing impaired apoptosis (Jeffers et al. 2003; Villunger et al. 2003). Of note and important for this discussion though, protection became apparent only after prolonged observation at 16-72 hrs, while little difference between wild type and PUMA-/- thymocytes existed within the first 8 hours after damage. However, as widely reported in the literature and reproduced in our own hands, γIR-induced thymocyte death in doses up to 10 Gy in vivo and in vitro is characterized by its surprising speed. In fact, thymocyte apoptosis in general is one of the fastest among mammalian cells, resulting in > 70% killing at 8-10 hrs (our unpublished work) (Clarke et al. 1993; Lowe et al. 1993). For example, we detect massive thymocyte death in the animal within the first 5 hrs after 5 Gy, in agreement with in vitro findings by Clarke et al who report 45% cultured thymocyte death at 8 hrs after 2.5 Gy and Lowe et al (Lowe et al. 1993) who report 60 % death at 10 hrs after 5 Gy and 90% death at 20 h after only 1 Gy. Intriguingly, PUMA deficiency also provides a much stronger protective effect against non-p53-mediated thymocyte death such as the PKC inhibitor staurosporine and most strikingly, the phorbol ester PMA, consistent with previous data (Villunger et al. 2003). Taken together, this data supports the notion that PUMA plays a significant role in later phases of thymocyte death. Furthermore, PUMA might be more critical for p53-independent than for p53-dependent thymocyte death. Nevertheless, with regard to transcriptional mechanisms of p53-mediated death in the thymus, PUMA clearly is the single most important target among all known p53 targets. Indeed, we find that in thymus in vivo, PUMA protein is the earliest product that is first induced at 2 hrs, while Noxa is only weakly and transiently induced at 4 hrs. This is in reasonable agreement with cultured thymocytes, where PUMA and Noxa mRNA transcripts are detected at 5 hrs and 6 hrs after γ-IR, respectively (Villunger et al. 2003).On the other hand, Bax protein appears only at 8 hrs, while Bid, Killer/DR5 and p53DinP1 fail to be induced altogether.

TRANSCRIPTION-INDEPENDENT PATHWAYS OF P53 APOPTOSIS

Evidence for transcription-independent pathways for p53- mediated apoptosis has been accumulating for years. In some cell types, p53-dependent apoptosis occurs in the absence of any gene transcription or protein synthesis (Mihara et al. 2003b; Sansome et al. 2001; Zou et al. 1999). Furthermore, inhibitors of protein phosphatases induce p53-dependent apoptosis in the absence of transcriptional activation (Flores et al. 2002). Moreover, the transcriptionally inactive p53 mutants del(1-214) and p53gln22; ser23, which fail to specifically bind to DNA or act as transcription factor of Waf1 and other target genes, act as potent inducers of apoptosis in several cell systems (Oren 2003; Sax et al. 2002). Interestingly, p53 protein from cell-free postnuclear extracts (which contain mitochondria), made from transformed fibroblasts that undergo p53-dependent apoptosis after γIR, directly mediates the activation of effector caspases. Immunodepletion of p53 protein from these extracts blocks this activity, suggesting direct protein-protein signaling from p53 to the Casp9/Casp3 activation cascade. This pathway requires caspase 8 (Okamura et al. 2001). Also, in cell-free cytoplasts, activation of cytosolic p53 can induce mitochondrial Cytochrome C release (Schuler et al. 2001). Together, these data indicate the co-existence of a transcription-independent pathway of p53-mediated apoptosis. However, the underlying mechanism of action remained unknown.

MITOCHONDRIA, PROTEIN TRANSLOCATION AND APOPTOSIS

Mitochondria are central integrators and transducers for pro-apoptotic signals, forming the nexus between the non-specific inducer phase and the final execution phase of apoptosis. This is particularly but not exclusively the case with those inducers of cell death that activate apoptosis from within the nucleus, independently of death receptor pathways on the cell surface. Such inducers include cell damage from γIR, anticancer drugs, hypoxia and growth factor withdrawal. A major reason for the central role of mitochondria is that these organelles store a host of critical apoptotic activators and effectors of cell death in their intermembranous space. These include Cytochrome C (activator of procaspase 9) (Liu et al. 1999); Smac Diablo inhibitor of cytosolic IAPs; Htra2 (inhibitor of cytosolic IAPs); apoptosis inducing factor AIF (a flavoprotein which induces chromatin condensation); endonuclease G (which degrades DNA) and fractions of cellular procaspases 2, 3 and 9. Permeabilization of the outer mitochondrial membrane (OMM), which is associated with mitochondrial dysfunction and collapse of the inner membrane gradient $\Delta\pi M$, causes the release of these pro-apoptogenic factors. This release constitutes the point of no return and triggers the execution phase of cell suicide because it directly activates the latent apoptotic machinery.

OMM permeabilization is regulated by the opposing actions of pro-and anti-apoptotic Bcl2 proteins, although the exact mechanism of how the Bcl2 family controls OMM permeability is unclear. The anti-apoptotic members, typified by Bcl2 and BclXL, constitutively reside at the OMM and mediate a critical pro-survival function by stabilizing the OMM and preventing the release of death factors. Overexpressed Bcl2 and BclXL suppress p53-dependent and -independent cell death. The pro-apoptotic members consist of the BH3-only class which regulate the protective Bcl2/XL proteins and the multidomain BH123 class. The type II BH3-only proteins Noxa, Puma, Bik, Bim and Bad couple death signals to mitochondria and in healthy cells are sequestered to cytosolic sites other than the OMM. Upon sensing death stimuli, BH3-only proteins undergo posttranslational modifications and mitochondrial translocation (Huang et al. 2000). Translocated BH3-only proteins then bind to Bcl2/XL via their BH3 domain, thereby inactivating their protective function (Cheng et al. 2001). In resting cells, BH123 proteins exist as inactive monomers in the cytosol (Bax) or at mitochondria (Bak) (Wolter et al. 1997) and can be induced to homo-oligomerize and insert into the OMM by tBid after death stimuli, leading to Cytochrome C release (Eskes et al. 2000; Wei et al. 2000). BH3-only proteins are upstream of BH123 proteins since Bax/Bak double-null cells are resistant to Bim and Bad-induced apoptosis (Zong et al. 2001).

One striking feature of apoptosis signaling is protein translocation of signalling and effector molecules between three major cellular compartments. This includes translocation to and from mitochondria, the cytoplasm and the nucleus. Of particular interest are a growing list of proapoptotic proteins that undergo translocation to mitochondria, where they exert their pro-apoptotic functions by inducing organellar dysfunction. The classic pro-apoptotic bcl family members Bax (Chen et al. 1996; Haupt et al. 1995), Bad, Bim (Regula et al. 2001) and truncated tBid (Chao et al. 2000; Kokontis et al. 2001) do this by increasing mitochondrial permeability which, among others, leads to the release of Cytochrome C. For example, inactive cytosolic Bid undergoes N-terminal cleavage by caspase 8 to tBid, leading to a newly exposed glycine residue which now becomes the target of posttranslational N-myristoylation (Chao et al. 2000; Jimenez et al. 2000). This modification promotes its mitochondrial targeting and pro-apoptotic effect. Moreover, ionizing IR induces translocation of the stress kinase SAPK/JNK to mitochondria which in turn causes phosphorylation and inactivation of anti-apoptotic Bcl-xL (Matas et al. 2001). Similarly, phorbol esters induce translocation of protein kinase C delta to mitochondria, altering a yet unknown substrate (Baptiste N. et al. 2002; Chao et al. 2000). The evolving theme here is that these mitochondrially translocating proteins belong to unrelated biochemical classes of molecules, most of which were previously not associated with mitochondrial functions. Most unexpectedly, this diversity is now even extending to transcription factors.

P53 AND MITOCHONDRIA

We searched for a basis of transcription-independent p53-mediated cell death. The key observation was the finding that in response to a death stimulus such as γIP, DNA damaging drugs or hypoxia, a fraction of the stabilized p53 rapidly translocates to mitochondria in primary, immortal and transformed cultured cells (Marchenko et al, 2000; Sansome et al, 2001; Mihara et al, 2003). To study the functional consequences of this phenomenon, we forcibly targeted exogenous p53 to mitochondria in p53-null cancer cells, and showed that mitochondrial p53 was sufficient to launch apoptosis and suppress colony formation in a transcription-independent fashion, (Mihara et al, 2003). Moreover, in response to death stimuli, endogenous mitochondrial p53 forms inhibitory complexes with endogenous anti-apoptotic BclXL and Bcl2 proteins. Purified wild type p53 protein induces oligomerization of Bak, permeabilization of the outer mitochondrial membrane and strongly promotes Cytochrome C release *in vitro*. Using computational and genetic approaches, we determined that the p53 DNA binding domain is involved in the p53-BclXL complex formation. Conversely, tumor-derived transactivation-deficient missense mutants concomitantly lose or compromise their ability to interact with BclXL and to promote cytochrome C release. Thus, tumor-derived p53 mutations may represent "double hits", eliminating the transcriptional as well as the direct mitochondrial functions of p53 (Mihara et al, 2003). Based on these tissue culture studies, we proposed that mitochondrial translocation of p53 triggers a rapid proapoptotic response that "jump-starts" and amplifies the slower transcriptionbased response, which requires a certain ramping time typically in the range of 4-8 hrs after p53 induction. The description that follows will detail these findings further.

A fraction of stress-induced endogenous p53 protein rapidly translocates to mitochondria in response to death stimuli in wtp53 harboring human and mouse malignant and non-malignant cell lines and primary cells (Mihara et al, 2003). Based on the average ratio between the total mitochondrial and nuclear volume, we estimate that the concentration of induced mitochondrial p53 is roughly equimolar to induced nuclear p53. This result was obtained using all methods available for determining subcellular localization including carefully controlled subcellular fractionation p53 translocation strictly occurs at the onset of p53-dependent apoptosis but not during p53-independent apoptosis triggered e. g. by the death receptor pathway via TNF-α nor does it occur during p53-mediated cell cycle arrest. The translocation of p53 to mitochondria is rapid (within 1 h after cell damage) and precedes changes in mitochondrial membrane potential, Cytochrome C release and procaspase-3 activation. Mitochondrial localization of endogenous p53 can be visualized by in situ immunofluorescence of whole cells after 5-6 h of hypoxic stress. Immuno-flow cytometry analysis of isolated mitochondria show that a significant amount of mitochondrial p53 localizes to the membranous compartment (Mihara et al, 2003). This result was confirmed by direct localization of p53 via immuno-electronmicroscopy of stressed mitochondria while untreated mitochondria failed to give a signal. Further suborganellar localization of p53 by limited trypsin digestion suggests that a significant amount of mitochondrial p53 is located at the surface of the organelle while a small subfraction appears to be intraorganellar trypsin-resistant. Mitochondrial association of p53 and its significant surface localization can be reproduced in vitro in an organellar pull-down assay with purified baculoviral p53 added to isolated mitochondria. Control recombinant PCNA protein fails to associate. Coimmunoprecipitation from stressed whole cells or mitochondria shows that mitochondrial p53 is found in an in vivo complex with the mitochondrial import motor mt hsp70. This p53 subpopulation possibly corresponds to the trypsin-resistant subfraction (Mihara et al, 2003). A similar in vivo complex between mt hsp 70 (also called mot- 2) and p53 was independently observed in NIH 3T3 cells. Interestingly, forced overexpression of ectopic mot-2 abolished the transcriptional ability of p53 in reporter assays and nuclear translocation suggesting that p53 can be completely siphoned off into mitochondria (Wadhwa et al, 1998).

Similar results were seen in clean ectopic systems in the absence of additional DNA damage. In p53-deficient EB cells which harbor a stable inducible wild-type p53 transgene mitochondrial p53 accumulation occurs concomitantly with nuclear p53 accumulation and precedes the onset of apoptosis after sole induction of ectopic p53 (Mihara et al, 2003). Recently we confirmed the participation of the direct mitochondrial p53 pathway in the apoptotic response in vivo (Erster et al, 2004). We find that in mice subjected to γIR or intravenous etoposide, radiosensitive target organs also undergo mitochondrial p53 translocation that can trigger a rapid first wave of caspase 3 activation and cell death, followed by a slower transcription-dependent p53 death wave. This is the first evidence that mitochondrial p53 indeed contributes to the physiological apoptotic response in the animal. This direct mitochondrial p53 death program jump-starts and amplifies the slower transcription-based p53 response and might be one of the distinguishing features between radiosensitive and radioresistant organs. It also suggests that the mitochondrial pathway may participate in tumor suppression in vivo.

An independent confirmation of mitochondrial p53 translocation during the apoptotic response was reported by M. Murphy's group (Dumont et al, 2003). The codon 72 polymorphic variants of human p53 -Arg72p53 and

Pro73p53 - have markedly different apoptotic potential. This paper links a polymorphism of wt p53 - which epidemiologically is associated with differential tumor susceptibility in several tumor types - to differential killing ability of the two variants. Critically, this functional difference was tightly and exclusively linked to the ability to translocate to mitochondria upon stress, while no difference was found in the variants' transactivation and transrepression ability or in their protein-protein interactions. This work confirms the translocation phenomenon and assigns further medical impact to its biology.

Importantly nuclear bypass experiments demonstrate that mitochondrial p53 localization is sufficient for launching p53-dependent apoptosis from the level of mitochondria. Deliberate targeting of ectopic p53 to mitochondria via fusions with either mitochondrial import leader peptide (designated Lp53) or the transmembrane domain of Bcl2 bypasses the nucleus and is sufficient to induce apoptosis and long-term colony suppression in variousp53-deficient tumor cell lines (Saos-2, H1299 and HeLa) (Mihara et al 2003 and unpublished observation). Targeting a control transcription factor - cRel - in the same way causes only a background rate of death. Mitochondrially targeted wtp53 lacks transactivation activity, as shown in sensitive reporter assays, compared to empty vector and nuclear wtp53. Interestingly, a mitochondrially targeted but truncated p53 protein which misses the entire C-terminus has a higher apoptotic activity than Lwtp53, indicating that the C-terminus containing the tetramerization domain is dispensable for mitochondrial action This is in contrast to the nuclear action of p53 which requires the C-terminus for tetramerization in order to optimally function as a transcription factor.

Mitochondrial regulators of apoptosis influence the induction of mitochondrial p53 accumulation. Overexpression of anti-apoptotic Bcl-2 or Bcl-xL abrogates stress-signal-mediated mitochondrial p53 accumulation *and* apoptosis, but does not abrogate total cellular p53 accumulation nor the ability to undergo stress-induced cell cycle arrest. Since this supports a functional link with protective Bcl members, we looked for and found physical complexes with p53. Endogenous p53/BclXL and p53/Bcl-2 complexes can be co-immunoprecipitated in both directions from mitochondria of stressed cells. The same complexes can also be precipitated with targeted *and* nuclear p53 in transfections and in vitro in GST-pull down assays. Of note, p53-BclXL/2 complexes are specific since p53-Bax complexes were not detectable despite many attempts (Mihara et al 2003). Next, we used protein modeling to determine the structural basis for the p53-BclXL complex and to predict interaction sites. Using the known crystallographic structures of the p53 DNA binding domain and BclXL, the model predicts that the contact surfaces involve the p53 DNA-binding domain, interacting with the alpha1/BH4 and the alpha2/BH3 domain of BclXL (Mihara et al 2003) (Figure 1). The structure of this complex was recently experimentally confirmed with NMR spectroscopy by S. Fesik's group (Petros et al, 2004). Of note, the interaction surface on p53 with BclXL indeed was determined to involve the same region as p53 uses to contact DNA. The Kd was estimated to be greater than 1 uM. Interaction of p53 with BclXL was blocked by the binding of a 25-residue peptide derived from Bad, which has a very high affinity (0.6 nM) to BclXL. On the other hand, it is predicted that Bak- and Bax-derived peptides, which bind much weaker to BclXL (480 nM and 13, 000 nM), may not be able to block p53 from binding to BclXL. This suggests a possible regulatory level of p53 binding by select but not all BH3 proteins (Petros et al, 2004).

This complex also implies that the p53 DNA binding domain is a dual function domain, mediating both the transactivation function and the mitochondrial apoptotic function. If that is so, do tumor-derived p53 mutations concomitantly loose the ability to interact with BclXL? To address this question, we looked at 4 randomly chosen breast cancer lines. They represent the classic structural and contact mutation hotspots H175 and H273, as well as Lys280 and Phe194. Of note, proportional to their abnormally stabilized p53, all 4 lines constitutively harbor mutant p53 at the mitochondria, in sharp contrast to wtp53 cells, whose mitochondrial p53 translocation depends on induction by a death stimulus. However, quantitative co-immunoprecipitation from comparable amounts of mitochondrial p53 showed that 3 of the 4 lines reproducibly showed no endogenous p53-BclXL complex, regardless of whether cells were subjected to prior DNA damage or not. The fourth line had greatly reduced complex, in stark contrast to Camptothecin-treated wild type cells. The BclXL levels of mutant and wild type cells were comparable. This data is very compelling and argues for an *in vivo* role of mitochondrial p53 in human tumors. Tumorigenic mutations that are selected during tumor formation might represent 'double-hit' mutations by abrogating both apoptotic activities of p53 (Mihara et al, 2003).

Purified p53 protein, when added *in vitro* to unstressed mitochondria freshly isolated from mouse liver, is sufficient to trigger the rapid release of Cytochrome C. In contrast, a C-terminal p53 fragment or bovine serum albumin yielded only minimal background levels of release. Moreover, this is a fast process. Within 5 minutes after adding p53, 69% of the total Cyto C release has already occurred and release is complete within 30 min (Mihara et al, 2003). What is the mechanism by which p53 triggers Cytochrome C release? There is a direct link between p53/BclXL complex formation and Cytochrome C release, since pellets of mitochondria *post* Cytochrome C release contain p53/BclXL complexes. Furthermore, since p53 is a direct inducer of Cytochrome C release and mitochondrial dysfunction, p53 might also indirectly activate the ultimate effectors for Cytochrome C release –Bak and Bax proteins – by inducing their conformational change and intramembraneous oligomerization. Indeed, using a chemial cross-linker, purified wtp53, when added to freshly isolated nascent mitochondria, readily induced cross-linked multimers of endogenous Bak (mouse liver mitochondria are poor in Bax). This p53 behaviour is similar to the prototypical BH3-only protein tBID that translocates to mitochondria upon a death stimulus to induce Bak oligomerization. Moreover, since overexpression of BclXL and Bcl2 completely block p53-dependent apoptosis in vivo, it predicts that excess BclXL also prevents p53-induced Cytochrome C release in vitro. To address this, we performed BclXL competition experiments. Indeed, the effectiveness of p53 in inducing Cytochrome C release is inhibited in a dose-dependent fashion by increasing amounts of excess GST-BclXL, indicating that the two proteins are in a rheostat-like relationship (Mihara et al, 2003).

Functional parallels between p53 and other pro-apoptotic proteins appear to be emerging (Figure 2). p53, Noxa, Puma, Bim, and Bad all rapidly translocate to mitochondria upon a death stimulus, bind and inhibit the protective BclXL and Bcl2 proteins and induce OMM permeabilization, leading to Cytochrome C release and caspase activation. For both p53 and the BH3 proteins, complex formation with BclXL/2 is critical for apoptogenicity, since interference with binding blocks their mitochondrial killing activity (our data; Cheng et al.,2001; Oda et al.,2000; Nakano et al.,2001; Yu et al.,2001). Other pro-apoptotic proteins such as Siva-1 and p53AIP1, which like p53 lack a BH3-domain, also bind and inactivate BclXL (Xue et al., 2002; Matsuda et al., 2002). While BH3-only proteins bind BclXL/2 through their BH3 domain, SIVA-1 binds through a unique amphipathic helix (Xue et al.,2002) and p53 binds through its central core effector domain (our data). Our finding of the ability of excess BclXL to block p53-mediated Cytochrome C release in vitro parallels that seen with Bax and tBid (Jurgensmeier et al., 1998; Desagher et al., 1999). Also, Puma and Siva-1 mediated apoptosis is suppressed by excess BclXL (Nakano et al., 2001; Xue et al., 2002).

The most persuasive evidence for the physiologic importance of the mitochondrial p53 pathway, however, would come from the behavior of mutant p53 proteins. Is there a direct functional link between the failure of p53 mutants to form BclXL complexes and a failure to release Cytochrome C in vitro? Indeed, in contrast to wtp53 that readily induces release, tumorderived mutant p53 proteins are defective in permeabilizing the outer membrane. The hotspot mutant R175H has completely lost the ability to release Cytochrome C, while the hotspot mutant R273H is severely suppressed in this activity. Six additional tumor-derived mutant proteins also have lost the ability to release Cytochrome C (Erster & Moll, unpublished).

Thus, this direct link supports the hypothesis that tumors select against both the nuclear and mitochondrial apoptotic function of p53. Moreover, it shows that the interaction with BclXL is required for p53-mediated Cytochrome C release (Mihara et al, 2003). Together, this demonstrates that p53 itself is able to directly permeabilize the OMM and trigger the release of pro-apoptotic activators. It does so by engaging in inhibitory complexes with protective Bcl2/XL, thereby indirectly activating pro-apoptotic BH123 proteins like Bak (Mihara et al, 2003a).

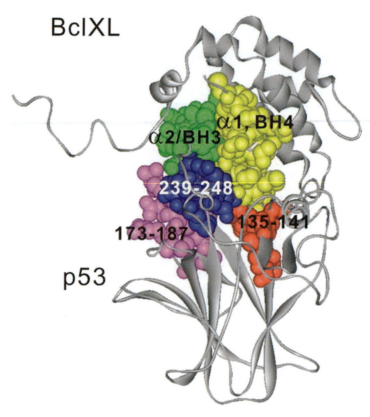

Figure 1. Structure/function analysis of the p53-BclXL complex. Predicted model of the complex (Mihara et al, 2003). The p53 DNA binding domain contacts BclXL. The protruding blue region of human p53 spanning residues 239-248, flanked by regions 135-141(red) and 173-187 (magenta), interact with a groove formed by the a1/BH4 and part of the a2/BH3 domains of human BclXL. In its essential elements, the structure of this complex was recently experimentally confirmed with NMR spectroscopy (Petros et al. 2004).

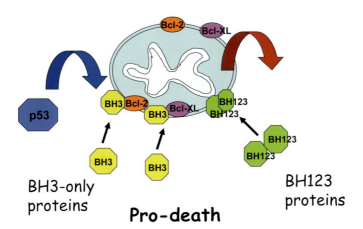

Figure 2. Functional parallels between p53 and other pro-apoptotic proteins. p53, Noxa, Puma, Bim, and Bad all rapidly translocate to mitochondria upon a death stimulus, bind and inhibit the protective BclXL and Bcl2 proteins and induce OMM permeabilization, leading to Cytochrome C release and caspase activation. For both p53 and the BH3 proteins, complex formation with BclXL/2 is critical for apoptogenicity, since interference with binding blocks their mitochondrial killing activity.

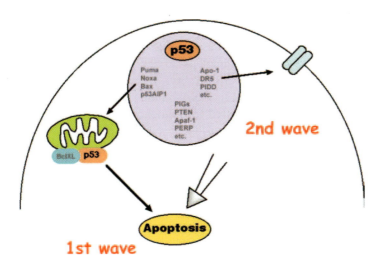

Figure 3. p53 exerts a rapid and direct pro-apoptogenic role at the mitochondria, thereby jump-starting and amplifying the transcription-based apoptotic action of p53.

Based on these results, we propose a model in which p53 can contribute to cell death by rapid direct signaling at the mitochondria. This pathway likely acts in synergy with the transcription-dependent mode of p53, thereby amplifying the apoptotic potency and speed of p53. Moreover, based on its implication in a broad spectrum of cell types and death signals, this enhancer pathway has the potential for being generic, accompanying the action of most or all p53-induced apoptotic genes (Fig. 3). Important questions for the future are: which are the protein(s) that mitochondrial p53 is talking to and what secondary apoptotic regulators are being activated as a result of this.

P53 AND BEYOND

Does mitochondrial accumulation of p53 represent a precedent for other transcription factors implicated in apoptosis? The answer is yes. The nuclear orphan steroid receptor TR3 (also called Nur77), a member of the steroid/thyroid receptor superfamily is a bona fide transcription factor with a zinc finger DNA-binding domain flanked by transactivation domains and a binding domain for an as yet unknown ligand. TR3 is induced and acts as a transcription factor in response to epidermal growth factor and all-transretinoic acid. On the other hand, TR3 mediates apoptosis in diverent cell types in vivo, e.g. in neurons, autoreactive developing T cells and human cancer cells (Villunger et al. 2003) and references within). TR3 is upregulated by apoptotic stimuli like seizures, neuronal ischemia and ligation of the T cell receptor. Unexpectedly, when TR3 works as an apoptotic factor, its transcriptional activation function is turned off (Villunger et al. 2003). Instead, in response to a wide variety of apoptotic stimuli, TR3 relocates from the nucleus to the surface of mitochondria, where it triggers membrane permeability, Cytochrome C release and apoptotic cell death. The TR3 DNA-binding domain, required for transcriptional activity, is not required for mitochondrial targeting. As is true for mitochondrial p53, mitochondrial TR3 is also sufficient to cause cell death. Moreover, as for mitochondrial p53, its action is blocked by bcl-2 (Villunger et al. 2003). Recently, the mechanism of TR3-mediated mitochondrial action was elucidated, and again it follows the p53 lead. TR3 binds to Bcl2 and this binding induces a Bcl-2 conformational change that exposes its BH3 domain, resulting in conversion of Bcl-2 from a protector to a killer (Lin et al. 2004). Thus, there are now two nuclear transcription factors that, by virtue of their subcellular relocalization, are capable of mediating lethal signaling directly through mitochondria. For one thing, these two examples add another level of complexity to the regulation of apoptosis. However, their greater significance might lie in the fact that it forces us to rethink our neat biochemical classification scheme of one protein - one mode of action. p53 stands as a prototype of a molecule that draws its power from fulfilling pleiotropic functions by biochemical flexibility. As a jack of more than one trade, it is both a transcription factor and a membrane permeability regulator.

ACKNOWLEDGMENTS

This work was supported by grants from the National Cancer Institute and the American Cancer Society".

REFERENCES

Attardi L.D., Reczek E.E., Cosmas C., Demicco E.G., McCurrach M.E., Lowe S.W., Jacks T. PERP, an apoptosis-associated target of p53, is a novel member of the PMP-22/gas3 family.Genes Dev. 2000; 14:704-18.

Baptiste N., Friedlander P, Chen X, C. P. The proline-rich domain of p53 is required for cooperation with anti-neoplastic agents to promote apoptosis of tumor cells.Oncogene. 2002; 21:9-21.

Caelles C., Helmberg A., Karin M. p53-dependent apoptosis in the absence of transcriptional activation of p53-target genes.Nature. 1994; 370:220-3.

Chao C., Saito S., Kang J., Anderson C.W., Appella E., Xu Y. p53 transcriptional activity is essential for p53-dependent apoptosis following DNA damage.Embo J. 2000; 19:4967-4975.

Chen X., Ko L.J., Jayaraman L., Prives C. p53 levels, functional domains, and DNA damage determine the extent of the apoptotic response of tumor cells.Genes Dev. 1996; 10:2438-2451.

Cheng E.H., Wei M.C., Weiler S., Flavell R.A., Mak T.W., Lindsten T., Korsmeyer S.J. BCL-2, BCL-X(L) sequester BH3 domain-only molecules preventing BAX- and BAK-mediated mitochondrial apoptosis.Mol Cell. 2001; 8:705-711.

Clarke A.R., Purdie C.A., Harrison D.J., Morris R.G., Bird C.C., Hooper M.L., Wyllie A.H. Thymocyte apoptosis induced by p53-dependent and independent pathways.Nature. 1993; 362:849-52.

Desagher S., Osen-Sand A., Nichols A., Eskes R., Montessuit S., Lauper S., Maundrell K., Antonsson B., Martinou J.C. Bid-induced conformational change of Bax is responsible for mitochondrial cytochrome c release during apoptosis.J Cell Biol. 1999; 144:891-901.

Dumont P., Leu J.I., Della Pietra A.C.3., George D.L., Murphy M. The codon 72 polymorphic variants of p53 have markedly different apoptotic potential.Nat Genet. 2003; 33:357-65.

Eskes R., Desagher S., Antonsson B., Martinou J.C. Bid induces the oligomerization and insertion of Bax into the outer mitochondrial membrane.Mol Cell Biol. 2000; 20:929-935.

Fei P., Bernhard E.J., El-Deiry W.S. Tissue-specific induction of p53 targets in vivo.Cancer Res. 2002; 62:7316-27.

Flores E.R., Tsai K.Y., Crowley D., Sengupta S., Yang A., McKeon F., Jacks T. p63 and p73 are required for p53-dependent apoptosis in response to DNA damage.Nature. 2002; 416:560-4.

Haupt Y., Rowan S., Shaulian E., Vousden K.H., Oren M. Induction of apoptosis in HeLa cells by trans-activation-deficient p53.Genes Dev. 1995; 9:2170-2183.

Huang D.C., Strasser A. BH3-Only proteins-essential initiators of apoptotic cell death.Cell. 2000; 103:839-842.

Jeffers J.R., Parganas E., Lee Y., Yang C., Wang J., Brennan J., MacLean K.H., Han J., Chittenden T., Ihle J.N., McKinnon P.J., Cleveland J.L., Zambetti G.P. Puma is an essential mediator of p53-dependent and -independent apoptotic pathways.Cancer Cell. 2003; 4:321-8.

Jimenez G.S., Nister M., Stommel J.M., Beeche M., Barcarse E.A., Zhang X.Q., O'Gorman S., Wahl G.M. A transactivation-deficient mouse model provides insights into Trp53 regulation and function.Nat Genet. 2000; 26:37-43.

Jurgensmeier J.M., Xie Z., Deveraux Q., Ellerby L., Bredesen D., Reed J.C. Bax directly induces release of cytochrome c from isolated mitochondria.Proc Natl Acad Sci U S A. 1998; 95:4997-5002.

Kokontis J.M., Wagner A.J., O'Leary M., Liao S., Hay N. A transcriptional activation function of p53 is dispensable for and inhibitory of its apoptotic function.Oncogene. 2001; 20:659-68.

Komarova E.A., Chernov M.V., Franks R., Wang K., Armin G., Zelnick C.R., Chin D.M., Bacus S.S., Stark G.R., Gudkov A.V. Transgenic mice with p53-responsive lacZ: p53 activity varies dramatically during normal development and determines radiation and drug sensitivity in vivo.EMBO J. 1997; 16:1391-400.

Lin B., Kolluri S.K., Lin F., Liu W., Han Y.H., Cao X., Dawson M.I., Reed J.C., Zhang X.K. Conversion of Bcl-2 from protector to killer by interaction with nuclear orphan receptor Nur77/TR3.Cell. 2004; 116:527-540.

Liu L., Scolnick D.M., Trievel R.C., Zhang H.B., Marmorstein R., Halazonetis T.D., Berger S.L. p53 sites acetylated in vitro by PCAF and p300 are acetylated in vivo in response to DNA damage.Mol Cell Biol. 1999; 19:1202-9.

Lowe S.W., Schmitt E.M., Smith S.W., Osborne B.A., Jacks T. p53 is required for radiation-induced apoptosis in mouse thymocytes.Nature. 1993; 362:847-9.

MacCallum D.E., Hall P.A., Wright E.G. The Trp53 pathway is induced in vivo by low doses of gamma radiation.Radiat Res. 2001; 156:324-7.

Matas D., Sigal A., Stambolsky P., Milyavsky M., Weisz L., Schwartz D., Goldfinger N., Rotter V. Integrity of the N-terminal transcription domain of p53 is required for mutant p53 interference with drug-induced apoptosis.Embo J. 2001; 20:4163-4172.

Mihara M., Erster S., Zaika A., Petrenko O., Chittenden T., Pancoska P., Moll U.M. p53 has a direct apoptogenic role at the mitochondria.Mol Cell. 2003a; 11:577-90.

Mihara M., Moll U.M. Detection of mitochondrial localization of p53.Methods Mol Biol. 2003b; 234:203-9.

Nakano K., Vousden K.H. PUMA, a novel proapoptotic gene, is induced by p53.Mol Cell. 2001; 7:683-94.

Oda E., Ohki R, Murasawa H, Nemoto J, Shibue T, Yamashita T, Tokino T, Taniguchi T, N. T. Noxa, a BH3-only member of the Bcl-2 family and candidate mediator of p53-induced apoptosis.Science. 2000; 288:1053-8.

Okamura S.L., Arakawa H., Tanaka T., Nakanishi H., Ng C.C., Taya Y., Monden M., Nakamura Y. p53DINP1, a p53-inducible gene, regulates p53-dependent apoptosis.Mol Cell. 2001; 8:85-94.

Oren M. Decision making by p53: life, death and cancer.Cell Death Differ. 2003; 10:431-42.

Petros A.M., Gunasekera A., Xu N., Olejniczak E.T., Fesik S.W. Defining the p53 DNA-binding domain/Bxl-xL binding interface using NMR.FEBS Letters. 2004; 28073:1-4.

Regula K.M., Kirshenbaum L.A. p53 activates the mitochondrial death pathway and apoptosis of ventricular myocytes independent of de novo gene transcription.J Mol Cell Cardiol. 2001; 33:1435-45.

Sansome C., Zaika A., Marchenko N.D., Moll U.M. Hypoxia death stimulus induces translocation of p53 protein to mitochondria. Detection by immunofluorescence on whole cells.FEBS Lett. 2001; 488:110-5.

Sax JK, Fei P, Murphy ME, Bernhard E, Korsmeyer SJ, WS. E.-D. BID regulation by p53 contributes to chemosensitivity.Nat Cell Biol. 2002; 4:842-9.

Schuler M., Green D.R. Mechanisms of p53-dependent apoptosis.Biochem Soc Trans. 2001; 29:684-8.

Shibue T., Takeda K., Oda E., Tanaka H., Murasawa H., Takaoka A., Morishita Y., Akira S., Taniguchi T., Tanaka N. Integral role of Noxa in p53-mediated apoptotic response.Genes Dev. 2003; 17:2233-8.

Villunger A., Michalak E.M., Coultas L., Mullauer F., Bock G., Ausserlechner M.J., Adams J.M., Strasser A. p53- and Drug-Induced Apoptotic Responses Mediated by BH3-Only Proteins Puma and Noxa.Science. 2003; [Epub ahead of print]-

Wagner AJ, Kokontis JM, N. H. Myc-mediated apoptosis requires wild-type p53 in a manner independent of cell cycle arrest and the ability of p53 to induce p21wafl/cip1.Genes Dev. 1994; 8:2817-30.

Wei M.C., Lindsten T., Mootha V.K., Weiler S., Gross A., Ashiya M., Thompson C.B., Korsmeyer S.J. tBID, a membrane-targeted death ligand, oligomerizes BAK to release cytochrome c.Genes Dev. 2000; 14:2060-2071.

Wolter K.G., Wolter K.G., Hsu Y.T., Smith C.L., Nechushtan A., Xi X.G., Youle R.J. Movement of Bax from the cytosol to mitochondria during apoptosis.J Cell Biol. 1997; 139:1281-1292.

Xue L., Chu F., Cheng Y., Sun X., Borthakur A., Ramarao M., Pandey P., Wu M., Schlossman S.F., Prasad K.V. Siva-1 binds to and inhibits BCL-X(L)-mediated protection against UV radiation-induced apoptosis.Proc Natl Acad Sci U S A. 2002; 99:6925-6930.

Yu J, Zhang L, Hwang PM, Kinzler KW, B. V. PUMA induces the rapid apoptosis of colorectal cancer cells.Mol Cell. 2001; 7:673-82.

Zhao R., Gish K., Murphy M., Yin Y., Notterman D., Hoffman W.H., Tom E., Mack D.H., Levine A.J. Analysis of p53-regulated gene expression patterns using oligonucleotide arrays.Genes Dev. 2000; 14:981-993.

Zong W.X., Lindsten T., Ross A.J., MacGregor G.R., Thompson C.B. BH3-only proteins that bind pro-survival Bcl-2 family members fail to induce apoptosis in the absence of Bax and Bak.Genes Dev. 2001; 15:1481-1486.

Zou H., Li Y., Liu X., Wang X. An APAF-1.cytochrome c multimeric complex is a functional apoptosome that activates procaspase-9.J Biol Chem. 1999; 274:11549-56.

Chapter 8

MANIPULATING THE P53 GENE IN THE MOUSE: ORGANISMAL FUNCTIONS OF A PROTOTYPE TUMOR SUPPRESSOR

Lawrence A. Donehower[1], Dora Bocangel[1], Melissa Dumble[1] and Guillermina Lozano[2]
[1]Baylor College of Medicine, Department of Molecular Virology and Microbiology, One Baylor Plaza, Houston, TX, USA; [2]Department of Molecular Genetics, Section of Cancer Genetics, The University of Texas M. D. Anderson Cancer Center, Houston, TX, USA

INTRODUCTION

The early discoveries elucidating p53 function were based on cell culture experiments. Most of our fundamental knowledge of the role of p53 in cell signaling, stress response, cell cycle control, and apoptosis are a result of these *in vitro* studies (Giaccia and Kastan, 1998; Ko and Prives, 1996; Levine, 1997; Vogelstein et al., 2000). However, a greater depth of understanding was facilitated by the advent first of transgenic mouse methodologies and then by embryonic stem (ES) cell-based genetic manipulations. The sequencing of the mouse genome (www.ensembl.org and www.myscience.appliedbiosystems.com) has greatly simplified and accelerated the generation of null alleles. Methods have been developed to generate single nucleotide substitutions in the germline of mice, and importantly, to generate somatic mutations in genes to study somatic inactivation as occurs in most human cancers. The availability of whole genome analysis at the RNA expression level (arrays) and at the genomic level (array CGH) provides another level of analysis that is sure to provide insights into the molecular changes that lead to the initiation, progression, and maintenance of the tumor phenotype.

P. Hainaut and K.G. Wiman (eds.), 25 Years of p53 Research, 183-207.

The ability to add, alter, and subtract *p53* genes in the context of a whole mammalian organism has significantly complemented and enriched our knowledge of the biological functions of p53 in tumor suppression, development, and aging. For example, many *p53* point mutations associated with human tumors contribute to transformation of fibroblasts in culture, illustrating the oncogenic potential of these types of mutations (Eliyahu et al., 1984; Finlay et al., 1989; Hinds et al., 1989). Such *in vitro* transformation assays tend to be binary in nature. A particular mutant either transforms or it does not. However, when *p53* mutations are introduced into the germline of mice, tumor phenotypes can vary significantly with each mutation, revealing new subtleties in an organismal context, such as an altered tumor spectrum, altered tumor pathology, more rapid tumor incidence, or increased metastatic potential. Thus, the mouse, since it mimics the development of human cancers, has become the *in vivo* system of choice to study the process of tumorigenesis. In this review, we will describe some of the major genetically engineered mice that have alterations in *p53* copy number or structure, and discuss some of the important insights that have been obtained from these mice. This review is not intended to be encyclopedic, but illustrative in nature, with emphases placed on several models that have been particularly informative in adding to or changing the p53 paradigm.

TRANSGENIC MICE WITH MUTANT P53 TRANSGENES

The first genetically engineered *p53* mutant mice were reported by the Bernstein laboratory in 1989 (Lavigueur et al., 1989). Two lines of mutant transgenic mice bearing multiple genomic copies of *p53* under the control of its own promoter were generated. The *p53* transgenes, which encoded the *p53* point mutants p53Pro193 or p53Val135, were expressed in many tissues, including the lung, spleen, lymph nodes, and thymus. Moreover, the levels of the transgene-derived p53 protein was as high as 50-fold that of endogenous wild type p53 in some tissues (Lavigueur et al., 1989). Despite the massive overexpression of mutant p53, these transgenic mice developed tumors at a relatively modest incidence of about 20-30% by 18 months of age. Multiple tumor types were reported, though lung adenocarcinomas, osteosarcomas, and lymphomas were particularly frequent. This was the first direct evidence that point mutants of *p53* are sufficient to induce tumor development *in vivo*. The tumorigenic effect was hypothesized to be due to the mutant p53 acting in a dominant-negative manner to inhibit the endogenous wild-type p53 protein. Subsequent support for this hypothesis was provided by the observation that the *p53Val135* transgene accelerated tumors only in the presence of endogenous wild type *p53*, but not in a homozygous null *p53* background (Harvey et al., 1995b). The lack of a pro-tumorigenic effect for this mutant in the absence of wild type *p53* does rule out an oncogenic gain-of-function effect as has been suggested for some p53 point mutants from *in vitro* transformation assays (Cadwell and Zambetti, 2001; Dittmer et al., 1993).

Since the description of these first *p53* transgenic models, a series of other mutant *p53* transgenic lines have been reported. Generally, while the majority of these models show augmented tumorigenesis, the degree of tumor promotion can be quite variable and is sometimes dependent on the presence of another oncogenic stimulus. Unlike the Bernstein transgenic mice, in which the *p53* transgene was expressed globally, subsequent transgenic mice have confined mutant *p53* expression to specific tissues by utilization of tissue-specific promoters. For example, Li et al. (1997) described a transgenic line expressing a codon 172 arg-to-his *p53* mutation driven by a mammary gland-specific whey acidic protein (WAP) promoter. This codon was identical to the human codon 175 arg-to-his mutation often associated with human breast carcinomas (Soussi and Beroud, 2003). Despite relatively high levels of *in vitro* transformation capability, the *p53-R172H* transgenic mice did not develop spontaneous mammary tumors. However, when this transgenic line was crossed into the mammary tumor-susceptible MMTV-ErbB2 transgenic line, mammary tumor latency was decreased significantly in the bitransgenic offspring, indicating cooperativity between the two transgenes (Li et al., 1997). Other *p53* mutant transgenic mouse strains have included one carrying a *p53-Δ236* codon deletion transgene driven by an astrocyte-specific promoter and a *p53*-codon R273H mutant transgene driven by a lung-specific promoter (Duan et al., 2002; Klein et al., 2000). While the *p53-Δ236* transgene accelerated carcinogen-induced brain tumor formation, the R273H *p53* transgene developed spontaneous lung tumors at a 23% incidence by 15 months of age. A *p53* mutant transgenic model for liver cancer was developed by the Sell laboratory (Ghebranious et al., 1995; Ghebranious and Sell, 1998). This mouse line contained a *p53* transgene encoding a codon 246 arg-to-ser mutation driven by a liver-specific promoter. This *p53* mutation (codon 249 arg-to-ser in humans) is particularly frequent in human hepatocellular carcinomas associated with exposure to the co-carcinogen Aflatoxin B1 (Shen and Ong, 1996). A 14% incidence of liver carcinomas was observed in transgenic males after treatment with aflatoxin B1, whereas untreated transgenic males and treated nontransgenic males did not develop liver carcinomas (Ghebranious et al., 1995; Ghebranious and Sell, 1998). When these *p53* transgenic mice were heterozygous for wild type endogenous *p53*

($p53^{+/-}$) or bitransgenic with a hepatitis B surface antigen gene, liver carcinoma incidence was further increased (Ghebranious and Sell, 1998). Thus, these studies nicely illustrated the cooperativity of known human hepatocellular carcinogens (aflatoxin B1, mutant *p53*, and hepatitis B surface antigen expression) in a *p53* altered mouse model.

As outlined above, mutant *p53* does not always promote rapid tumors independent of co-carcinogens. This relatively weak induction of tumors despite often very high expression levels of mutant p53 protein indicates that inhibition of wild type p53 by mutant forms of p53 may be relatively inefficient *in vivo*. This contrasts with a number of *in vitro* transformation and transcription assays that suggest that mutant forms of p53 efficiently inhibit wild type p53 activity. The variability of the tumorigenic responses in the various mutant *p53* transgenic lines are likely due to the complex interactions of many factors, including the nature of the *p53* mutation and its dominant-negative effects, the expression level of the transgene, the sensitivity of a particular organ to mutant p53, and the genetic background of the transgenic animal. As discussed in the next section, mutant transgenes rarely display the tumor-inducing ability that the loss-of-function knockout mutations of *p53* confer.

P53 KNOCKOUT MICE

The development of embryonic stem cell gene targeting methodologies for mice provided an ideal way to directly test tumor suppressor function in the context of an intact mammalian organism. At best, the *p53* mutant transgenics could only indirectly show the tumor suppressor function of *p53* *in vivo*. The novel gene targeting techniques allowed the deletion and inactivation of resident genes in the mouse germline. Because tumor suppressors are recessive in function, their deletion and inactivation through gene targeting can more directly mimic the genetics of human cancer predisposition syndromes in which a tumor suppressor gene is mutationally inactivated. The discovery in 1990 that the Li-Fraumeni inherited cancer syndrome in humans was often due to a germline *p53* mutation provided further impetus to model the syndrome in mice (Malkin et al., 1990; Srivastava et al., 1990). Affected members of Li-Fraumeni families carry a defective *p53* allele and have a 50% incidence of cancer by the age of 30. Cancer types in these patients vary widely, but frequent tumors include brain tumors, breast tumors, osteosarcomas, soft tissue sarcomas and leukemias (Malkin et al., 1990). An important question was whether mice with *p53* germline mutations would mimic the human Li-Fraumeni syndrome.

The first reported tumor suppressor knockout was the *p53* knockout mouse (Donehower et al., 1992). Four groups generated mice with inactivating germline mutations in the endogenous *p53* allele, leading to an unequivocal demonstration of the fundamental role of *p53* as a tumor suppressor. Homologous recombination was used to disrupt the *p53* gene by replacing intron 4 and exon 5 (Donehower et al., 1992), exons 2-6 (Jacks et al., 1994; Purdie et al., 1994), or exon 2 (Tsukada et al., 1993) with a selectable marker cassette. Surprisingly, given the known role of p53 in cell cycle regulation in cell culture models, *p53*-null mice were found to be viable. A small fraction of female null embryos display exencephaly, a condition resulting from a failure of neural tube closure (Armstrong et al., 1995; Sah et al., 1995). As expected, the *p53*-null mice displayed early onset tumors, within 2-9 months after birth. Median time to tumor was about 18 weeks (Donehower, 1996; Donehower et al., 1992). The *p53*$^{-/-}$ mice in two of these studies were on a mixed C57BL/6 and 129/Sv genetic background, and developed primarily lymphomas and sarcomas of various types (Donehower et al., 1992; Jacks et al., 1994). The lymphomas in the null mice were generally CD4/CD8 double positive T-cell lymphomas. The sarcomas that were observed included hemangiosarcomas, osteosarcomas, fibrosarcomas, rhabdomyosarcomas, and anaplastic sarcomas (Donehower et al., 1992; Jacks et al., 1994).

The heterozygous *p53*$^{+/-}$ mice are also susceptible to spontaneous tumors, but these develop later and with a different tumor spectrum than observed in the *p53*-null mice (Donehower, 1996; Harvey et al., 1993a; Jacks et al., 1994; Purdie et al., 1994). The heterozygous mice display tumors beginning at nine months of age and by 18 months, half of them have succumbed to a variety of tumors (Donehower, 1996; Harvey et al., 1993a). The most frequently observed tumor type is osteosarcoma, observed at particularly high frequency in females. Soft tissue sarcomas, such as hemangiosarcomas, fibrosarcomas, and rhabdomyosarcomas are also relatively frequently observed. Lymphomas occur in about a third of the *p53*$^{+/-}$ mice, but these tend to be B-cell in origin and are found more often in non-thymic lymphoid organs, such as the spleen and mesenteric lymph nodes. The reduced incidence of thymic lymphomas is hypothesized to be due to the limited developmental window during which loss of heterozygosity (LOH) of the remaining wild-type *p53* allele can occur, because the number of thymocytes decreases substantially during thymic involution within several weeks after birth (Attardi and Jacks, 1999). Finally, carcinomas are observed in the *p53*$^{+/-}$ mice, but at considerably lower frequency than the lymphomas and sarcomas.

Since one of the goals in developing the *p53*-deficient mouse is to model Li-Fraumeni syndrome, an important question is whether the knockout mice indeed mimic the human syndrome. Fortunately, in some respects the mouse model could be considered accurate. For example, the *p53*$^{+/-}$ mice (the better model, since the human patients are *p53* heterozygotes) have a 50% tumor incidence at mid life span (18 months), roughly equivalent to the 50% incidence in humans by age 30 (Donehower, 1996; Malkin et al., 1990). Moreover, the tumor spectra have some similarities. Both mice and humans have a high incidence of osteosarcomas, soft tissue sarcomas and leukemias/lymphomas. However, Li-Fraumeni patients also exhibit relatively high levels of brain tumors and breast tumors, neoplasms not observed in the mixed C57BL/6-129/Sv strain of *p53* heterozygotes. However, this absence of mammary cancers in the mice can be circumvented through alteration of the strain background (see below). Thus, the *p53*$^{+/-}$ mice might be considered at least a partial success as a model for their human Li-Fraumeni syndrome counterparts.

The fate of the remaining wild type *p53* allele in the *p53*$^{+/-}$ tumors is of particular interest. According to Knudson's two hit hypothesis for tumor suppressors, tumor suppressor genes are recessive, and loss of function of both alleles is a prerequisite for tumor formation (Knudson, 1986). Surprisingly, only about half of the various tumors in heterozygous mice display LOH at the *p53* locus, suggesting that complete loss of *p53* is important for tumor development in many, but not all cases (Venkatachalam et al., 1998). In those tumors that apparently retained wild type *p53*, numerous functional assays for wild type p53 activity were performed (Venkatachalam et al., 1998). The p53 in these tumors exhibits normal activation of p53 target genes and a radiation-induced apoptotic response, while such responses are not observed in *p53*$^{+/-}$ tumors with LOH. These assays confirm that p53 in these non-LOH *p53*$^{+/-}$ tumors is both structurally and functionally intact. These results indicate that while loss of the remaining wild type *p53* allele is certainly a tumor promoting event in the *p53*$^{+/-}$ mice, it is not necessarily a prerequisite for tumor formation. Thus, a 50% reduced dosage of *p53* may be sufficient to promote tumorigenesis, at least in the mouse. This idea is consistent with a number of cell-based assays showing that *p53* heterozygous cells have cell cycle, apoptotic, and DNA damage response phenotypes intermediate between wild-type and *p53*-null cells (Venkatachalam et al., 2001). Whether this *p53* haploinsufficiency for tumor suppression holds true for humans is unclear, since in sporadically arising tumors with *p53* mutations, both *p53* alleles are usually observed to be mutated or lost (Baker et al., 1990; Hollstein et al., 1991). However, it has been reported that roughly half of the tumors arising in Li-Fraumeni patients exhibit retention of the wild type *p53* allele (Varley et al., 1997). It could be argued that many of these Li-Fraumeni patients have point mutations in *p53* and produce a protein with dominant-negative activity on wild type p53, but in fact patients with germline *p53* frameshift mutations, unlikely to express dominant-negative p53 proteins, may also exhibit no LOH in their tumors (Varley et al., 1997). Recent studies on many tumor suppressor genes indicate that there may be numerous other exceptions to the two-hit model (Paige, 2003).

Further genetic analyses of the *p53*-deficient tumors have provided some additional insights into the *in vivo* role of p53 in cancer prevention. For example, it has long been known that absence of *p53* in cultured cells leads to increases in karyotypic instability (Bischoff et al., 1990; Harvey et al., 1993b). Recently, it has been shown that *p53* mutations in human tumors also correlate with genomic instability (Overholtzer et al., 2003). Examination of *p53$^{+/-}$* and *p53$^{-/-}$* tumors by comparative genomic hybridization revealed that tumors without *p53* (*p53$^{-/-}$* tumors) or tumors that had lost *p53* (*p53$^{+/-}$* with LOH) had five-fold more copy number gains and losses than did tumors retaining *p53* (*p53$^{+/-}$* without LOH) (Venkatachalam et al., 1998). Moreover, the chromosomal copy number changes are both random and non-random. The non-random changes indicate the likelihood of other genetic alterations that collaborate with *p53*-deficiency to promote tumorigenesis.

The multiple anti-proliferative functions of p53 have often raised the question as to which of these functions plays the critical role in preventing tumor formation. Ever since p53 was discovered in cell culture assays to exhibit both cell cycle arrest and apoptosis-inducing capabilities, there has been conflicting evidence over which role takes precedence in an *in vivo* cancer preventative context. Early evidence using knockout/transgenic mouse models supported apoptosis induction as being critically important, particularly in models where retinoblastoma (Rb) gene function was inhibited or c-myc levels were augmented (Schmitt et al., 2002; Symonds et al., 1994). In these models, the presence of p53 in tumors often correlated with high levels of apoptosis and slowed tumor growth, while the absence of *p53* was associated with low levels of apoptosis and high rates of tumor growth. However, some mouse models suggested the possibility that in other *in vivo* contexts the cell cycle inhibitory role of p53 was dominant in suppressing tumors (Hundley et al., 1997; Jones et al., 1997). For example, comparison of *p53$^{+/+}$*, *p53$^{+/-}$*, and *p53$^{-/-}$* tumors for apoptotic cell levels found few differences among the different *p53* genotypes, but did see a pronounced inverse correlation between *p53* dosage and cell cycle progression, tumor growth rates, and genomic instability (Tyner et al., 1999). Crosses of *p53*-deficient mice to other tumor-susceptible transgenic models also showed a *p53*-dependent inhibition of tumor growth predominantly through cell cycle regulation (Hundley et al., 1997; Jones et al., 1997). Perhaps the most direct experiment to address this issue was through the generation of a *p53* mutant mouse that was deficient for p53-induced apoptosis, but retained cell cycle arrest capacity (described below). The likely conclusion to these conflicting studies, however, is that p53 contributes to tumor suppression through both its apoptotic and cell cycle regulatory functions, and the degree to which each role contributes may depend on the affected cell types and the oncogenic signaling pathways that are dysregulated.

BACKGROUND MATTERS

As discussed above, analysis of transgenic models that overexpress mutant p53 and mice with deletion of the *p53* gene indicate a crucial role of *p53* as a tumor suppressor in many cell types. The tumor spectrum of *p53*$^{+/-}$ mice in a mixed C57BL/6 and 129/Sv background resembles, but is not identical to that of Li-Fraumeni syndrome patients. The most important tumor types missing are the breast and brain tumors that are common in humans inheriting a defective *p53* gene. One possibility is that the kind of *p53* mutation inherited is important to transformation of a specific cell type. The mutation in mice described above represents absence of *p53*, while approximately 80% of patients with Li-Fraumeni syndrome inherit a missense mutation in *p53*. Since numerous *in vitro* data indicate that missense mutations have a gain-of-function and dominant-negative phenotype, disease progression *in vivo* with absence of *p53* may not be the same as that of mutant *p53*. Another possibility is simply that mice are not humans. The tumor spectrum of the *p53*-null allele was examined in a mixed C57BL/6 and 129/Sv cohort because of the nature of the methodology. Mutations or deletions are generated in 129/Sv embryonic stem (ES) cells, injected into C57BL/6 host blastocysts, and usually maintained by mating with C57BL/6 mice because these mice are docile, breed efficiently, and are very well characterized.

To examine potential *in vivo* effects of *p53* loss in a different strain, the *p53*-null allele was backcrossed to the 129/Sv strain of mice. In a pure 129/Sv background, mice heterozygous for the *p53*-null allele succumb to tumors faster than in a mixed C57BL/6 and 129/Sv strain. Importantly, the incidence of germ cell tumors was magnified and occurred in about half of the 129/Sv *p53*-null mice. This kind of cancer is rarely seen in *p53*-null mice in the C57BL/6-129/Sv mixed background (Donehower et al., 1995). Importantly, these data suggest that strain differences exist and that these differences contribute to the tumor phenotype. In order to identify the different chromosomal regions and eventually the specific genes that contribute to germ cell tumor development, genomic differences between strains can be exploited. Mice are generated using different mating schemes such that chromosomes from the two strains are mixed and matched and correlated with the onset of germ cell tumors. Using such a scheme, Muller et al. (2000) mapped one of the loci that contributes to the onset of germ cell tumors in the 129/Sv strain to chromosome 13 with a significant LOD score of 7.82. Thus, the mouse models facilitate the identification and cloning of genes that modify the tumor phenotype.

The *p53*-null allele was also introduced into another strain of mice, Balb/c. Approximately 20% of Balb/c mice normally succumb to breast cancer with a latency of 16-17 months (Heston and Vlahakis, 1971). In this background, 55% of the Balb/c mice heterozygous for *p53* develop breast tumors compared to about 1% in the mixed C57BL/6-129/Sv strain (Kuperwasser et al., 2000). Latency is also drastically decreased to 8-14 months in Balb/c *p53*$^{+/-}$ mice. These data suggest a modifier for breast cancer exists in the susceptible Balb/c strain. Such modifiers are more likely to be identified in the mouse and can then be examined for their contribution to human cancers.

Numerous other strains of mice have tendencies toward tumors at low frequencies and late in life (jax.org). Augmenting the tumor phenotype by deletion of *p53* makes them amenable models for cloning some of the genes that contribute to cancers. The CE/J inbred strain of mice for example, is susceptible to ovarian and breast carcinomas, adrenocortical carcinomas, and sarcomas. Approximately, 33% of virgin females die of ovarian tumors, and adrenocortical carcinomas occur in gonadectomized mice (Heston, 1963; Murphy, 1966). All tumors, however, develop with long latency and low penetrance. To increase the incidence and speed the onset of tumors, CE/J mice were mated with *p53*-null 129/SvJ mice. In the F2 generation containing a mix of CE/J and 129/Sv backgrounds, the combination with *p53* heterozygosity or homozygosity, resulted in embryonic lethality of half the *p53* heterozygote mice and half the *p53*-null mice. These data suggest that a recessive modifier of *p53* is present in CE/J mice. Linkage analysis identified a genetic locus on chromosome 11 (Evans et al., 2004). These mice also show differences in tumor spectrum (Evans, unpublished data). These data suggest that modifiers increase the likelihood of specific developmental defects or diseases in inbred strains of mice. The use of different mouse strains that differ in their susceptibility to specific kinds of cancers will aid in the identification of genes that modify the disease (Dragani, 2003; Mao and Balmain, 2003). The variations seen in inbred strains of mice may account for the heterogeneity seen in humans with Li-Fraumeni syndrome.

ONCOGENIC COOPERATIVITY OF P53

Even in the total absence of *p53*, isolated cancers arise with some latency, indicating the need for other cooperating oncogenic lesions to produce a tumor. To assess potential tumorigenic cooperativity, a large number of crosses between *p53* knockout mice and other tumor-susceptible strains have been performed. Such tumor-susceptible strains include transgenic mice overexpressing oncogenes in specific tissues or knockout mice lacking tumor suppressor genes. By revealing genetic interactions between p53 and other proteins involved in tumorigenesis, such crosses can help to define the pathways governing proliferation. Furthermore, these models can provide a powerful approach to generate more accurate mimics for human carcinogenesis. Finally, as briefly indicated above, these crosses can reveal those p53 functions that are critical for tumor suppression in particular *in vivo* contexts.

Due to space constraints, it would be impossible to detail the many crosses that have been performed between *p53*-deficient mice and other tumor-susceptible lines, so the reader is referred to more comprehensive reviews on this subject (Attardi and Jacks, 1999; Donehower, 1996; Lozano and Liu, 1998). In assessing the results of these cooperativity experiments, a few common themes emerge. First, in the great majority of crosses between *p53*-deficient mice and either tumor-prone tumor suppressor knockout mice or transgenic oncogene-containing mice, the tumor latency is invariably decreased in comparison to both parental lines. The basis for this acceleration is likely due to true cooperativity, i.e. a non-overlapping contribution by dysregulation of two differing growth or death signaling pathways in the nascent cancer cell. Another component contributing to the acceleration of tumors in these crosses could be pathway independent. We have hypothesized that *p53* deficiency and the consequent defects in cell cycle control and apoptosis could create a cellular environment more conducive to the initiation and fixation of oncogenic mutations and the emergence of tumorigenic clones (Venkatachalam et al., 2001). This hypothesis is consistent with the observation that 88% of genotoxic carcinogens reduce tumor latency in $p53^{+/-}$ mice, while non-genotoxic carcinogens rarely produce a reduction in tumor latency (Storer et al., 2001). We theorize that the oncogenic lesions induced by the genotoxic carcinogen are more likely to be preserved and not eliminated in the $p53^{+/-}$ cell, resulting in a higher rate of oncogenic conversion. Thus, in the *p53*-deficient crosses, random oncogenic lesions that cooperate with either *p53* deficiency or the second genetically altered allele are likely to occur at an increased rate and may contribute to the accelerated tumorigenesis.

Another surprisingly frequent occurrence in these *p53* crosses is the appearance of novel tumor types. For example, while *Rb* heterozygous mice are susceptible to pituitary and thyroid tumors, on a *p53$^{-/-}$* background, they now develop novel tumor types, including pinealoblastomas and pancreatic islet cell carcinomas (Harvey et al., 1995a; Williams et al., 1994). Similarly, *Nf1$^{+/-}$*; *p53$^{+/-}$* compound mice develop malignant peripheral nerve sheath tumors and astrocytomas, neither of which is observed in the singly heterozygous parents (Cichowski et al., 1999; Reilly et al., 2000). Breeding some of these tumor suppressor knockout strains to *p53*-deficient mice may also be helpful in mimicking a predisposition to the tumor type associated with the corresponding human familial cancer syndrome. This is illustrated by mice lacking the *Brca2* gene, which were generated in part to model human breast cancer-susceptible families with *BRCA2* germline mutations. The *Brca2*+/- mice are not predisposed to breast cancer until bred onto a *p53*-deficient background (Jonkers et al., 2001). Since *p53* is inactivated in *Brca2*-associated human breast cancers as well, these compound mutants provide good models for inherited human breast cancers. Additional crosses in the future will assist in generating better mouse models of human cancer.

An example of such a cross may be the telomerase-deficient, *p53*-deficient compound mice (Artandi et al., 2000). *p53*-deficient mice develop relatively few carcinomas, tumors of epithelial origin. Yet carcinomas are the most common tumor type observed in humans. One possible explanation for this difference is that the telomeres of inbred mice are very long (>50 kb). Humans have considerably shorter telomeres, which become progressively shorter with age and the number of cell divisions completed in the absence of telomerase, the enzyme that maintains telomere length. When telomerase-deficient mice are crossed for 4-6 generations, their telomeres become short enough to activate the p53 damage response, leading to increased p53 activity, apoptosis, and chromosomal instability, accompanied by premature aging phenotypes (Artandi et al., 2000). When these mice are crossed into a *p53$^{+/-}$* background, the resultant reduction in *p53* dosage leads to an increase in survival of the chromosomally unstable cells, but these animals now display a much increased incidence of carcinomas of the mammary gland, colon and skin (Artandi et al., 2000). Thus, the telomerase-deficient, *p53* heterozygous mice have been useful for providing a more humanized cancer model and for potentially illustrating the importance of telomeres in the genesis of epithelial cancers.

SOMATIC INACTIVATION OF *P53*

The generation of mice with a deletion of one *p53* allele mimics the inheritance of *p53* loss-of-function mutations in Li-Fraumeni syndrome. However, these mice do not represent the majority of human cancers that arise from somatic inactivation of *p53*. To develop a mouse in which *p53* can be deleted almost at will, Dr. Berns' group developed an allele of *p53* flanked by loxP sites in introns 1 and 10 (Jonkers et al., 2001). Expression of the Cre recombinase allows recombination at the loxP sites and deletion of DNA sequences in between, thus generating a *p53*-null allele. The expression of Cre then regulates deletion of *p53*. This *p53* conditional allele has been used to study loss of *Brca2* and *p53* in the genesis of mammary tumors in mice using the *K14cre* transgene. *Brca2* and *p53* deletion cooperated in the genesis of mammary tumors and skin tumors, but not for example in salivary gland tumors and tumors in other cell types expressing Cre. This experiment highlights the tissue specific nature of tumorigenesis caused by deletions in *Brca2*. Numerous transgenic mice expressing Cre in different cell types and at different times in development have been generated and can be used to delete *p53* in specific tissues. Deletion of *p53* in specific cell types also offsets the problem of multiple tumors in *p53* heterozygous mice. Additionally, mice with a tamoxifen inducible Cre gene allow transient expression of Cre by addition of tamoxifen providing temporal control of gene expression (Hayashi and McMahon, 2002). The Flp recombinase is yet another recombinase that has more recently been developed for use in mice (Possemato et al., 2002). The concept is identical to that of Cre, but involves a different DNA sequence and a different recombinase protein. These developments will allow us to delete different tumor suppressors at different times to monitor the effects of timing and the order of events on the tumorigenic process.

MICE WITH *P53* MISSENSE MUTATIONS

While the development of the *p53*-null mice has tremendously expanded our knowledge of the role of p53 in tumorigenesis, the most common type of alteration seen in human cancers and in Li-Fraumeni syndrome is a missense mutation in *p53*. *In vitro* overexpression systems suggest that *p53* missense mutants would be more detrimental than absence of *p53*. Gain-of-function mutations are those missense mutations in which mutant p53 has additional functions not seen in wild-type p53 (Sigal and Rotter, 2000). For example, the *p53R175H* mutant was overexpressed in a nontransformed cell line lacking *p53*. In a test for tumorigenicity, the cell lines expressing the *p53R175H* mutant yielded tumors in nude mice, while the parental cell lines did not (Dittmer et al., 1993). The dominant-negative hypothesis is strongly supported by the observations that many mutant p53 proteins have an increased half-life (Slingerland et al., 1993) and that they oligomerize with wild type p53, inhibiting its function (Jeffrey et al., 1995; Milner and Medcalf, 1991; Sturzbecher et al., 1992). The *in vivo* relevance of missense mutations remains, largely unexplored. Several missense mutations in *p53* have been generated and studied in embryonic stem (ES) cells (previously reviewed by Parant and Lozano, 2003). This part of the review will focus on mice established from ES cells with missense mutations in *p53*.

The first mouse to be described with a missense mutation in *p53* encoded an arg-to-his substitution at amino acid172 which corresponds to the arg-to-his mutation at amino acid 175 in human *p53* (Liu et al., 2000). The p53R175H mutation alters the conformation of the p53 protein and is a common hot spot mutation present in 6% human cancers and in patients with Li-Fraumeni syndrome (Cho et al., 1994; Frebourg et al., 1995; Hussain and Harris, 1998). This mouse, however, expresses low levels of the mutant p53 protein due to the additional mutation at a splice junction. Mice heterozygous for this hypomorphic allele differed from $p53^{+/-}$ mice in tumor spectrum, with a significant increase in the number of carcinomas and a slight decrease in the number of lymphomas. Moreover, 69% of the osteosarcomas and 40% of the carcinomas that developed in $p53^{R172H\Delta g}$ heterozygous mice metastasized, a rare occurrence in *p53* heterozygous mice, providing the first *in vivo* evidence that a p53 missense mutation has a gain-of-function phenotype. An LOH study of tumors indicated that only 1 of 11 had lost the wild type *p53* allele. These data highlight clear differences between a *p53* missense mutation and a *p53*-null allele in tumorigenesis *in vivo* and suggest that the $p53^{R172H\Delta g}$ mutant represents a gain-of-function allele. Mice containing a *p53R172H* allele without secondary mutations have been developed, but the results have not yet been published.

As mentioned above, the ability of p53 to induce apoptosis is important in prevention of tumorigenesis. p53, however, also has less well studied roles in cell cycle arrest and senescence. Another human tumor specific mutation at amino acid 175, an arg-to-pro substitution results in a p53 protein that still binds and activates the *p21* promoter, but not the *bax* or *IGF-BP3* promoters, for example. In various cell lines tested, p53R175P can induce cell cycle arrest, but not apoptosis (Ludwig et al., 1996; Rowan et al., 1996). Thus, this important mutant separates the cell cycle arrest and apoptotic functions of p53. To distinguish the importance of apoptosis versus cell cycle arrest in the development of tumors, a mouse was generated containing this point mutation (Liu et al., 2004).

Mice inheriting the equivalent mutation in mice, an arg-to-pro substitution at amino acid 172, and mouse embryo fibroblasts (MEFs) homozygous for the alteration were characterized in detail. The initial experiments were aimed at determining whether these $p53^{515C/515C}$ mice lacked the ability to induce apoptosis yet retained cell cycle arrest function. Apoptosis was measured using MEFs sensitized to induce apoptosis with E1A and treated with adriamycin or deprived of serum. p53-dependent apoptosis was also measured *in vivo* in thymocytes and in embryonic brains after IR. In all assays, in all cell types examined, the p53-R172P protein was unable to induce apoptosis reminiscent of cells lacking *p53*. With regards to cell cycle arrest, cells homozygous for $p53^{515C}$ retained a partial cell cycle arrest function in response to IR, and initiated expression of *p21*, albeit at lower levels that wild type *p53* cells. Thus, *in vivo*, the p53-R172P protein retained a partial p53 cell cycle arrest function, but was devoid of apoptotic activity and allowed testing of the hypothesis that the cell cycle arrest function was critical for tumor development *in vivo*.

Mice homozygous for the $p53^{515C}$ mutation were monitored for tumor development. When compared to $p53^{-/-}$ mice, $p53^{515C/515C}$ mice show delayed onset of tumorigenesis. At 6 months of age whereby approximately 90% of the *p53*-null mice have developed tumors, only 10% of the $p53^{515C/515C}$ mice have died. If apoptosis was the only critical function of p53 involved in tumor suppression, no difference in survival should have been noted. The delay in tumorigenesis in $p53^{515C/515C}$ mice is strong evidence that p53-dependent apoptosis is not the sole determinant of p53 tumor suppression. The majority of tumors that developed in these mice are lymphomas and sarcomas. Approximately 38% of the mice had highly disseminated lymphoma, only two of which were T cell lymphomas. Most lymphomas stained positive for both CD4 and B cell markers suggesting that they arose from an early progenitor or histiocytic cell. In contrast, 75% of *p53*-null mice develop thymic lymphomas (Donehower et al., 1992; Jacks et al., 1994). 43% of $p53^{515C}$ homozygous mice had different types of high-grade sarcomas including osteosarcomas, angiosarcomas, and rhabdomyosarcomas. Atypical hyperplasia in the spleen was commonly observed in 43% of $p53^{515C/515C}$ mice. Thus, the tumor spectrum of mice that retain a cell cycle arrest function is different from those with deletion of *p53*.

To begin to get at the function of the p53-R172P protein that delayed tumorigenesis, tumor samples were analyzed for ploidy. Amazingly, tumors from $p53^{515C}$ homozygous mice were diploid in sharp contrast to tumors derived from $p53^{-/-}$ mice, which are aneuploid. This phenotype was also visible in MEFs. Whereas *p53*-null MEFs become aneuploid within a few passages in culture, cells homozygous for the $p53^{515C}$ allele remain diploid. This observation led to the intriguing possibility that the cell cycle arrest function of p53 is responsible for maintaining a normal genome. Further experiments will be initiated to prove or disprove this hypothesis. However, the ability of p53 to arrest the cell cycle can no longer be ignored as a mechanism of tumor suppression.

p53 PHOSPHORYLATION MUTANTS

Phosphorylation plays a key role in regulating p53 activity. p53 is phosphorylated at multiple sites by numerous kinases in response to DNA damage (Appella and Anderson, 2001). The functional consequences of phosphorylation vary. Phosphorylation at the amino terminus activates p53 by dislodging its inhibitor, MDM2 (Shieh et al., 1997; Unger et al., 1999). Phosphorylation and dephosphorylation of different amino acids at the carboxyl terminus activate the intrinsic DNA binding activity of p53 (Hupp et al., 1992; Waterman et al., 1998). The role of phosphorylation is further complicated by the finding that the phosphorylation status varies as a function of the stimulus. For example, treatment of cells with UV but not IR causes phosphorylation of p53 serine 392 (Kapoor and Lozano, 1998; Lu et al., 1998) and phosphorylation of serine 20 is absent in UV treated cells (Shieh et al., 1999). Additional experiments mutating every serine and threonine yielded a p53 protein that could not be phosphorylated, yet retained transcriptional activity in transient transfection assays (Ashcroft et al., 1999), suggesting that mechanisms other than phosphorylation can also activate p53. *In vivo* analysis of different phosphorylation mutants is likely to provide insight into the importance of p53 phosphorylation.

To date, several ES cell lines have been characterized with specific mutations of phosphorylated amino acids, but only one has been used to generate mice. Jones and colleagues have generated mice containing a ser-to-ala mutation at murine p53 amino serine 18 that corresponds to serine 15 in human p53 (Sluss et al., 2004). Phosphorylation at this amino acid upregulates p53 activity by interfering with Mdm2 binding and degradation of p53. Thus, a mutation at p53 serine 18 should result in decreased p53 activity due to the inability of Mdm2 to be dislodged by phosphorylation. In response to IR, thymocytes homozygous for *p53ala18* showed a decreased apoptotic response. In all other assays, however, cells homozygous for *p53ala18* looked identical to wild type cells. *p53ala18* homozygous mice also did not exhibit a tumor phenotype at 40 weeks of age. The data from this experiment is contrary to expectations, but emphasizes the complex nature of phosphorylation in p53 regulation and perhaps indicates a role for other phosphorylation or regulatory events in regulating p53 activity.

A MOUSE MODEL WITH A HUMAN/MOUSE CHIMERIC *p53* GENE

One other fascinating allele has been generated in mice. The human *p53* knock-in (hupki) allele replaces exons 4-9 of the mouse with human *p53* sequences (Luo et al., 2001b). The region encompassing exons 4-9 (amino acids 33-331 in human p53) encodes the DNA binding domain where most p53 mutations occur and contains 4 of the 5 highly conserved domains. The hupki protein seems identical to the mouse wild type p53 protein in several assays. The *p53* and *p53hupki* alleles show no difference in RNA expression levels, induction of p53 targets *p21* and *bax* in response to IR, DNA binding activity, and p53-dependent apoptosis in thymocytes. Importantly, no spontaneous tumor formation was observed in *p53hupki* homozygous mice as might have been expected if the human/mouse chimeric protein malfunctioned. Thus, the *p53hupki* homozygous mice appear biologically equivalent to *p53* wild type mice and provide a unique tool for studying spontaneous or chemical induced tumorigenesis. For example, aflatoxin exposure causes a common mutation at codon 249 altering AGG to AGT (an arg to ser mutation) in human p53. Since the murine sequence differs at this nucleotide, aflatoxin exposure does not alter this amino acid in the mouse. A related approach was tested by UV irradiation of the skin of the *p53hupki* mice. DNA from the irradiated cells exhibited *p53* mutations at the same hotspot sites identified in human skin cancers (Luo et al., 2001a). Additionally, the *p53hupki* mouse will allow modeling of the proline polymorphism at human p53 amino acid 72.

MOUSE MODELS OF ELEVATED P53 EXPRESSION AND ACTIVITY

Much work has been done generating mice that lack *p53* or contain various mutant forms of the protein. In the past, many have attempted to generate a mouse with elevated levels of p53 (Allemand et al., 1999; Choi and Donehower, 1999; Godley et al., 1996; Nakamura et al., 1995). These attempts, for the most part, have been of mixed success, hampered by difficulties obtaining the appropriate transcriptional regulation of the transgene. Most of the attempts have resulted in faulty differentiation or lethality due to overly high levels of p53.

Recently the production of a p53 mutant mouse with elevated p53 activity that did not result in lethality was reported (Tyner et al., 2002). The $p53^{+/m}$ mouse contains one wild type *p53* allele and one mutant allele. The *p53* mutant allele consists of exons 7 to 11 with a point mutant at position

245. A promoter region, now known to belong to the *Vamp2* gene, upstream of the mutant *p53* allele drives expression of a short mRNA, which is translated into a 150 amino acid C-terminal p53 protein. The *m* allele mRNA (Tyner et al., 2002) and protein (unpublished results) have been detected in differing amounts in many tissues of the mouse. It was demonstrated that wild type *p53* in the presence of the m protein is more abundant, has increased stability and increased transactivation activity. Thus, the $p53^{+/m}$ mouse model is one of constitutively hyperactivated p53.

Consistent with the expectation of elevated p53 activity, $p53^{+/m}$ mice are tumor resistant. These mice were monitored over their lifespan and none of the 35 mice developed overt, life-threatening tumors, while 2 of 35 developed localized tumor lesions (Tyner et al., 2002). In comparison, 48% of their wild type counterparts developed a range of tumor lesions, predominantly lymphomas. However, it seemed that the mutant mouse was not trouble free. The $p53^{+/m}$ mouse, when aged, died earlier than the wild type; in fact, the median lifespan of a $p53^{+/m}$ mouse is 96 weeks versus 118 weeks for a wild type mouse. Close characterization revealed the mutant mouse to have several phenotypes consistent with premature aging. These included reduced body weight, reduced size and cellularity of various organs, osteopenia and associated kyphosis of the spine, and reduced regenerative ability following tissue ablation. It should be noted that there were a few aging traits the $p53^{+/m}$ mouse did not possess, including atherosclerosis, joint diseases, cataracts, hair graying, alopecia, autoimmune diseases, liver or kidney pathologies, brain atrophy or amyloid plaques and atrophy of intestinal villi. Thus the $p53^{+/m}$ mouse model is one of tumor resistance and "partial" accelerated aging characteristics. It is unclear what the cause of death was for these mice, since they were not shown to display any obvious lethal pathologies; their death is most consistent with the human frailty syndrome of organ functional decline.

How the m protein increases wild type p53 activity and stability is not known. It is possible that upon binding to p53, the m protein alters p53 latent conformation rendering it more active. Another unknown is how elevated p53 contributes to the aging process. p53 is a transcriptional regulator of many genes which have an ultimate effect on cellular fate decisions including: cell cycle arrest, both transient and terminal (senescence), and apoptosis. Obviously these cellular outcomes if augmented could have a drastic effect on tissue homeostasis and ultimately lifespan. In addition, the $p53^{+/m}$ mouse phenotypes of reduced organ cellularity and tissue regenerative ability are consistent with a potential stem cell aberration. It is possible that elevated p53 has a negative effect on tissue stem cells resulting in a reduced functionality. These processes and many more are being studied carefully in an attempt to elucidate p53's role in orchestrating the aging process.

Garcia-Cao and co-workers (Garcia-Cao et al., 2002) recently reported the production of yet another mouse with elevated p53 levels, named the super-p53 mouse (Garcia-Cao et al., 2002). The group bypassed traditional problems associated with making a *p53* transgenic by utilizing bacterial artificial chromosomes (BACS) to introduce *p53* into the genome. They isolated 2 genomic sequences containing *p53* from a mouse BAC library, containing each one (*p53*-tg) and two (*p53*-tg^b) copies of *p53*, and made two transgenic mice. The transgenes were shown to be functional; when the *p53*-tg mice (one copy of *p53*) were crossed to *p53*-null mice, the offspring had p53 responses (transactivation ability, apoptosis, and tumorigenesis) similar to the *p53*$^{+/-}$ mouse. Their work concentrated on the "super-p53" mouse, which contains wild type genomic *p53* and a single copy of the *p53*-tg BAC. The mouse was demonstrated to be developmentally normal (fertile with no obvious nor histological phenotypes), but possessed elevated p53 responses to cellular stress. Consistent with elevated active p53, the mice were shown to be tumor resistant when exposed to two different carcinogenic agents. In light of the *p53*$^{+/m}$ mouse, Garcia-Cao *et al* (2002) analyzed the super-p53 mouse for evidence of accelerated aging. The super-p53 mice have a similar life span to wild type mice and do not exhibit any phenotypes consistent with aging. Tumorigenesis was not the main cause of death in the animals since only 16% of old super-53 mice died of tumors, versus 47% of wild type mice. It seemed the most obvious lethal pathology was renal dysfunction caused by glomerulonephritis, also a common cause of death in wild type mice.

Discrepancies clearly exist between these two described models of p53 over expression. Both models strongly support the role of p53 as a tumor suppressor and demonstrate increased tumor resistance with elevated p53. However, the *p53*$^{+/m}$ mouse prematurely ages resulting in decreased longevity, while the super-p53 mouse has a normal lifespan. It has been proposed the reason for this difference is again due to p53 regulation (Donehower, 2002; Klatt and Serrano, 2003). It is hypothesized that the *p53*$^{+/m}$ mouse has constitutively higher levels of p53 whereas the super-p53 mouse has normal regulation of p53 and levels are only increased following cellular stress. It is possible that the more constitutively active p53 in the *p53*$^{+/m}$ model is responsible for the aging phenotype seen.

Another possible interpretation is provided by a recent report from Maier et al. (2004) A transgenic mouse containing a truncated transgene of 44 kDa missing an amino terminal p53 segment was shown to result in reduced growth and shortened life span accompanied by some premature aging phenotypes such as osteoporosis. However, in this case, the phenotypes were ascribed to hyperactivation effects of the transgene product on the IGF signaling pathway. This IGF signaling in turn produced more activated ERK signaling, which ultimately increased activation of the p21 cyclin dependent kinase inhibitor. The growth retardation and aging phenotypes were thus attributed to growth inhibitory effects mediated by p21 through p53 upregulation of IGF-1 signaling. However, whether other p53 activated pathways were also involved in the observed phenotypes was unclear. When the p44 transgene product could interact with wild type p53 and when it was crossed into a *p53*-null background, there was no significant effect on growth rates in the absence of p53, indicating the interaction of p44 with wild type p53 is likely to be critical for the p44 effects. Thus, it now appears that two different lines of mice with truncated forms of p53 can affect longevity and aging-associated phenotypes, implicating wild type p53 as a regulator of the aging process.

SUMMARY

This review was intended to illustrate some of the insights gained from the various genetic manipulations of the *p53* gene in the germline of the mouse. It should be clear from the models presented that alteration of p53 structure and/or expression levels can have amazingly diverse effects on cancer, development, and aging in the mouse. Many of the phenotypes observed in the *p53* mutant mice could not have been predicted from cell culture studies or even analysis of *p53* mutations in human tumors. Thus, *p53* mutant mouse models have provided a rich resource for further understanding of p53 functions in an organismal context. Future refinements of genetic engineering techniques in the mouse should lead to even more insights and perhaps even more surprises.

REFERENCES

Allemand, I., Anglo, A., Jeantet, A. Y., Cerutti, I., and May, E. (1999). Testicular wild-type p53 expression in transgenic mice induces spermiogenesis alterations ranging from differentiation defects to apoptosis. Oncogene *18*, 6521-6530.

Appella, E., and Anderson, C. W. (2001). Post-translational modifications and activation of p53 by genotoxic stresses. Eur J Biochem *268*, 2764-2772.

Armstrong, J. F., Kaufman, M. H., Harrison, D. J., and Clarke, A. R. (1995). High-frequency developmental abnormalities in p53-deficient mice. Curr Biol *5*, 931-936.

Artandi, S. E., Chang, S., Lee, S. L., Alson, S., Gottlieb, G. J., Chin, L., and DePinho, R. A. (2000). Telomere dysfunction promotes non-reciprocal translocations and epithelial cancers in mice. Nature *406*, 641-645.

Ashcroft, M., Kubbutat, M. H., and Vousden, K. H. (1999). Regulation of p53 function and stability by phosphorylation. Mol Cell Biol *19*, 1751-1758.

Attardi, L. D., and Jacks, T. (1999). The role of p53 in tumour suppression: lessons from mouse models. Cell Mol Life Sci *55*, 48-63.

Baker, S. J., Preisinger, A. C., Jessup, J. M., Paraskeva, C., Markowitz, S., Willson, J. K., Hamilton, S., and Vogelstein, B. (1990). p53 gene mutations occur in combination with 17p allelic deletions as late events in colorectal tumorigenesis. Cancer Res *50*, 7717-7722.

Bischoff, F. Z., Yim, S. O., Pathak, S., Grant, G., Siciliano, M. J., Giovanella, B. C., Strong, L. C., and Tainsky, M. A. (1990). Spontaneous abnormalities in normal fibroblasts from patients with Li-Fraumeni cancer syndrome: aneuploidy and immortalization. Cancer Res *50*, 7979-7984.

Cadwell, C., and Zambetti, G. P. (2001). The effects of wild-type p53 tumor suppressor activity and mutant p53 gain-of-function on cell growth. Gene *277*, 15-30.

Cho, Y., Gorina, S., Jeffrey, P. D., and Pavletich, N. P. (1994). Crystal structure of a p53 tumor suppressor-DNA complex: understanding tumorigenic mutations. Science *265*, 346-355.

Choi, J., and Donehower, L. A. (1999). p53 in embryonic development: maintaining a fine balance. Cell Mol Life Sci *55*, 38-47.

Cichowski, K., Shih, T. S., Schmitt, E., Santiago, S., Reilly, K., McLaughlin, M. E., Bronson, R. T., and Jacks, T. (1999). Mouse models of tumor development in neurofibromatosis type 1. Science *286*, 2172-2176.

Dittmer, D., Pati, S., Zambetti, G., Chu, S., Teresky, A. K., Moore, M., Finlay, C., and Levine, A. J. (1993). Gain of function mutations in p53. Nat Genet *4*, 42-46.

Donehower, L. A. (1996). The p53-deficient mouse: a model for basic and applied cancer studies. Semin Cancer Biol *7*, 269-278.

Donehower, L. A. (2002). Does p53 affect organismal aging? J Cell Physiol *192*, 23-33.

Donehower, L. A., Harvey, M., Slagle, B. L., McArthur, M. J., Montgomery, C. A., Jr., Butel, J. S., and Bradley, A. (1992). Mice deficient for p53 are developmentally normal but susceptible to spontaneous tumours. Nature *356*, 215-221.

Donehower, L. A., Harvey, M., Vogel, H., McArthur, M. J., Montgomery, C. A., Jr., Park, S. H., Thompson, T., Ford, R. J., and Bradley, A. (1995). Effects of genetic background on tumorigenesis in p53-deficient mice. Mol Carcinog *14*, 16-22.

Dragani, T. A. (2003). 10 years of mouse cancer modifier loci: human relevance. Cancer Res *63*, 3011-3018.

Duan, W., Ding, H., Subler, M. A., Zhu, W. G., Zhang, H., Stoner, G. D., Windle, J. J., Otterson, G. A., and Villalona-Calero, M. A. (2002). Lung-specific expression of human mutant p53-273H is associated with a high frequency of lung adenocarcinoma in transgenic mice. Oncogene *21*, 7831-7838.

Eliyahu, D., Raz, A., Gruss, P., Givol, D., and Oren, M. (1984). Participation of p53 cellular tumour antigen in transformation of normal embryonic cells. Nature *312*, 646-649.

Evans, S. C., Liang, M., Amos, C., Gu, X., and Lozano, G. (2004). A novel genetic modifier of p53, mop1, results in embryonic lethality. In press, Mammalian Genome, 2004.

Finlay, C. A., Hinds, P. W., and Levine, A. J. (1989). The p53 proto-oncogene can act as a suppressor of transformation. Cell *57*, 1083-1093.

Frebourg, T., Barbier, N., Yan, Y. X., Garber, J. E., Dreyfus, M., Fraumeni, J., Jr., Li, F. P., and Friend, S. H. (1995). Germ-line p53 mutations in 15 families with Li-Fraumeni syndrome. Am J Hum Genet *56*, 608-615.

Garcia-Cao, I., Garcia-Cao, M., Martin-Caballero, J., Criado, L. M., Klatt, P., Flores, J. M., Weill, J. C., Blasco, M. A., and Serrano, M. (2002). "Super p53" mice exhibit enhanced DNA damage response, are tumor resistant and age normally. Embo J *21*, 6225-6235.

Ghebranious, N., Knoll, B. J., Wu, H., Lozano, G., and Sell, S. (1995). Characterization of a murine p53ser246 mutant equivalent to the human p53ser249 associated with hepatocellular carcinoma and aflatoxin exposure. Mol Carcinog *13*, 104-111.

Ghebranious, N., and Sell, S. (1998). The mouse equivalent of the human p53ser249 mutation p53ser246 enhances aflatoxin hepatocarcinogenesis in hepatitis B surface antigen transgenic and p53 heterozygous null mice. Hepatology *27*, 967-973.

Giaccia, A. J., and Kastan, M. B. (1998). The complexity of p53 modulation: emerging patterns from divergent signals. Genes Dev *12*, 2973-2983.

Godley, L. A., Kopp, J. B., Eckhaus, M., Paglino, J. J., Owens, J., and Varmus, H. E. (1996). Wild-type p53 transgenic mice exhibit altered differentiation of the ureteric bud and possess small kidneys. Genes Dev *10*, 836-850.

Harvey, M., McArthur, M. J., Montgomery, C. A., Jr., Butel, J. S., Bradley, A., and Donehower, L. A. (1993a). Spontaneous and carcinogen-induced tumorigenesis in p53-deficient mice. Nat Genet *5*, 225-229.

Harvey, M., Sands, A. T., Weiss, R. S., Hegi, M. E., Wiseman, R. W., Pantazis, P., Giovanella, B. C., Tainsky, M. A., Bradley, A., and Donehower, L. A. (1993b). In vitro growth characteristics of embryo fibroblasts isolated from p53-deficient mice. Oncogene *8*, 2457-2467.

Harvey, M., Vogel, H., Lee, E. Y., Bradley, A., and Donehower, L. A. (1995a). Mice deficient in both p53 and Rb develop tumors primarily of endocrine origin. Cancer Res *55*, 1146-1151.

Harvey, M., Vogel, H., Morris, D., Bradley, A., Bernstein, A., and Donehower, L. A. (1995b). A mutant p53 transgene accelerates tumour development in heterozygous but not nullizygous p53-deficient mice. Nat Genet *9*, 305-311.

Hayashi, S., and McMahon, A. P. (2002). Efficient recombination in diverse tissues by a tamoxifen-inducible form of Cre: a tool for temporally regulated gene activation/inactivation in the mouse. Dev Biol *244*, 305-318.

Heston, W. E. (1963). Genetics of neoplasia, in Methodology in mammalian genetics. (Burdette W J, ed), Holden-Day, San Francisco, 247-268.

Heston, W. E., and Vlahakis, G. (1971). Mammary tumors, plaques, and hyperplastic alveolar nodules in various combinations of mouse inbred strains and the different lines of the mammary tumor virus. Int J Cancer *7*, 141-148.

Hinds, P., Finlay, C., and Levine, A. J. (1989). Mutation is required to activate the p53 gene for cooperation with the ras oncogene and transformation. J Virol *63*, 739-746.

Hollstein, M., Sidransky, D., Vogelstein, B., and Harris, C. C. (1991). p53 mutations in human cancers. Science *253*, 49-53.

Hundley, J. E., Koester, S. K., Troyer, D. A., Hilsenbeck, S. G., Subler, M. A., and Windle, J. J. (1997). Increased tumor proliferation and genomic instability without decreased apoptosis in MMTV-ras mice deficient in p53. Mol Cell Biol *17*, 723-731.

Hupp, T. R., Meek, D. W., Midgley, C. A., and Lane, D. P. (1992). Regulation of the specific DNA binding function of p53. Cell *71*, 875-886.

Hussain, S. P., and Harris, C. C. (1998). Molecular epidemiology of human cancer: contribution of mutation spectra studies of tumor suppressor genes. Cancer Res *58*, 4023-4037.

Jacks, T., Remington, L., Williams, B. O., Schmitt, E. M., Halachmi, S., Bronson, R. T., and Weinberg, R. A. (1994). Tumor spectrum analysis in p53-mutant mice. Curr Biol *4*, 1-7.

Jeffrey, P. D., Gorina, S., and Pavletich, N. P. (1995). Crystal structure of the tetramerization domain of the p53 tumor suppressor at 1.7 angstroms. Science *267*, 1498-1502.

Jones, J. M., Attardi, L., Godley, L. A., Laucirica, R., Medina, D., Jacks, T., Varmus, H. E., and Donehower, L. A. (1997). Absence of p53 in a mouse mammary tumor model promotes tumor cell proliferation without affecting apoptosis. Cell Growth Differ *8*, 829-838.

Jonkers, J., Meuwissen, R., van der Gulden, H., Peterse, H., van der Valk, M., and Berns, A. (2001). Synergistic tumor suppressor activity of BRCA2 and p53 in a conditional mouse model for breast cancer. Nat Genet *29*, 418-425.

Kapoor, M., and Lozano, G. (1998). Functional activation of p53 via phosphorylation following DNA damage by UV but not gamma radiation. Proc Natl Acad Sci U S A *95*, 2834-2837.

Klatt, P., and Serrano, M. (2003). Engineering cancer resistance in mice. Carcinogenesis *24*, 817-826.

Klein, M. A., Ruedi, D., Nozaki, M., Dell, E. W., Diserens, A. C., Seelentag, W., Janzer, R. C., Aguzzi, A., and Hegi, M. E. (2000). Reduced latency but no increased brain tumor penetrance in mice with astrocyte specific expression of a human p53 mutant. Oncogene *19*, 5329-5337.

Knudson, A. G., Jr. (1986). Genetics of human cancer. Annu Rev Genet *20*, 231-251.

Ko, L. J., and Prives, C. (1996). p53: puzzle and paradigm. Genes Dev *10*, 1054-1072.

Kuperwasser, C., Hurlbut, G. D., Kittrell, F. S., Dickinson, E. S., Laucirica, R., Medina, D., Naber, S. P., and Jerry, D. J. (2000). Development of spontaneous mammary tumors in BALB/c p53 heterozygous mice. A model for Li-Fraumeni syndrome. Am J Pathol *157*, 2151-2159.

Lavigueur, A., Maltby, V., Mock, D., Rossant, J., Pawson, T., and Bernstein, A. (1989). High incidence of lung, bone, and lymphoid tumors in transgenic mice overexpressing mutant alleles of the p53 oncogene. Mol Cell Biol *9*, 3982-3991.

Levine, A. J. (1997). p53, the cellular gatekeeper for growth and division. Cell *88*, 323-331.

Li, B., Rosen, J. M., McMenamin-Balano, J., Muller, W. J., and Perkins, A. S. (1997). neu/ERBB2 cooperates with p53-172H during mammary tumorigenesis in transgenic mice. Mol Cell Biol *17*, 3155-3163.

Liu, G., McDonnell, T. J., Montes de Oca Luna, R., Kapoor, M., Mims, B., El-Naggar, A. K., and Lozano, G. (2000). High metastatic potential in mice inheriting a targeted p53 missense mutation. Proc Natl Acad Sci U S A *97*, 4174-4179.

Liu, G., Parant, J. M., Lang, G., Chau, P., Chavez-Reyes, A., El-Naggar, A. K., Multani, A., Chang, S., and Lozano, G. (2004). Chromosome stability, in the absence of apoptosis, is critical for suppression of tumorigenesis in Trp53 mutant mice. Nat Genet *36*, 63-68.

Lozano, G., and Liu, G. (1998). Mouse models dissect the role of p53 in cancer and development. Semin Cancer Biol *8*, 337-344.

Lu, H., Taya, Y., Ikeda, M., and Levine, A. J. (1998). Ultraviolet radiation, but not gamma radiation or etoposide-induced DNA damage, results in the phosphorylation of the murine p53 protein at serine-389. Proc Natl Acad Sci U S A *95*, 6399-6402.

Ludwig, R. L., Bates, S., and Vousden, K. H. (1996). Differential activation of target cellular promoters by p53 mutants with impaired apoptotic function. Mol Cell Biol *16*, 4952-4960.

Luo, J. L., Tong, W. M., Yoon, J. H., Hergenhahn, M., Koomagi, R., Yang, Q., Galendo, D., Pfeifer, G. P., Wang, Z. Q., and Hollstein, M. (2001a). UV-induced DNA damage and mutations in Hupki (human p53 knock-in) mice recapitulate p53 hotspot alterations in sun-exposed human skin. Cancer Res *61*, 8158-8163.

Luo, J. L., Yang, Q., Tong, W. M., Hergenhahn, M., Wang, Z. Q., and Hollstein, M. (2001b). Knock-in mice with a chimeric human/murine p53 gene develop normally and show wild-type p53 responses to DNA damaging agents: a new biomedical research tool. Oncogene *20*, 320-328.

Maier, B., Gluba, W., Bernier, B., Turner, T., Mohammad, K., Guise, T., Sutherland, A., Thorner, M., and Scrable, H. (2004). Modulation of mammalian life span by the short isoform of p53. Genes Dev *18*, 306-319.

Malkin, D., Li, F. P., Strong, L. C., Fraumeni, J. F., Jr., Nelson, C. E., Kim, D. H., Kassel, J., Gryka, M. A., Bischoff, F. Z., and Tainsky, M. A. (1990). Germ line p53 mutations in a familial syndrome of breast cancer, sarcomas, and other neoplasms. [see comments.]. Science *250*, 1233-1238.

Mao, J. H., and Balmain, A. (2003). Genomic approaches to identification of tumour-susceptibility genes using mouse models. Curr Opin Genet Dev *13*, 14-19.

Milner, J., and Medcalf, E. A. (1991). Cotranslation of activated mutant p53 with wild type drives the wild-type p53 protein into the mutant conformation. Cell *65*, 765-774.

Muller, A. J., Teresky, A. K., and Levine, A. J. (2000). A male germ cell tumor-susceptibility-determining locus, pgct1, identified on murine chromosome 13. Proc Natl Acad Sci U S A *97*, 8421-8426.

Murphy, E. D. (1966). Characteristic tumors, in Biology of the laboratory mouse, 2nd ed. (Green, E L, ed), McGraw-Hill, New York, 521-562.

Nakamura, T., Pichel, J. G., Williams-Simons, L., and Westphal, H. (1995). An apoptotic defect in lens differentiation caused by human p53 is rescued by a mutant allele. Proc Natl Acad Sci U S A *92*, 6142-6146.

Overholtzer, M., Rao, P. H., Favis, R., Lu, X. Y., Elowitz, M. B., Barany, F., Ladanyi, M., Gorlick, R., and Levine, A. J. (2003). The presence of p53 mutations in human osteosarcomas correlates with high levels of genomic instability. Proc Natl Acad Sci U S A *100*, 11547-11552.

Paige, A. J. (2003). Redefining tumour suppressor genes: exceptions to the two-hit hypothesis. Cell Mol Life Sci *60*, 2147-2163.

Parant, J. M., and Lozano, G. (2003). Disrupting TP53 in mouse models of human cancers. Hum Mutat *21*, 321-326.

Possemato, R., Eggan, K., Moeller, B. J., Jaenisch, R., and Jackson-Grusby, L. (2002). Flp recombinase regulated lacZ expression at the ROSA26 locus. Genesis *32*, 184-186.

Purdie, C. A., Harrison, D. J., Peter, A., Dobbie, L., White, S., Howie, S. E., Salter, D. M., Bird, C. C., Wyllie, A. H., and Hooper, M. L. (1994). Tumour incidence, spectrum and ploidy in mice with a large deletion in the p53 gene. Oncogene *9*, 603-609.

Reilly, K. M., Loisel, D. A., Bronson, R. T., McLaughlin, M. E., and Jacks, T. (2000). Nf1;Trp53 mutant mice develop glioblastoma with evidence of strain- specific effects. Nat Genet *26*, 109-113.

Rowan, S., Ludwig, R. L., Haupt, Y., Bates, S., Lu, X., Oren, M., and Vousden, K. H. (1996). Specific loss of apoptotic but not cell-cycle arrest function in a human tumor derived p53 mutant. Embo J *15*, 827-838.

Sah, V. P., Attardi, L. D., Mulligan, G. J., Williams, B. O., Bronson, R. T., and Jacks, T. (1995). A subset of p53-deficient embryos exhibit exencephaly. Nat Genet *10*, 175-180.

Schmitt, C. A., Fridman, J. S., Yang, M., Baranov, E., Hoffman, R. M., and Lowe, S. W. (2002). Dissecting p53 tumor suppressor functions in vivo. Cancer Cell *1*, 289-298.

Shen, H. M., and Ong, C. N. (1996). Mutations of the p53 tumor suppressor gene and ras oncogenes in aflatoxin hepatocarcinogenesis. Mutat Res *366*, 23-44.

Shieh, S. Y., Ikeda, M., Taya, Y., and Prives, C. (1997). DNA damage-induced phosphorylation of p53 alleviates inhibition by MDM2. Cell *91*, 325-334.

Shieh, S. Y., Taya, Y., and Prives, C. (1999). DNA damage-inducible phosphorylation of p53 at N-terminal sites including a novel site, Ser20, requires tetramerization. Embo J *18*, 1815-1823.

Sigal, A., and Rotter, V. (2000). Oncogenic mutations of the p53 tumor suppressor: the demons of the guardian of the genome. Cancer Res *60*, 6788-6793.

Slingerland, J. M., Jenkins, J. R., and Benchimol, S. (1993). The transforming and suppressor functions of p53 alleles: effects of mutations that disrupt phosphorylation, oligomerization and nuclear translocation. Embo J *12*, 1029-1037.

Sluss, H. K., Armata, H., Gallant, J., and Jones, S. N. (2004). Phosphorylation of serine 18 regulates distinct p53 functions in mice. Mol Cell Biol *24*, 976-984.

Soussi, T., and Beroud, C. (2003). Significance of TP53 mutations in human cancer: a critical analysis of mutations at CpG dinucleotides. Hum Mutat *21*, 192-200.

Srivastava, S., Zou, Z. Q., Pirollo, K., Blattner, W., and Chang, E. H. (1990). Germ-line transmission of a mutated p53 gene in a cancer-prone family with Li-Fraumeni syndrome. Nature *348*, 747-749.

Storer, R. D., French, J. E., Haseman, J., Hajian, G., LeGrand, E. K., Long, G. G., Mixson, L. A., Ochoa, R., Sagartz, J. E., and Soper, K. A. (2001). P53+/- hemizygous knockout mouse: overview of available data. Toxicol Pathol *29 Suppl*, 30-50.

Sturzbecher, H. W., Brain, R., Addison, C., Rudge, K., Remm, M., Grimaldi, M., Keenan, E., and Jenkins, J. R. (1992). A C-terminal alpha-helix plus basic region motif is the major structural determinant of p53 tetramerization. Oncogene *7*, 1513-1523.

Symonds, H., Krall, L., Remington, L., Saenz-Robles, M., Lowe, S., Jacks, T., and Van Dyke, T. (1994). p53-dependent apoptosis suppresses tumor growth and progression in vivo. Cell *78*, 703-711.

Tsukada, T., Tomooka, Y., Takai, S., Ueda, Y., Nishikawa, S., Yagi, T., Tokunaga, T., Takeda, N., Suda, Y., Abe, S., and et al. (1993). Enhanced proliferative potential in culture of cells from p53-deficient mice. Oncogene *8*, 3313-3322.

Tyner, S. D., Choi, J., Laucirica, R., Ford, R. J., and Donehower, L. A. (1999). Increased tumor cell proliferation in murine tumors with decreasing dosage of wild-type p53. Mol Carcinog *24*, 197-208.

Tyner, S. D., Venkatachalam, S., Choi, J., Jones, S., Ghebranious, N., Igelmann, H., Lu, X., Soron, G., Cooper, B., Brayton, C., *et al.* (2002). p53 mutant mice that display early ageing-associated phenotypes. Nature *415*, 45-53.

Unger, T., Juven-Gershon, T., Moallem, E., Berger, M., Vogt Sionov, R., Lozano, G., Oren, M., and Haupt, Y. (1999). Critical role for Ser20 of human p53 in the negative regulation of p53 by Mdm2. Embo J *18*, 1805-1814.

Varley, J. M., Thorncroft, M., McGown, G., Appleby, J., Kelsey, A. M., Tricker, K. J., Evans, D. G., and Birch, J. M. (1997). A detailed study of loss of heterozygosity on chromosome 17 in tumours from Li-Fraumeni patients carrying a mutation to the TP53 gene. Oncogene *14*, 865-871.

Venkatachalam, S., Shi, Y. P., Jones, S. N., Vogel, H., Bradley, A., Pinkel, D., and Donehower, L. A. (1998). Retention of wild-type p53 in tumors from p53 heterozygous mice: reduction of p53 dosage can promote cancer formation. Embo J *17*, 4657-4667.

Venkatachalam, S., Tyner, S. D., Pickering, C. R., Boley, S., Recio, L., French, J. E., and Donehower, L. A. (2001). Is p53 haploinsufficient for tumor suppression? Implications for the p53+/- mouse model in carcinogenicity testing. Toxicol Pathol *29*, 147-154.

Vogelstein, B., Lane, D., and Levine, A. J. (2000). Surfing the p53 network. Nature *408*, 307-310.

Waterman, M. J., Stavridi, E. S., Waterman, J. L., and Halazonetis, T. D. (1998). ATM-dependent activation of p53 involves dephosphorylation and association with 14-3-3 proteins. Nat Genet *19*, 175-178.

Williams, B. O., Remington, L., Albert, D. M., Mukai, S., Bronson, R. T., and Jacks, T. (1994). Cooperative tumorigenic effects of germline mutations in Rb and p53. Nat Genet *7*, 480-484.

Chapter 9

P53, P63, AND P73: INTERNECINE RELATIONS?

Frank McKeon[1] and Annie Yang[2]
[1]*Department of Cell Biology,* [2]*Department of Biological Chemistry and Molecular Pharmacology, Harvard Medical School, Boston, MA, USA*

INTRODUCTION

The discoveries of p63 and p73 as genes related to the vaunted tumor suppressor p53 launched questions that remain largely unanswered, and fuel controversy and debate. Do these homologs behave like p53? Do they also act in tumor suppression? What were their origins – spin-offs of an ancestral p53 gene, or, in fact, predecessors of this famed 'guardian of the genome'? *In vivo* studies clearly reveal distinct physiological roles for p53, p63, and p73. But do these belie cooperative or antagonistic interactions within the p53 gene family?

Answers to these questions will be key to understanding the biological activities of the p53 family, the mechanisms by which each member directs unique physiological programs, as well as the extent of their functional interactions. Beyond any significance they may have in cancer biology, the p53 homologs are also distinguishing themselves as pivotal mediators of stem cell maintenance, homeostasis, and development. We discuss the various and often contrasting views in studies to date, as they underscore the complexity and challenges presented by this intriguing gene family.

P. Hainaut and K.G. Wiman (eds.), 25 Years of p53 Research, 209-222.
© 2005 *Springer. Printed in the Netherlands.*

STARTING AT THE BEGINNING: ORIGINS OF THE P53 FAMILY

It had been known prior to the discoveries of p73 and p63 that genes distantly related to p53 were present in mollusks (Barker et al., 1997; van Beneden et al., 1997), suggesting that p53 functions arose in invertebrates. This notion was both supported yet challenged by the cloning of p73 (Kaghad et al., 1997) and p63 (Yang et al., 1998; Osada et al., 1998; Augustin et al., 1998; Senoo et al., 1998) – which demonstrated that these homologs were actually much more related to the mollusk p53-like gene than was mammalian p53 itself. Yet upon the sequencing of the *Drosophila melanogaster* and *C. elegans* genomes, a single p53-like gene was identified in each species and purported to be, in fact, the ancestral p53 (Brodsky et al., 2000; Ollmann et al., 2000; Derry et al., 2001; Schumacher et al., 2001). Further experiments in each of these organisms supported p53-like functions, including apoptosis and chromosome maintenance. In particular, Drosophila p53 (Dmp53) was generally linked to apoptosis in response to DNA damage in a process that could involve its ability to transactivate *Reaper* (Brodsky et al., 2000; Ollmann et al., 2000). In more genetically defined experiments, *C. elegans* p53 (Cep53 or CEP-1) was shown to be important for apoptosis of germ cells following DNA damage, and separate experiments tied Cep53 to meiotic chromosome segregation (Derry et al., 2001; Schumacher et al., 2001). Curiously, somatic cells did not appear dependent on Cep53 for DNA damage-induced apoptosis, suggesting the possibility that this early p53 activity served to protect genome fidelity rather than to marshal against tumor growth. More recent work has shown that Dmp53 is a central feature of the response to DNA damage through the simultaneous activation of genes involved in apoptosis, such as *Hid*, *Reaper*, and *Sickle*, as well as DNA repair genes including Ku70 and Ku80 (Peters et al., 2002; Jassim et al., 2003; Sogame et al., 2003; Brodsky et al., 2004). The sum of these data argues strongly that at least some of the basic functions we attribute to mammalian p53 evolved in invertebrates and was executed by a p53-like protein.

What may be incorrect about this picture is the identity, or designation, of these orthologs as p53 *per se*. Certainly BLAST searches reveal both Dmp53 and Cep53 to be more closely related to p63 than to p73 or p53. But an even stronger argument for the priority of p63 in the evolution of this gene family comes from the analysis of p53-like genes in metazoan evolution. As noted, the only p53-like protein reported in mollusks is highly homologous with mammalian p63 (Barker et al., 1997; van Beneden et al., 1997; Yang et al., 1998). This observation is significant because phylogenic assemblies of metazoan evolution indicate a bifurcation, occurring some 400 million years ago, of protostomes (including mollusks, insects, and nematodes) and deuterostomes (represented by hemi-chordates and chordates) (Gerhart and Kirschner, 1997; Aguinaldo et al., 1997). The presence of p63, and not p73 or p53, on both divergent paths therefore highlights its identity as the primitive ancestor of the ensuing gene family (Yang et al., 2002; Figure 1).

Arguments about the phylogeny of the p53 family may be more than academic. The capacity of *Drosophila* and *C. elegans* as powerful model systems from which to deduce gene function in higher organisms is less than clear: do the data from flies and worms tell us more about p53, or do they reflect the functions of its true ancestor, p63?

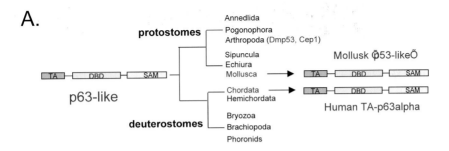

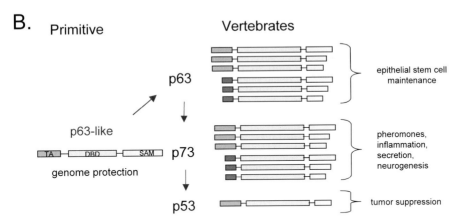

Figure 1. Evolution of the p53 Family of Genes (A) A plausible scenario for the evolution of the p53 family members based on comparisons of p53-like sequences derived from invertebrates and vertebrates. A major divergence of metazoan phylogeny giving rise to protostomes, including mollusks, and to deuterostomes, including vertebrates, is shown. A schematic of the solitary p53-like gene in mollusks is depicted, encoding a product with strong homology to the human TA-p63alpha protein. Similarly, the single p53-like molecule in flies and nematode worms is also more similar to p63 than to p73 or p53, suggesting the likely ancestral gene was similar to TAp63alpha. TA, transactivation domain, DBD, DNA binding domain, SAM, sterile alpha motif, a protein-protein interaction domain. (B)

Schematic depicting the ancestral member of the p53 family having a transactivation domain (TA), a DNA-binding domain (DBD), and a C-terminal domain (C-term). The arrows indicate the most likely sequence of events leading to the p53 family seen in vertebrates. First, there appears to be the gain of a separate promoter that gives rise to ΔN isoforms of the original gene, as well as splicing strategies that yield C-terminal variations. During the genome quadruplication (Holland, 2003) events that led up to vertebrate evolution, the p73 gene is thought to arise from a duplication of the p63 gene, and the mammalian p53 is shown to arise from a duplication of the p73 gene. The functions listed for each gene is a partial list and mostly derived from analyses of mouse gene knockout models.

TYPECASTING THE HOMOLOGS: DYING TO BE P53?

In light of their striking sequence and structural similarities to p53, the homologs were widely anticipated to share p53's role in growth control, cell death, and even tumor suppression. Indeed, p73 is localized to a region on chromosome 1p36 commonly lost in human cancers, a trademark of tumor suppressor genes. p63 and p73 also passed a bevy of litmus tests for p53 function: they can bind to and transactivate canonical p53 targets (Jost et al., 1997; Kaghad et al., 1997; Yang et al., 1998; Chen et al., 2003) induce apoptosis when overexpressed in cells (Jost et al., 1997; Osada et al., 1998; Yang et al., 1998), and appear to respond to DNA damage in some instances (Agami et al., 1999; Gong et al., 1999; Yuan et al., 1999; Flores et al., 2002).

Yet there were hints early on that p63 and p73 would not fit so easily into the p53 tumor suppressor mold. For one, mutations in p63 or p73 are exceedingly rare in human cancers (Kaghad et al., 1997; Osada et al., 1998), though epigenetic modulation of these genes and links between expression and disease prognosis have been reported (Hibi et al., 2000; Davis and Dowdy, 2001; Lindstrom and Wiman, 2002; Puig et al., 2003; Koga et al., 2003). These clinical findings were verified by mouse knockout (KO) studies which revealed no predisposition for tumors in p73-deficient mice (Yang et al., 2000). The neonatal lethality in p63 KO mice precluded an analysis of tumor susceptibility (Yang et al., 1999; Mills et al., 1999), but no evidence of a link between p63 deficiency and tumorigenesis has been established *in vivo*. Indeed, mutations in p63 have been found in human patients, but these are associated with severe developmental syndromes without a corresponding increase in cancer rates (Celli et al., 1999; Duijf et al., 2002; van Bokhoven and McKeon, 2002).

A closer examination of p63 and p73 provides a likely explanation for their deviation from p53 paradigms: unlike the p53 gene, which encodes a single, major transcript, the p63 and p73 genes share a more complex structure (Figure 2). Two separate promoters drive the expression of fundamentally different classes of p63/p73 proteins. The distal promoter gives rise to p63/p73 proteins containing an acidic, N-terminal transactivation (TAp63/p73) domain analogous to that found in p53. A second, internal promoter generates products lacking this N-terminus (ΔNp63/p73), but having the same DNA-binding and oligomerization sequences. Notably, these ΔN isoforms are capable of acting as dominant negatives toward their TA counterparts, and even toward p53 itself (Yang et al., 1998; Yang et al., 2000, Grob et al., 2001). The physiological significance of these findings is suggested by the apparent need to down-regulate ΔN isoforms of p63 or p73 in some p53-dependent processes of cell death (Liefer et al., 2000; Pozniak et al., 2000). Alternative splicing events at the C-terminus offer even more variety to p63 and p73's form and functionality (Figure 2). For instance, p63/p73 alpha isoforms contain a sterile-alpha-motif (SAM) that may be important for protein-protein interactions. A stretch of 70 amino acids at the end of TAp63/p73 has also been shown to bind the transactivation domain in a mechanism of intramolecular inhibition (Serber et al., 2002).

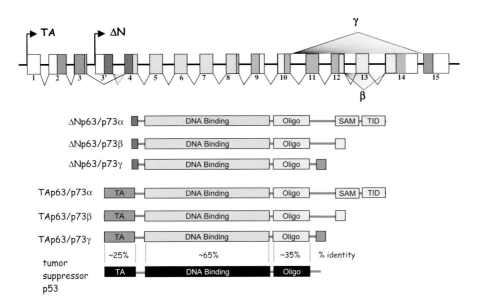

Figure 2. Structure of the p53 Homologs. Schematic depicts the genomic organization of the p63 and p73 genes, located on human chromsome 3q27 and 1p36.3, respectively. Complex alternative splicing and the use of two distinct promoters (TA, ΔN) result in a wide array of transcripts. The main functional domains essential for p53's tumor suppressor and growth control activities - an N-terminal transactivation (TA) region, a core DNA-binding domain, and a C-terminal oligomerization (oligo) domain - are conserved in p63 and p73. Additional domains, including a sterile alpha motif (SAM) and a transactivation inhibitory domain (TID) at the extreme C-terminus, may serve as regulatory elements not found in p53.

Despite obvious differences between p53 and its homologs, and the nuances posed by the multitude of gene products, functional studies on p63 and p73 have displayed a decidedly p53-centric theme. The two major activities of p53 – growth arrest and induction of apoptosis – can each be executed by p63 and p73 in experimental settings. Several groups have thus championed the notion that p73 – and, to an extent, p63 – play fundamental roles in regulating proliferation and cell death. p73 is reportedly an essential component of E2F-mediated apoptosis, both in response to DNA damage as well as in activated T cells (Lissy et al., 2000; Irwin et al., 2000; Alexander et al., 2003; Pediconi et al., 2003). Studies in E1A-transformed fibroblasts go on to demonstrate a requirement for p63 and p73 in the apoptosis functions of p53 itself (Flores et al., 2002). Finally, it has been reported that inactivation of p73 contributes to chemoresistance in human cancer cells (Irwin et al., 2003), leading some to suggest p73's importance in clinical outcome and response to therapy (Bergamaschi et al., 2003; Gasco and Crook, 2003).

While tantalizing, the above results are tempered by several lines of evidence against the notion that p63 and p73 serve predominantly p53-like roles in apoptosis and tumor suppression. For one, most of these experiments fail to address or explain the activity of ΔNp63/p73 proteins, which are often expressed at much higher levels than their TA counterparts under basal conditions. In fact, numerous studies point to the over-expression of wild type p63 or p73 in tumor cells (Kaghad et al., 1997; Yang et al., 1998; Hibi et al., 2000), thereby challenging the notion that these homologs have anti-tumor properties. Indeed, it may be telling that p63 and p73 do not appear to be targets for mutagenesis or inactivation in human cancers (Ikawa et al., 1999; Melino et al., 2002; Bernard et al., 2003). Arguably the most convincing proof that p53 is a tumor suppressor comes from its high rates of mutation and functional alteration in tumors. If p63 or p73 plays such an essential role in cell death, one might expect the clinical data to reflect selective pressure against these genes in cancer. And, as noted above, mouse models of p73 deficiency do not reveal an increased susceptibility to spontaneous tumorigenesis (Yang et al., 2000). Recent work from our laboratory has also found no evidence for p63 or p73 in p53-dependent or independent apoptosis in genetically defined (with respect to p53/p63/p73 status) T cells (M. Senoo and F.M., unpublished observations). Together, these data argue against a general requirement for p63 and p73 in mechanisms of cell death, and caution against the prevailing belief that the p53 homologs play pivotal roles in tumor suppression.

P63 IN EPITHELIAL PROGENITORS – THE CONTROVERSY CONTINUES

If p63 and p73 ultimately disappoint as tumor suppressors, they have nevertheless emerged as critical regulators of several fundamental physiological processes. Of the two, p63 is perhaps more accessible with regard to efforts toward understanding its biological function. Our first clues came with the observation that anti-p63 monoclonal antibodies detected abundant expression in the progenitor cells of many glandular and squamous epithelial tissues, including skin, breast, prostate, and urothelia. In all these epithelial tissues, p63 decreased dramatically with cellular differentiation. To further probe p63 function, we and others engineered mice with a targeted disruption of the p63 locus (Yang et al., 1999; Mills et al., 1999). Strikingly, p63-deficient mice were born without limbs, epidermis, breast, prostate, urothelia, and all other tissues which normally harbored p63 in their progenitor cells. While the overall phenotype and cause of death of these mice were stipulated by the two groups, the mechanism underlying the loss of epithelial tissues was the subject of considerable debate (Figure 3). Mills et al. argued that p63 was actually a "commitment" factor required for the conversion of an undifferentiated ectoderm to, for example, an epidermal lineage (Mills et al., 1999; Koster et al., 2004). The basis for this argument was their failure to observe either a stratified epidermis or epidermal differentiation markers on the surface of the embryos. In contrast, we observed clear evidence of stratification and differentiation markers, although most of this differentiated epidermis appeared to be in the process of lifting from the surface of the embryo (Yang et al., 1999). We therefore argued that neither commitment nor differentiation was affected by the loss of p63. Instead, the catastrophic loss of stratifying epidermis, coupled with the high expression of p63 in progenitor cells, led us to conclude that the defect was in 'proliferation potential'-- these tissues simply ran out of stem, or reserve, cells needed to maintain such regenerative epithelia. At the time, this seemed a fairly novel proposition -- corresponding factors essential for the proliferative potential of embryonic, hematopoietic, or neural stem cells had yet to be determined. Additional lines of evidence can be put forth to support a role for p63 in proliferative potential rather than commitment, differentiation, or proliferation *per se*. For one, DeLuca and colleagues used the epithelial stem cell cloning techniques of Howard Green (Barrandon and Green, 1985) to show that cells with the highest proliferative potential have the highest expression of p63 (Pelligrini et al., 2001). In addition, we have examined the thymic epithelium in p63-deficient mice, and have shown that the defects in thymus development are attributed to the loss of proliferative potential. In these animals, thymic epithelial commitment and differentiation

-- as judged by their ability to support T cell development and proliferation –
appear unaffected (M. Senoo and F.M., unpublished).

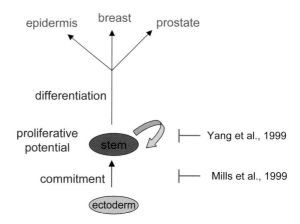

Figure 3. Function in Epithelial Stem Cells Deduced from Mouse Mutants. Epithelial tissues
including epidermis, breast, and prostate are depicted arising from a committed stem cell that
in turn is derived from an uncommitted ectodermal (or endodermal) cell. Mills et al. (1999)
argued that the loss of p63 prevents commitment and therefore differentiation of ectoderm to
stem cells. Yang et al. (1999) argued that differentiation of epidermis was unimpeded in the
p63-null mouse, and therefore that neither differentiation nor commitment was affected.
Instead, the latter view would propose that p63 is required to maintain the proliferative
potential of epithelial stem cells.

    If p63 is required to maintain the proliferative potential of the progenitor
cells of the various epithelial tissues, how does it do it? Whereas one simple
explanation would be that p63 transactivates a number of "stem cell genes"
that maintain the cell in some immature state, virtually all of the p63
expressed in these cells is of the "ΔN" form, and therefore largely lacking
any obvious means of driving gene expression. An alternative is that ΔNp63
is acting to repress genes that would otherwise be engaged in programs of
differentiation or senescence.   This hypothesis is somewhat appealing
because it brings p63 into analogy with Bmi-1, a gene recently shown to be
essential for maintaining hematopoietic and neural stem cells (Lessard and
Sauvageau, 2003; Park et al., 2003; Molofsky et al., 2003). More detailed
experiments have now narrowed its function to proliferative potential of
blood cells and subsets of neurons. Bmi-1 is also of tremendous interest
because it is homologous to proteins in the polycomb complex known to
repress *Hox* gene expression during development and other genes in adults.
Given that p63 and Bmi-1 are the only known adult stem cell factors
required for proliferative potential, and both appear to act as repressors, an
attractive possibility is that they function in analogous pathways to maintain

stem cells. One of the target genes for Bmi-1 repression is the INK4A locus implicated in pathways of cellular senescence (Jacobs et al., 1999; Jacobs and van Lohuizen, 2002). Presumably, suppressing pathways of senescence is one of many required to maintain a cellular state where immortality, without loss of proliferation control, is achieved. It remains to be seen whether p63 achieves stem cell control by similar mechanisms, perhaps even via direct interactions with polycomb complexes or other chromatin remodeling factors.

SENSE AND SENSIBILITY: THE MANY FUNCTIONS OF P73

p73's biological activities are probably the most enigmatic of the p53 family members because they appear so distributed (Kaghad et al., 1997; Yang et al., 2000). Indeed, p73 seems to be recruited for everything from pheromone transduction to inflammation control to the development of key portions of the brain required for memory acquisition. Again, the mouse knockout phenotype has been instrumental to uncovering the remarkably diverse roles assumed by p73 (Yang et al., 2000). Unlike the p63-null mouse, which dies at birth, p73-deficient mice are viable and appear similar to littermates at birth. Within the next few days, however, they assume a runted appearance, show inflammation in their nasal passages, and develop gastrointestinal hemorrhages leading to high rates of mortality. Survivors gain weight but show varying degrees of hydrocephalus, likely due to an inappropriate secretion and/or absorption of cerebrospinal fluid (CSF). In adulthood, additional defects become apparent as p73-deficient mice show limited interactions with other mice and fail to respond to social and mating cues. Further analysis linked these behavioral defects to the absence of pheromone receptors in neurosensory cells of the vomeronasal organ. Together, these aspects of the knockout phenotype pointed to an essential role for p73 in various pathways of sensory and homeostatic control. Such activities are somewhat analogous to the DNA damage-sensing abilities of p53, and may reflect a conserved biological function in members of this gene family.

p73-deficient mice also display an unusual defect in the hippocampus, thought to be the center of learning and memory in higher mammals. Marked by a 'wavy' formation of pyramidal neurons in the CA2 region and a highly distorted dentate gyrus, this hippocampal dysgenesis is accompanied by the absence of Cajal-Retizus (CR) cells, a distinct population of bipolar "pioneer" neurons that occupy the marginal zone of the cortex and the molecular zone of the hippocampus (Sarnat and Flores-Sarnat,

2002). Importantly, the CR neurons are known to secrete reelin, a key regulator of neuronal migration, and have thus been linked to catastrophic cortical lamination defects in the Reeler mouse (Frotscher, 1998). p73-deficient mice, however, do not exhibit any defects in cortical lamination such as those seen in *reeler* mutants. This finding challenged previously held notions on the cortical lamination function of the Cajal-Retzius cells, and instead focused efforts on understanding how these pioneer neurons affect the organization of the hippocampus (Yang et al., 2000; Meyer at al., 2002).

The apparent requirement for p73 in the maintenance of CR neurons suggests a pro-survival function for this p53 homolog. In a separate study of the peripheral nervous system, Kaplan and colleagues reported the loss of nearly fifty percent of the sympathetic neurons in the p73-null mouse (Pozniak et al., 2000). They attributed this loss to the requirement of dominant-negative forms of p73 to offset the actions of p53 that would lead to apoptosis. We are far from a mechanistic understanding of p73 function in Cajal-Retzius neurons, other than knowing that the loss of p53 does not rescue the hippocampal dysgenesis in p53/p73 double mutant mice (A.Y. & F.M., unpublished observations).

It is also unclear if common themes for p73 function can be derived from the litany of p73-dependent physiological systems affected in the knockout mouse. Part of the problem in making functional links between these seemingly diverse systems is that some might use p73 as a transcriptional activator and others as transcriptional repressors in the form of ΔNp73. As with p63, it will be important to determine p73's isoform specificity and function in the signaling pathways of pheromone detection, inflammation, CSF dynamics, and neuronal survival.

CONCLUSIONS

The p53 family has undergone a remarkable series of structural and functional metamorphoses in its progression from protector of germ cells in invertebrates to regulators of tumor suppression, epithelial stem cells, homeostasis, and neurogenesis in vertebrates. The accumulated complexity of the p53 homologs is at once daunting and provocative. The various debates and contradictions seem unavoidable given the functional dichotomy inherent to the TA and ΔN isoforms of the p63 and p73 genes. In the area of cancer biology, battle lines are still being drawn with respect to the roles of these genes in tumor suppression versus tumor promotion, both of which are conceivable in light of the pro- and anti-p53 properties of the TA and ΔN isoforms. The mouse knockout models have offered intriguing insights into the individual functions of the p53 family members, but have left us with no clear impression of overlapping or synergistic functional interactions within the family. Given the high conservation of their DNA binding domains, the lack of phenotypic similarities in the p53, p63, and p73 knockout mice is surprising. An outstanding question is how these transcription factors distinguish among their physiological targets, particularly in cells where two or more family members are co-expressed. Obviously the levels of control of genetic programs are more subtle and intricate than imagined at present, as are the extent and consequence of interactions among the p53 family members. The challenge facing the field is to understand how these intriguing molecules work -- alone or together -- to effect intricate, subtle, and mysterious processes in the cell.

REFERENCES

Agami, R., Blandino, G., Oren, M., and Shaul, Y. 1999. Interaction of c-Abl and p73alpha and their collaboration to induce apoptosis. Nature 399, 809-813.

Aguinaldo, A.M., Turbeville, J.M., Linford, L.S., Rivera, M.C., Garey, J.R., Raff, R.A., and Lake, J.A., 1997. Evidence for a clade of nematodes, arthropods and other moulting animals. Nature 387, 489-493.

Alexander, K., Yang, H.S., and Hinds, P.W. 2003. pRb inactivation in senescent cells leads to an E2F-dependent apoptosis requiring p73. Mol. Cancer Res. 1, 716-728.

Augustin, M., Bamberger, C., Paul, D., and Schmale, H. 1998. Cloning and chromosomal mapping of the human p53-related KET gene to chromosome 3q27 and its murine homolog Ket to mouse chromosome 16. Mamm. Genome 9, 899-902.

Barker, C.M., Calvert, R.J., Walker, C.W., and Reinisch, C.L. 1997. Detection of mutant p53 in clam leukemia cells. Exp. Cell Res. 232, 240-245.

Barrandon, Y., and Green, H. 1985. Cell size as a determinant of the clone-forming ability of human keratinocytes. Proc. Natl. Acad. Sci. USA 82, 5390-5394.

Benard J, Douc-Rasy S, Ahomadegbe JC. 2003. TP53 family members and human cancers. Hum Mutat. 21(3):182-91.

Bergamaschi, D., Gasco, M., Hiller, L., Sullivan, A., Syed, N., Trigiante, G., Yulug, I., Merlano, M., Numico, G., Comino, A., Attard, M., Reelfs, O., Gusterson, B., Bell, A.K., Heath, V., Tavassoli, M., Farrell, P.J., Smith, P., Lu, X., and Crook, T. 2003. p53 polymorphism influences response in cancer chemotherapy via modulation of p73-dependent apoptosis. Cancer Cell 3, 387-402.

Brodsky, M.H., Nordstrom, W., Tsang, G., Kwan, E., Rubin, G.M., and Abrams, J.M. 2000. Drosophila p53 binds a damage response element at the reaper locus. Cell 101, 103-113.

Brodsky, M.H., Weinert, B.T., Tsang, G., Rong, Y.S., McGinnis, N.M., Golic, K.G., Rio, D.C., and Rubin, G.M. 2004. Drosophila melanogaster MNK/Chk2 and p53 regulate multiple DNA repair and apoptotic pathways following DNA damage. Mol. Cell. Biol. 24, 1219-1231.

Celli, J., Duijf, P., Hamel, B.C., Bamshad, M., Kramer, B., Smits, A.P., Newbury-Ecob, R., Hennekam, R.C., van Buggenhout, G., van Haeringen, A., Woods, C.G., van Essen, A.J., de Waal, R., Vriend, G., Haber, D.A., Yang, A., McKeon, F., Brunner, H.G., van

Bokhoven, H. 1999. Heterozygous germline mutations in the p53 homolog p63 are the cause of EEC syndrome. Cell 99, 143-153.

Chen, X., Liu, G., Zhu, J., Jiang, J., Nozell, S., and Willis, A. 2003. Isolation and characterization of fourteen novel putative and nine known target genes of the p53 family. Cancer Biol. Ther. 2, 55-62.

Davis, P.K., and Dowdy, S.F. 2001. p73. Int. J. Biochem. Cell Biol. 33, 935-939.

Derry, W.B., Putzke, A.P., and Rothman, J.H. 2001. Caenorhabditis elegans p53: role in apoptosis, meiosis, and stress resistance. Science 294, 591-595.

Duijf, P.H., Vanmolkot, K.R., Propping, P., Friedl, W., Krieger, E., McKeon, F., Dotsch, V., Brunner, H.G., and van Bokhoven, H. 2002. Gain-of-function mutation in ADULT syndrome reveals the presence of a second transactivation domain in p63. *Hum Mol Genet.* **11:** 799-804.

Flores, E.R., Tsai, K.Y., Crowley, D., Sengupta, S., Yang, A., McKeon, F., and Jacks T. 2002. p63 and p73 are required for p53-dependent apoptosis in response to DNA damage. Nature 416, 560-564.

Frotscher, M. 1998. Cajal-Retzius cells, Reelin, and the formation of layers. Curr. Opin. Neurobiol. 8, 570-575.

Gasco, M., and Crook, T. 2003. p53 family members and chemoresistance in cancer: what we know and what we need to know. Drug Resist Updat. 6, 323-328.

Gerhart, J., and Kirschner, M. 1997. Cells, Embryos, and Evolution. Blackwell Science, Malden,MA,USA.

Gong, J.G., Costanzo, A., Yang, H.Q., Melino, G., Kaelin, W.G. Jr., Levrero, M., Wang, J Y 1999. The tyrosine kinase c-Abl regulates p73 in apoptotic response to cisplatin-induced DNA damage. Nature 399, 806-809.

Grob, T.J., Novak, U., Maisse, C, Barcaroli, D., Luthi, A.U., Pirnia, F., Hugli, B., Graber, H.U., De Laurenzi, V., Fey, M.F., Melino, G., and Tobler, A. 2001. Human delta Np73 regulates a dominant negative feedback loop for TAp73 and p53. Cell Death Differ. 8, 1213-1223.

Hibi, K., Trink, B., Patturajan, M., Westra, W.H., Caballero, O.L., Hill, D.E., Ratovitski, E.A., Jen, J., and Sidransky, D. 2000. AIS is an oncogene amplified in squamous cell carcinoma. Proc. Natl. Acad. Sci. USA 97, 5462-5467.

Holland, P.W. 2003. More genes in vertebrates? J. Struct. Funct. Genomics 3, 75-84.

Ikawa, S., Nakagawara, A., and Ikawa, Y. 1999. p53 family genes: structural comparison, expression and mutation. Cell Death Differ. 6, 1154-1161.

Irwin, M., Marin, M.C., Phillips, A.C., Seelan, R.S., Smith, D.I., Liu, W., Flores, E.R., Tsai, K.Y., Jacks, T., Vousden, K.H., and Kaelin, W.G. Jr. 2000. Role for the p53 homologue p73 in E2F-1-induced apoptosis. Nature 407, 645-648.

Irwin, M.S., Kondo, K., Marin, M.C., Cheng, L.S., Hahn, W.C., and Kaelin, W.G. Jr. 2003. Chemosensitivity linked to p73 function. Cancer Cell 3, 403-410.

Ishioka, C., Frebourg, T., Yan, Y.X., Vidal, M., Friend, S.H., Schmidt, S., and Iggo, R. 1993. Screening patients for heterozygous p53 mutations using a functional assay in yeast. Nat. Genet. 5, 124-129.

Jacobs, J.J., Kieboom, K., Marino, S., DePinho, R.A., van Lohuizen, M. 1999. The oncogene and Polycomb-group gene bmi-1 regulates cell proliferation and senescence through the ink4a locus. Nature 397, 164-168.

Jacobs, J.J., and van Lohuizen, M. 2002. Polycomb repression: from cellular memory to cellular proliferation and cancer. Biochim. Biophys. Acta. 1602, 151-161.

Jassim, O.W., Fink, J.L., and Cagan, R.L. 2003. Dmp53 protects the Drosophila retina during a developmentally regulated DNA damage response. EMBO J. 22, 5622-5632.

Jost, C.A., Marin, M.C., and Kaelin, W.G. Jr. 1997. p73 is a simian [correction of human] p53-related protein that can induce apoptosis. Nature 389, 191-194.

Kaghad, M., Bonnet, H., Yang, A., Creancier, L., Biscan, J.C., Valent, A., Minty, A., Chalon, P., Lelias, J.M., Dumont, X., Ferrara, P., McKeon, F., and Caput D. 1997. Monoallelically expressed gene related to p53 at 1p36, a region frequently deleted in neuroblastoma and other human cancers. Cell 90, 809-819.

Koga, F., Kawakami, S., Fujii, Y., Saito, K., Ohtsuka, Y., Iwai, A., Ando, N., Takizawa, T., Kageyama, Y., and Kihara, K. 2003. Impaired p63 expression associates with poor prognosis and uroplakin III expression in invasive urothelial carcinoma of the bladder. Clin. Cancer Res. 9, 5501-5507.

Koster, M.I., Kim, S., Mills, A.A., DeMayo, F.J., and Roop, D.R. 2004. p63 is the molecular switch for initiation of an epithelial stratification program. *Genes Dev.* 18: 126-131.

Lessard, J., and Sauvageau, G. 2003. Bmi-1 determines the proliferative capacity of normal and leukaemic stem cells. Nature 423, 255-260.

Liefer, K.M., Koster, M.I., Wang, X.J., Yang, A., McKeon, F., and Roop, D.R. 2000. Down-regulation of p63 is required for epidermal UV-B-induced apoptosis. Cancer Res. 60, 4016-4020.

Lindstrom, M.S., and Wiman, K.G. 2002. Role of genetic and epigenetic changes in Burkitt lymphoma. Semin. Cancer Biol. 12, 381-387.

Lissy, N.A., Davis, P.K., Irwin, M., Kaelin, W.G., and Dowdy, S.F. 2001. A common E2F-1 and p73 pathway mediates cell death induced by TCR activation. Nature 407, 642-645.

Melino, G., De Laurenzi, V., and Vousden, K.H. 2002. p73: Friend or foe in tumorigenesis. Nat. Rev. Cancer. 2, 605-615.

Melino, G., Lu, X., Gasco, M., Crook, T., and Knight, R.A. 2003. Functional regulation of p73 and p63: development and cancer. *Trends Biochem. Sci.* 28: 663-670.

Meyer, G., Perez-Garcia, C.G., Abraham, H., and Caput, D. 2002. Expression of p73 and Reelin in the developing human cortex. J. Neurosci. 22, 4973-4986.

Mills, A.A., Zheng, B., Wang, X.J., Vogel, H., Roop, D.R., and Bradley, A. 1999. p63 is a p53 homologue required for limb and epidermal morphogenesis. *Nature.* 398: 708-713.

Molofsky, A.V., Pardal, R., Iwashita, T., Park, I.K., Clarke, M.F., and Morrison, S.J. 2003. Bmi-1 dependence distinguishes neural stem cell self-renewal from progenitor proliferation. Nature 425, 962-967.

Ollmann, M., Young, L.M., Di Como, C.J., Karim, F., Belvin, M., Robertson, S., Whittaker, K., Demsky, M., Fisher, W.W., Buchman, A., Duyk, G., Friedman, L., Prives, C., and Kopczynski, C. 2000. Drosophila p53 is a structural and functional homolog of the tumor suppressor p53. Cell 101, 91-101.

Osada, M., Ohba, M., Kawahara, C., Ishioka, C., Kanamaru, R., Katoh, I., Ikawa, Y., Nimura, Y., Nakagawara, A., Obinata, M., and Ikawa, S. 1998. Cloning and functional analysis of human p51, which structurally and functionally resembles p53. Nat. Med. 4, 839-843.

Park, I.K., Qian, D., Kiel, M., Becker, M.W., Pihalja, M., Weissman, I.L., Morrison, S.J., and Clarke, M.F. 2003. Bmi-1 is required for maintenance of adult self-renewing haematopoietic stem cells. Nature 423, 302-305.

Pediconi, N., Ianari, A., Costanzo, A., Belloni, L., Gallo, R., Cimino, L., Porcellini, A., Screpanti, I., Balsano, C., Alesse, E., Gulino, A., and Levrero, M. 2003. Differential regulation of E2F1 apoptotic target genes in response to DNA damage. Nat. Cell Biol. 5, 552-558.

Pellegrini, G., Dellambra, E., Golisano, O., Martinelli, E., Fantozzi, I., Bondanza, S., Ponzin, D., McKeon, F., and De Luca, M. 2001. p63 identifies keratinocyte stem cells. *Proc. Natl. Acad. Sci. U. S. A.* 98: 3156-3161.

Peters, M., DeLuca, C., Hirao, A., Stambolic, V., Potter, J., Zhou, L., Liepa, J., Snow, B., Arya, S., Wong, J., Bouchard, D., Binari, R., Manoukian, A.S., and Mak, T.W. 2002. Chk2 regulates irradiation-induced, p53-mediated apoptosis in Drosophila. Proc. Natl. Acad. Sci. USA. 99,11305-11310.

Pozniak, C.D., Radinovic, S., Yang, A., McKeon, F., Kaplan, D.R., and Miller, F.D. 2000. An anti-apoptotic role for the p53 family member, p73, during developmental neuron death. Science 289, 304-306.

Puig, P., Capodieci, P., Drobnjak, M., Verbel, D., Prives, C., Cordon-Cardo, C., Di Como, C.J. 2003. p73 Expression in human normal and tumor tissues: loss of p73alpha expression is associated with tumor progression in bladder cancer. Clin. Cancer Res. 9, 5642-5651.

Sarnat, H.B., and Flores-Sarnat, L. 2002. Role of Cajal-Retzius and subplate neurons in cerebral cortical development. Sem. Ped. Neurol. 9, 302-308.

Schumacher, B., Hofmann, K., Boulton, S., and Gartner, A. 2001. The C. elegans homolog of the p53 tumor suppressor is required for DNA damage-induced apoptosis. Curr. Biol. 11, 1722-1727.

Senoo, M., Seki, N., Ohira, M., Sugano, S., Watanabe, M., Inuzuka, S., Okamoto, T., Tachibana, M., Tanaka, T., Shinkai, Y., and Kato, H. 1998. A second p53-related protein, p73L, with high homology to p73. Biochem. Biophys. Res. Commun. 248, 603-607.

Serber, Z., Lai, H.C., Yang, A., Ou, H.D., Sigal, M.S., Kelly, A.E., Darimont, B.D., Duijf, P.H, van Bokhoven, H., McKeon, F., and Dotsch, V. 2002. A C-terminal inhibitory domain controls the activity of p63 by an intramolecular mechanism. Mol. Cell. Biol. 22, 8601-8611.

Sogame, N., Kim, M., and Abrams, J.M. 2003. Drosophila p53 preserves genomic stability by regulating cell death. Proc. Natl. Acad. Sci. U S A 100, 4696-4701.

Van Beneden, R.J., Walker, C.W., Laughner, E.S. 1997. Characterization of gene expression of a p53 homologue in the soft-shell clam (Mya arenaria). Mol. Mar. Biol. Biotechnol. 6, 116-122.

van Bokhoven, H., and McKeon, F. 2002. Mutations in the p53 homolog p63: allele-specific developmental syndromes in humans. Trends Mol. Med. 8, 133-139.

Yang, A., Kaghad, M., Wang, Y., Gillett, E., Fleming, M.D., Dotsch, V., Andrews, N.C., Caput, D., and McKeon, F. 1998. p63, a p53 homolog at 3q27-29, encodes multiple products with transactivating, death-inducing, and dominant-negative activities. Mol. Cell 2:305-316.

Yang, A., Schweitzer, R., Sun, D., Kaghad, M., Walker, N., Bronson, R.T., Tabin, C., Sharpe, A., Caput, D., Crum, C., and McKeon, F. 1999. p63 is essential for regenerative proliferation in limb, craniofacial and epithelial development. *Nature*. **398**, 714-718.

Yang, A., Walker, N., Bronson, R., Kaghad, M., Oosterwegel, M., Bonnin, J., Vagner, C., Bonnet, H., Dikkes, P., Sharpe, A., McKeon, F., and Caput, D. 2000. p73-deficient mice have neurological, pheromonal and inflammatory defects but lack spontaneous tumours. Nature 404, 99-103.

Yang, A., Kaghad, M., Caput, D., and McKeon, F. 2002. On the shoulders of giants: p63, p73 and the rise of p53. Trends Genet. 18, 90-95.

Yuan, Z.M., Shioya, H., Ishiko, T., Sun, X., Gu, J., Huang, Y.Y., Lu, H., Kharbanda, S., Weichselbaum, R., and Kufe, D. 1999. p73 is regulated by tyrosine kinase c-Abl in the apoptotic response to DNA damage. Nature 399, 814-817.

Chapter 10

P73, P63 AND MUTANT P53: MEMBERS OF PROTEIN COMPLEXES FLOATING IN CANCER CELLS

Olimpia Monti, Alexander Damalas, Sabrina Strano and Giovanni Blandino
Department of Experimental Oncology, Regina Elena Cancer Institute, Rome, Italy

INTRODUCTION

Approximately half of human tumors bear p53 mutations (Hollestein et al., 1997). The most prevalent type consists of missense mutations that are frequently accompanied by loss of the remaining wild-type p53 (wt-p53) allele (Hainaut et al., 1997; Levine, 1997). The major site of the p53 mutations is the highly conserved DNA binding core domain (Hussain et al., 1998; Prives et al., 1999). Thus, mutant p53 (mt-p53) proteins are unable to specifically bind DNA and to activate specific wt-p53 target genes. Unlike wt-p53, whose half-life is short, mutant p53 proteins are quite stable and abundantly present in cancer cells. One certain outcome of p53 mutations is the loss of wild type activities such as growth arrest, apoptosis, and differentiation (Michalovitz et al., 1990; Yonish-Rouach et al., 1991; Soddu et al.,1996; Almog et al.,1997). However, at variance with other tumor suppressor genes, cells with p53 mutations maintain expression of the full-length protein. This may suggest that, at least certain mutant forms of p53 can gain additional functions through which actively contribute to cancer progression (Prives et al., 1999; Sigal et al., 2000; Strano et al., 2001; Bullock et al., 2001). Such evidence is provided by several *in vitro* and *in vivo* studies (Haley et al., 1990; Dittmer et al., 1993; Gualberto et al., 1998; Frazier et al., 1998; Li et al., 1998; Blandino et al., 1999; Aas et al., 1996; Irwin et al., 2003; Strano et al., 2003)

P. Hainaut and K.G. Wiman (eds.), 25 Years of p53 Research, 223-253.
© 2005 *Springer. Printed in the Netherlands.*

Most genes are members of a family. It is generally believed that a gene family derives from an ancestral gene by duplication and divergence. The tumor suppressor p53 was a striking exception to this established rule. The recent discovery of two new p53 family members, p73 and p63, has solved such exception (Kaghad et al., 1997; Jost, et al., 1997; Osada et al., 1998; Trink et al., 1998; Mills et al., 1999; Levrero et al., 2000). At the sequence level, *p63* and *p73* are more similar to each other than each is to *p53*, suggesting the possibility that the ancestral gene is a gene resembling *p63/p73*, while *p53* is phylogenetically younger (Kaelin, 1999; Yang et al., 2000; Strano et al., 2001). The complexity of the family has also been enriched by the alternatively spliced forms of p63 and p73, which give rise to a complex network of proteins involved in the control of cell proliferation, apoptosis, and development (De Laurenzi et al., 1998; Levrero et al., 2000).

In this chapter we will mainly discuss the existence and the functional implications of protein complexes involving mt-p53, p73 and p63 in cancer cells.

PROTEIN-PROTEIN INTERACTIONS BETWEEN P53 FAMILY MEMBERS IN CANCER CELLS

Cells containing mt-p53 proteins have turned out to be a context in which protein complexes involving the diverse members of the p53 family exist (Fig. 1). Unlike wt-p53/p73 and wt-p53/p63 protein complexes, whose existence in living cells has not been found despite extensive efforts, mt-p53/p73 and/or mt-p53/p63 complexes have been evidenced in different tumor cell lines (Di Como et al., 1999; Marin et al., 2000; Strano et al., 2000; Gaiddon et al., 2001; Strano et al., 2002; Irwin et al., 2003) (Figure 1).

According to the type of structural alteration, mt-p53 proteins can be divided in two main classes: DNA-contact defective and conformational mutants (Cho et al., 1994; Bullock et al., 2001). DNA contact defective mutants include p53 proteins whose missense mutations impact on the region of that binds DNA, while conformational mutants are marked by mutations that reside in the L2 and L3 loops. Do diverse p53 mutant proteins bind differently to p73 and p63? A first glance reply has come from published results raised by different labs (Di Como et al., 1999; Marin et al., 2000; Strano et al., 2000; Gaiddon et al., 2001; Strano et al., 2002; Irwin et al., 2003; Strano et al., 2003). They have shown that both types of mt-p53 proteins can engage in a physical interaction with the long, short, and truncated forms of p73 and p63 (Figure 1).

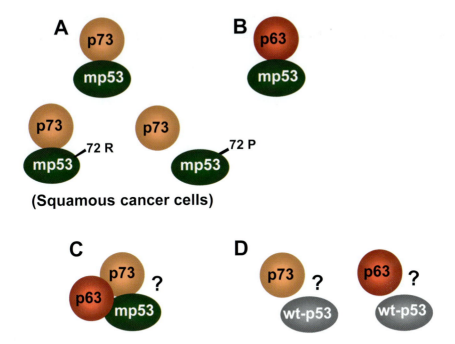

Figure 1. p53 family members protein complexes. (A) Protein-protein interaction between mutant p53 and p73. p53 polymorphism in squamous cancer cells is a determinant of the association with p73; (B) Protein complex involving mutant p53 and p63; (C) Triple complex between mutant p53, p73, and p63; (D) Protein-protein interaction between wt-p53, p73 or p63; ? Protein complexes that have not been found in vivo.

Experiments aimed to measure potential differences in the binding affinity to p73 and p63 of diverse mt-p53s are still missing. It was reported that the mutant p53Gly281 binds weakly to p73 and does not bind to p63 (Strano et al., 2000; Marin et al., 2000; Strano et al., 2002). This mt-p53 protein was shown to exert gain of function activity *in vitro* and *in vivo* (Dittmer et al., 1994). Furthermore, mt-p53Gly281, whose critical residues (22,23) were mutated, lost gain of function activity either as increased tumorigenicity in nude mice or as the ability to transactivate target genes, suggesting that the transactivation domain plays an important role in the oncogenic activity of this type of p53 mutant (Lin et al., 1995; Gualberto et al. 1998; Frazier et al., 1998) Altogether these finding might allow the definition of two classes of gain of function p53 mutants. The first one may account for mutants (p53His175 and p53His273) whose gain of function activity relies on protein-protein interactions with p63 and p73, while the second one includes p53 mutants (p53Gly281) that exert such activity independently from the physical association with the p53 family members.

Recent evidence showed that association between human tumor-derived mt-p53 and p73 is governed by a common polymorphism at codon 72 of p53 that encodes Arg or Pro. The 72R forms of p53Ala143 and p53His175 mutants bind to p73 and impairs p73-mediated gene target transcriptional activation more efficiently than the equivalent 72P mutants (Marin et al., 2000; Marin et al., 2003, Bergamaschi et al., 2003) (Figure 1).

What domains of mt-p53, p73 and p63 are directly involved in the protein-protein interaction? Co-precipitation assays have revealed that the protein-protein interaction surface involves the core domain of mtp53 and the specific DNA-binding domain of p73 and p63, respectively (Strano et al., 2000; Gaiddon et al.,2001; Strano et al., 2002). The minimal sequence of each domain involved in the interaction still needs to be identified. A recent report has shown that change in the conformation of the DNA binding domain of p53 is a key element for the binding to and interference with p73 (Bensaad et al., 2003). Interestingly, the core domains of mt-p53 proteins have been regarded as "dead" domains since they cannot bind and activate p53 target genes. However, these core domains acquire a protein-protein interaction capacity that might contribute to gain of function activities of mt-p53 by sequestering and inactivating proteins required for anti-tumor functions. Indeed, colony suppression assays have shown that the core domain of mutant is as efficient as full-length mutant p53 in inhibiting growth suppression activity of both p63 and p73 (Strano et al., 2002; Bensaad et al., 2003): A rather speculative hypothesis might suggest that pressure for the selection of p53 mutations in the core domain is partially related to its ability to gain new protein-protein interaction properties. The latter could play a central role in the gain of function activity of mutant p53.

FUNCTIONAL IMPLICATIONS OF THE PRESENCE OF MT-P53/P73 AND MT-P53/P63 PROTEIN COMPLEXES IN CANCER CELLS.

Despite the developmental phenotypes of p73 and p63 deficient mice, several *in vitro* and *in vivo* studies have clearly shown the involvement of the p53 family members in apoptosis (Yang et al., 1999; Mills et al., 1999; Yang et al., 2000). The TA isoforms of p73 and p63 can transactivate p53 target genes and induce apoptosis, whereas the DN isoforms are dominant inhibitors of p53-responsive gene expression (Yang et al., 2000) (Figure 2). Thus, p73 and p63 can have both pro-apoptotic and anti-apoptotic effects. It was originally shown that p73 is not induced by DNA damage (Kaghad et al, 1997; Jost et al., 1997). Subsequently, it was reported that cisplatin and ionizing radiation could regulate protein levels of p73 through tyrosine phosphorylation (Agami et al., 1999; Gong et al., 1999; Yuan et al., 1999). This modification of p73 in response to DNA damage is quite peculiar and has never been found for p53. This would imply that cells exposed to DNA damage recruit a p73-dependent pathway distinct from that activated by p53. Thus, specific protein-protein interactions, as well as activation or repression of specific target genes different from those recruited by wt-p53 could be hallmarks for p73-dependent pathways in response to DNA damage (Strano et al., 2001; Fontemaggi et al., 2002). For instance, the transcriptional co-activator Yes-associated protein (YAP) depicts two levels of specificity in the binding to p53 family members. It binds to long but not to short forms of p73 and p63 and does not bind to p53 at all (Strano et al., 2001). The binding of YAP to p73 or p63 results in a strong transcriptional co-activation (Strano et al., 2001; Basu et al., 2003, S.S. and G.B., unpublished observations). It has been shown that most of p73 post-translational modifications occur through its physical interaction with the active c-Abl kinase and promote the apoptotic activity of p73 (Agami et al., 1999; Gong et al., 1999; Yuan et al., 1999) (Figure 2). Furthermore, p73 can also be acetylated by p300 upon treatment with cisplatin and doxorubicin (Costanzo et al., 2002) (Figure2). Acetylation of p73 in response to DNA damage was found to be a determinant in the activation of pro-apoptotic target genes (Costanzo et al., 2002). Two reports have recently shown that TAp73 is induced by a wide variety of chemotherapeutic agents including camptothectin, etoposide, taxol, cisplatin, and doxorubicin (Irwin et al., 2003; Bergamaschi et al., 2003). Of note, blocking TAp73 function either by overexpressing specific dominant negative proteins or using specific RNA interfering oligonucleotides leads to enhanced chemoresistance (Irwin et al., 2003; Bergamaschi et al., 2003). Altogether these data demonstrate that chemosensitivity of cancer cells can also be ascribed to p73 function. There is no evidence whether p73 posttranslational modifications in response to DNA can positively or negatively impinge on p73 binding to mt-p53.

Is p73-mediated apoptosis impaired in cancer cells bearing mt-p53? Unlike p73 inactivation that results in increased chemoresistance, downregulation of mt-p53 enhances chemosensitivity of cancer cells (Irwin et al., 2003). Interestingly, patients whose head and neck cancers express mt-p53 with the 72R polymorphism have a worse response to therapy than those expressing the 72P. Thus, expression of mt-p53 with 72R, that efficiently inhibit p73, is associated with a particular poor outcome to therapy (Bergamaschi et al., 2003). These very intriguing findings establish a strong link between polymorphism in p53 and response to therapy of patients affected by specific tumors, but they still leave unsolved the question whether the protein-protein interaction between mutant p53 and p73 or p63 plays a role in the chemoresistance of tumor cells (Strano et al., 2003).

Indeed, mt-p53/p73 complexes have been found in either untreated or camptothecin treated tumor cells (Irwin et al., 2003). It was recently shown that both p73 and p63 are necessary for the efficient wt-p53-mediated apoptosis. Furthermore, wt-p53, p73, and p63 can be concomitantly recruited on the regulatory regions of p53-apoptotic target genes (Flores et al., 2002). By using stable transfected lung cancer cell lines, whose expression of p53His175 mutant was tightly regulated, we found that the recruitment *in vivo* of both p73 and p63 to their target genes is impaired by the concomitant expression of mt-p53 (Strano et al., 2002) (Figure 2). Similar experiments need to be done either in response to DNA damage or upon inactivation of mt-p53 by specific RNA interference. The first set of experiments might show that after the exposure of mt-p53 cells to DNA damaging agents the recruitment of both p73 and p63 to their target genes remains similar to that of the untreated cells. Conversely, interference of mt-p53 might allow p73 and p63 to recruit a larger repertoire of target genes respect to cells with intact mt-p53 and consequently promote efficient apoptosis in response to anticancer treatments.

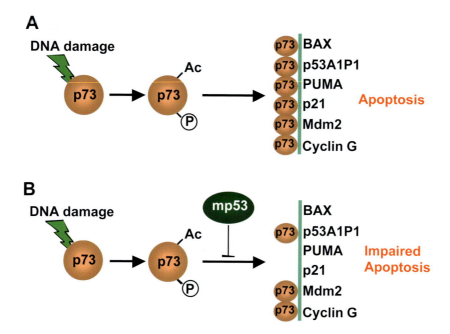

Figure 2. Cross-talk between p73 and mutant p53 proteins in response to DNA damage. (A) In response to some DNA damaging agents p73 is stabilized, phosphorylated, and acetylated. The net biological output of these posttranslational modifications is the increased pro-apoptotic activity of p73 that occurs mainly through the activation of specific target genes. (B) Mutant p53 proteins binding to and sequestering p73 impair its recruitment to specific target gene promoters. This might reduce the rate of p73-mediated apoptosis.

PERSPECTIVES

Basic research in cancer is required not only to identify and dissect molecular mechanisms but also to design new therapeutic approaches. Despite the rapid progress in the biochemical and functional characterization of the mt-p53/p73 and mt-p53/p63 interactions, direct evidence that disruption of such interactions rescues gain of function activity of mt-p53 by making both p73 and p63 available to promote anti-tumor effects is still missing. The design of small peptides aimed to disassemble the mt-p53/p73 and mt-p53/p63 protein complexes could potentially address this unsolved issue. Future work will tell us if these peptides can be reasonably used to treat tumors bearing mutant p53.

REFERENCES

Agami R., Blandino G., Oren M., and Shaul. Y. Interaction of c-Abl and p73 and their collaboration to induce apoptosis. Nature 1999; 399: 809-13.

Almog N., and Rotter V. Involvement of p53 in cell differentiation and development. Biochem. Biophys. Acta. 1997; 1333: F1-27.

Aas T., Borresen A. L., Geisler S., Smith-Sorensen B., Johnsen H., Varhaug J. E., Akslen L. A., and Lonning P. E. Specific P53 mutations are associated with de novo resistance to doxorubicin in breast cancer patients. Nat. Med. 1996; 2:811-4.

Basu S., Totty N.F., Irwin M.S., Sudol M., Downward J. Akt phosphorylates the Yes-associated protein, YAP, to induce interaction with 14-3-3 and attenuation of p73-mediated apoptosis. Mol. Cell. 2003; 11:11-23.

Blandino G., Levine A. J., and Oren M. Mutant p53 gain of function: differential effects of different p53 mutants on resistance of cultured cells to chemotherapy. Oncogene, 1999, 18: 477-85.

Bensaad, K., Le Bras M., Unsal K., Strano S., Blandino G., Tominaga O., Rouillard D., and Soussi T. Change of conformation of the DNA binding domain of p53 is the only key element for binding of and interference with p73. J. Biol. Chem. 2003; 278:10546-10555.

Bergamaschi D., Gasco M., Hiller L., Sullivan A., Syed N., Trigiante G., Yulug I., Merlano M., Numico G., Comino A., Attard M., Reelfs O., Gusterson B., Bell A.K., Heath V., Tavassoli M., Farrel P.J., Smith P., Lu X., and Crook T. p53 polymorphism influences response in cancer chemotherapy via modulation of p73-dependent apoptosis. Cancer Cell 2003; 3: 387-402.

Bullock A.N., and Fersht A.R. Rescuing the function of mutant p53. Nat. Rev. Cancer 2001; 1: 68-76.

Cho, Y. J., S. Gorina, P. D. Jeffrey, and N. P. Pavletich. Crystal structure of a p53 tumor suppressor DNA complex: understanding tumorigenic mutations. Science 1994; 265:346-355.

Costanzo A., Merlo P., Pediconi N., Fulco M., Sartorelli V., Cole P. A., Fontemaggi G., Fanciulli M., Schiltz, L., Blandino, G., Balsano C., and Levrero M. DNA damage-dependent acetylation of p73 dictates the selective activation of apoptotic target genes. Mol. Cell 2002; 9: 175-86.

De Laurenzi V., Costanzo A., Barcaroli D., Terrinoni A., Falco, M., Annichiarico-Petruzzelli M., Levrero M., and Melino G. Two new p73 splice variants, gamma and delta, with different transcriptional activity. J. Exp. Med. 1998; 188: 1763-68.

De Laurenzi V., Raschellà G., Barcaroli D., Annichiarico-Petruzzelli, M., Ranalli, M., Catani M. V., Tanno B., Costanzo A., Levrero M., and Melino G. Induction of neuronal differentiation by p73 in a neuroblastoma cell line. J. Biol. Chem. 2000; 275: 15226-231.

Di Como C. J., Gaiddon C., and Prives C. p73 function is inhibited by tumor-derived p53 mutants in mammalian cells. Mol. Cell. Biol. 1999; 19: 1438-49.

Dittmer D., Pati S., Zambetti G., Chu S., Teresky A. K., Moore M., Finlay C., .and Levine A. J. Gain of function mutations in p53. Nat. Genet. 1993; 4:42-6.

Flores E.R., Tsai K.Y., Crowley D., Sengupta S., Yang A., McKeon F., and Jacks T. p63 and p73 are required for p53-dependent apoptosis in response to DNA damage. Nature 2002; 416: 560-564.

Fontemaggi G., Gurtner A., Strano S., Higashi Y., Sacchi A., Piaggio G., and Blandino G. The transcriptional repressor ZEB regulates p73 expression at the cross-road between proliferation and differentiation. Mol. Cell. Biol. 2001; 24: 8461-470.

Fontemaggi G., Kela I., Amariglio N., Rechavi G., Krishnamurthy J., Strano S., Sacchi A., Givol D., and Blandino G. Identification of direct p73 target genes combining DNA microarray and chromatin immunoprecipitation analyses. J. Biol Chem. 2002; 277: 43359-368.

Frazier M. W., He X., Wang J., Gu Z., Cleveland J. L., and Zambetti G. P. Activation of c-myc gene expression by tumor-derived p53 mutants requires a discrete C-terminal domain. Mol. Cell. Biol. 1998; 18: 3735-43.

Gaiddon C., Lokshin, M., Ahn, J., Zhang T., and Prives. C. A subset of tumor-derived mutant forms of p53 down-regulate p63 and p73 through a direct interaction with the p53 core domain. Mol. Cell. Biol. 2001; 21: 1874-87.

Gong J. G., Costanzo A., Yang, H. Q., Melino G., Kaelin W. G., Levrero M., and Wang, J. Y. J. The tyrosine kinase c-Abl regulates p73 in apoptotic response to cisplatin-induced DNA damage. Nature 1999; 39: 806-9.

Gualberto A., Aldape K., Kozakiewicz K., and Tlsty, T. D. An oncogenic form of p53 confers a dominant, gain-of-function phenotype that disrupts spindle checkpoint control. Proc. Natl. Acad. Sci. U S A 1998; 95: 5166-71.

Hainaut P., Soussi T., Shomer B., Hollstein M., Greenblatt M., Hovig E., Harris C.C., and Montesano R. Database of p53 gene somatic mutations in human tumors and cell lines: updated compilation and future prospects. Nucleic Acids Res. 1997; 25: 151-7.

Haley O., Michalovitz D., and Oren M. Different tumor-derived p53 mutants exhibit distinct biological activities. Science 1990; 250: 113-6.

Hollstein M., Soussi T., Thomas G., von Brevern M. C., and Bartsch H. p53 gene alterations in human tumors: perspectives for cancer control. Recent Results Cancer Res. 1997; 143: 369-89.

Hussain, S. P., and Harris C. C.. Molecular epidemiology of human cancer: contribution of mutation spectra studies of tumor suppressor genes. Cancer Res. 1998; 58: 4023-37.

Jost C. A., Marin M. C., and Kaelin W. G.. p73 is a human p53 related protein that can induce apoptosis. Nature 1997; 389: 191-4.

Irwin M., Kondo K., Marin M.C., Cheng L.S., Hahn W.C., and Kaelin W.G.jr. Chemosensitivity linked to p73 function. Cancer Cell 2003; 3: 403-10.

Kaelin W. G., Jr. The emerging p53 gene family. J. Natl. Cancer Inst. 1999; 91: 594-8.

Kaghad M., Bonnet H., Yang A., Creancier L., Biscan J.C., Valent A., Minty A., Chalon P., Lelias J.M., Dumont, X., Ferrara, P., McKeon F., and Caput D. Monoallelically expressed gene related to p53 at 1p36, a region frequently deleted in neuroblastoma and other human cancers. Cell 1997; 90: 809-19.

Levine A. J. p53, the cellular gatekeeper for growth and division. Cell 1997; 88: 323-31.

Levrero M., De Laurenzi V., Costanzo A., Gong J., Wang J. Y., and Melino G. The p53/p63/p73 family of transcription factors: overlapping and distinct functions. J. Cell. Sci. 2000; 113: 1661-70.

Li R., Sutphin P. D., Schwartz D., Matas D., Almog, N., Wolkowicz R., Goldfinger N., Pei H., Prokocimer M., and Rotter. V. Mutant p53 protein expression interferes with p53-independent apoptotic pathways. Oncogene 1998; 16: 3269-77.

Lin, J., A. K. Teresky, and A. J. Levine. Two critical hydrophobic amino acids in the N-terminal domain of the p53 protein are required for the gain of function phenotypes of human p53 mutants. Oncogene 1995; 10: 2387-90.

Marin M. C., Jost C. A., Brooks L. A., Irwin M. S., O'Nions J., Tidy J. A., James, N., McGregor J. M., Harwood C. A.,. Yulug I. G., Vousden, K. H., Allday M. J., Gusterson B., Ikawa S,. Hinds P. W., Crook T., and Kaelin W. G. Jr. A common polymorphism acts as an intragenic modifier of mutant p53 behaviour. Nat. Genet. 2000; 25: 47-54.

Matas D., Sigal A., Stambolsky P., Milyavsky M., Veisz L., Schwartz D., Goldfinger N., and Rotter V. Integrity of the N-terminal transcription domain of p53 is required for mutant p53 interference with drug-induced apoptosis. EMBO J. 2001; 20: 4163-72.

Michalovitz D., Halevy O., Oren M. Conditional inhibition of transformation and of cell proliferation by a temperature-sensitive. Cell 1990; 62: 671-80.

Mills A. A., Zheng B., Wang, X. J., Vogel H., Roop D. R., and Bradley A. p63 is a p53 homologue required for limb and epidermal morphogenesis. Nature 1999; 398: 708-13.

Morena A.R., Riccioni S., Marchetti A., Tartaglia Polcini A., Mercurio A.M., Blandino G., Sacchi A., and Falcioni R. Expression of β4 integrin subunit induces monocytic differentiation of 32D/v-Abl cells. Blood 2002; 100: 96-106.

Osada M., Ohba M., Kawahara C., Ishioka C., Kanamaru R., Katoh I., Ikawa Y., Nimura Y., Nakagawara A., Obinata, M., and Ikawa S. Cloning and functional analysis of human p51, which structurally and functionally resembles p53. Nat. Med. 1998; 4: 839-43.

Prives C., and Hall P. A.. The p53 pathway. J Pathol. 1999; 187: 112-26.

Sigal A., and Rotter V. Oncogenic mutations of the p53 tumor suppressor: the demons of the guardian of the genome. Cancer Res. 2000; 60: 6788-93.

Soddu S., Blandino G., Scardigli R., Coen S., Marchetti A., Rizzo M.G., Bossi G., Cimino L., Crescenzi M., and Sacchi A. Interference with p53 protein inhibits hematopoietic and muscle differentiation. The Journal of Cell Biology 1996; 134: 193-204.

Strano S., Munarriz E., Rossi, M., Cristofanelli B., Shaul Y., Castagnoli, L., Levine, A. J., Sacchi, A., Cesareni, G., Oren M. and Blandino. G. Physical and functional interaction between p53 mutants and different isoforms of p73. J. Biol. Chem. 2000; 275: 29503-12.

Strano S., Fontemaggi G., Costanzo A., Rizzo M.G., Monti O., Baccarini A., Del Sal G., Levrero M., Sacchi A., Oren M. and Blandino G. Physical interaction with human tumor derived p53 mutants inhibits p63 activities. J. Biol Chem. 2002; 277: 18817-826.

Strano S., Munarriz E., Rossi M., Cristofanelli B., Castagnoli L., Shaul Y., Sacchi A., Oren M., Sudol M., Cesareni G., and Blandino G. Physical interaction with Yes-associated protein (YAP) enhances p73 transcriptional activity. J. Biol. Chem. 2001; 276: 15164-173.

Strano S., Rossi M., Fontemaggi, G., Munarriz, E., Soddu, S., Sacchi, A., and Blandino G. From p63 to p53 across p73. FEBS Lett. 2001; 490: 163-70.

Strano S., and Blandino G. p73-mediated chemosensitivity: a preferential target of oncogenic mutant p53. Cell Cycle. 2003; 2: 348-9.

Sudol M.. Yes-associated protein (YAP65) is a proline-rich phosphoprotein that binds to the SH3 domain of the Yes proto-oncogene product. Oncogene 1994; 9: 2145-52.

Trink B., Okami K., Wu L., Sriuranpong V., Jen J., and Sidransky. D. A new human p53 homologue. Nat. Med. 1998; 4: 747-8.

Yang, A., N. Walker, R. Bronson, M. Kaghad, M. Oosterwegel, J. Bonnin, C. Vagner, H. Bonnet, P. Dikkes, A. Sharpe, F. McKeon, and D. Caput. p73-deficient mice have neurological, pheromonal and inflammatory defects but lack spontaneous tumours. Nature 2000; 404: 99-103.

Yang A., and McKeon. F. p63 and p73: p53 mimics, menaces and more. Nat. Rev. Mol. Cell. Biol. 2000; 1: 199-207.

Yang A., Schweitzer R., Sun D., Kaghad M., Walker N., Bronson R. T., Tabin C., Sharpe A., Caput D., Crum C., and McKeon F. p63 is essential for regenerative proliferation in limb, craniofacial and epithelial development. Nature 1999; 398: 714-8.

Yang A., Kaghad M., Wang, Y., Gillett E., Fleming M. D., Dotsch V., Andrews N.C., Caput D., and McKeon. F. p63, a p53 homolog at 3q27-29, encodes multiple products with transactivating, death-inducing, and dominant-negative activities. Mol. Cell. 1998; 2: 305-16.

Yonish-Rouach E., Resnitzky D., Lotem J., Sachs L., Kimchi A., and Oren M. Wild type p53 induces apoptosis of myeloid leukaemic cells that is inhibited by interleukin-6. Nature 1991, 352: 345-47.

Yuan Z. M., Shioya H,. Ishiko T, Sun X., Gu J., Huang Y. Y., Lu H., Kharbanda S., Weichselbaum R., and Kufe. D. p73 is regulated by tyrosine kinase c-Abl in the apoptotic response to DNA damage. Nature1999; 399: 814-17.

Chapter 11

P53: GATEKEEPER, CARETAKER OR BOTH?

Carlos P. Rubbi and Jo Milner
YCR p53 Laboratory, Biology Department, University of York, UK

P53 AS A TUMOUR SUPPRESSOR

Soon after the discovery of the p53 protein in 1979, the p53 gene was found to be mutated in about half of all human cancers. However, mainly due to the fact that the normally low levels of the p53 protein can be elevated in many cancers, it took some time to realise that p53 was in fact a tumour suppressor gene (Finlay et al., 1989; Levine et al., 1991). This concept was further consolidated by the discovery that the familial cancer predisposition known as Li-Fraumeni syndrome[1] is linked to germ-line mutation of the p53 gene (for a review see Varley et al., 1997), and by the clear tumour propensity shown by p53 knock-out mice (for a review see Venkatachalam and Donehower, 1998).

A mechanism for the tumour suppressor activity of p53 soon started to emerge: under conditions of DNA damage or oncogene activation p53 will drive cells into cell cycle arrest and/or apoptosis thus acting as a 'gatekeeper of cell growth' (Levine, 1997). However, since DNA damage (cause of mutation) and oncogene activation (consequence of mutation) can both be present in the majority of oncogenic processes, the capacity of p53 to react to both stresses confounds its mode and place of action as a tumour suppressor.

---

[1] The classic human Li-Fraumeni syndrome is defined by a proband aged under 45 years with a sarcoma having a first-degree relative aged under 45 years with any cancer and an additional first- or second-degree relative aged under 45 years in the same lineage with any cancer or a sarcoma at any age (taken from Varley et al., 1997).

P. Hainaut and K.G. Wiman (eds.), 25 Years of p53 Research, 233-253.
© 2005 *Springer. Printed in the Netherlands.*

In an elegant set of experiments Symonds et al. (1994) managed to bypass the need for a source of genomic instability in oncogenic development by creating mice carrying a transgenic fragment of the large T antigen of the SV40 virus. These mice develop choroid plexus tumours through the inactivation of proteins of the Rb family. Importantly, the T antigen fragment used inhibits Rb function while not interfering with p53 function (Symonds et al., 1994). In this system, loss of Rb function leads to de-repression of E2F (the oncogenic event) resulting in oncogene activation and entry into S-phase, which should cause p53-mediated apoptosis and elimination of the potential tumour cells (for a detailed description of the pathway see for example Sherr and Weber, 2000 - see Figure 1).

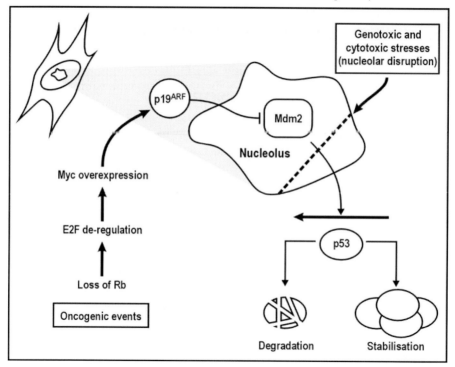

Figure 1. p53 stabilisation A central player in p53 stabilisation in response to oncogenic events is an increase in the p19ARF protein. Several oncogenic events can cascade to increase p19ARF expression. Immediate to it is the overexpression of myc. This, in turn may be caused by loss of repression of the transcription factor E2F. And this E2F de-repression may be caused by inactivation (e.g. adenovirus E1A protein), phosphorylation (e.g. by cyclin-dependent kinases) or loss expression (e.g. retinoblastoma) of the Rb protein. p19ARF inhibits mdm2-mediated degradation of p53 probably by sequestering mdm2 in the nucleolus (for reviews see Sherr and Weber, 2000; Vousden and Lu 2002). While the p19ARF pathway does not appear to mediate p53 stabilisation in response to genotoxic stresses (Stott et al., 1998), most genotoxic/cytotoxic stresses will disrupt nucleolar function which, in turn appears to lead to p53 stabilisation (Rubbi and Milner, 2003c).

By crossing these transgenic mice into a p53$^{-/-}$ background Symonds et al. observed that loss of p53 caused rapid tumour progression and that p53-dependent apoptosis was required to suppress this tumour development. Importantly, the increased tumour progression showed by p53$^{+/-}$ mice was associated to loss of the remaining wild-type p53 allele (Loss of Heterozygosity - LOH). Hence, p53 as a suppressor of oncogenic development appeared completely to agree with Knudson's 'two-hit' hypothesis for tumour development in that a germline loss of a tumour suppressor allele has to be followed by a somatic loss of the second wild-type allele (Knudson, 1993).

THE LOSS-OF-HETEROZYGOSITY PROBLEM

However, in the "real life" situation where the oncogenic insult is likely to derive from a mutation, p53 LOH is unusually low in tumours developed in cancer-prone Li-Fraumeni individuals (around 57% - see Varley et al., 1997). This low LOH is mirrored in p53 heterozygous mice: Venkatachalam et al. (1998) reported 50% wild-type allele retention in tumours in p53$^{+/-}$ mice under 18 months of age and 85% retention in older mice, and concluded that loss of both p53 alleles is no prerequisite for tumour formation. In both humans and mice the functional status of the wild-type p53 alleles retained in tumours has been analysed and while in some cases a reduction in some biological activities has been reported, the expected p53 functions are still present in heterozygous individuals. Camplejohn et al. (1995) reported that peripheral blood lymphocytes of Li-Fraumeni donors retained the capacity to arrest in G_1 in response to ionising radiation (IR), while apoptosis induction was reduced but still present. Williams et al. (1997), on the other hand, observed that Li-Fraumeni fibroblasts had similar transient G_1 arrest compared with normal fibroblasts with a reduced permanent G_1 arrest. Parallel experiments showed no arrest in p53$^{-/-}$ mice. Varley et al. (1998) compared cells from normal and carrier individuals of a Li-Fraumeni-like family harbouring a p53 splice donor mutation (for a definition of Li-Fraumeni-like syndrome see Varley et al., 1997), and reported no differences in IR-induced apoptosis and a small but statistically significant reduction in cell cycle arrest (from ~97% to ~83%). Interestingly, they found a significant increase in radiation-induced chromosome damage. Similarly, Venkatachalam et al. (1998) carried out exhaustive functional studies on the status of the remaining wild-type p53 allele in tumours from p53$^{+/-}$ mice. They found no reduction in radiation-induced apoptosis, lower but significant p21 induction, normal PCNA downregulation and normal *in vitro* binding of p53 to its response elements (EMSA). Importantly, they found no mdm2 amplification that might cause inhibition of p53 function. These observations indicate that the occurrence of dominant-negative p53 mutations, while possible, are not the norm. Thus, the high dominance of the cell cycle arrest and apoptotic functions of p53 indicates that if they are the sole basis for its tumour suppressor capacity, then the remaining wild-type allele should be inactivated with a high frequency, as predicted by the 'two-hit' model and observed for other tumour suppressor genes (Knudson, 1993).

IS P53 AN EXCEPTION TO THE 'TWO-HIT' MODEL?

Is this low LOH paradox telling us something about the way p53 suppresses tumours? While the results of Symonds et al. (1994) discussed above point to a clear correlation between p53 LOH and tumour development under direct oncogene activation (hence implying that p53 is a dominant tumour suppressor), a 'full' oncogenic process suggests that p53 is haploinsufficient for tumour suppression. Thus, haploinsufficiency appears to be revealed in conditions where somatic mutations are required for driving tumour progression. In fact, p53$^{+/-}$ mice are highly sensitive to genotoxic carcinogens while maintaining a normal sensitivity to non-genotoxic carcinogens (French et al. 2001b; Storer et al., 2001; Venkatachalam, et al., 2001). Furthermore, Venkatachalam, et al. (2001) showed that p53 LOH can be tissue- and carcinogen-dependent, since the clastogenic agent benzene induces a high proportion of thymic lymphomas with high LOH. Interestingly, while p53$^{+/-}$ mice show a tumour spectrum similar to that of p53$^{+/+}$ mice, p53$^{-/-}$ mice develop mainly lymphomas (Venkatachalam and Donehower, 1998; Venkatachalam, et al., 2001 and see below), suggesting that p53 loss and p53 haploinsufficiency may cause tumour permissiveness through different mechanisms.

GATEKEEPERS, CARETAKERS AND PATHWAYS FOR TUMOUR DEVELOPMENT

Although definitions do not introduce new information, they can be helpful in organising and clarifying the existing one. Thus, we shall recall here the classification of tumour suppressor genes as gatekeepers and caretakers according to the definitions used by Macleod (2000). For a tumour suppressor gene to be classified as a gatekeeper first its loss of function has to be rate-limiting for a particular step in multi-stage tumorigenesis; second, it must act directly to prevent tumour growth; and third, its restoration into a tumour cell must suppress neoplasia. On the other hand, a caretaker tumour suppressor gene acts by limiting genomic instability through effective DNA repair, and restoration of its function into a developing tumour will not limit tumour growth if a gatekeeper mutation has already occurred (Macleod, 2000).

In the present article we discuss the dual capacity of p53 as a gatekeeper and a caretaker and argue that it is simultaneously a dominant gatekeeper and a haploinsufficient caretaker. This classification is not a trivial language game: it sheds light into the aetiology of p53-related tumours and, whilst it may not tell us anything new about p53, by organising what we know about p53, it helps us understand the relative weight that a particular p53 function may have in an oncogenic process.

It should be noted that a dual caretaker/gatekeeper character of p53 has been previously pointed out, for instance by Macleod herself (Macleod, 2000), but it is important to bear in mind that in the majority of cases this 'caretaker' function of p53 has been invoked in relation to its capacity to induce apoptosis and/or cell cycle arrest in response to DNA damage. In fact, this capacity has led to its appellative 'guardian of the genome' (Lane, 1992). However, as we shall discuss later, the proapoptotic capacity of p53 in response to DNA damage is not fundamentally different from its response to oncogenic stress (see Figure 1) and it also consists of the elimination of genetically damaged cells rather than the prevention or the correction of the damage (at the risk of introducing a further definition, it may be regarded as a *'pseudo-caretaker'* function). In this article we shall reserve the term caretaker for those functions of p53 directly related to either the enhancement or the modulation of DNA repair. Moreover, as discussed above, while it may display some gene dosage effect, the capacity of wtp53 to induce apoptosis in response to DNA damage is essentially present in heterozygous cells and thus has to be regarded as dominant. The role of p53 in DNA repair, we shall argue, is likely to be haploinsufficient.

In the description of the mutator phenotype Loeb (1991) argues that the mutation frequencies in normal cells, even in the presence of environmental carcinogens, can hardly account for the cancer frequencies observed in humans and proposes that cancer progression must involve genetic alterations that increase the mutation rate in the developing neoplasm. However, as Loeb also points out, mutation rate analysis indicates that if a growth advantage is introduced by an early mutation, the ensuing clonal expansion increases the *probability* of subsequent mutations, even without a significant increase in the overall *frequency* of mutation (Loeb, 1991). We are thus presented with two extreme models of tumour development: a mutation-driven process where a tumour-committed cell lineage has an increased mutation frequency with no significant increase in growth rate until the accumulated mutations cause the malignant transformation of a clone, and a selection-driven process where a tumour-committed clone does not have a significantly higher mutation rate compared with normal tissue, but due to its increased growth rate has an increased mutation probability (Temin, 1988, see scheme in Figure 2).

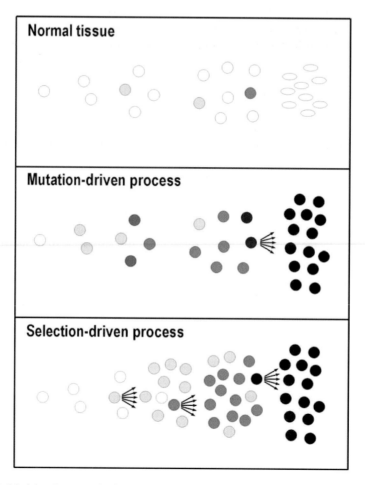

Figure 2. Models of oncogenic development.Normal cells grow at normal rate and mutate at normal frequency until either terminally differentiated or eliminated by tissue renewal. Mutation-driven process: cells grow at normal rate with higher mutation frequency, until malignant transformation. Selection-driven process: clones grow at an accelerated rate with normal mutation frequency, until malignant transformation. Grey shading indicates accumulation of mutations.

In their discussion of colon carcinogenesis Kinzler and Vogelstein (1996) introduced the analogous concepts of diseases of tumour initiation (exemplified by Familial Adenomatous Polyposis - FAP - with a germline deficiency in the adenomatous polyposis coli - *APC* - gatekeeper) and diseases of tumour progression (such as the Hereditary Nonpolyposis Colorectal Carcinoma - HNPCC - driven by a deficiency in mismatch repair - MMR). As also noted by Kinzler and Vogelstein (1996) the progression of a tumour initiation disease may require the introduction of mutations in a particular order. Selection-driven and mutation-driven processes are extreme models and in real situations both mechanisms will co-exist to varying degrees, as a selection-driven only process would not exist without an underlying increase in mutation rate (Cahill et al., 1999). While it is unlikely that a tumour suppressor will strictly fit only one of them, this classification is useful to exemplify the different roles of gatekeeper and caretaker tumour suppressors and the expected consequences of their failure. Caretakers, by limiting the mutation rate, will suppress a mutation-driven process, while gatekeepers by controlling the entrance into cell division will suppress a selection-driven process.

In general, defects in gatekeeper genes are associated with a particular cancer type which can display high penetrance in affected families (see for example Venkatachalam and Donehower, 1998). Since growth control networks can be specific to cell lineages it is reasonable to expect a high correlation between gatekeeper germline defect and tumour type. Thus, a specific defect may be coupled to a specific tumour type as, for example the *APC* gene mentioned above and adenomatous polyposis coli. Interestingly, this is not necessarily valid across species: the retinoblastoma (*Rb*) gene for example, which derives its name from the high penetrance tumour that it causes in humans, in mice is found to develop almost exclusively pituitary tumours (Venkatachalam and Donehower, 1998). The association between tumour suppressor genes and high penetrance tumours thus agrees with a selection-driven oncogenic process. However, p53 heterozygosity, both in Li-Fraumeni patients and in p53$^{+/-}$ mice, causes a broad range of tumors. In fact, while p53$^{+/-}$ mice have a decrease in tumour latency under carcinogen treatment, the spectrum of tumours is similar to that of normal, wild-type p53 mice. In contrast, p53$^{-/-}$ mice are strikingly different in that they develop primarily lymphomas. This suggests that the tumour promotion mechanisms in the p53 heterozygous and nullizygous cases are fundamentally different[2].

P53 AND GENOMIC STABILITY

Fibroblasts derived from Li-Fraumeni patients display an unusual capacity for immortalisation *in vitro*, in stark contrast with normal human

---

[2] In his 1993 update of the 'two-hit' hypothesis Knudson (1993) also points out that Li-Fraumeni syndrome patients do not display high incidence of cancers such as colon carcinoma and small-cell lung carcinoma where p53 mutations are common.

fibroblasts, and even with fibroblasts from other cancer-prone syndromes, which are rarely immortalised (see for example Bischoff et al., 1990, and references therein). This growth characteristic was studied in detail by Bischoff et al. (1990), who reported that it is accompanied by random chromosome loss and various chromosomal anomalies, noticeable even at early passages. The *in vitro* immortalised cells, however, do not necessarily become tumorigenic as demonstrated by injection into nude mice (Bischoff et al., 1990). It thus became apparent that chromosomal instability and *in vitro* immortalisation were distinctive features of Li-Fraumeni-derived fibroblasts which suggested that genomic instability might drive the cancer propensity of Li-Fraumeni syndrome.

With the realisation that Li-Fraumeni syndrome is linked to a germline mutation of the p53 gene several groups in the early 1990's decided to test the newly emerging tumour suppressor for any involvement on genomic stability. Yin et al. (1992) demonstrated that Li-Fraumeni fibroblasts, with chromosomal instability at low passage number, would show loss or mutation of the remaining wild-type p53 allele at post-crisis passages and that this LOH always accompanied immortalisation. This p53 LOH also resulted in loss of the cell cycle arrest response to PALA[3] treatment (which causes depletion of the nucleotide triphosphate pool and, as we now know, p53-dependent cell cycle arrest - Linke et al., 1996), which could be restored by re-introduction of wtp53 (Yin et al., 1992). Similar behaviour was observed in mouse cells by Livingstone et al. (1992). With passage number, p53$^{+/-}$ (and also p53$^{-/-}$) mouse embryonic fibroblasts displayed a high spontaneous conversion to aneuploidy *in vitro* which was not observed in wild-type p53 homozygous cells (Livingstone et al., 1992). These observations suggest that while a reduction in p53 gene dosage compromises genomic stability, loss of p53 is required for loss of proliferation control (and in this genotype the instability will persist). This dual capacity of p53 is a likely explanation for the similar and wide spectrum of tumours induced by carcinogens in p53$^{+/+}$ and p53$^{+/-}$ mice, as opposed to the predominance of thymic lymphomas observed in p53$^{-/-}$ mice (see above).

Interestingly, Lalle et al. (1995) reported that the chromosomal instability found in Li-Fraumeni fibroblasts was not present in immortalised lymphocytes from the same syndrome, suggesting some degree of tissue-dependence in the p53-associated genomic instability phenotype. This behaviour is suggestive of mutation-driven oncogenesis in p53 heterozygosity, but selection-driven oncogenesis in p53 nullizygosity. But because the chromosomal instability caused by p53 haploinsufficiency can be expected occasionally to cause the loss of the remaining wild-type allele (Yin et al., 1992), the relative contributions to an oncogenic development of

---

[3] N-phosphonoacetyl-L-aspartate.

increased genomic instability versus loss of proliferation control (and loss of apoptotic capacity) are difficult to assign.

P53 AND NUCLEOTIDE EXCISION REPAIR

Despite the above considerations, when dealing with the mechanism of tumour suppression by p53 in conditions of DNA damage the majority of the bibliography refers to the capacity of p53 to induce cell cycle arrest or apoptosis. As we have seen above, these 'pseudo-caretaker' DNA damage responses, while important, fail to account for the low p53 LOH observed in tumours developed in p53 heterozygous individuals. Another tumour suppressive DNA damage response of p53 has to be invoked that (i) is gene dose dependent, (ii) is different form induction of apoptosis and/or cell cycle arrest, and (iii) can affect genomic stability. The answer may lay in the large body of evidence indicating that p53 directly participates in DNA repair.

In the mid 1990's several groups reported evidence for the involvement of p53 in nucleotide excision repair (NER - Wang et al. 1995; Smith et al. 1995; Ford and Hanawalt, 1995). This mode of repair is responsible for the removal of bulky adducts and UV photoproducts from DNA (usually called 'NER lesions'), and comprises two pathways: transcription-coupled repair (TCR) responsible for the removal of lesions from the transcribed strand of active genes, and global genomic repair (GGR) which removes lesions from the rest of the chromatin (Friedberg et al., 1995). Removal of DNA adducts from the global chromatin is crucial for cancer protection, as demonstrated by the cancer propensity of individuals inheriting the autosomal recessive disorder Xeroderma pigmentosum, which suffer from different kinds of impairments of NER (Friedberg, 2001). In particular, the cancer propensity of Xeroderma pigmentosum complementation group C individuals, where a mutated XPC gene confers specific impairment of GGR, indicates that this pathway is especially important in carcinogenesis. This cancer-prone phenotype associated to the loss of GGR has now been demonstrated in XPC mice (Friedberg et al., 2000).

Detailed studies by Ford and Hanawalt (1995, 1997) demonstrated that the role of p53 in NER consists in the enhancement of GGR of UV photoproducts. Later studies confirmed and extended this observation to chemically induced DNA adducts (Lloyd and Hanawalt 2000; Lloyd and Hanawalt 2002; Wani et al. 2002a). Two crucial features of the participation of p53 in GGR then emerged. First, the expression of p21 (a downstream effector of p53-mediated cell cycle arrest - see Sherr and Weber, 2000) has no effect on NER of either UV photoproducts or BPDE[4] adducts (BPDE is a carcinogenic polycyclic aromatic hydrocarbon associated with cigarette smoking - Adimoolam et al. 2001, Wani et al., 2002a,b) indicating that p53-mediated cell cycle arrest (which arguably might enhance genomic stability by allowing time for DNA repair before entering S-phase) has no role in GGR. Second, studies of NER efficiency in Li-Fraumeni cells indicate that p53 is haploinsufficient for GGR (Ford and Hanawalt, 1997; Abrahams et al., 1998). In fact, many of the genotoxic carcinogens against which p53 is haploinsufficient produce NER lesions (French et al. 2001a, and see van Oostrom et al., 1999 for BPDE). In order for p53 fully to fit the role of caretaker its direct participation in DNA repair has to be demonstrated. As a transcription factor which can also be involved in regulatory protein-protein interactions (Levine, 1997) it is often difficult to discriminate p53 functions exerted by direct action from indirect functions mediated by its downstream transactivation products (Vousden and Lu, 2002). In fact, Hwang et al. (1999) reported that p53 controls the expression of the DDB2 gene which encodes the p48 component of the DNA damage sensor DDB. This offers a potential explanation for the effect of p53 in GGR, since DDB is required for global detection of UV photoproducts (Tang et al., 2000). While this indirect mode of action of p53 on GGR appears plausible, its significance outside the specific cell system used in these studies is yet to be demonstrated, since a survey of the gene expression database (GEO - http://www.ncbi.nlm.nih.gov/geo/) indicates that in general DDB2 expression does not correlate with p53 status (see also Yoon et al., 2002) and furthermore, Tang et al. (2000) also demonstrated strong DDB activity in HeLa cell extracts[5].

We have recently shown that p53 can participate directly in GGR, independently of whether it transactivates repair genes or not (Rubbi and Milner, 2003a). In our model, p53 acts as a chromatin relaxation factor providing the chromatin accessibility necessary for the detection of global DNA lesions. It has long been known that the NER machinery has to overcome the hindrance caused by chromatin for the detection of certain DNA lesions such as chemical adducts and some UV-induced photolesions (Smerdon and Thoma, 1998; Friedberg, 2001; Green and Almouzni, 2002). Not all NER lesions are equally hindered by chromatin conformation: UV-induced pyrimidine (6-4) pyrimidon photoproducts (6-4PPs), which produce a significant DNA distortion and can only be accommodated in the inter-nucleosome linker, are more accessible (and hence more rapidly repaired) than *cis-syn* cyclobutane pyrimidine dimers (CPDs - Thoma, 1999).

---

[4] benzo[a]pyrene 7,8-diol-9,10-epoxide.
[5] In HeLa cells endogenous p53 is degraded through the action of the E6 ubiquitinase from human papilloma virus.

Crucially, Ford and Hanawalt (1997) demonstrated that p53 is required for GGR removal of CPDs, with no significant effect on GGR of 6-4PPs. This characteristic of p53, together with its capacity to promote chromatin relaxation by recruitment of histone acetyltransferases (HAT) to chromatin (for example p300, as demonstrated by Espinosa and Emerson, 2001) was the basis for our chromatin accessibility factor model (Rubbi and Milner, 2003a,b). We showed that p53 senses bulky DNA lesions through the blockage of RNA polymerase II transcription that they cause (in agreement with the localisation of p53 in active transcription sites - Rubbi and Milner, 2000) and triggers global histone acetylation and chromatin relaxation through mobilisation of HATs (Rubbi and Milner, 2003a). Hence, DNA lesions whose accessibility is assured by the transcription machinery induce, through p53, global chromatin relaxation that grants accessibility to lesions in the rest of the genome. Interestingly, the need for transcription-associated lesion detection for efficient global removal had been previously suggested by van Oosterwijk et al. (1996) who observed that lesions which are capable of stalling RNA polymerase II are also very efficiently recognised and removed by GGR, even when TCR may not necessarily make an important contribution to their removal.

An important aspect of our model is that p53-mediated global chromatin relaxation operates efficiently at levels of UV DNA damage too low to cause p53 protein stabilisation (e.g. $4J/m^2$) and thus to induce p53-mediated cell-cycle arrest or apoptosis (Yamaizumi and Sugano, 1994; Ljungman and Zhang, 1996; Rubbi and Milner, 2003a,c). This is consistent with the fact that GGR removes a large fraction of lesions before significant p53 stabilisation (and transactivation of downstream genes - see for example Ford and Hanawalt, 1997). In fact, maximal p53-mediated chromatin relaxation is observed within 1h after irradiation (Rubbi and Milner, 2003a). Thus, the chromatin relaxation function of p53 can be effected at normal (resting) levels of p53 well before any stabilisation occurs. Since p53 is normally expressed at low levels and has a very short half life (20min - Levine, 1997), gene dosage is likely to have a significant effect on any p53 function that operates in the absence of stabilisation (see for example Ross et al., 1994; Cook et al., 1998; and note that increases in p53 protein levels are generally caused by reduced degradation, rather that increased synthesis - Levine, 1997).

On the other hand, it has now become clear that p53 stabilisation is a general requirement for the induction of transactivation-dependent cell cycle arrest or apoptosis (Ljungman, 2000; Jaks et al.; 2001; Vousden and Lu, 2002; Kaeser and Iggo, 2002). It is reasonable to assume that the involvement of p53 protein stabilisation (which is achieved by inhibition of degradation) in the expression of these transactivation-dependent functions makes them largely independent from gene dosage: this independence would account for their dominant phenotype discussed above. Thus, the finding that a stabilisation-independent function of p53 (chromatin accessibility for GGR) correlates with a haploinsufficient behaviour (increased genomic instability in heterozygous under genotoxic carcinogens, particularly those inducing NER lesions - see above) is one possible explanation as to why p53 can show characteristics of both a haploinsufficient caretaker and a dominant gatekeeper.

P53 AND REPAIR OF DOUBLE STRAND BREAKS

In 1997 the groups of Lopez and Powell independently reported that inactivation of p53 or expression of mutant p53 increased the rate of spontaneous homologous recombination (Bertrand et al., 1997; Mekeel et al., 1997). This is highly significant because homologous recombination (HR) is one of the major pathways for the repair of DNA double strand breaks (DSB). The alternative pathway is direct non-homologous end-joining (NHEJ) (for a review see Karran 2000; Jackson, 2001). In HR a DNA strand is resected from the DSB and the free single strand, through the action of the Rad51 protein, searches for a region of homology in undamaged DNA and through subsequent DNA synthesis, ligation and resolution of crossovers the sequence integrity is restored to the original damaged DNA (Karran 2000; Jackson, 2001). However, the accuracy of this homology-driven process can be lost when direct repeats are used as a source of homology. Mammalian cells repair DSB mainly through NHEJ, in contrast to *Saccharomyces cerevisiae* where HR predominates (Karran, 2000).

The search for an involvement of p53 in HR was inspired by previous reports that p53 can interact with Rad51 (Bertrand et al. 1997; Mekeel et al., 1997), but so far this interaction has not been demonstrated to suppress HR *in vivo* in favour of NHEJ, and the molecular partners of p53 in this mechanism are not yet defined. Moreover, the studies of Tang et al. (1999) on rejoining of double strand breaks in linearised plasmid substrates suggest that p53 can promote NHEJ of cohesive but not blunt ends. A possible lead into the molecular mechanism of HR suppression, however, comes from the recent report by Sengupta et al. (2003) that p53 recruits the Bloom syndrome helicase (BLM) to Rad51 foci induced by stalled replication forks. This recruitment, which appears to be necessary for an interaction between p53 and Rad51, correlates with a reduction in the rate of sister chromatid exchanges.

Intriguingly, the characterisation of the suppressive effect of p53 on HR produced a picture similar to that observed for NER. Hence, it was shown to be independent of cell cycle arrest in G_1 and of transactivation (Saintigny et al., 1999; Dudenhoffer et al., 1999; Willers et al., 2000; Lu et al., 2003). Dissociation from G_1 arrest is crucial to the demonstration of a direct action of p53 in HR suppression, since HR operates preferably in S-phase (where homologous sequences are readily available and are used by HR for rescuing stalled replication forks - see Helleday, 2003).

As in NER, p53 heterozygosity affects suppression of HR. Saintigny and Lopez (2002) demonstrated that HR (induced by inhibition of replication) was stimulated by the introduction of mutant p53 in a p53 wild-type background. Lu et al. (2003) reported a higher HR frequency in embryo fibroblasts from p53$^{+/-}$ mice compared with p53$^{+/+}$ mice (but in both cases lower than in cells from p53$^{-/-}$ mice). If an interaction between p53 and Rad51 mediates suppression of HR, it is conceivable that gene dosage, by limiting the basal levels of p53, might affect this suppression. However, careful studies by Kumari et al. on HR induced by replication inhibition suggest that rather than reducing Rad51-dependent HR, p53 limits the onset of recombinogenic lesions (Kumari et al., 2004).

Willers et al. (2001) have shown that p53 fails to suppress HR when the recombination substrate is located in an extrachromosomal plasmid, even for a Rad51-dependent process (gene conversion). These observations suggest that the chromatin environment plays a role in p53 modulation of DSB repair. Indeed, there is evidence that chromatin condensation arises in response to DSB and that it may suppress resolution of the DNA breaks by HR in favour of NHEJ. Strand breaks in human centromeric heterochromatin appear to be resolved almost exclusively by NHEJ (Rief and Lobrich, 2002) and Tsukamoto et al. (1997) showed that chromatin silencing by Sir proteins promotes DSB resolution by NHEJ. Critchlow and Jackson (1998, see also Jackson, 1997, 2001) have pointed out several ways in which heterochromatin formation may facilitate NHEJ: it may prevent the transcription and replication machineries as well as nucleases from accessing the damaged chromatin; it may hinder strand invasion for HR; and it may reduce chromosomal flexibility and help juxtapose the two broken DNA ends facilitating ligation and effectively preventing the DSB from turning into a full chromosomal break. With regard to the latter effect, the experiments of Tumbar et al. (1999) provide a striking example of the changes in spatial distribution of DNA that chromatin condensation/relaxation can cause. They showed that a 90-Mbp tract of repeated LacI-binding sites, normally visualised as a single spot under fluorescence microscopy (LacI-GFP labelling), after transcriptional activation would appear as an extended 25 - 40μm ribbon (Tumbar et al., 1999). Thus, following on our work on p53-dependent chromatin relaxation in response to UV irradiation, we have now tested whether the introduction of DSB can cause p53-mediated chromatin modifications. We found that within 1h of bleomycin treatment human fibroblasts undergo a transient chromatin condensation which is dependent on p53 and is marked by deacetylation at Lys9 of histone H3 (Rubbi and Milner, manuscript in preparation). The mechanism responsible for this deacetylation appears to be p53 dependent mobilisation of PML from the nuclear bodies into the nucleoplasm (already described by us and others - Carbone et al. 2002; Seker et al. 2003), which in turn causes spread of histone deacetylase 1 (HDAC1) from PML bodies to the nucleoplasm (Rubbi and Milner, manuscript in preparation).

Whether through interaction with proteins of the recombination repair machinery, or through chromatin condensation, or both, it is evident that p53 can directly modulate the activity of HR. This modulation appears to be transactivation-independent and gene dosage-dependent, indicating that loss of p53, even heterozygous, adds a level of genomic instability in response to DSB and replication blockage.

P53 AND BASE-EXCISION REPAIR

In recent years evidence has accumulated that indicates that p53 also has a role in base excision repair (BER - Offer et al. 1999; Offer et al., 2001a; Offer et al., 2001b; Zhou et al., 2001). BER removes bases that have been chemically modified by alkylation, oxidation, etc. In general a glycosylase removes the modified base leaving an abasic (apurinic or apyrimidinic - AP) deoxyribose. These AP sites are removed by the sequential action of an AP endonuclease (APE) followed by a DNA deoxyribophosphodiesterase and then the gap is filled by a DNA polymerase and a ligase (Friedberg et al., 1995).

p53 has been shown to promote BER activity *in vitro* in a transactivation-independent manner (Offer et al., 1999, 2001a). A possible mechanism for this action has been suggested by Zhou et al. (2001) who showed that p53 stimulation of BER correlates with its capacity to bind both APE and DNA polymerase β and stabilises the interaction between DNA polymerase β and abasic DNA. Thus, p53 appears to be directly involved in this important DNA repair pathway as well. However, it is not clear if heterozygosity will affect the participation of p53 in BER.

P53 IN EMBRYOGENESIS

There remains one more p53 paradox, and this is evident in two ways. First, p53-null mice develop normally, but, if p53 induces cell cycle arrest or apoptosis in response to endogenous DNA damage surely there should be developmental problems associated to loss of p53 (Choi and Donehower, 1999). Second, p53 is highly conserved in vertebrates, but if its main function is tumour suppression and if the onset of cancer is generally post-reproductive maturity, how could it be evolutionarily selected? (Hall and Lane, 1997; Sansom and Clarke, 2000). The simple answer is: p53 *does* have a role in development, although the phenotype of its failure has a surprisingly low penetrance. This evidence has been aptly reviewed (see Hall and Lane, 1997; Choi and Donehower, 1999; Sansom and Clarke, 2000) and here we shall highlight the most important conclusions.

Some heterozygous crosses in mice show a frequency of nullizygous lower (~20%) than the Mendelian ratio with a lower number of females. In particular, a fraction of female p53-null embryos show excencephaly (a defect in neural tube closure) which is increased when embryos are subjected to IR. Choi and Donehower (1999) propose that p53 is a teratological suppressor: p53 heterozygous and nullizygous mice have a larger embryo lethality by treatment with BPDE, and show developmental anomalies after IR, again pointing to a role of p53 in DNA repair. In fact, the appellative 'guardian of the babies' has been applied to p53 (Hall and Lane, 1997; Choi and Donehower, 1999). Thus, the same pattern of enhancement of DNA repair and induction of apoptosis by p53 that is observed during tumour development, is also likely to operate during embryo development, again, protecting against DNA damage and eliminating excessively damaged cells.

CONCLUSIONS

As discussed throughout this article, a large body of evidence indicates that p53 has independent roles in several DNA repair pathways. Its participation in DNA repair appears to be sensitive to gene dosage and to have a direct impact in oncogenesis. Thus, p53 failure contributes to a mutator phenotype and makes p53 a *caretaker* gene. In a tumour development process, however, mutation will lead to an oncogenic event which will trigger the capacity of p53 to induce cell cycle arrest and/or apoptosis through its stabilisation, that is to say, its *gatekeeper* function will be brought into play. Hence in a typical oncogenic development both p53 functions are likely to be intertwined. In humans, however, an oncogenic process is followed by an anti-cancer therapy and this most commonly involves the introduction of a high level of DNA damage, expected to induce apoptosis even in the presence of mutant p53 (Weinstein et al., 1997). Thus, the roles of p53 in DNA repair and, as a consequence, the increased mutagenicity associated with p53 failure have to be taken into account in the presence of this 'iatrogenic' DNA damage. The ideal situation of triggering gatekeeper activation without fuelling the mutator phenotype could be achieved through therapies that promote tumour cell death without resorting to DNA damage.

We have recently showed that nucleolar function appears to be crucial for the maintenance of low levels of p53 and that impairment of this function abrogates the capacity of mdm2 to promote p53 degradation leading to its stabilisation (Rubbi and Milner, 2003c and see Figure 1). Others have also indicated that the link between p53 stabilisation and loss of nucleolar function can be mediated by small p53-binding molecules released from the nucleolus under stress (Colombo et al. 2002; Lohrum et al., 2003). Thus, the nucleolus appears to be a sensor of cellular stress that can integrate a variety of stresses, including DNA damage, into a p53 response. With hundreds of proteins already identified in the nucleolar proteome (Andersen et al., 2002; Scherl et al., 2002) there is great potential for the identification of targets for non-DNA damaging anti-cancer therapies.

REFERENCES

Abrahams P.J., Houweling A., Cornelissen-Steijger P.D., Jaspers N.G., Darroudi F., Meijers C.M., Mullenders L.H., Filon R., Arwert F., Pinedo H.M., Natarajan A.P., Terleth C., Van Zeeland A.A., van der Eb A.J. Impaired DNA repair capacity in skin fibroblasts from various hereditary cancer-prone syndromes. Mutat Res 1998; 407:189-201.

Adimoolam S., Lin C.X., Ford J.M. The p53-regulated cyclin-dependent kinase inhibitor, p21 (cip1, waf1, sdi1), is not required for global genomic and transcription-coupled nucleotide excision repair of UV-induced DNA photoproducts. J Biol Chem 2001; 276:25813-25822.

Andersen J.S., Lyon C.E., Fox A.H., Leung A.K., Lam Y.W., Steen H., Mann M., Lamond A.I. Directed proteomic analysis of the human nucleolus. Curr Biol 2002; 12:1-11.

Bertrand P., Rouillard D., Boulet A., Levalois C., Soussi T., Lopez B.S. Increase of spontaneous intrachromosomal homologous recombination in mammalian cells expressing a mutant p53 protein. Oncogene 1997; 14:1117-1122.

Bischoff F.Z., Yim S.O., Pathak S., Grant G., Siciliano M.J., Giovanella B.C., Strong L.C., Tainsky M.A. Spontaneous abnormalities in normal fibroblasts from patients with Li-Fraumeni cancer syndrome: aneuploidy and immortalization. Cancer Res 1990; 50:7979-7984.

Cahill D.P., Kinzler K.W., Vogelstein B., Lengauer C. Genetic instability and darwinian selection in tumours. Trends Cell Biol 1999; 9:M57-60.

Camplejohn R.S., Perry P., Hodgson S.V., Turner G., Williams A., Upton C., MacGeoch C., Mohammed S., Barnes D.M. A possible screening test for inherited p53-related defects based on the apoptotic response of peripheral blood lymphocytes to DNA damage. Br J Cancer 1995; 72:654-662.

Carbone R., Pearson M., Minucci S., Pelicci P.G. PML NBs associate with the hMre11 complex and p53 at sites of irradiation induced DNA damage. Oncogene 2002; 21:1633-1640.

Choi J., Donehower L.A. p53 in embryonic development: maintaining a fine balance. Cell. Mol Life Sci 1999; 55:38-47.

Colombo E., Marine J.C., Danovi D., Falini B., Pelicci P.G. Nucleophosmin regulates the stability and transcriptional activity of p53. Nat Cell Biol 2002; 4:529-533.

Cook D.L., Gerber A.N., Tapscott S.J. Modeling stochastic gene expression: implications for haploinsuficiency. Proc. Natl Acad Sci USA 1998; 95:15641-15646.

Critchlow S.E., Jackson S.P. DNA end-joining: from yeast to man. Trends Biochem Sci 1998; 23:394-398.

Dudenhoffer C., Kurth M., Janus F., Deppert W., Wiesmuller L. Dissociation of the recombination control and the sequence-specific transactivation function of p53. Oncogene 1999; 18:5773-5784.

Espinosa J.M., Emerson B.M. Transcriptional regulation by p53 through intrinsic DNA/chromatin binding and site-directed cofactor recruitment. Mol Cell 2001; 8:57-69.

Finlay C.A., Hinds P.W., Levine A.J. The p53 proto-oncogene can act as a suppressor of transformation. Cell 1989; 57:1083-1093.

Ford J.M., Hanawalt P.C. Li-Fraumeni syndrome fibroblasts homozygous for p53 mutations are deficient in global DNA repair but exhibit normal transcription-coupled repair and enhanced UV resistance. Proc Natl Acad Sci. USA 1995; 92:8876-8880.

Ford J.M., Hanawalt P.C. Expression of wild-type p53 is required for efficient global genomic nucleotide excision repair in UV-irradiated human fibroblasts. J. Biol. Chem. 1997; 272:28073-28080.

French J.E., Lacks G.D., Trempus C., Dunnick J.K., Foley J., Mahler J., Tice R.R., Tennant R.W. Loss of heterozygosity frequency at the Trp53 locus in p53-deficient (+/-) mouse tumors is carcinogen-and tissue-dependent. Carcinogenesis 2001a; 22:99-106.

French J., Storer R.D., Donehower L.A. The nature of the heterozygous Trp53 knockout model for identification of mutagenic carcinogens. Toxicol Pathol 2001b; 29 Suppl:24-29.

Friedberg E.C. How nucleotide excision repair protects against cancer. Nature Rev Cancer 2001; 1:22-33.

Friedberg E.C., Walker G.C., and Siede, W. *DNA repair and mutagenesis.* Washington, DC: ASM Press, 1995.

Friedberg E.C., Bond J.P., Burns D.K., Cheo D.L., Greenblatt M.S., Meira L.B., Nahari D., Reis A.M. Defective nucleotide excision repair in xpc mutant mice and its association with cancer predisposition. Mutat. Res. 2000; 459:99-108.

Green C.M., Almouzni G. When repair meets chromatin. EMBO Rep 2002; 3:28-33.

Hall P.A., Lane D.P. Tumor suppressors: a developing role for p53?. Curr Biol 1997; 7:R144-147.

Helleday T. Pathways for mitotic homologous recombination in mammalian cells. Mutat Res 2003; 532:103-115.

Hwang B.J., Ford J.M., Hanawalt P.C., Chu G. Expression of the p48 xeroderma pigmentosum gene is p53-dependent and is involved in global genomic repair. Proc Natl Acad Sci USA 1999; 96:424-428.

Jackson S.P. Silencing and DNA repair connect. Nature 1997; 388:829-830.

Jackson S.P. Detecting, signalling and repairing DNA double-strand breaks. Biochem Soc Trans 2001; 29:655-661.

Jaks V., Joers A., Kristjuhan A., Maimets T. p53 protein accumulation in addition to the transactivation activity is required for p53-dependent cell cycle arrest after treatment of cells with camptothecin. Oncogene 2001; 20:1212-1219.

Kaeser M.D., Iggo R.D. Chromatin immunoprecipitation analysis fails to support the latency model for regulation of p53 DNA binding activity in vivo. Proc. Natl. Acad. Sci. USA 2002; 99:95-100.

Karran P. DNA double strand break repair in mammalian cells. Curr Opin Genet Dev 2000; 10:144-150.

Kinzler K.W., Vogelstein B. Lessons from hereditary colorectal cancer. Cell 1996; 87:159-170.

Knudson A.G. Antioncogenes and human cancer. Proc Natl Acad Sci USA 1993; 90:10914-10921.

Kumari A., Schultz N., Helleday T. p53 protects from replication-associated DNA double-strand breaks in mammalian cells. Oncogene 2004; 23:2324-2329.

Lalle P., Moyret-Lalle.C., Wang Q., Vialle J.M., Navarro C., Bressac-de Paillerets B., Magaud J.P., Ozturk M. Genomic stability and wild-type p53 function of lymphoblastoid cells with germ-line p53 mutation. Oncogene 1995; 10:2447-2454.

Lane D.P. p53, guardian of the genome. Nature 1992; 358:15-16.

Levine A.J. p53, the cellular gatekeeper for growth and division. Cell 1997; 88:323-331.

Levine A.J., Momand J., Finlay C.A. The p53 tumour suppressor gene. Nature 1991; 351:453-456.

Linke S.P., Clarkin K.C., Di Leonardo A., Tsou A., Wahl G.M. A reversible, p53-dependent G0/G1 cell cycle arrest induced by ribonucleotide depletion in the absence of detectable DNA damage. Genes Dev 1996; 10:934-947.

Livingstone L.R., White A., Sprouse J., Livanos E., Jacks T., Tlsty T.D. Altered cell cycle arrest and gene amplification potential accompany loss of wild-type p53. Cell 1992; 70:923-935.

Ljungman M. Dial 9-1-1 for p53: mechanisms of p53 activation by cellular stress. Neoplasia 2000; 2:208-225.

Ljungman M., Zhang F. Blockage of RNA polymerase as a possible trigger for u.v. light-induced apoptosis. Oncogene 1996; 13:823-831.

Lloyd D.R., Hanawalt P.C. p53-dependent global genomic repair of benzo[a]pyrene-7,8-diol-9,10-epoxide adducts in human cells. Cancer Res 2000; 60:517-521.

Lloyd D.R., Hanawalt P.C. p53 controls global nucleotide excision repair of low levels of structurally diverse benzo(g)chrysene-DNA adducts in human fibroblasts. Cancer Res 2002; 62:5288-5294.

Loeb L.A. Mutator phenotype may be required for multistage carcinogenesis. Cancer Res 1991; 51:3075-3079.

Lohrum M.A., Ludwig R.L., Kubbutat M.H., Hanlon M., Vousden K.H. Regulation of HDM2 activity by the ribosomal protein L11. Cancer Cell 2003; 3:577-587.

Lu X., Lozano G., Donehower L.A. Activities of wildtype and mutant p53 in suppression of homologous recombination as measured by a retroviral vector system. Mutat Res 2003; 522:69-83.

Macleod K. Tumor suppressor genes. Curr Opin Genet Dev 2000; 10:81-93.

Mekeel K.L., Tang W., Kachnic L.A., Luo C.M., DeFrank J.S., Powell S.N. Inactivation of p53 results in high rates of homologous recombination. Oncogene 1997; 14:1847-1857.

Offer H., Wolkowicz R., Matas D., Blumenstein S., Livneh Z., Rotter V. Direct involvement of p53 in the base excision repair pathway of the DNA repair machinery. FEBS Lett 1999; 450:197-204.

Offer H., Milyavsky M., Erez N., Matas D., Zurer I., Harris C.C., Rotter V. Structural and functional involvement of p53 in BER in vitro and in vivo. Oncogene 2001a; 20:581-589.

Offer H., Zurer I., Banfalvi G., Reha'k M., Falcovitz A., Milyavsky M., Goldfinger N., Rotter V. p53 modulates base excision repair activity in a cell cycle-specific manner after genotoxic stress. Cancer Res 2001b; 61:88-96.

Rief N., Lobrich M. Efficient rejoining of radiation-induced DNA double-strand breaks in centromeric DNA of human cells. J Biol Chem 2002; 277:20572-20582.

Ross I.L., Browne C.M., Hume D.A. Transcription of individual genes in eukaryotic cells occurs randomly and infrequently. Immunol Cell Biol 1994; 72:177-185.

Rubbi C.P., Milner J. Non-activated p53 co-localizes with sites of transcription within both the nucleoplasm and the nucleolus. Oncogene 2000; 19:85-96.

Rubbi C.P., Milner J. p53 is a chromatin accessibility factor for nucleotide excision repair of DNA damage. EMBO J 2003a; 22:975-986.

Rubbi C.P., Milner J. p53: Guardian of a genome's guardian?. Cell Cycle 2003b; 2:20-21.

Rubbi C.P., Milner J. Disruption of the nucleolus mediates stabilization of p53 in response to DNA damage and other stresses. EMBO J 2003c; 22:6068-6077.

Saintigny Y., Lopez B.S. Homologous recombination induced by replication inhibition, is stimulated by expression of mutant p53. Oncogene 2002; 21:488-492.

Saintigny Y., Rouillard D., Chaput B., Soussi T., Lopez B.S. Mutant p53 proteins stimulate spontaneous and radiation-induced intrachromosomal homologous recombination independently of the alteration of the transactivation activity and of the G1 checkpoint. Oncogene 1999; 18:3553-3563.

Sansom O.J., Clarke A.R. p53 null mice: damaging the hypothesis?. Mutat Res 2000; 452:149-162.

Scherl A., Coute Y., Deon C., Calle A., Kindbeiter K., Sanchez J.C., Greco A., Hochstrasser D., Diaz J.J. Functional proteomic analysis of human nucleolus. Mol Biol Cell 2002; 13:4100-4109.

Seker H., Rubbi C., Linke S.P., Bowman E.D., Garfield S., Hansen L., Borden K.L., Milner J., Harris C.C. UV-C-induced DNA damage leads to p53-dependent nuclear trafficking of PML. Oncogene 2003; 22:1620-1628.

Sengupta S., Linke S.P., Pedeux R., Yang Q., Farnsworth J., Garfield S.H., Valerie K., Shay J.W., Ellis N.A., Wasylyk B., Harris C.C. BLM helicase-dependent transport of p53 to sites of stalled DNA replication forks modulates homologous recombination. EMBO J 2003; 22:1210-1222.

Sherr C.J., Weber J.D. The ARF/p53 pathway. Curr Opin Genet Dev 2000; 10:94-99.

Smerdon M.J., Thoma F. "Modulations in chromatin structure during DNA damage formation and DNA repair". In *DNA Damage and Repair. Volume II: DNA repair in Higher Eukaryotes*, J. A. Nickoloff and M. F. Hoekstra, ed. Totowa, N.J.: Humana Press, 1998.

Smith M.L., I.T.Chen Q.Zhan P.M.O'Connor and A.J. Fornace. Involvement of the p53 tumor suppressor in repair of u.v.-type DNA damage. Oncogene 1995; 10:1053-1059.

Storer R.D., French J.E., Haseman J., Hajian G., LeGrand E.K., Long G.G., Mixson L.A., Ochoa R., Sagartz J.E., Soper K.A. p53+/- hemizygous knockout mouse: overview of available data. Toxicol Pathol 2001; 29 Suppl:30-50.

Stott F.J., Bates S., James M.C., McConnell B.B., Starborg M., Brookes S., Palmero I., Ryan K., Hara E., Vousden K.H., Peters G. The alternative product from the human CDKN2A locus, p14(ARF), participates in a regulatory feedback loop with p53 and MDM2. EMBO J 1998; 17:5001-5014.

Symonds H., Krall L., Remington L., Saenz-Robles M., Lowe S., Jacks T., Van Dyke T. p53-dependent apoptosis suppresses tumor growth and progression in vivo. Cell 1994; 78:703-711.

Tang W., Willers H., Powell S.N. p53 directly enhances rejoining of DNA double-strand breaks with cohesive ends in gamma-irradiated mouse fibroblasts. Cancer Res 1999; 59:2562-2565.

Tang J.Y., Hwang B.J., Ford J.M., Hanawalt P.C., Chu G. Xeroderma pigmentosum p48 gene enhances global genomic repair and suppresses UV-induced mutagenesis. Mol Cell 2000; 5:737-744.

Temin H.M. Evolution of cancer genes as a mutation-driven process. Cancer Res 1988; 48:1697-1701.

Thoma F. Light and dark in chromatin repair: repair of UV-induced DNA lesions by photolyase and nucleotide excision repair. EMBO J 1999; 18:6585-6598.

Tsukamoto Y., Kato J., Ikeda H. Silencing factors participate in DNA repair and recombination in Saccharomyces cerevisiae. Nature 1997; 388:900-903.

Tumbar T., Sudlow G., Belmont A.S. Large-scale chromatin unfolding and remodeling induced by VP16 acidic activation domain. J Cell Biol 1999; 145:1341-1354.

van Oosterwijk M.F., Filon R., Kalle W.H., Mullenders L.H., van Zeeland A.A. The sensitivity of human fibroblasts to N-acetoxy-2-acetylaminofluorene is determined by the extent of transcription-coupled repair, and/or their capability to counteract RNA synthesis inhibition. Nucleic Acids Res 1996; 24:4653-4659.

van Oostrom C.T., Boeve M., van Den Berg J., de Vries A., Dolle M.E., Beems R.B., van Kreijl C.F., Vijg J., van Steeg H. Effect of heterozygous loss of p53 on benzo[a]pyrene-induced mutations and tumors in DNA repair-deficient XPA mice. Environ Mol Mutagen 1999; 34:124-130.

Varley J.M., Evans D.G., Birch J.M. Li-Fraumeni syndrome - a molecular and clinical review. Br J Cancer 1997; 76:1-14.

Varley J.M., Chapman P., McGown G., Thorncroft M., White G.R., Greaves M.J., Scott D., Spreadborough A., Tricker K.J., Birch J.M., Evans D.G., Reddel R., Camplejohn R.S., Burn J., Boyle J.M. Genetic and functional studies of a germline TP53 splicing mutation in a Li-Fraumeni-like family. Oncogene 1998; 16:3291-3298.

Venkatachalam S., Donehower L.A. Murine tumor suppressor models. Mutat Res 1998; 400:391-407.

Venkatachalam S., Shi Y.P., Jones S.N., Vogel H., Bradley A., Pinkel D., Donehower L.A. Retention of wild-type p53 in tumors from p53 heterozygous mice: reduction of p53 dosage can promote cancer formation. EMBO J 1998; 17:4657-4667.

Venkatachalam S., Tyner S.D., Pickering C.R., Boley S., Recio L., French J.E., Donehower L.A. Is p53 haploinsufficient for tumor suppression? Implications for the p53+/- mouse model in carcinogenicity testing. Toxicol Pathol 2001; 29 Suppl:147-154.

Vousden K.H., Lu X. Live or let die: the cell's response to p53. Nat Rev Cancer 2002; 2:594-604.

Wang X.W., Yeh H., Schaeffer L., Roy R., Moncollin V., Egly J.M., Wang Z., Friedberg E.C., Evans M.K., Taffe B.G., Bohr V.A., Weeda G., Hoeijmakers J.H., Forrester K., Harris C.C. p53 modulation of TFIIH-associated nucleotide excision-repair activity. Nat Genet 1995; 10:188-195.

Wani M.A., El-Mahdy M.A., Hamada F.M., Wani G., Zhu Q., Wang Q.E., Wani A.A. Efficient repair of bulky anti-BPDE DNA adducts from non-transcribed DNA strand requires functional p53 but not p21(waf1/cip1) and pRb. Mutat Res 2002a; 505:13-25.

Wani M.A., Wani G., Yao J., Zhu Q., Wani A.A. Human cells deficient in p53 regulated p21(waf1/cip1) expression exhibit normal nucleotide excision repair of UV-induced DNA damage. Carcinogenesis 2002b; 23:403-410.

Weinstein J.N., Myers T.G., O'Connor P.M., Friend S.H., Fornace A.J., Kohn K.W., Fojo T., Bates S.E., Rubinstein L.V., Anderson N.L., Buolamwini J.K., van Osdol W.W., Monks

A.P., Scudiero D.A., Sausville E.A., Zaharevitz D.W., Bunow B., Viswanadhan V.N., Johnson G.S., Wittes R.E., Paull K.D. An information-intensive approach to the molecular pharmacology of cancer. Science 1997; 275:343-349.

Willers H., McCarthy E.E., Wu B., Wunsch H., Tang W., Taghian D.G., Xia F., Powell S.N. Dissociation of p53-mediated suppression of homologous recombination from G1/S cell cycle checkpoint control. Oncogene 2000; 19:632-639.

Willers H., McCarthy E.E., Hubbe P., Dahm-Daphi J., Powell S.N. Homologous recombination in extrachromosomal plasmid substrates is not suppressed by p53. Carcinogenesis 2001; 22:1757-1763.

Williams K.J., Boyle J.M., Birch J.M., Norton J.D., Scott D. Cell cycle arrest defect in Li-Fraumeni Syndrome: a mechanism of cancer predisposition?. Oncogene 1997; 14:277-282.

Yamaizumi M., Sugano T. U.V.-induced nuclear accumulation of p53 is evoked through DNA damage of actively transcribed genes independent of the cell cycle. Oncogene 1994; 9:2775-2784.

Yin Y., Tainsky M.A., Bischoff F.Z., Strong L.C., Wahl G.M. Wild-type p53 restores cell cycle control and inhibits gene amplification in cells with mutant p53 alleles. Cell 1992; 70:937-948.

Yoon H., Liyanarachchi S., Wright F.A., Davuluri R., Lockman J.C., de la Chapelle A., Pellegata N.S. Gene expression profiling of isogenic cells with different TP53 gene dosage reveals numerous genes that are affected by TP53 dosage and identifies CSPG2 as a direct target of p53. Proc. Natl Acad Sci USA 2002; 99:15632-15637.

Zhou J., Ahn J., Wilson S.H., Prives C. A role for p53 in base excision repair. EMBO J 2001; 20:914-923.

Chapter 12

ANALYSIS OF P53 GENE ALTERATIONS IN CANCER: A CRITICAL VIEW

Thierry Soussi

Laboratoire de Génotoxicologie des tumeurs, EA 3493 Service de Pneumologie, Hôpital Tenon, Paris, France

> *He threw himself into the water and swam out in search of the swans, who caught sight of him and hurried towards him, their feathers ruffled. "Kill me" cried the poor animal, hanging his head towards the surface of the water, awaiting death. But what did he see in the transparent water? He saw his own image beneath him, no longer that of an ugly, dirty grey duckling, but that of a majestic swan. There's no harm in being born in a farmyard when you hatch from a swan's egg.*
>
> Hans Christian Andersen, *The Ugly Duckling*

INTRODUCTION

"Guardian of the genome" (Lane, 1992), "Death star" (Vousden, 2000), "Good and bad cop" (Sharpless and DePinho, 2002), "An acrobat in tumorigenesis" (Moll and Schramm, 1998), are just a few of the names that have been attributed to the p53 gene over recent years. However, the cameras (and funding) were certainly not present at the time of the discovery of p53 in 1979 (Crawford, 1983). It was only when the first alterations of the p53 gene in human cancers were discovered 10 years later, in 1989, that p53 started to become really popular, with the title of "molecule of the year" attributed by *Science*, in 1993 (Harris, 1993). This title was certainly justified, as the observation that more than one half of human cancers expressed a mutant p53 raised extensive clinical possibilities both for

P. Hainaut and K.G. Wiman (eds.), 25 Years of p53 Research, 255-292.

diagnosis and treatment. As always, during the rapid growth phase of a new field of investigation, great hopes were raised and the pharmaceutical industry became actively involved. Although, from a scientific point of view, research has clearly shown the importance of p53 signalling pathways in the surveillance of the cell after a genotoxic stress (Vogelstein et al., 2000), clinical applications are nevertheless limited at the present time. This situation is not specific to p53, as technology transfer to clinical applications is always a difficult process. The contingencies required to validate a new marker are identical to those used to validate a new therapeutic molecule (see below). The field of diagnosis is also currently undergoing a major revolution with the development of high throughput technologies, such as DNA biochips, proteomic analyses or TMA (tissue microarray). Most of these technologies cannot be applied in routine clinical practice and therefore remain confined to the field of "gene discovery". On the other hand, their discoveries can have applications in routine clinical practice by means of technologies such as high throughput quantitative PCR, detection of mutations on DNA biochips or any other support allowing exhaustive analysis of a large number of samples. In the case of p53, a biochip could be developed allowing the simultaneous definition of the mutational profile of the gene, while also verifying the expression status of the genes involved in upstream or downstream signalling pathways.

In this article, I will review the analysis of p53 gene alterations in human cancers. The first part will discuss the specific properties of the p53 gene which make it an atypical tumour suppressor gene. The second part will evaluate the situations in which the diagnosis of p53 mutations can be useful and the third part will present a critical analysis of the various technological aspects of p53 analysis in human cancers.

THE P53 GENE: A TUMOUR SUPPRESSOR GENE OR AN ONCOGENE? A CARETAKER OR A GATEKEEPER?

The history of p53 is a chaotic voyage from the world of oncogenes to the world of tumour suppressor genes, while retaining a certain degree of individuality (Lane and Benchimol, 1990). Apart from artefactual problems related to involuntary cloning of mutant p53, this ambiguity is also due to our propensity to over-categorize in order to satisfy our Cartesian and oversimplistic view of science .

The idea that some p53 mutations can actively participate in cellular transformation was already postulated in 1990 and several arguments are in favour of such a model (Eliyahu et al., 1990; Lane and Benchimol, 1990). First of all, the mode of "inactivation" of wild-type p53. Unlike most other tumour suppressor genes that are inactivated by frameshift or nonsense mutations leading to disappearance or aberrant synthesis of the gene product, almost 90% of p53 gene mutations are missense mutations leading to the synthesis of a stable protein, lacking its specific DNA binding function and accumulating in the nucleus of tumour cells (Soussi and Béroud, 2001). This particular selection for accumulation of p53 mutations in tumour cells can have two consequences: i) a dominant negative role by hetero-oligomerization with wild-type p53 expressed by the second allele, or ii) a specific gain of function of mutant p53. Many studies have tried to distinguish between these two hypotheses, with no clear-cut conclusions (Michalovitz et al., 1991; Milner, 1995). This task is further complicated by the fact that not all p53 mutations appear to be equivalent and present a marked heterogeneity of structure or loss of function. Transfection of various p53 mutations into cells devoid of endogenous p53 leads to an increase in their carcinogenicity, which varies according to the type of mutation (Dittmer et al., 1993; Halevy et al., 1990). This research into the oncogenic potential of certain p53 mutations is not purely theoretical, but has obvious clinical implications, as it could explain the marked disparity of the results of studies trying to demonstrate a relationship between the presence of a p53 gene mutation and various clinical parameters, such as survival or response to treatment. In breast cancer patients, the response to adriamycin is very strongly correlated with the presence of a mutation specifically localized in the loop domains L2 or L3 of the p53 protein (Aas et al., 1996). *In vitro*, the expression of p53 mutations in position 175 (R175H) specifically induces resistance of cells to etoposides compared to other p53 mutations (Blandino et al., 1999).

The two homologous genes of p53, p63 and p73, discovered 6 years ago, express many isoforms due to alternating use of transcription promoters and alternative splicing (Yang et al., 2002). Long isoforms (TA-p73 or TA-p63) are able to transactivate the same target genes as p53 and induce apoptosis, while short forms (DN-P63 or Dnp73) have an opposite activity via dominant negative mechanisms. p63 and p73 are able to cooperate with p53 to induce apoptosis, suggesting the existence of a complex network of interactions between the products of these three genes (Melino et al., 2002). Although wt p53 does not interact with p73 or p63, some mutant p53 bind strongly to the two p53 homologs via their DNA binding domains. This interaction leads to the inactivation of p73 and p63 function (DiComo et al., 1999; Gaiddon et al., 2001; Marin et al., 2000) (Strano et al., 2000). Recent studies by T. Crook and B. Kaelin show that the activity of resistance to anticancer agents involves inactivation of the apoptotic function of p73 protein by a subset of mutant p53 that have sustained a change of conformation (Bergamaschi et al., 2003) (Irwin et al., 2003).

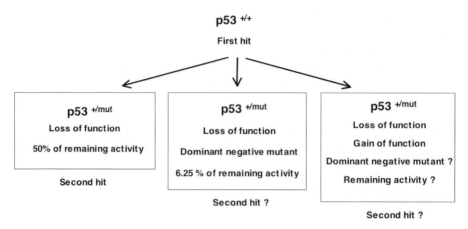

Figure 1. p53 and cancer: in the classical situation (left), the first hit leads to inactivation of p53 without affecting the activity of the second allele. However, this loss of function may have certain consequences for the cell. The second hit leads to complete loss of function of p53. In the case of a dominant negative mutant (middle), an estimated 1/16 of the tetramers theoretically have 4 wild-type monomers if the two proteins are expressed in identical quantities and tetramerization is not affected by the mutation. If the dominant negative effect of a single monomer is sufficient to inactivate p53, then the remaining activity will be 6.25%. In the case of partial penetrance of the dominant negative effect, the activity gradient will be between 6% and 50%. Loss of the second allele may not be mandatory, depending on the remaining activity. In the case of a mutant with a gain of function (right), the situation is more complex with an important combinatorial effect. Loss of the second allele may also be unnecessary in this case. This particular situation of the p53 gene results from: i) its particular mode of inactivation by missense mutations and ii) its tetrameric structure.

It is therefore likely that although the <u>wild-type</u> p53 gene is effectively a tumour suppressor gene, some p53 <u>mutants</u> can be considered to be oncogenes. The distinction between oncogene and tumour suppressor gene, although simple and able to account for the mechanisms of activation or inactivation of these various genes, is probably imperfect and does not allow characterization of all genes involved in the processes of carcinogenesis.

The genes involved can also be classified as a function of the various roles that they play in malignant transformation of a cell when they are altered. Three types of genes are distinguished at the present time(Kinzler and Vogelstein, 1997; Kinzler and Vogelstein, 1998). Gatekeeper genes regulate cellular homeostasis and the cell cycle by controlling the entry of the cell in the various phases of the cell cycle (Rb, VHL or APC). Caretaker genes participate in maintenance of the integrity of the genome and allow the cell to transmit an identical genome during successive cell divisions; they act as caretakers of the genome (MLH1, MSH2, or gene XP). Finally, landscaper genes maintain the integrity and equilibrium of the various cellular components of a tissue (PTEN, Smad4). Once again, the situation is more ambiguous for the p53 gene, which can be classified as both a gatekeeper and a caretaker gene. Its apoptotic and antiproliferative activity make it an important gatekeeper and re-introduction of a wild-type p53 gene into a tumour cell effectively restores the cell cycle control properties. On the other hand, its possible activity in the control of chromosomal stability would classify p53 as a caretaker gene, whose phenotype cannot be corrected by functional supplementation. p53 probably has a very heterogeneous role in tumour processes as a function of the tissue considered, the chronology of the p53 alteration (early or late), the p53 function targeted (cell cycle or apoptosis) and finally the nature of the other alterations already present in the cell.

WHY ANALYSE P53 GENE ALTERATIONS IN CANCERS?

Before discussing the various approaches to the analysis of p53 gene status in human tumours, it is useful to ask the question of the practical value of this analysis. At the dawn of an age devoted to "high throughput" and "global analyses", one wonders whether a diagnosis based on a single gene analysis, either p53 or another gene, has any future in clinical practice. Furthermore, as I will discuss below, the p53 protein belongs to a complex signalling network with a high tissue specificity. This has several important consequences in terms of diagnosis: i) the p53 pathway can be inactivated in many ways; ii) the behaviour of p53 mutants can be very heterogeneous according to the organ and the p53 gene status therefore probably does not always reflect the activity of the p53 signalling pathway

Multiple pathways of inactivation of the p53 gene

Inactivation of the p53 gene is essentially due to small mutations (missense and nonsense mutations or insertions/deletions of several nucleotides), which lead to either expression of a mutant protein (90% of cases) or absence of protein (10% of cases)(see the special issue of Human Mutation devoted to p53 published in January 2002 for more information). No inactivation of p53 gene expression by hypermethylation of transcription promoters has been demonstrated at the present time, which supports the hypothesis of a function for p53 mutants. In many cases, these mutations are associated with loss of the wild-type allele of the p53 gene located on the short arm of chromosome 17, which is why the p53 gene is said to behave like a classical tumour suppressor gene with a mutation on one allele and loss of heterozygosity (LOH) of the second allele. This is in line with the good concordance observed between cancers with a high frequency of p53 gene mutations and the frequency of LOH of the short arm of chromosome 17. However, another tumour suppressor gene, located adjacent to p53, could also be the target of these deletions (Makos Walles et al., 1995). In fact, many studies on murine models show that the haplo-insufficiency of p53 is sufficient to lead to an abnormal cell phenotype (Venkatachalam et al., 2001). Almost 50% of tumours in these mice still express the wild-type allele. It is therefore perfectly possible that inactivation of the remaining wild-type allele is not systematically necessary. As indicated above, the dominant negative activity of some mutants may not always require LOH of p53.

Although the presence of a p53 gene mutation is generally unambiguous, other situations are much more equivocal (Table 1). The mdm2 protein regulates the stability of the p53 protein by ubiquitination and transport towards the proteasome (Iwakuma and Lozano, 2003; Moll and Petrenko, 2003). Abnormal accumulation of the mdm2 protein is observed in many tumours, especially sarcomas (Onel and Cordon-Cardo, 2004). This accumulation can be due to amplification of the mdm2 gene, enhanced transcription of the gene or enhanced translation of its messenger RNA (Michael and Oren, 2002). Although these tumours would be expected to no longer express p53, the opposite situation is generally observed, with a large number of tumours overexpressing both p53 and mdm2. The reasons for this apparent paradox have not been elucidated. No formal exclusion between p53 gene mutation and mdm2 accumulation has been clearly demonstrated, suggesting that this situation could be due to an oncogenic activity of mdm2 independent of p53.

The situation is somewhat clearer for cervical cancer. The E6 viral protein expressed by HPV specifically binds to the p53 protein and induces its degradation (Scheffner et al., 1990). This observation explains the rarity of p53 mutations in cervical cancers (Crook et al., 1992). p53 inactivation by a viral protein has not been formally demonstrated in other human cancers associated with viral infection, such as HCC (associated with HBV) or Burkitt lymphoma (associated with EBV). In inflammatory breast cancers or neuroblastomas, molecular and immunohistochemical analyses demonstrate accumulation of wild-type p53 in the cytoplasm of tumour cells, leading to functional inactivation of p53 (Moll et al., 1995; Moll et al., 1996; Moll et al., 1992).

Table 1. Multiple pathways of inactivation of p53.

Inactivation	p53 gene status	p53 protein status	Type of cancer	Possible method of analysis*
p53 gene mutation	Mutant	Stable, nuclear	50% of all cancers	Molecular Immunohistochemical Functional
mdm2 amplification	Wild-type	Stable, nuclear	Sarcoma	Molecular Immunohistochemical
Nuclear exclusion (overexpression of the Parc protein)	Wild-type	Cytoplasmic	Neuroblastoma Inflammatory breast cancer	Immunohistochemical**
chk2 gene mutation	Wild-type	Not induced?	LFS (rare)	
Alteration of p14ARF	Wild-type***	Nuclear	20% of cancers, melanomas	
Alteration of Apaf1	Wild-type	Nuclear	Melanomas	
HPV 16 or 18 infection	Wild-type	Degraded	Cervical cancer	

* for p53 status only. For the other genes, the methodology depends on the modes of inactivation (see text for more details);

** the mechanisms leading to accumulation of parc are unknown at the present time;

*** the relationship between p14ARF mutation and absence of p53 mutation is not 100%

Nikolaev et al. recently isolated a new protein, parc, which sequesters p53 in the cytoplasm of cells in the absence of any lesions (Nikolaev et al., 2003). Abnormal accumulation of this protein is observed in neuroblastoma cells and could therefore account for functional inactivation of p53 in this cancer.

p53 mutations are very rare in malignant melanoma, a highly chemoresistant tumour, but are much more frequent in other skin cancers, such as BCC and SCC (Brash and Ponten, 1998). This situation could be due to an alteration of apoptotic pathways upstream and downstream to p53 signals. Alteration of upstream signals corresponds to inactivation of the CDKN2 locus, which expresses the p14^{ARF} protein, an activator of p53 in response to an oncogenic stress and the cyclin kinase inhibitor p16. The interaction of p14^{ARF} with mdm2 blocks MDM2 shuttling between the nucleus and cytoplasm via the nucleolus (Iwakuma and Lozano, 2003; Vousden and Woude, 2000). Sequestration of MDM2 in the nucleolus thus results in activation of p53. Germline alterations (point mutations) and somatic alterations (point mutations and hypermethylation of the promoter) of this gene are frequent in malignant melanoma and impair the induction of p53 after an oncogenic stress (Chin et al., 1998). Alteration of downstream signals corresponds to a marked reduction of expression of the pro-apoptotic apaf1 gene, the absence of which is correlated with resistance to chemotherapeutic agents (Soengas et al., 1999). Decreased apaf1 expression is due to hypermethylation of its promoter.

The role of the hchk2 gene, a kinase activated by ATM following irradiation, is more ambiguous. Many studies have shown that hchk2 is necessary for phosphorylation and stabilization of p53 after genotoxic lesions (Chehab et al., 2000; Shieh et al., 2000). These studies also found a strong supprt with the description of hchk2 germline mutations in families with Li-Fraumeni syndrome (LFS) not presenting any p53 mutations (Bell et al., 1999). However, these results probably need to be interpreted differently, as more recent studies have shown that, in human cells devoid of hchk2, p53 stabilization and induction of its target genes and arrest of the cell cycle after irradiation are perfectly normal (Ahn et al., 2003; Jallepalli et al., 2003).

Many mechanisms are therefore involved in the inactivation of signalling pathways regulated by p53, but the clinical consequences of each of these mechanisms needs to be more clearly elucidated.

Molecular epidemiology

The greatest contribution to the study of p53 mutations has been provided by molecular epidemiology and its applications (Harris, 1991; Soussi, 1996). We will not discuss these epidemiological studies in more detail, as they have been the subject of many detailed reviews and are discussed in another chapter of this book. The most important findings of molecular epidemiology are summarized in table 2.

These studies demonstrate a link between exposure to various types of carcinogens and the development of specific cancers. The most striking example is that of tandem mutations, specifically induced by ultraviolet radiation, which are only observed in skin cancers. The relationships between G->T transversion and lung cancer in smokers or mutation of codon 249 observed in aflatoxin B1-induced liver cancers are also very demonstrative. From a cognitive point of view, these findings are important in that they confirm that a large number of mutations are exogenous and therefore avoidable. On the other hand, these studies will not have any impact in terms of public health as they will not be followed by any political or administrative decisions to modify exposure situations. The only finding that could possibly have a major application concerns the mutation of codon 249 in liver cancer in individuals exposed to aflatoxin B1, a hepatocarcinogenic molecule expressed by a fungus, which frequently contaminates certain agricultural products harvested in tropical or subtropical regions. Contamination of food by this fungus and exposure to aflatoxin B1 are documented facts in several developing countries (Wild et al., 1993). The singularity of the unique mutation is due to the binding specificity of substances derived from aflatoxin B1 on codon 249 of the p53 gene (Puisieux et al., 1991). This specificity has allowed the development of extremely sensitive screening methods for this mutation (1 mutant copy per 10^5 wild-type copies) (Aguilar et al., 1993). By using these approaches, Aguilar et al. demonstrated that the mutation occurs very early and can be demonstrated in the liver of asymptomatic healthy subjects derived from regions exposed to aflatoxin B1 (Aguilar et al., 1994). No mutation was detected in subjects derived from non-exposed regions. This test has subsequently been successfully applied to the detection of p53 mutations in serum DNA from individuals living in high-risk regions (Kirk et al., 2000). This type of analysis should allow very early detection of individuals at high risk of developing hepatocellular carcinoma. However, it can only be effective when it is associated with an infrastructure able to follow these individuals and propose early intervention.

Table 2. Relationship between p53 gene mutations and exposure to carcinogens. Only the most striking observations are summarized in this table. For more details, the reader can refer to the recent review by Vähäkangas et al. (Vahakangas, 2003)

Type of cancer	Particularity of mutations p53*	Genotoxic agent incriminated	Comments	References
Lung cancer	High frequency of G->T transversions** Hot spot on codons 157 and 158	Benzo(a)pyrene (cigarette smoke)	Benzo(a)pyrene has a particular affinity for codons 157 and 158	(Denissenko et al., 1996; Toyooka et al., 2003)
Hepatocellular carcinoma	Specific G->T transversions in codon 249	Aflatoxin B1	Aflatoxin B1 binds specifically to codon 249	(Puisieux et al., 1991; Staib et al., 2003)
Skin cancer (BCC and SCC)	Very high frequency of mutations on pyrimidine dimers High frequency of tandem mutations	Ultraviolet radiation	Photo-induced mutations	(Brash et al., 1991; Tornaletti et al., 1993)
Hepatic angiosarcoma	High frequency of A:T -> T:A transversions	Vinyl chloride		(Hollstein et al., 1994)
Wilson's disease, haemochromatosis	Specific G->T transversions in codon 249	In these diseases related to iron or copper overload, overproduction of free radicals leads to high oxidative stress	Exposure of cells to a carcinogen derived from lipid peroxidation leads to alterations on codon 249 of the p53 gene.	(Hussain et al., 2000; Marrogi et al., 2001)

* compared to p53 mutations observed in the absence of exposure to the agent incriminated.
** this high frequency of transversion is also observed in cancers of the oesophagus and head and neck cancers associated with drinking and smoking.

p53 mutations: a new clinical marker?

This is certainly one of the most chaotic subjects in the field of p53. This confusion is due to a number of explanations: diversity of methodologies and strategies used to analyse p53 status, marked heterogeneity in study populations, a very heterogeneous behaviour of mutant p53 and especially our current ignorance concerning all branches of p53 signalling pathways. Research has focussed on the prognostic value of p53 and also on the response to therapy, as it has now been clearly demonstrated that a large number of molecules used in cancer chemotherapy induce p53-dependent apoptosis.

It is essential to avoid confusion about the terms prognostic and predictive. A prognostic marker can be defined as any factor that, at the time of diagnosis, can provide information on the clinical outcome of the patient, such as survival or disease-free survival. The most powerful prognostic factors are tumour size, clinical spread (stage) and histological grade. Among the molecular markers that have been tested during the past decade, *N-MYC* amplification in neuroblastoma remains the best prognostic marker. A predictive factor is defined as any marker that gives information regarding the response to a specific treatment. Prototype predictive markers are the oestrogen and progesterone receptors that mediate the response to the hormone therapy tamoxifen. With a few exceptions, none of the potentially useful prognostic or predictive markers have led to any consistent results in independent clinical studies. Factors that influence these studies include inadequate patient recruitment (sample size, diagnostic entry criteria, heterogeneous treatment) and methodological problems (quality of starting tissue, assay variability). This unsatisfactory situation has led several authors to propose a hierarchy of prognostic and predictive studies, analogous to the hierarchical study design in drug trials (Sullivan Pepe et al., 2001). Such an approach allows logical exploration and step-by-step validation of potential markers. Phase I studies are early exploratory studies of the association between a prognostic marker and important disease characteristics. They should also lead to the definition of a standardized assay. Phase II studies should define the clinical utility of the marker by identifying the optimal cut-off value between high-risk and low-risk patients. Both of these retrospective phases should be performed in carefully controlled (preferably case-controlled) cohorts of well-defined patients. Phase III studies are large, prospective, confirmatory studies in which the marker is evaluated and compared with other well-defined factors. The *TP53* status in human cancer could be considered at the end of Phase I (Bray et al., 1998). Several meta-analyses have indicated that, despite disagreement in the literature, *p53* status could have prognostic significance in non-small-cell lung cancer

(Mitsudomi et al., 2000; Steels et al., 2001) or in breast cancer (Pharoah et al., 1999), so the time is ripe to begin Phase II studies to unravel the true potential of using *p53* status for clinical decision-making. This topic will be extensively developed by Borensen et al. in another chapter of this book.

p53 germline mutations

The discovery of germline mutations in families with LFS was a major argument leading to the classification of p53 as a tumour suppressor gene (Malkin et al., 1990; Srivastava et al., 1990). LFS, with autosomal dominant transmission, is a rare disease of young subjects who present a predisposition to various tumours (Li et al., 1988). The classical and historical definition is based on familial criteria, essentially the observation of a sarcoma in an affected subject before the age of 45 years with a first-degree relative who developed any type of cancer before the age of 45 years or with a second-degree relative with a cancer or sarcoma before the age of 45 years. It is difficult to estimate the incidence of this rare syndrome due to its poorly defined diagnostic criteria. The most characteristic tumours are osteosarcomas, soft tissue sarcomas, breast cancers in young subjects, leukaemias/lymphomas, brain tumours and adrenal cortical carcinomas. However, any type of tumours can be observed. There are an estimated 400 families with LFS in the world at the present time. A germline mutation of the p53 gene is detected in about 70% of LFS families and in certain families or cases suggestive of the syndrome, without strictly satisfying all of the criteria (LFL) (Varley, 2003). A germline mutation of the *hCHK2* gene has also been described in rare families (Bell et al., 1999). A subject with a deleterious mutation of the *TP53* gene has a 15% risk of developing a cancer at the age of 15 years, with a risk of 80% for 50-year-old women and 40% for 50-year-old men; the significant difference between the sexes can be almost completely explained by breast cancer. These patients also present a high risk of second cancers, especially radio-induced cancers.

p53 germline mutations have also been demonstrated in almost 50% of children with adrenal cortical carcinoma with no family history of cancer (Varley et al., 1999b; Wagner et al., 1994). This cancer of the adrenal cortex accounts for 0.2% of all childhood tumours with an international incidence of 0.5/million, and occurs more frequently in girls than in boys (ratio of 1.5:1). The incidence is higher in patients with isolated hemihypertrophy, Wiedemann-Beckwith syndrome, congenital adrenal hyperplasia and LFS. The biological causes of this specific association are unknown.

No international consensus has yet been reached concerning the management of LFS families, which remains subject to local guidelines. These patients have a high risk of developing various types of other primary malignant tumours, which, because of their very broad spectrum, raise difficult screening problems. There are no data, at the present time, to suggest that a particular constitutional mutation is associated with a specific tumour type, but our knowledge of the biology of p53 suggests that DNA lesions induced by treatment of a first cancer in these patients with a constitutional p53 mutation can induce the emergence of cells with an abnormal genotype and therefore the appearance of other tumours.

No consensus has been reached to define the methodological approach to screening of the p53 gene status in affected families. The most advanced studies have been conducted by J. Varley's team, which clearly showed that only an approach involving exhaustive DNA (exons and introns) and RNA analysis can provide rigorous results (Varley et al., 1999a). The use of the FASAY test can also be considered (see below).

Contribution of p53 gene mutation analysis to basic research into the p53 protein.

This contribution is rarely mentioned, but, in the case of p53, basic research provided an unexpected wealth of data. Historically, the first description of p53 mutations was published in 1989 (Nigro et al., 1989; Takahashi et al., 1989), while the transactivation and the specific DNA binding activity of p53 was only discovered several years later (Bargonetti et al., 1993; Fields and Jang, 1990; Pavletich et al., 1993; Raycroft et al., 1990). Analysis of the mutations detected in human cancers not only allowed definition of the functionally important regions of the protein, but their heterogeneity also allowed us to more precisely dissect the various functions of p53. The establishment of databases combining all of this information provided valuable tools for this research. I would like to cite 3 examples, which appear to be particularly representative of this type of study.

Proline 175 (H175P) mutation

Despite the fact that the H175P mutation is situated on a mutation hot spot (codon 175), it is very rarely detected in human cancers (4 times) in contrast with its little sister (H175R), reported 650 times (Soussi and Béroud, 2003). The H175P mutation has a normal cell cycle arrest and gene p21 induction behaviour (Ory et al., 1994), but is deficient for apoptotic activity and does not transactivate bax or PIG3 genes (Friedlander et al., 1996; Rowan et al., 1996). The reasons for this heterogeneity are unknown at the present time, but could be related to a difference of interaction with various co-activating molecules. A strain of transgenic mice specifically expressing this mutation was recently produced (Liu et al., 2004). These mice have a very low predisposition to develop tumours compared to mice not expressing p53. However, these tumours do not present the chromosomal instability revealed in p53 -/- mice. These results derived from purely basic research, but based on a clinical observation, suggest that the apoptotic activity cannot be the primary activity targeted by p53 gene alterations (Attardi and DePinho, 2004). This represents a major challenge in relation to current models, which define apoptosis as being the fundamental activity of p53. Only the future will clarify the real role of p53.

Histidine 337 (R337H) mutation

The R337H mutation was found as a germline mutation specifically associated with paediatric adrenal cortical carcinoma in southern Brazil in several independent families that were not predisposed to other tumours (Latronico et al., 2001). In every transactivation assay, this mutant showed a wt behaviour. Precise chemical analysis revealed that this R337H mutant is highly sensitive to pH in the physiological range leading to folding changes depending on the protonated state of the protein (DiGiammarino et al., 2002). The observation that the syndrome associated with this mutation has been predominantly found in Brazil suggests that it could be linked to other modifier genes that could influence folding of p53 or cell pH. This type of observation emphasizes an important aspect of the p53 protein, its *in vitro* and *in vivo* flexibility and the influence of this flexibility on its properties. This type of observation must be considered in the light of the fact that almost 100 p53 mutants have a temperature-sensitive behaviour, i.e. active at 30°C and inactive at 37°C (Shiraishi et al., 2004).

Construction of a p53 mutation library

Many studies have emphasized the quantitative and qualitative heterogeneity of p53 mutations (Forrester et al., 1995; Ory et al., 1994; Resnick and Inga, 2003). In order to obtain a global view of this heterogeneity, Kato et al. constructed a library of 2,500 p53 mutations (Kato et al., 2003). Their transcriptional activity was tested on 8 transcription promoters characteristic of the various activities of p53 (cell cycle arrest, repair or apoptosis). Apart from the technical feat achieved by these authors, this study provides us with a wealth of valuable data to understand the individual activity of each p53 mutation, which can be classified according to the decreasing order of their remaining activity. It remains to be seen whether this classification has any correlation with clinical and laboratory parameters, such as response to therapy.

These three examples clearly illustrate this ping pong game, which can and must exist between basic research and clinical observations.

p53 gene mutations and therapy

Several chapters of this book are devoted to new therapeutic approaches applied to p53 and will not be discussed here. When these approaches are used in clinical trials, the patients' p53 status must be precisely defined in order to measure the efficacy of the new treatment.

Detection of p53 polymorphism in codon 72

Discovered in 1986, this intragenic polymorphism leads to the expression of two different p53 proteins, with Arginine or Proline in codon 72 in a region rich in proline residues (Harris et al., 1986). This region could be involved in the apoptotic activity of p53 (Walker and Levine, 1996). The distribution of this polymorphism in the general population is heterogeneous with a frequency of the Pro/Pro haplotype of 16% in Scandinavian populations and 63% in Nigerian populations (Beckman et al., 1994). The reason for this North/South gradient is unknown at the present time. Many studies have investigated whether one of the haplotypes could be associated with a higher susceptibility to develop cancers. The results of these studies are very contradictory and have not demonstrated any highly significant findings. In 1998, Storey et al. showed that p53^{Arg} was very sensitive to the degradation activity induced by papillomavirus protein E6, while p53^{Pro} was more resistant. This observation, that has not been contested, was associated with an epidemiological study on cervical cancer, which showed an over-representation of women presenting the p53^{Arg} allele (Makni et al., 2000; Storey et al., 1998; Thomas et al., 1999b). Unfortunately, this second part of the study was strongly contested by many other studies (Helland et al., 1998; Hildesheim et al., 1998). Nevertheless, these two wild-type p53, p53^{Arg} and p53^{Pro}, do not have exactly the same properties (table 3).

Some p53 mutants interact with p73 protein and inactivate its apoptotic function. Recently, T. Crook's team showed that this activity is specific to the p53^{Arg} mutant (Bergamaschi et al., 2003). Analysis of a homogeneous population of patients with head and neck cancer demonstrates that the majority of patients expressing a mutant p53 associated with this polymorphism have a poor response to chemotherapy and a shorter survival. If this result is confirmed by other studies, it is therefore possible that analysis of this polymorphism could have a clinical value, but, as discussed below, this analysis is difficult to perform.

Table 3. Comparison of the biological activities of the two polymorphic p53.

	p53Arg72	p53Pro72	Reference
Sensitivity to HPV protein E6	Sensitive	Resistant	(Thomas et al., 1999b)
Induction of apoptosis	High*	Moderate	(Dumont et al., 2003)
Interaction with p73 (in the case of mutant p53)	High	Low	(Marin et al., 2000)
Association with response to treatment	Poor*	Better	(Bergamaschi et al., 2003)
Interaction with transcriptional machinery	Low	High	(Thomas et al., 1999b)
Transactivation	Moderate	Higher	(Thomas et al., 1999b)
DNA binding	identical	identical	(Thomas et al., 1999b)

* In each case, the concept of high/low or poor/better is relative and only concerns the comparison between the two forms of p53.

HOW TO ANALYSE P53 GENE ALTERATIONS IN CANCERS

Molecular analysis

Direct sequencing of the p53 gene after PCR amplification remains the "Gold Standard" of molecular analysis. For the p53 gene, this approach is facilitated by the fact that the 10 coding exons are smaller than 350 bp and can therefore be easily amplified individually. Mutations involving partial or total gene deletions are relatively rare.

Unfortunately, although considerable progress has been made in the field of DNA sequencing in terms of throughput, its sensitivity still remains limited. The major problem of molecular analysis of tumour specimens is the presence of normal cells (lymphocytes, stromal cells) that contaminate the tumour samples. According to the type of tumour or the type of sample, the rate of contamination can range from several percent (surgical tumour sample) to 50% (biopsies) or even more than 95% (urine, stools or bronchial lavage). It is generally accepted that direct sequencing requires at least 20% of mutant alleles, but this can vary considerably according to the quality of the sample. This qualitative aspect is generally underestimated. The quantity and quality of DNA obtained varies considerably according to the origin of the sample (frozen tumour, formalin- or paraffin-embedded tissues). This variability can lead to the generation of PCR artefacts, which can be falsely interpreted as mutations. In the case of heavily contaminated samples, microdissection can be performed in order to enrich the tumour cell content, but this complicates the manipulations and cannot be performed routinely at the present time. The application of molecular technologies to routine analysis in hospital is a very important aspect. Many extremely sensitive molecular analysis methodologies have been developed, but their clinical application is generally limited because of the complex installation, their low throughput, the use of radioactivity or the need for highly qualified personnel.

Up until now, molecular analyses have been performed on exons 5-8 of the p53 gene, as the majority of mutations are located in these regions. It is generally established that 90% of mutational events are missense mutations leading to the synthesis of an abnormal protein that is not degraded and which accumulates in the nucleus of tumour cells. The remaining 10% of mutational events are nonsense mutations or small deletions that do not lead to accumulation of p53. This type of mutation excludes the possibility of using molecular methodologies such as PTT (Protein truncature test) based on expression of truncated proteins. More recently, molecular studies have been extended to the other exons, as exons 4, 9 and 10 have been found to contain a considerable number of mutations (about 15%) (Soussi and Béroud, 2001). Analysis of molecular events also shows a high proportion of nonsense mutations in these exons. Analysis of the latest version of the p53 gene mutation database shows that about 20% to 25% of mutations do not lead to the synthesis of a p53 protein. These mutations also present a marked variability as a function of the type of cancer: they are more frequent in lung cancers and breast cancers than in colon cancers (Soussi and Béroud, 2001). About 280 of the 393 codons of the p53 gene can be affected by a mutation. Furthermore, as each codon comprises 3 bases, which can each be altered generating a different amino acid, there is a very large number of theoretical combinations. 1,300 different variants have been identified in the p53 mutation database, which comprises more than 15,000 mutations derived from as many tumours (Béroud et al., 2000).

Many prescreening methodologies have been used to increase the sensitivity of detection of mutations and to concentrate the sequencing exclusively on the mutant exon. Unfortunately, many of these methods, possibly with the exception of DHPLC (denaturing high-performance liquid chromatography), remain confined to specialized laboratories and the sensitivity of detection of some of them is incompatible with the needs of clinical diagnosis. However, they present the advantage of being able to detect mutations in samples heavily contaminated by normal DNA.

An important point, rarely discussed, concerns the initial genetic material that can be used for molecular analysis. DNA is obviously a material of choice, as it is robust and easy to handle, but many studies use RNA and its derived product cDNA for the detection of mutations. This is particularly the case when studying haematopoietic tumours, which are easier to manipulate than solid tumours. In the case of p53, there are several advantages to using RNA as starting material: i) the coding region is small (1,200 bp), allowing analysis of the entire gene; ii) several comparative studies of DNA versus RNA have demonstrated a higher sensitivity of detection of mutations with RNA (Forslund et al., 2002; Williams et al., 1998); iii) the use of RNA can detect the presence of splicing variants not identified on DNA. This last point is important, as the frequency of these splicing mutants of p53 has probably been underestimated. The most striking case is that of the mutation at codon 125 (ACG->ACA) described by many authors as a polymorphic variant. RNA analysis of this mutation, located at the 3' extremity of exon 4, shows that the splicing of the p53 gene is totally aberrant (Warneford et al., 1992).

No major progress in Sanger's sequencing technique has been made over recent years. The only new robust approach that has been developed is pyrosequencing, which does not require electrophoretic separation (Ronaghi, 2001). This approach has been successfully evaluated for the analysis of the p53 gene. However, it has a number of disadvantages at the present time, particularly the unreliable detection of frameshift mutations and the fact that it can only be applied to DNA fragments smaller than 100 bp.

On the other hand, considerable progress has been made in electrophoretic separation methods with the development of multiplexing and automation. Capillary electrophoresis (CE) now allows simultaneous separation of 96 reactions and this number will probably be increased over the next few years. Miniaturization of this approach and the development of new separation matrices and new solid phases have also led to the development of microchannel electrophoresis. The analysis of p53 gene mutations has often been used as a paradigm to evaluate the efficacy of these new approaches (table). We will not describe all of these methods in this review, but it should be noted that they can improve the sensitivity of conventional methods. The combined use of SSCP and HA techniques associated with solid-phase electrophoresis ensures a very high sensitivity (close to 100%) (Kourkine et al., 2002).

Table 4. Comparison of techniques to identify p53 mutation. Adapted from Kirk et al. with the kind permission of the author (Kirk et al., 2002).

	Advantages	Current Limitations	References
Direct sequencing	1) Detects any mutation up to 600 bp /reaction. 2) As rapid as SSCP and DGGE but more accurate. 3) High throughput with capillary electrophoresis	1) Electrophoresis of some specific DNA sequences can be difficult, leading to compression 2) Cannot be applied to tissues with less than 20% of tumour tissue (low level mutations)	many
Pyrosequencing	1) No problem of template compression during electrophoresis 2) High throughput 3) Identify the mutation	1) Does not detect low level mutations. 2) Can only be applied to short DNA fragments (< 100 bp) 3) Poor detection of deletions and insertions	(Garcia et al., 2000)
SSCP	1) Detects low level mutations. 2) Rapid, does not require extra enzymatic steps	1) Misses 30% of possible mutations. 2) Sequencing is necessary to identify the mutation 3) Does not locate the position of polymorphisms. 4) Can miss mutations adjacent to common polymorphisms.	many
DGGE, CDGE, DHPLC	1) Detects low level mutations. 2) Rapid, does not require extra enzymatic steps.	1) Large-scale screen missed 13% of mutations. 2) Sequencing is necessary to identify the mutation 3) Requires GC clamp; limited to small fragments. 4) Some regions can be refractory to analysis	(Breton et al., 2003; Cottu et al., 1996; Smith-Sorensen et al., 1993)
SSCP-HA	1) High throughput 2) Detects low level mutations 3) High sensitivity (close to 100%)	1) Sequencing is necessary to identify the mutation 2) New technique	(Kourkine et al., 2002)
Primer extension	1) Identify the mutation 2) High throughput when combined with a hybridization array	1) Some sequences can be refractory to primer extension 2) Cannot detect mononucleotide repeat insertions and deletions	(Shumaker et al., 2001; Tonisson et al., 2002)
ddF, REF	1) Detects virtually all possible mutations.	1) Does not detect low level mutations. 2) Sequencing is necessary to identify the mutation	(Feng et al., 1999; Kovach et al., 1996)
Cleavase	1) Heteroduplex not required.	1) High background. 2) Sequencing is necessary to identify the mutation 3) Requires optimization for each mutation.	(O. Connell et al., 1999)

	Advantages	Current Limitations	References
T4 endoVII, MutY, CEL I	1) Identifies approximate position of most mutations. 2) Identifies missense, frameshift and nonsense mutations.	1) Difficult to detect low level mutations. 2) High background for T4 endoVII and MutY. 3) Sequencing is necessary to identify the mutation	(Chakrabarti et al., 2000; Zhang et al., 2002)
Thermostable Endonuclease V -	1) Identifies the approximate position of the mutation. 2) Identifies missense, frameshift and nonsense mutations, up to 1,750 bp/reaction. 3) Detects low level mutations; 1 in 20. 4) In combination with sequencing, most rapid screen to directly identify mutation.	1) Does not detect transition mutations in GGCG or RCGC sequences. 2) New technique	(Huang et al., 2002)
LDR	1) Identifies the mutation 2) Detects point mutations, small insertions and deletions 3) High multiplexing capabilities 4) Detects low level mutations; 1 in 20 5) High throughput when combined with a capture array	1) Each mutat on requires a specific pair of primers 2) New technique	(Favis et al., 2003; Fouquet et al., 2004)
Padlock probes, rolling circle amplification	1) No PCR amplification required 2) Compatible with in situ applications	1) Requires very long probes 2) Limited multiplexing capabilities	(Thomas et al., 1999a)
Hybridization array	1) Scans for mutations in thousands of positions. 2) Detects some small insertions/deletions 3) High throughput	1) Does not detect low level mutations. 2) Some arrays miss frameshift mutations	(Ahrendt et al., 1999; Okamoto et al., 2000; Wen et al., 2000)
(MALDI-TOF MS)	1) Identifies the mutation 2) High throughput	1) Each mutation requires a specific primer 2) New technique	(Kim et al., 2003)

This table provides information for the detection of small mutations (missense, nonsense and frameshift mutations). These technologies obviously do not detect large deletions or gene rearrangements. Reference are only given for the diagnosis of p53 mutations. The respective advantages and limitations can be different for other genes. Technologies such as PTT (protein truncation test), devoted exclusively to the detection of frameshift mutations, are not included in this table.

CDGE: Constant Denaturing Gel Electrophoresis; ddf: dideoxy Fingerprinting; DGGE: Denaturing Gradient Gel Electrophoresis; DHPLC: Denaturing High Pressure Liquid Chropatography; LDR: Ligase Detection Reaction; MALDI-TOF MS: Matrix-assisted laser rescrption/ ionization time-of-flight mass spectrometry; REF: Restriction Endonuclease Fingerprinting; SSCP: Single Stranded Conformation Polymophism. SSCA-HA: SSCP Heteroduplex Analysis.

Multiplexing approaches have also been improved with the development of DNA biochips. They can use various approaches such as mismatch detection, primer extension or hybridization. Many studies have compared the sensitivities of these biochip approaches to those of more conventional methods and have generally shown a good concordance between the various methods (Ahrendt et al., 1999; Schaefer et al., 2002; Wen et al., 2000; Wikman et al., 2000). The only problem concerns deletions greater than 1 bp that are not analysed by these approaches, which can raise a problem in certain cancers such as ovarian or head and neck cancers, presenting a high rate of frameshift mutations.

The LDR technique was recently linked to a biochip binding approach. This combination allows high sensitivity and high throughput. Analysis of bronchial biopsies identified p53 gene mutations not detected by conventional approaches (Favis et al., 2003; Fouquet et al., 2004).

FASAY functional assay

Originally described for the detection of germline mutations in patients with LFS, this methodology has been improved and its current sensitivity allows it to be used for the detection of mutations in tumour samples (Flaman et al., 1995; Ishioka et al., 1993). The initial material used for FASAY (Functional Assay in Yeast) is cDNA obtained from tumour RNA. PCR amplification of this cDNA, using primers corresponding to codons 52 to 364 (68% of exons 4 to 10), followed by introduction of the PCR product into an indicator yeast, where it recombines with an expression vector, can be used to define the transactivating activity of the protein expressed. Red yeast colonies express mutant p53, while white colonies express wild-type p53 (Fronza et al., 2000). The amplified region corresponds to 95 % of the mutations identified to date, which makes FASAY a very good approach for exhaustive analysis of p53 gene mutations. All alterations leading to absence of RNA expression will obviously not be detected, but this is a relatively rare situation for p53. The only criticism that can be formulated in relation to this methodology is that it provides no information about the type of mutation, so that sequencing must always be performed subsequently. As sequencing is performed on DNA extracted from red colonies (mutant p53), problems of sensitivity are eliminated. This methodology can also be used to demonstrate splicing alterations. Waridel et al. have modified the FASAY technique to increase its sensitivity and robustness (Waridel et al., 1997). Recently, we used this approach to detect p53 gene mutations in biopsies containing only 5% of tumour cells (Fouquet et al., 2004). These mutations could not be detected by direct sequencing. In addition to this high level of sensitivity, the FASAY technique also presents the advantage of being simple and robust. FASAY avoids selecting active variants (see below for problems related to this biological activity). It also has the advantage of being the only method able to rigorously demonstrate codon 72 polymorphism linked to the mutant allele (see beginning of the chapter for the significance of this polymorphism) (Tada et al., 2001). The cloning and sequencing of cDNA in the indicator yeast provide a non-fragmented molecule corresponding to the initial RNA expressed by the p53 gene. Direct sequencing of individual exons would not allow this type of analysis. Contamination of the sample by normal cells and the possibility of loss of heterozygosity prevents any interpretation. At the present time, FASAY remains the only approach suitable for the demonstration of associations between p53 gene mutations and codon 72 polymorphism (Tada et al., 2001).

The first generation FASAY test was based on the transcriptional activity of p53 on the RGC response element exclusively recognized by wild-type p53. This test is generally performed at 30°C, the growth temperature of the yeast. The observation that some p53 mutations are temperature-sensitive (active at 30°C, but inactive at 37°C), has led to the addition of a phase of yeast culture at 37°C to verify the presence of such mutants (Shiraishi et al., 2004). Various authors have also shown that, depending on the response element used (RGC, bax, WAF or PIG3), p53 mutants may have a heterogeneous behaviour (Campomenosi et al., 2001; Flaman et al., 1998). Some mutants that have only lost their apoptotic activity are inactive in relation to the PIG3 or bax elements, but continue to transactivate the WAF or mdm2 genes. On the other hand, the demonstration of pink colonies on the FASAY test shows that not all mutations have the same penetrance. New indicator yeasts allowing more accurate evaluation of p53 activity have been developed. It remains to be seen whether the use of these yeasts can help to increase the sensitivity of this test.

Immunohistochemical analysis

As the p53 protein is the end-product of gene expression, it therefore appears logical to try to directly visualize protein expression by immunohistochemical analysis linked with morphological analysis allowing qualitative evaluation of the cells presenting these defects (Dowell et al., 1994). p53 is a paradigm for this type of study, as most point mutations lead to the synthesis of a stable but "inactive" protein in the nucleus of tumour cells. Although this phenomenon was identified more than 10 years ago, no truly satisfactory explanation for this phenomenon has yet been proposed. Several non-exclusive possibilities have been proposed: i) absence of induction of mdm2 which can no longer regulate p53; ii) conformational change and decreased sensitivity to degradation; iii) stability or over-translation of messenger RNA. It is interesting to note that normal cells from LFS patients do not overexpress p53, in contrast with the tumours observed in these same patients and in the absence of LOH. It is therefore possible that the tumour context is also important for stabilization of the p53 protein. This remains an unexplored field of investigation, which would certainly provide useful information.

Immunohistochemical studies concerning p53, as for other markers, suffer from a lack of standardization, leading to very heterogeneous results. The sources of heterogeneity are multiple: i) the various antibodies used; ii) methodological aspects (amplification, epitope unmasking); iii) the initial material (paraffin block, frozen tumour) and storage conditions; iv) the positive cut-off value, which can vary from 1% to 20% according to the authors; and v) individual variability of interpretation of the results (McShane et al., 2000; Schmitz-Drager et al., 2000).

Nonsense or frameshift mutations do not lead to accumulation of p53 protein. This is certainly due to instability of truncated proteins, which are generally not detectable despite the use of monoclonal antibodies which recognize an epitope situated in the amino-terminal domain of p53.

The correlation between missense mutations and nuclear accumulation appears to be 80% with variations from one type of cancer to another (Casey et al., 1996). The accumulation of wild-type p53 is more complicated to understand and its clinical and biological significance is unknown. One of the important parameters not revealed by immunohistochemical analysis is the variability of behaviour of p53 mutants (Hashimoto et al., 1999; van Oijen and Slootweg, 2000). Some monoclonal antibodies are described as being specific for p53 mutants (Pab240, HO15.4). However, they must be used cautiously, as: i) not all p53 mutants are recognized by these antibodies; and ii) denaturation of p53 during binding or epitope unmasking procedures leads to recognition of all forms of p53 (Legros et al., 1994b).

It is beyond the scope of this chapter to present an exhaustive review of the literature concerning immunohistochemical analysis of p53 and its clinical applications. The reader can refer to reviews already published on the subject or other chapters of this book (Hall and Lane, 1994; Save et al., 1998).

Like molecular methodologies, immunohistochemistry has also taken a great leap forward with the development of tissue microarrays (TMA), which allow simultaneous analysis of several hundred tumours with a single antibody (Kallioniemi et al., 2001). These TMA can also be used for in situ hybridization or FISH analyses. However, this approach cannot be used for routine diagnostic analyses, but it is very useful for immunohistochemical studies which can analyse several dozen different antibodies on sections derived from the same TMA block. It would therefore be very easy to analyse combinations of antibodies and to evaluate their clinical significance in terms of prognosis or prediction of response to treatment. It is also possible to analyse the relationships between various antibodies and to study tumours in terms of "pathways" with multiple antibodies in order to determine the clinical significance and independence of each antibody (Hoos et al., 2001; Simon and Sauter, 2003). Analysis of 288 Hodgkin's lymphomas for the expression of 28 proteins involved in regulation of the cell cycle demonstrated that expression of genes involved in apoptosis is strongly correlated with poor prognosis (Garcia et al., 2003). Analysis of 852 small breast cancers (T1N0M0) showed that p53 expression is not correlated with survival, in contrast with c-erb2 overexpression (Joensuu et al., 2003).

Serological analysis of p53 gene alterations

In 1979, DeLeo et al. showed that the humoral response of mice to some methylcholanthrene-induced tumour cells such as MethA was directed against the p53 protein (De Leo et al., 1979). It was subsequently found that animals bearing several types of tumours elicited an immune response specific for p53 (Kress et al., 1979; Melero et al., 1979; Rotter et al., 1980). In 1982, Crawford et al. first described antibodies directed against human p53 protein in the serum of 9% of breast cancer patients (Crawford et al., 1982). No significant clinical correlation was reported, and at that time, no information was available concerning p53 gene mutations. Caron de Fromentel et al. later found that these antibodies were present in serum of children with a wide variety of cancers. The average frequency was 12%, but a frequency of 20% is observed in Burkitt lymphoma (Caron de Fromentel et al., 1987).

Since 1992, a new series of studies has shown that p53-Abs can be found in the serum of patients with various types of cancer, whereas the prevalence of these antibodies in the normal population remains very low. To date, the majority of published studies suggest that most patients with p53 antibodies have a p53 mutation leading to p53 accumulation. It is also clear that not all patients with a p53 alteration develop p53 antibodies. Comparison of the frequency of p53 alterations in the literature indicates that 30% to 40% of patients with an alteration of the p53 gene develop p53 antibodies (Lubin et al., 1995a).

It was initially believed that these antibodies were directed against the central region of the p53 protein, which is the target for the various mutations. Surprisingly, a careful study of the epitope recognized by these p53-Abs indicates that they bind both wild-type and mutant p53 (Labrecque and Matlashewski, 1995; Lubin et al., 1993; Schlichtholz et al., 1992;

Schlichtholz et al., 1994). Using either truncated p53 or synthetic peptides, it has been demonstrated that the epitopes recognized by the p53-Abs are mainly located in the amino and carboxy terminal regions of the protein, regions not corresponding to the p53 mutation hot spots (Lubin et al., 1993; Schlichtholz et al., 1992). These immunodominant epitopes have also been detected in the serum of mice and rabbits hyperimmunized with wild-type p53 (Legros et al., 1994a). Taken together, i) the presence of immunodominant epitopes outside the hot spot region of p53 mutations, ii) the correlation between p53 accumulation (and p53 gene mutation) in tumour cells and p53 antibody responses, iii) the similarity of humoral responses in patients independent of the cancer type and iv) the similarity of antigenic site profiles in patients and hyperimmunized animals, all suggest that p53 accumulation is a major component of the humoral response in patients with cancer. This accumulation could lead to a self-immunization process culminating in the appearance of p53 antibodies. As stated above, the level of p53 proteins in a normal organism is very low, suggesting very weak (if any) tolerance to endogenous p53 (Soussi, 2000).

Numerous studies have tried to determine the clinical value of p53-Abs (Bourhis et al., 1996; Cabelguenne et al., 2000; Lenner et al., 1999; Peyrat et al., 1995; Werner et al., 1997). As for p53 mutations and p53 immunohistochemical analysis, these studies have reported contradictory results and have been recently reviewed (Soussi, 2000). There is a trend towards an association between p53-Abs and poorly differentiated tumours, a feature already observed with p53 mutations.

As p53 accumulation is the main trigger of this humoral response, it was interesting to examine the behaviour of these p53-Abs during therapy to see whether there is a relationship between tumour disappearance and a decrease in p53-Abs. Several studies have addressed this question in various types of cancer (Angelopoulou and Diamandis, 1997; Angelopoulou et al., 1994; Hammel et al., 1999; Saffroy et al., 1999; Zalcman et al., 1998). Zalcman et al. showed that there is a good correlation between the specific time-course of p53-Abs titres and the response to therapy in patients with lung cancer (Zalcman et al., 1998). A similar situation was described in colorectal and ovarian cancer. In several patients, the disappearance of p53-Abs was very rapid, nearly as rapid as the half-life of human IgG (Lubin et al., 1995a). In breast cancer, it is possible to detect the reappearance of p53-Abs two years after initial therapy. This increase in p53-Abs was detected 3 months before the detection of a relapse. In these tumour types, p53-Abs could therefore be a useful tool to determine response to therapy and to monitor certain early relapses before they are clinically detectable.

p53 accumulation is the major component in the appearance of these p53-Abs. It is therefore reasonable to assume that p53-Abs could be used as an early indicator of p53 mutations in tumours in which such alterations occur early during tumour progression. A good model to test this hypothesis is that of lung cancer and heavy smokers. It is well established that p53 accumulation is an early event in lung cancer and that this cancer is strongly associated with tobacco smoking. In 1994, p53-Abs were found in a heavy smoker in whom no lung cancer could be detected. Two years later, lung cancer was detected in this patient prior to any clinical manifestations of the disease (Lubin et al., 1995b; Schlichtholz et al., 1994). The patient showed good response to therapy in parallel with complete disappearance of p53-Abs (Lubin et al., 1995b). To our knowledge, this is the only prospective study addressing the importance of p53-Abs in individuals at high risk for cancer and using p53-Ab assays for clinical management of the patient. Since this publication, several studies have demonstrated the presence of p53-Abs in the serum of high-risk individuals (Trivers et al., 1995; Trivers et al., 1996).

A similar situation is observed in subjects with premalignant oral lesions (leukoplakia) due to tobacco or betel nut chewing. Such individuals are at high risk of developing oral cancer (5-10%). A high frequency of p53-Abs has been found in patients with premalignant and malignant lesions, suggesting that these antibodies could be used for early detection of cancer (Ralhan et al., 1998). Unfortunately, no follow-up has been performed on these patients. Due to the high frequency of this type of cancer in countries such as India or Pakistan, this type of diagnosis could be particularly useful. The recent discovery that p53-Abs can be found in saliva indicates that a simple screening method could be organized to verify the value of these antibodies (Tavassoli et al., 1998).

The majority of the literature clearly demonstrates the specificity of this serological analysis, as such antibodies are very rare in the normal population. The specificity of this assay can be estimated to be 95%. This high specificity is supported by the fact that p53 specifically accumulates in the nucleus of tumour cells after gene mutation. One of the disadvantages of this assay is its lack of sensitivity, as only 20% to 40% of patients with p53 mutations develop p53-Abs. This lack of sensitivity totally precludes the use of the assay to evaluate p53 alterations in human tumour. Nevertheless, if we estimate that there are 8 million patients with various types of cancer throughout the world, and 50% of them have a mutation in their p53 gene, then we can deduce that about 1 million of these patients would have p53-Abs.

There are several situations in which p53-Abs could be clinically useful. The first situation is that of serum monitoring during therapy. Only prospective studies on various types of cancer in which relapses occur several months or years after treatment would be able to validate this assay.

The use of standardized assays which have been validated for quantitative analysis would be useful for such studies.

The second situation concerns p53-Abs in high-risk individuals. One of the challenges of the next decade is the early detection of tumours using highly sensitive assays with gene probes specific for tumour genetic alterations. These approaches are still under development and remain costly. I believe that there is still room for serological assays of p53 antibody tumour markers. In developing countries, there is an increased burden of tumours due to carcinogen exposure as a result of increasing cigarette consumption, higher pollution caused by political laxity, uncontrolled industrial development and the absence of regulations concerning waste elimination. There may be a high incidence of p53 mutations in cancers related to this type of exposure and the use of an inexpensive assay for the detection of p53-Abs could be of public health benefit in these countries.

CONCLUSIONS

Fifteen years after the first description of mutations of the p53 gene, the clinical situation of p53 is still uncertain. The multiplicity of mutations and their properties makes the situation complex in terms of the clinical, biological and diagnostic significance. Unless the p53 status can provide new data, independent of the information provided by the other markers used at the present time, implementation of routine p53 diagnostic screening would be useless. On the other hand, as indicated throughout this chapter, the main problems related to the study of p53 are: i) heterogeneity of functions of mutant p53 and ii) a far from perfect correlation between p53 mutations and inactivation of the p53 signalling pathway. It is likely that, in the near future, global analysis of this signalling pathway could be evaluated more exhaustively by a mutational approach associated with an expression profile.

REFERENCES

Aas, T., Borresen, A. L., Geisler, S., Smithsorensen, B., Johnsen, H., Varhaug, J. E., Akslen, L. A., and Lonning, P. E. (1996). Specific p53 mutations are associated with de novo resistance to doxorubicin in breast cancer patients. Nature Med *2*, 811-814.

Aguilar, F., Harris, C. C., Sun, T., Hollstein, M., and Cerutti, P. (1994). Geographic variation of p53 mutational profile in nonmalignant human liver. Science *264*, 1317-1319.

Aguilar, F., Hussain, S. P., and Cerutti, P. (1993). Aflatoxin-B(1) induces the transversion of G->T in codon 249 of the p53 tumor suppressor gene in human hepatocytes. Proc Natl Acad Sci USA *90*, 8586-8590.

Ahn, J., Urist, M., and Prives, C. (2003). Questioning the role of checkpoint kinase 2 in the p53 DNA damage response. J Biol Chem *278*, 20480-20489.

Ahrendt, S. A., Halachmi, S., Chow, J. T., Wu, L., Halachmi, N., Yang, S. C., Wehage, S., Jen, J., and Sidransky, D. (1999). Rapid p53 sequence analysis in primary lung cancer using an oligonucleotide probe array. Proc Natl Acad Sci U S A *96*, 7382-7387.

Angelopoulou, K., and Diamandis, E. P. (1997). Detection of the TP53 tumour suppressor gene product and p53 auto-antibodies in the ascites of women with ovarian cancer. Eur J Cancer *33*, 115-121.

Angelopoulou, K., Diamandis, E. P., Sutherland, D. J. A., Kellen, J. A., and Bunting, P. S. (1994). Prevalence of serum antibodies against the p53 tumor suppressor gene protein in various cancers. Int J Cancer *58*, 480-487.

Attardi, L. D., and DePinho, R. A. (2004). Conquering the complexity of p53. Nat Genet *36*, 7-8.

Bargonetti, J., Manfredi, J. J., Chen, X. B., Marshak, D. R., and Prives, C. (1993). A proteolytic fragment from the central region of p53 has marked Sequence-Specific DNA-Binding activity when generated from Wild-Type but not from oncogenic mutant p53-Protein. Gene Develop *7*, 2565-2574.

Beckman, G., Birgander, R., Sjalander, A., Saha, N., Holmberg, P. A., Kivela, A., and Beckman, L. (1994). Is p53 polymorphism maintained by natural selection? Hum Hered *44*, 266-270.

Bell, D. W., Varley, J. M., Szydlo, T. E., Kang, D. H., Wahrer, D. C., Shannon, K. E., Lubratovich, M., Verselis, S. J., Isselbacher, K. J., Fraumeni, J. F., et al. (1999). Heterozygous germ line hCHK2 mutations in Li-Fraumeni syndrome. Science *286*, 2528-2531.

Bergamaschi, D., Gasco, M., Hiller, L., Sullivan, A., Syed, N., Trigiante, G., Yulug, I., Merlano, M., Numico, G., Comino, A., et al. (2003). p53 polymorphism influences response in cancer chemotherapy via modulation of p73-dependent apoptosis. Cancer Cell *3*, 387-402.

Béroud, C., Collod-Béroud, G., Boileau, C., Soussi, T., and Junien, C. (2000). UMD (Universal Mutation Database): A generic software to build and analyze locus-specific databases. Hum Mutat *15*, 86-94.

Blandino, G., Levine, A. J., and Oren, M. (1999). Mutant p53 gain of function: differential effects of different p53 mutants on resistance of cultured cells to chemotherapy. Oncogene *18*, 477-485.

Bourhis, J., Lubin, R., Roche, B., Koscielny, S., Bosq, J., Dubois, I., Talbot, M., Marandas, P., Schwaab, G., Wibault, P., et al. (1996). Analysis of p53 serum antibodies in patients with head and neck squamous cell carcinoma. J Nat Cancer Inst *88*, 1228-1233.

Brash, D. E., and Ponten, J. (1998). Skin precancer. Cancer Surv *32*, 69-113.

Brash, D. E., Rudolph, J. A., Simon, J. A., Lin, A., Mckenna, G. J., Baden, H. P., Halperin, A. J., and Ponten, J. (1991). A Role for sunlight in skin cancer - UV-induced p53 mutations in squamous cell carcinoma. Proc Natl Acad Sci USA *88*, 10124-10128.

Bray, S. E., Schorl, C., and Hall, P. A. (1998). The challenge of p53: Linking biochemistry, biology, and patient management. Stem Cells *16*, 248-260.

Breton, J., Sichel, F., Abbas, A., Marnay, J., Arsene, D., and Lechevrel, M. (2003). Simultaneous use of DGGE and DHPLC to screen TP53 mutations in cancers of the esophagus and cardia from a European high incidence area (Lower Normandy, France). Mutagenesis *18*, 299-306.

Cabelguenne, A., Blons, H., de Waziers, I., Carnot, F., Houllier, A. M., Soussi, T., Brasnu, D., Beaune, P., Laccourreye, O., and Laurent-Puig, P. (2000). p53 alterations predict tumor response to neoadjuvant chemotherapy in head and neck squamous cell carcinoma: a prospective series. J Clin Oncol *18*, 1465-1473.

Campomenosi, P., Monti, P., Aprile, A., Abbondandolo, A., Frebourg, T., Gold, B., Crook, T., Inga, A., Resnick, M. A., Iggo, R., and Fronza, G. (2001). p53 mutants can often transactivate promoters containing a p21 but not Bax or PIG3 responsive elements. Oncogene *20*, 3573-3579.

Caron de Fromentel, C., May-Levin, F., Mouriesse, H., Lemerle, J., Chandrasekaran, K., and May, P. (1987). Presence of circulating antibodies against cellular protein p53 in a notable proportion of children with B-cell lymphoma. Int J Cancer *39*, 185-189.

Casey, G., Lopez, M. E., Ramos, J. C., Plummer, S. J., Arboleda, M. J., Shaughnessy, M., Karlan, B., and Slamon, D. J. (1996). DNA sequence analysis of exons 2 through 11 and immunohistochemical staining are required to detect all known p53 alterations in human malignancies. Oncogene *13*, 1971-1981.

Chakrabarti, S., Price, B. D., Tetradis, S., Fox, E. A., Zhang, Y., Maulik, G., and Makrigiorgos, G. M. (2000). Highly selective isolation of unknown mutations in diverse DNA fragments: toward new multiplex screening in cancer. Cancer Res *60*, 3732-3737.

Chehab, N. H., Malikzay, A., Appel, M., and Halazonetis, T. D. (2000). Chk2/hCds1 functions as a DNA damage checkpoint in G(1) by stabilizing p53. Genes Dev *14*, 278-288.

Chin, L., Merlino, G., and DePinho, R. A. (1998). Malignant melanoma: modern black plague and genetic black box. Genes Dev *12*, 3467-3481.

Cottu, P. H., Muzeau, F., Estreicher, A., Flejou, J. F., Iggo, R., Thomas, G., and Hamelin, R. (1996). Inverse correlation between RER(+) status and p53 mutation in colorectal cancer cell lines. Oncogene *13*, 2727-2730.

Crawford, L. (1983). The 53,000-dalton cellular protein and its role in transformation. Int Rev Exp Path *25*, 1-50.

Crawford, L. V., Pim, D. C., and Bulbrook, R. D. (1982). Detection of antibodies against the cellular protein p53 in sera from patients with breast cancer. Int J Cancer *30*, 403-408.

Crook, T., Wrede, D., Tidy, J. A., Mason, W. P., Evans, D. J., and Vousden, K. H. (1992). Clonal p53 mutation in primary cervical cancer - association with human-papillomavirus-negative tumours. Lancet *339*, 1070-1073.

De Leo, A. B., Jay, G., Appella, E., Dubois, G. C., Law, L. W., and Old, L. J. (1979). Detection of a transformation-related antigen in chemically induced sarcomas and other transformed cells of the mouse. Proc Natl Acad Sci USA *76*, 2420-2424.

Denissenko, M. F., Pao, A., Tang, M. S., and Pfeifer, G. P. (1996). Preferential formation of benzo[a]pyrene adducts at lung cancer mutational hotspots in P53. Science *274*, 430-432.

DiComo, C. J., Gaiddon, C., and Prives, C. (1999). p73 function is inhibited by tumor-derived p53 mutants in mammalian cells. Mol Cell Biol *19*, 1438-1449.

DiGiammarino, E. L., Lee, A. S., Cadwell, C., Zhang, W., Bothner, B., Ribeiro, R. C., Zambetti, G., and Kriwacki, R. W. (2002). A novel mechanism of tumorigenesis involving pH-dependent destabilization of a mutant p53 tetramer. Nat Struct Biol *9*, 12-16.

Dittmer, D., Pati, S., Zambetti, G., Chu, S., Teresky, A. K., Moore, M., Finlay, C., and Levine, A. J. (1993). Gain of function mutations in p53. Nature genetics *4*, 42-46.

Dowell, S. P., Wilson, P. O. G., Derias, N. W., Lane, D. P., and Hall, P. A. (1994). Clinical utility of the immunocytochemical detection of p53 protein in cytological specimens. Cancer Res *54*, 2914-2918.

Dumont, P., Leu, J. I., Della Pietra, A. C., 3rd, George, D. L., and Murphy, M. (2003). The codon 72 polymorphic variants of p53 have markedly different apoptotic potential. Nat Genet *33*, 357-365.

Eliyahu, D., Michalovitz, D., Eliyahu, S., Pinhasikimhi, O., and Oren, M. (1990). p53 - Oncogene or anti-oncogene. Oncogenes in Cancer Diagnosis *39*, 125-134.

Favis, R., Huang, J., Gerry, N. P., Culliford, A., Paty, P., Soussi, T., and Barany, F. (2003). Harmonized microarray mutation scanning analysis of p53 mutation in Undissected Colorectal Tumors. Hum Mutat *in press*.

Feng, J., Buzin, C. H., Tang, S. H., Scaringe, W. A., and Sommer, S. S. (1999). Highly sensitive mutation screening by REF with low concentrations of urea: A blinded analysis of a 2-kb region of the p53 gene reveals two common haplotypes. Hum Mutat *14*, 175-180.

Fields, S., and Jang, S. K. (1990). Presence of a potent transcription activating sequence in the p53 protein. Science *249*, 1046-1049.

Flaman, J. M., Frebourg, T., Moreau, V., Charbonnier, F., Martin, C., Chappuis, P., Sappino, A. P., Limacher, J. M., Bron, L., Benhattar, J., *et al.* (1995). A simple p53 functional assay for screening cell lines, blood, and tumors. Proc Natl Acad Sci USA *92*, 3963-3967.

Flaman, J. M., Robert, V., Lenglet, S., Moreau, V., Iggo, R., and Frebourg, T. (1998). Identification of human p53 mutations with differential effects on the bax and p21 promoters using functional assays in yeast. Oncogene *16*, 1369-1372.

Forrester, K., Lupold, S. E., Ott, V. L., Chay, C. H., Band, V., Wang, X. W., and Harris, C. C. (1995). Effects of p53 mutants on wild-type p53-mediated transactivation are cell type dependent. Oncogene *10*, 2103-2111.

Forslund, A., Kressner, U., Lonnroth, C., Andersson, M., Lindmark, G., and Lundholm, K. (2002). p53 mutations in colorectal cancer assessed in both genomic DNA and cDNA as compared to the presence of p53 LOH. Int J Oncol *21*, 409-415.

Fouquet, C., Antoine, M., Tisserand, P., Favis, R., Wislez, M., Como, F., Rabbe, N., Carette, M. F., Milleron, B., Barany, F., *et al.* (2004). Rapid and sensitive p53 alteration analysis in biopsies from lung cancer patients using a functional assay and a universal oligonucleotide array: a prospective study. Clin Cancer Res *in press*.

Friedlander, P., Haupt, Y., Prives, C., and Oren, M. (1996). A mutant p53 that discriminates between p53-responsive genes cannot induce apoptosis. Mol Cell Biol *16*, 4961-4971.

Fronza, G., Inga, A., Monti, P., Scott, G., Campomenosi, P., Menichini, P., Ottaggio, L., Viaggi, S., Burns, P. A., Gold, B., and Abbondandolo, A. (2000). The yeast p53 functional assay: a new tool for molecular epidemiology. Hopes and facts. Mutat Res *462*, 293-301.

Gaiddon, C., Lokshin, M., Ahn, J., Zhang, T., and Prives, C. (2001). A Subset of Tumor-Derived Mutant Forms of p53 Down-Regulate p63 and p73 through a Direct Interaction with the p53 Core Domain. Mol Cell Biol *21*, 1874-1887.

Garcia, C. A., Ahmadian, A., Gharizadeh, B., Lundeberg, J., Ronaghi, M., and Nyren, P. (2000). Mutation detection by pyrosequencing: sequencing of exons 5-8 of the p53 tumor suppressor gene. Gene *253*, 249-257.

Garcia, J. F., Camacho, F. I., Morente, M., Fraga, M., Montalban, C., Alvaro, T., Bellas, C., Castano, A., Diez, A., Flores, T., *et al.* (2003). Hodgkin and Reed-Sternberg cells harbor alterations in the major tumor suppressor pathways and cell-cycle checkpoints: analyses using tissue microarrays. Blood *101*, 681-689.

Halevy, O., Michalovitz, D., and Oren, M. (1990). Different tumor-derived p53 mutants exhibit distinct biological activities. Science *250*, 113-116.

Hall, P. A., and Lane, D. P. (1994). p53 in tumour pathology - can we trust immunohistochemistry - revisited. J Pathol *172*, 1-4.

Hammel, P., LeroyViard, K., Chaumette, M. T., Villaudy, J., Falzone, M. C., Rouillard, D., Hamelin, R., Boissier, B., and Remvikos, Y. (1999). Correlations between p53-protein accumulation, serum antibodies and gene mutation in colorectal cancer. Int J Cancer *81*, 712-718.

Harris, C. C. (1991). Chemical and physical carcinogenesis: Advances and perspectives for the 1990s. Cancer Res *51*, 5023s-5044s.

Harris, C. C. (1993). p53 - at the crossroads of molecular carcinogenesis and risk assessment. Science *262*, 1980-1981.

Harris, N., Brill, E., Shohat, O., Prokocimer, M., Wolf, D., Arai, N., and Rotter, V. (1986). Molecular basis for heterogeneity of the human p53 protein. Mol Cell Biol *6*, 4650-4656.

Hashimoto, T., Tokuchi, Y., Hayashi, M., Kobayashi, Y., Nishida, K., Hayashi, S., Ishikawa, Y., Tsuchiya, S., Nakagawa, K., Hayashi, J., and Tsuchiya, E. (1999). p53 null mutations undetected by immunohistochemical staining predict a poor outcome with early-stage non-small cell lung carcinomas. Cancer Res *59*, 5572-5577.

Helland, A., Langerod, A., Johnsen, H., Olsen, A. O., Skovlund, E., and BorresenDale, A. L. (1998). p53 polymorphism and risk of cervical cancer. Nature *396*, 530-531.

Hildesheim, A., Schiffman, M., Brinton, L. A., Fraumeni, J. F., Herrero, R., Bratti, M. C., Schwartz, P., Mortel, R., Barnes, W., Greenberg, M., *et al.* (1998). p53 polymorphism and risk of cervical cancer. Nature *396*, 531-532.

Hollstein, M., Marion, M. J., Lehman, T., Welsh, J., Harris, C. C., Martelplanche, G., Kusters, I., and Montesano, R. (1994). p53 mutations at A:T base pairs in angiosarcomas of vinyl Chloride-Exposed factory workers. Carcinogenesis *15*, 1-3.

Hoos, A., Urist, M. J., Stojadinovic, A., Mastorides, S., Dudas, M. E., Leung, D. H., Kuo, D., Brennan, M. F., Lewis, J. J., and Cordon-Cardo, C. (2001). Validation of tissue microarrays for immunohistochemical profiling of cancer specimens using the example of human fibroblastic tumors. Am J Pathol *158*, 1245-1251.

Huang, J., Kirk, B., Favis, R., Soussi, T., Paty, P., Cao, W., and Barany, F. (2002). An endonuclease/ligase based mutation scanning method especially suited for analysis of neoplastic tissue. Oncogene *21*, 1909-1921.

Hussain, S. P., Raja, K., Amstad, P. A., Sawyer, M., Trudel, L. J., Wogan, G. N., Hofseth, L. J., Shields, P. G., Billiar, T. R., Trautwein, C., *et al.* (2000). Increased p53 mutation load in nontumorous human liver of wilson disease and hemochromatosis: oxyradical overload diseases [In Process Citation]. Proc Natl Acad Sci U S A *97*, 12770-12775.

Irwin, M. S., Kondo, K., Marin, M. C., Cheng, L. S., Hahn, W. C., and Kaelin Jr, W. J. (2003). Chemosensitivity linked to p73 function. Cancer cell *in press.*

Ishioka, C., Frebourg, T., Yan, Y., Vidal, M., Friend, S. H., Schmidt, S., and Iggo, R. (1993). Screening patients for heterozygotous p53 mutations using a functional assay in yeast. Nature Genetics *5*, 124-129.

Iwakuma, T., and Lozano, G. (2003). MDM2, an introduction. Mol Cancer Res *1*, 993-1000.

Jallepalli, P. V., Lengauer, C., Vogelstein, B., and Bunz, F. (2003). The Chk2 tumor suppressor is not required for p53 responses in human cancer cells. J Biol Chem *278*, 20475-20479.

Joensuu, H., Isola, J., Lundin, M., Salminen, T., Holli, K., Kataja, V., Pylkkanen, L., Turpeenniemi-Hujanen, T., von Smitten, K., and Lundin, J. (2003). Amplification of erbB2 and erbB2 expression are superior to estrogen receptor status as risk factors for distant recurrence in pT1N0M0 breast cancer: a nationwide population-based study. Clin Cancer Res *9*, 923-930.

Kallioniemi, O. P., Wagner, U., Kononen, J., and Sauter, G. (2001). Tissue microarray technology for high-throughput molecular profiling of cancer. Hum Mol Genet *10*, 657-662.

Kato, S., Han, S. Y., Liu, W., Otsuka, K., Shibata, H., Kanamaru, R., and Ishioka, C. (2003). Understanding the function-structure and function-mutation relationships of p53 tumor suppressor protein by high-resolution missense mutation analysis. Proc Natl Acad Sci U S A *100*, 8424-8429.

Kim, S., Ruparel, H. D., Gilliam, T. C., and Ju, J. (2003). Digital genotyping using molecular affinity and mass spectrometry. Nat Rev Genet *4*, 1001-1008.

Kinzler, K. W., and Vogelstein, B. (1997). Cancer-susceptibility genes - Gatekeepers and caretakers. Nature *386*, 761.

Kinzler, K. W., and Vogelstein, B. (1998). Landscaping the cancer terrain. Science *280*, 1036-1037.

Kirk, B. W., Feinsod, M., Favis, R., Kliman, R. M., and Barany, F. (2002). Single nucleotide polymorphism seeking long term association with complex disease. Nucleic Acids Res *30*, 3295-3311.

Kirk, G. D., Camus-Randon, A. M., Mendy, M., Goedert, J. J., Merle, P., Trepo, C., Brechot, C., Hainaut, P., and Montesano, R. (2000). Ser-249 p53 mutations in plasma DNA of patients with hepatocellular carcinoma from The Gambia. J Natl Cancer Inst *92*, 148-153.

Kourkine, I. V., Hestekin, C. N., Buchholz, B. A., and Barron, A. E. (2002). High-throughput, high-sensitivity genetic mutation detection by tandem single-strand conformation polymorphism/heteroduplex analysis capillary array electrophoresis. Anal Chem *74*, 2565-2572.

Kovach, J. S., Hartmann, A., Blaszyk, H., Cunningham, J., Schaid, D., and Sommer, S. S. (1996). Mutation detection by highly sensitive methods indicates that p53 gene mutations in breast cancer can have important prognostic value. Proc Natl Acad Sci USA *93*, 1093-1096.

Kress, M., May, E., Cassingena, R., and May, P. (1979). Simian Virus 40-transformed cells express new species of proteins precipitable by anti-simian virus 40 serum. J Virol *31*, 472-483.

Labrecque, S., and Matlashewski, G. J. (1995). Viability of wild type p53-containing and p53-deficient tumor cells following anticancer treatment: the use of human papillomavirus e6 to target p53. Oncogene *11*, 387-392.

Lane, D. (1992). p53, guardian of the genome. Nature *358*, 15-16.

Lane, D. P., and Benchimol, S. (1990). p53: oncogene or anti-oncogene? Genes and Development *4*, 1-8.

Latronico, A. C., Pinto, E. M., Domenice, S., Fragoso, M. C., Martin, R. M., Zerbini, M. C., Lucon, A. M., and Mendonca, B. B. (2001). An inherited mutation outside the highly conserved DNA-binding domain of the p53 tumor suppressor protein in children and adults with sporadic adrenocortical tumors. J Clin Endocrinol Metab *86*, 4970-4973.

Legros, Y., Lafon, C., and Soussi, T. (1994a). Linear antigenic sites defined by the B-cell response to human p53 are localized predominantly in the amino and carboxy-termini of the protein. Oncogene *9*, 2071-2076.

Legros, Y., Meyer, A., Ory, K., and Soussi, T. (1994b). Mutations in p53 produce a common conformational effect that can be detected with a panel of monoclonal antibodies directed toward the central part of the p53 protein. Oncogene *9*, 3689-3694.

Lenner, P., Wiklund, F., Emdin, S. O., Arnerlov, C., Eklund, C., Hallmans, G., Zentgraf, H., and Dillner, J. (1999). Serum antibodies against p53 in relation to cancer risk and prognosis in breast cancer: a population-based epidemiological study. Brit J Cancer *79*, 927-932.

Li, F. P., Fraumeni Jr., J. F., Mulvihill, J. J., Blatner, W. A., Dreyfus, M. G., Tucker, M. A., and Miller, R. W. (1988). A cancer family syndrome in twenty-four kindreds. Cancer Res *48*, 5358-5362.

Liu, G., Parant, J. M., Lang, G., Chau, P., Chavez-Reyes, A., El-Naggar, A. K., Multani, A., Chang, S., and Lozano, G. (2004). Chromosome stability, in the absence of apoptosis, is critical for suppression of tumorigenesis in Trp53 mutant mice. Nat Genet *36*, 63-68.

Lubin, R., Schlichtholz, B., Bengoufa, D., Zalcman, G., Tredaniel, J., Hirsch, A., Caron de Fromentel, C., Preudhomme, C., Fenaux, P., Fournier, G., *et al.* (1993). Analysis of p53 antibodies in patients with various cancers define B-Cell epitopes of human p53 - distribution on primary structure and exposure on protein surface. Cancer Res *53*, 5872-5876.

Lubin, R., Schlichtholz, B., Teillaud, J. L., Garay, E., Bussel, A., Wild, C., and Soussi, T. (1995a). p53 antibodies in patients with various types of cancer: assay, identification and characterization. Clinical Cancer Res *1*, 1463-1469.

Lubin, R., Zalcman, G., Bouchet, L., Trédaniel, J., Legros, Y., Cazals, D., Hirsh, A., and Soussi, T. (1995b). Serum p53 antibodies as early markers of lung cancer. Nature Med *1*, 701-702.

Makni, H., Franco, E. L., Kaiano, J., Villa, L. L., Labrecque, S., Dudley, R., Storey, A., and Matlashewski, G. (2000). p53 polymorphism in codon 72 and risk of human papillomavirus-induced cervical cancer: effect of inter-laboratory variation. Int J Cancer *87*, 528-533.

Makos Walles, M., Biel, M. A., El-Deiry, W. S., Nelkin, B. D., Issa, J. P., Cavenee, W. K., Kuerbitz, S. J., and Baylin, S. B. (1995). p53 activates expression of HIC-1, a new candidate tumor suppressor gene on 17p13.3. Nature Med *1*, 570-577.

Malkin, D., Li, F. P., Strong, L. C., Fraumeni, J. F., Nelson, C. E., Kim, D. H., Kassel, J., Gryka, M. A., Bischoff, F. Z., Tainsky, M. A., and Friend, S. H. (1990). Germ line p53 mutations in a familial syndrome of breast cancer, sarcomas, and other neoplasms. Science *250*, 1233-1238.

Marin, M. C., Jost, C. A., Brooks, L. A., Irwin, M. S., O'Nions, J., Tidy, J. A., James, N., McGregor, J. M., Harwood, C. A., Yulug, I. G., *et al.* (2000). A common polymorphism acts as an intragenic modifier of mutant p53 behaviour. Nat Genet *25*, 47-54.

Marrogi, A. J., Khan, M. A., van Gijssel, H. E., Welsh, J. A., Rahim, H., Demetris, A. J., Kowdley, K. V., Hussain, S. P., Nair, J., Bartsch, H., *et al.* (2001). Oxidative stress and p53 mutations in the carcinogenesis of iron overload-associated hepatocellular carcinoma. J Natl Cancer Inst *93*, 1652-1655.

McShane, L. M., Aamodt, R., Cordon-Cardo, C., Cote, R., Faraggi, D., Fradet, Y., Grossman, H. B., Peng, A., Taube, S. E., and Waldman, F. M. (2000). Reproducibility of p53 immunohistochemistry in bladder tumors. National Cancer Institute, Bladder Tumor Marker Network [In Process Citation]. Clin Cancer Res *6*, 1854-1864.

Melero, J. A., Stitt, D. T., Mangel, W. F., and Carroll, R. B. (1979). Identification of new polypeptide species (48-55K) immunoprecipitable by antiserum to purified large T antigen and present in simian virus 40-infected and transformed cells. J Virol *93*, 466-480.

Melino, G., De Laurenzi, V., and Vousden, K. H. (2002). p73: Friend or foe in tumorigenesis. Nat Rev Cancer *2*, 605-615.

Michael, D., and Oren, M. (2002). The p53 and Mdm2 families in cancer. Curr Opin Genet Dev *12*, 53-59.

Michalovitz, D., Halevy, O., and Oren, M. (1991). p53 mutations - gains or losses. J Cell Biochem *45*, 22-29.

Milner, J. (1995). Flexibility: the key to p53 function? Trends Biochem Sci *20*, 49-51.

Mitsudomi, T., Hamajima, N., Ogawa, M., and Takahashi, T. (2000). Prognostic significance of p53 alterations in patients with non-small cell lung cancer: a meta-analysis [In Process Citation]. Clin Cancer Res *6*, 4055-4063.

Moll, U. M., Laquaglia, M., Benard, J., and Riou, G. (1995). Wild-type p53 protein undergoes cytoplasmic sequestration in undifferentiated neuroblastomas but not in differentiated tumors. Proc Natl Acad Sci USA *92*, 4407-4411.

Moll, U. M., Ostermeyer, A. G., Haladay, R., Winkfield, B., Frazier, M., and Zambetti, G. (1996). Cytoplasmic sequestration of wild-type p53 protein impairs the G1 checkpoint after DNA damage. Mol Cell Biol *16*, 1126-1137.

Moll, U. M., and Petrenko, O. (2003). The MDM2-p53 interaction. Mol Cancer Res *1*, 1001-1008.

Moll, U. M., Riou, G., and Levine, A. J. (1992). Two distinct mechanisms alter p53 in breast cancer - mutation and nuclear exclusion. Proc Natl Acad Sci USA *89*, 7262-7266.

Moll, U. M., and Schramm, L. M. (1998). p53 - An acrobat in tumorigenesis. Crit Rev Oral Biol Med *9*, 23-37.

Nigro, J. M., Baker, S. J., Preisinger, A. C., Jessup, J. M., Hostetter, R., Cleary, K., Bigner, S. H., Davidson, N., Baylin, S., Devilee, P., *et al.* (1989). Mutations in the p53 gene occur in diverse human tumour types. Nature *342*, 705-708.

Nikolaev, A. Y., Li, M., Puskas, N., Qin, J., and Gu, W. (2003). Parc: a cytoplasmic anchor for p53. Cell *112*, 29-40.

O. Connell, C. D., Atha, D. H., Oldenburg, M. C., Tian, J. X., Siebert, M., Handrow, R., Grooms, K., Heisler, L., and deArruda, M. (1999). Detection of p53 gene mutation: Analysis by single-strand conformation polymorphism and Cleavase fragment length polymorphism. Electrophoresis *20*, 1211-1223.

Okamoto, T., Suzuki, T., and Yamamoto, N. (2000). Microarray fabrication with covalent attachment of DNA using bubble jet technology. Nat Biotechnol *18*, 438-441.

Onel, K., and Cordon-Cardo, C. (2004). MDM2 and prognosis. Mol Cancer Res *2*, 1-8.

Ory, K., Legros, Y., Auguin, C., and Soussi, T. (1994). Analysis of the most representative tumour-derived p53 mutants reveals that changes in protein conformation are not correlated with loss of transactivation or inhibition of cell proliferation. EMBO J *13*, 3496-3504.

Pavletich, N. P., Chambers, K. A., and Pabo, C. O. (1993). The DNA-Binding domain of p53 contains the 4 conserved regions and the major mutation hot spots. Gene Develop *7*, 2556-2564.

Peyrat, J. P., Bonneterre, J., Lubin, R., Vanlemmens, L., Fournier, J., and Soussi, T. (1995). Prognostic significance of circulating p53 antibodies in patients undergoing surgery for locoregional breast cancer. Lancet *345*, 621-622.

Pharoah, P. D. P., Day, N. E., and Caldas, C. (1999). Somatic mutations in the p53 gene and prognosis in breast cancer: a meta-analysis. Brit J Cancer *80*, 1968-1973.

Puisieux, A., Lim, S., Groopman, J., and Ozturk, M. (1991). Selective targeting of p53 gene mutational hotspots in human cancers by etiologically defined carcinogens. Cancer Res *51*, 6185-6189.

Ralhan, R., Nath, N., Agarwal, S., Mathur, M., Wasylyk, B., and Shukla, N. K. (1998). Circulating p53 antibodies as early markers of oral cancer: correlation with p53 alterations [In Process Citation]. Clin Cancer Res *4*, 2147-2152.

Raycroft, L., Wu, H., and Lozano, G. (1990). Transcriptional activation by wild-type but not transforming mutants of the p53 anti-oncogene. Science *249*, 1049-1051.

Resnick, M. A., and Inga, A. (2003). Functional mutants of the sequence-specific transcription factor p53 and implications for master genes of diversity. Proc Natl Acad Sci U S A *100*, 9934-9939.

Ronaghi, M. (2001). Pyrosequencing sheds light on DNA sequencing. Genome Res *11*, 3-11.

Rotter, V., Witte, O. N., Coffman, R., and Baltimore, D. (1980). Abelson murine leukemia virus-induced tumors elicit antibodies against a host cell protein, p50. J Virol *36*, 547-555.

Rowan, S., Ludwig, R. L., Haupt, Y., Bates, S., Lu, X., Oren, M., and Vousden, K. H. (1996). Specific loss of apoptotic but not cell-cycle arrest function in a human tumor derived p53 mutant. EMBO J *15*, 827-838.

Saffroy, R., Lelong, J. C., Azoulay, D., Salvucci, M., Reynes, M., Bismuth, H., Debuire, B., and Lemoine, A. (1999). Clinical significance of circulating anti-p53 antibodies in European patients with hepatocellular carcinoma. Brit J Cancer *79*, 604-610.

Save, V., Nylander, K., and Hall, P. A. (1998). Why is p53 protein stabilized in neoplasia? Some answers but many more question! J Pathol *184*, 348-350.

Schaefer, K. L., Wai, D., Poremba, C., Diallo, R., Boecker, W., and Dockhorn-Dworniczak, B. (2002). Analysis of TP53 germline mutations in pediatric tumor patients using DNA microarray-based sequencing technology. Med Pediatr Oncol *38*, 247-253.

Scheffner, M., Werness, B. A., Huibregtse, J. M., Levine, A. J., and Howley, P. M. (1990). The E6 oncoprotein encoded by human papillomavirus type-16 and type-18 promotes the degradation of p53. Cell *63*, 1129-1136.

Schlichtholz, B., Legros, Y., Gillet, D., Gaillard, C., Marty, M., Lane, D., Calvo, F., and Soussi, T. (1992). The immune response to p53 in breast cancer patients is directed against immunodominant epitopes unrelated to the mutational hot spot. Cancer Res *52*, 6380-6384.

Schlichtholz, B., Tredaniel, J., Lubin, R., Zalcman, G., Hirsch, A., and Soussi, T. (1994). Analyses of p53 antibodies in sera of patients with lung carcinoma define immunodominant regions in the p53 protein. Br J Cancer *69*, 809-816.

Schmitz-Drager, B. J., Goebell, P. J., Ebert, T., and Fradet, Y. (2000). p53 immunohistochemistry as a prognostic marker in bladder cancer. playground for urology scientists? [In Process Citation]. Eur Urol *38*, 691-700.

Sharpless, N. E., and DePinho, R. A. (2002). p53: good cop/bad cop. Cell *110*, 9-12.

Shieh, S. Y., Ahn, J., Tamai, K., Taya, Y., and Prives, C. (2000). The human homologs of checkpoint kinases Chk1 and Cds1 (Chk2) phosphorylate p53 at multiple DNA damage-inducible sites [published erratum appears in Genes Dev 2000 Mar 15;14(6):750]. Genes Dev *14*, 289-300.

Shiraishi, K., Kato, S., Han, S. Y., Liu, W., Otsuka, K., Sakayori, M., Ishida, T., Takeda, M., Kanamaru, R., Ohuchi, N., and Ishioka, C. (2004). Isolation of temperature-sensitive p53 mutations from a comprehensive missense mutation library. J Biol Chem *279*, 348-355.

Shumaker, J. M., Tollet, J. J., Filbin, K. J., Montague-Smith, M. P., and Pirrung, M. C. (2001). APEX disease gene resequencing: mutations in exon 7 of the p53 tumor suppressor gene. Bioorg Med Chem *9*, 2269-2278.

Simon, R., and Sauter, G. (2003). Tissue microarray (TMA) applications: implications for molecular medicine. Expert Rev Mol Med *2003*, 1-12.

Smith-Sørensen, B., Gebhardt, M. C., Kloen, P., McIntyre, J., Aguilar, F., Cerutti, P., and Børresen, A.-L. (1993). Screening for TP53 mutations in osteosarcomas using constant denaturant gel electrophoresis (CDGE). Human Mutation *2*, 274-285.

Soengas, M. S., Alarcon, R. M., Yoshida, H., Giaccia, A. J., Hakem, R., Mak, T. W., and Lowe, S. W. (1999). Apaf-1 and caspase-9 in p53-dependent apoptosis and tumor inhibition. Science *284*, 156-159.

Soussi, T. (1996). The p53 tumour suppressor gene: a model for molecular epidemiology of human cancer. Mol Med Today *2*, 32-37.

Soussi, T. (2000). p53 Antibodies in the sera of patients with various types of cancer: a review. Cancer Res *60*, 1777-1788.

Soussi, T., and Béroud, C. (2001). Assessing TP53 status in human tumours to evaluate clinical outcome. Nat Rev Cancer *1*, 233-240.

Soussi, T., and Béroud, C. (2003). Significance of TP53 mutations in human cancer: A critical analysis of mutations at CpG dinucleotides. Hum Mutat *21*, 192-200.

Srivastava, S., Zou, Z. Q., Pirollo, K., Blattner, W., and Chang, E. H. (1990). Germ-line transmission of a mutated p53 gene in a cancer-prone family with li-fraumeni syndrome. Nature *348*, 747-749.

Staib, F., Hussain, S. P., Hofseth, L. J., Wang, X. W., and Harris, C. C. (2003). TP53 and liver carcinogenesis. Hum Mutat *21*, 201-216.

Steels, E., Paesmans, M., Berghmans, T., Branle, F., Lemaitre, F., Mascaux, C., Meert, A. P., Vallot, F., Lafitte, J. J., and Sculier, J. P. (2001). Role of p53 as a prognostic factor for survival in lung cancer: a systematic review of the literature with a meta-analysis. Eur Respir J *18*, 705-719.

Storey, A., Thomas, M., Kalita, A., Harwood, C., Gardiol, D., Mantovani, F., Breuer, J., Leigh, I. M., Matlashewski, G., and Banks, L. (1998). Role of a p53 polymorphism in the development of human papillomavirus-associated cancer. Nature *393*, 229-234.

Strano, S., Munarriz, E., Rossi, M., Cristofanelli, B., Shaul, Y., Castagnoli, L., Levine, A. J., Sacchi, A., Cesareni, G., Oren, M., and Blandino, G. (2000). Physical and functional interaction between p53 mutants and different isoforms of p73. J Biol Chem *275*, 29503-29512.

Sullivan Pepe, M., Etzioni, R., Feng, Z., Potter, J. D., Thompson, M. L., Thornquist, M., Winget, M., and Yasui, Y. (2001). Phases of biomarker development for early detection of cancer. J Natl Cancer Inst *93*, 1054-1061.

Tada, M., Furuuchi, K., Kaneda, M., Matsumoto, J., Takahashi, M., Hirai, A., Mitsumoto, Y., Iggo, R. D., and Moriuchi, T. (2001). Inactivate the remaining p53 allele or the alternate p73? Preferential selection of the Arg72 polymorphism in cancers with recessive p53 mutants but not transdominant mutants. Carcinogenesis *22*, 515-517.

Takahashi, T., Nau, M. M., Chiba, I., Birrer, M. J., Rosenberg, R. K., Vinocour, M., Levitt, M., Pass, H., Gazdar, A. F., and Minna, J. D. (1989). p53 - a frequent target for genetic abnormalities in lung cancer. Science *246*, 491-494.

Tavassoli, M., Brunel, N., Maher, R., Johnson, N. W., and Soussi, T. (1998). p53 antibodies in the saliva of patients with squamous cell carcinoma of the oral cavity. Int J Cancer *78*, 390-391.

Thomas, D. C., Nardone, G. A., and Randall, S. K. (1999a). Amplification of padlock probes for DNA diagnostics by cascade rolling circle amplification or the polymerase chain reaction. Arch Pathol Lab Med *123*, 1170-1176.

Thomas, M., Kalita, A., Labrecque, S., Pim, D., Banks, L., and Matlashewski, G. (1999b). Two polymorphic variants of wild-type p53 differ biochemically and biologically. Mol Cell Biol *19*, 1092-1100.

Tonisson, N., Zernant, J., Kurg, A., Pavel, H., Slavin, G., Roomere, H., Meiel, A., Hainaut, P., and Metspalu, A. (2002). Evaluating the arrayed primer extension resequencing assay of TP53 tumor suppressor gene. Proc Natl Acad Sci U S A *99*, 5503-5508.

Tornaletti, S., Rozek, D., and Pfeifer, G. P. (1993). The distribution of UV photoproducts along the human p53 gene and its relation to mutations in skin cancer. Oncogene *8*, 2051-2057.

Toyooka, S., Tsuda, T., and Gazdar, A. F. (2003). The TP53 gene, tobacco exposure, and lung cancer. Hum Mutat *21*, 229-239.

Trivers, G. E., Cawley, H. L., Debenedetti, V. M. G., Hollstein, M., Marion, M. J., Bennett, W. P., Hoover, M. L., Prives, C. C., Tamburro, C. C., and Harris, C. C. (1995). Anti-p53 antibodies in sera of workers occupationally exposed to vinyl chloride. J Nat Cancer Inst *87*, 1400-1407.

Trivers, G. E., De Benedetti, V. M. G., Cawley, H. L., Caron, G., Harrington, A. M., Bennet, W. P., Jett, J. R., Colby, T. V., Tazelaar, H., Pairolero, P., et al. (1996). Anti-p53 antibodies in sera from patients with chronic obstructive pulmonary disease can predate a diagnosis of cancer. Clin Cancer Res *2*, 1767-1775.

Vahakangas, K. (2003). TP53 mutations in workers exposed to occupational carcinogens. Hum Mutat *21*, 240-251.

van Oijen, M. G., and Slootweg, P. J. (2000). Gain-of-function mutations in the tumor suppressor gene p53 [In Process Citation]. Clin Cancer Res *6*, 2138-2145.

Varley, J. M. (2003). Germline TP53 mutations and Li-Fraumeni syndrome. Hum Mutat *21*, 313-320.

Varley, J. M., Attwooll, C., White, G., McGown, G., Thorncroft, M., Kelsey, A. M., Greaves, M., Boyle, J., and Birch, J. M. (1999a). Characterization of germline TP53 splicing mutations and their genetic and functional analysis. Oncogene *20*, 2647-2654.

Varley, J. M., McGown, G., Thorncroft, M., James, L. A., Margison, G. P., Forster, G., Evans, D. G. R., Harris, M., Kelsey, A. M., and Birch, J. M. (1999b). Are there low-penetrance TP53 alleles? Evidence from childhood adrenocortical tumors. Amer J Hum Genet *65*, 995-1006.

Venkatachalam, S., Tyner, S. D., Pickering, C. R., Boley, S., Recio, L., French, J. E., and Donehower, L. A. (2001). Is p53 haploinsufficient for tumor suppression? Implications for the p53+/- mouse model in carcinogenicity testing. Toxicol Pathol *29 Suppl*, 147-154.

Vogelstein, B., Lane, D., and Levine, A. J. (2000). Surfing the p53 network. Nature *408*, 307-310.

Vousden, K. H. (2000). p53. Death star [In Process Citation]. Cell *103*, 691-694.

Vousden, K. H., and Woude, G. F. (2000). The ins and outs of p53 [In Process Citation]. Nat Cell Biol *2*, E178-180.

Wagner, J., Portwine, C., Rabin, K., Leclerc, J. M., Narod, S. A., and Malkin, D. (1994). High frequency of germline p53 mutations in childhood adrenocortical cancer. J Nat Cancer Inst *86*, 1707-1710.

Walker, K. K., and Levine, A. J. (1996). Identification of a novel p53 functional domain that is necessary for efficient growth suppression. Proc Natl Acad Sci USA *93*, 15335-15340.

Waridel, F., Estreicher, A., Bron, L., Flaman, J. M., Fontolliet, C., Monnier, P., Frebourg, T., and Iggo, R. (1997). Field cancerisation and polyclonal p53 mutation in the upper aerodigestive tract. Oncogene *14*, 163-169.

Warneford, S. G., Witton, L. J., Townsend, M. L., Rowe, P. B., Reddel, R. R., Dalla-Pozza, L., and Symonds, G. (1992). Germ-line splicing mutation of the p53 gene in a cancer-prone family. Cell Growth Differ *3*, 839-846.

Wen, W. H., Bernstein, L., Lescallett, J., Beazer-Barclay, Y., Sullivan-Halley, J., White, M., and Press, M. F. (2000). Comparison of TP53 mutations identified by oligonucleotide microarray and conventional DNA sequence analysis. Cancer Res *60*, 2716-2722.

Werner, J. A., Gottschlich, S., Folz, B. J., Goeroegh, T., Lippert, B. M., Maass, J. D., and Rudert, H. (1997). p53 serum antibodies as prognostic indicator in head and neck cancer. Cancer Immunol Immunother *44*, 112-116.

Wikman, F. P., Lu, M. L., Thykjaer, T., Olesen, S. H., Andersen, L. D., Cordon-Cardo, C., and Orntoft, T. F. (2000). Evaluation of the performance of a p53 sequencing microarray chip using 140 previously sequenced bladder tumor samples [In Process Citation]. Clin Chem *46*, 1555-1561.

Wild, C. P., Jansen, L. A., Cova, L., and Montesano, R. (1993). Molecular dosimetry of aflatoxin exposure: contribution to understanding the multifactorial etiopathogenesis of primary hepatocellular carcinoma with particular reference to hepatitis B virus. Environ Health Perspect *99*, 115-122.

Williams, C., Norberg, T., Ahmadian, A., Ponten, F., Bergh, J., Inganas, M., Lundeberg, J., and Uhlen, M. (1998). Assessment of sequence-based p53 gene analysis in human breast cancer: messenger RNA in comparison with genomic DNA targets. Clin Chem *44*, 455-462.

Yang, A., Kaghad, M., Caput, D., and McKeon, F. (2002). On the shoulders of giants: p63, p73 and the rise of p53. Trends Genet *18*, 90-95.

Zalcman, G., Schlichtholz, B., Trédaniel, J., Urban, T., Lubin, R., Dubois, I., Milleron, B., Hirsh, A., and Soussi, T. (1998). Monitoring of p53 auto antibodies in lung cancer during therapy: relationship to response to treatment. Clin Cancer Res *4*, 1359-1366.

Zhang, Y., Kaur, M., Price, B. D., Tetradis, S., and Makrigiorgos, G. M. (2002). An amplification and ligation-based method to scan for unknown mutations in DNA. Hum Mutat *20*, 139-147.

Chapter 13

PATTERNS OF TP53 MUTATIONS IN HUMAN CANCER: INTERPLAY BETWEEN MUTAGENESIS, DNA REPAIR AND SELECTION

Hong Shi, Florence Le Calvez, Magali Olivier and Pierre Hainaut
International Agency for Research on Cancer, Lyon, France

INTRODUCTION

Somatic mutations are the cornerstone of cancer (Hanahan et al. 2000). The development of cancer involves the contributions of many heritable genetic events as well as of a large number of epigenetic changes, but what makes the turning point between untransformed and transformed cell irreversible is the acquisition of targeted, somatic mutations, conferring to cells a selective advantage for clonal proliferation. These mutations can occur in many different genes, but only a handful of them are frequently mutated in a wide variety of human cancers. They include genes of the *RAS* family (mainly *KRAS*), *BRAF1, APC, β–Catenin, p16/INK4a, PTEN* and *TP53*. After over 20 years of research on mutation detection in cancers, *TP53* remains the world champion of somatic mutations, with over 70% of all the mutations described so far in human cancers (Hainaut et al. 2000). The database of *TP53* mutations maintained at IARC (http://www.iarc.fr/p53) contains close to 20,000 mutations identified in primary human cancers, cell lines and pre-cancerous lesions (Olivier et al. 2002). The overall spectrum of these mutations has several unique features. As for many tumor suppressor genes, mutations are scattered along large parts of the coding sequence, and also occur at splice junctions, in agreement with the notion that their primary consequence is to disrupt p53 protein function. In contrast to most other suppressors, however, the vast majority of *TP53* mutations are missense rather than nonsense or frameshift mutations.

P. Hainaut and K.G. Wiman (eds.), 25 Years of p53 Research, 293-319.

This predominance of missense mutations is only matched by oncogenes such as *KRAS*. In the latter case, missense mutations fall at a small number of codons (12, 13, 59 and 61), encoding residues involved in GTP binding, thus resulting in gain-of-signaling function. Thus, the overall profile of *TP53* mutations carries similarities with those of oncogenes as well as of other tumor suppressor genes, reflecting in its alterations the functional ambivalence of the protein.

In 1991, Monica Hollstein and colleagues published the first compilation of mutation and popularized the concept that mutations could occur at many different positions, with a few "hotspots" that were common irrespective of cancer site or histology (Hollstein et al. 1991). By 1994, Curtis Harris and his collaborators had correctly pinpointed most of the tumor specific *TP53* mutation patterns, based on a dataset of less than 2500 mutations, and it became evident that carcinogens leave fingerprints in the human genome (Greenblatt et al. 1994). These observations contributed to establish the principle that *TP53* mutation patterns could be read as "reporters" of carcinogenic exposures. However, in recent years, it has emerged that, mutagenesis alone could not fully explain the distribution of mutations along the coding sequence of *TP53*, and that other mechanisms, in particular preferential repair or biological selection of mutants with special properties, may operate as an additional filter to eliminate some of the expected mutations. This chapter reviews and discusses the case of "mutagenesis versus selection" in shaping *TP53* mutation patterns in human cancers.

INTERPRETING MUTATION PATTERNS

What is a mutation pattern?

There is a wide diversity in the types of mutations that arise in the human genome and, in many instances, they provide clues on the mechanisms for their generation. The Human Gene Mutation Database (HGMD, http://www.uwcm.ac.uk/uwcm/mg/hgmd0.html) compiles mutations reported in the coding regions of human genes causing genetic diseases (Antonarakis et al. 2002; Stenson et al. 2003). In this database, single base-pair substitutions account for about 50% of all reported mutations. Other common changes include deletions, insertions, duplications, inversions, and alterations of unstable repeated sequences. Among single base-pair substitutions, there are major differences in frequency according to the type of nucleotide change. This observation has led to the concept of "mutation pattern" or "mutation spectrum", both terms being used to identify a particular distribution of types of mutations within a selected set of mutation data. However, the notion of "mutation pattern" is somewhat ambiguous and often corresponds to different concepts from one publication to another.

Basically, six elements may come into consideration in defining a mutation pattern: type of mutation, nucleotide change, sequence context, strand distribution, occurrence of the mutation at a position of known structure or function (e.g. mutations in exons, introns, at mRNA splice junctions or other structures involved in mRNA processing, within promoter regions, etc..), and consequence of the mutation on the gene structure and its coding potential (e.g. silent, missense, nonsense, mutations affecting exon processing or expression levels) (Antonarakis et al. 2002). A "mutation pattern" occurs when there is a significant difference in any combination of these elements, between a set of "test" mutations (e.g. mutations identified in a particular type of cancer) and a set of "reference" mutations. In many instances in studies on *TP53*, the reference set is represented by mutations found in all cancers, minus the ones with the same criteria as in the "test" set. This type of analysis is prone to many methodological biases depending on selection criteria for both "test" and "reference" sets. Ideally, such comparisons should be performed in a strict, case-control study design (e.g. comparing mutations between patients with the same cancers, matched for individual parameters, and exposed or not to a particular risk factor). This is however not the case in most published studies. The *TP53* mutation database maintained at IARC has been developed to provide a repository for all published *TP53* mutation data, with extensive annotations that allows the selection by users of the best possible reference datasets. However, the *TP53* mutation database does not provide a population-based reference and is open to many biases since it compiles data from studies that differ in size, methods, design, case selection criteria and annotations (reviewed in (Hernandez-Boussard et al. 1999)).

The formation of a mutation pattern in cancer can be seen as the result of a complex process of elimination of candidate lesions through a succession of filters (Table 1). The first filter is determined by carcinogen metabolism and by the chemistry of DNA damage. Mutagens can damage DNA in a specific way, generating promutagenic lesions that, to some extend, reflect the chemistry of DNA damage (Essigmann et al. 1993). Moreover, base position, accessibility and sequence context are the main factors that influence the type and form of DNA damage, forming a second filter (Antonarakis et al. 2002). The third filter consists into DNA repair, a complex set of processes that removes the large majority of lesions, but does so in a selective manner so that all types of lesions are not eliminated with the same efficiency (Hanawalt et al. 2003).

Table 1. Formation of a mutation pattern through a succession of "filters": the example of Benzo(a)pyrene from tobacco smoke

Exposure	**Tobacco smoke contains over 60 substances classified as carcinogenic to human by IARC, including 1 to 40 ng Benzo[a]Pyrene (B[a]P)/cigarette**		
	Filter	**Example**	**Type of lesion**
Filter 1	Chemistry of DNA damage	B[a]P is metabolized by CYP450 to generate B[a]P-7,8-diol-9,10-epoxide (BPDE) that binds on N2 position of guanine	BPDE-N^2-dG adduct
Filter 2	Base position and sequence context	Adduct preferentially form at G adjacent to methylated cytosines at mCpG sites	Major adducts at codons 156, 157, 245, 248, 273
Filter 3	DNA repair	Transcription-coupled repair preferentially removes lesions on the transcribed DNA strand (TS)	Strand bias with persistence of adducts on G on the NTS
Filter 4	DNA replication	Lesion bypass of an adducted template by Pol eta misincorporates A instead of C; replication results in substitition of G to T opposite to misincoporated A	Formation of G to T transversions
Filter 5	Protein filter	Only mutations that inactivate p53 protein contribute to the clonal expansion of cancer cells and are detectable in cancer lesions	Selection of mutations at codons 157, 245, 248, 273; counter-selection of mutation at codon 156, which is silent.
Mutation pattern in cancer	Excess of G to T transversions on the Non-transcribed strand at specific codons in lung cancers of smokers		

In particular, transcription-coupled excision repair preferentially eliminates polymerase-blocking lesions occurring on the transcribed DNA strand and is responsible for DNA strand bias in mutation distribution (Vrieling et al. 1998). A fourth filter is represented by DNA replication. Although a very accurate system, DNA replication is also error-prone and critically dependent on polymerase fidelity (Kunkel 2004). Finally, once DNA alterations have been selected through these various filters, they become targets for biological selection through the so-called "protein filter" that confines the detectable mutations to genetic alterations that produce a functionally altered gene product (Dogliotti et al. 1998; Hollstein et al.

1999). However, detection of a mutation in a clone of cancer cells does not necessarily imply that mutation is causal for cancer growth. In many cases, mutations can arise as by-products of the transformation process or may occur before the initiation of clonal expansion. Thus, a number of mutations may be hitch-hikers, rather than drivers, in the process of clonal expansion. This is particularly true for the mutations that occur at rare codon positions ("coldspots") in the IARC *TP53* mutation database.

Mechanisms of "carcinogen fingerprints"

The primary DNA structure is constantly under attack by metabolites resulting from cellular processes as well as from exogenous DNA-damaging agents. These metabolites can induce covalent and non-covalent anomalies in DNA structures, the most common ones being various forms of base damage, single- or double DNA-strand nicks and gaps, intrastrand or interstrand as well as protein-DNA crosslinks (Wogan et al. 2004). All these forms of damage elicit specific DNA repair reactions. However, some of these reactions can also be triggered by the formation of DNA forks, Holliday structures and other non-paired DNA structures during replication and recombination. The main forms of base damage are oxidized, reduced, fragmented bases as well as covalent adducts formed of small chemical groups (such as alkyl adducts formed by alkylating agents) or large compounds (the so-called "bulky adducts" induced by metabolites of polycyclic aromatic hydrocarbons, arylamines or mycotoxins). Imperfect repair of these various types of lesions has the potential to induce irreversible changes in the primary DNA structure. A "carcinogen fingerprint" arises when the frequent occurrence of a specific type of mutation can be taken as evidence of DNA attack by a specific type of carcinogen. Table 2 lists a series of chemicals that induce defined types of promutagenic DNA lesions, and describes the major types of mutations that result from these lesions in experimental systems.

When confronted to carcinogen exposures, human cells develop responses similar to those for any other foreign compound or drug. Many carcinogens are lipophilic compounds that readily cross plasma membranes to accumulate in the cytoplasm and the nucleus. To counterbalance their immediate toxic effects, cytochrome P (CYP) 450 enzymes initiate a cascade of metabolic detoxification reactions by catalyzing the addition of an oxygen to the carcinogen, increasing its solubility in water and converting it to a more readily excretable form (Guengerich 2000). This process is amplified by conjugation enzymes such as Glutathione-S-Transferases, which convert the oxygenated carcinogen to a form that is highly soluble in water. These detoxification reactions are highly efficient and provide a first line of metabolic protection against the immediate, toxic injury inflicted by these chemicals (Burchell et al. 1997). However, during this process, cytochrome P450 enzymes modify the carcinogens to form reactive compounds that often contain an electrophilic (electron-deficient) center. Such metabolites can attack DNA bases on specific N positions (depending upon their chemical structure and the target base), resulting in the formation of stable DNA adducts (Guengerich 2003). This process is known as metabolic activation (Miller 1994). The genes involved in metabolic detoxification and activation reactions are multiple, redundant and polymorphic. Therefore, the balance between metabolic activation and detoxification varies among populations and individuals, and is one of the determinants of genetic susceptibility to cancer.

Table 2. Frequent Mutations Induced by Some Exogenous and Endogenous DNA-Damaging Agents.

Site of premutagenic lesion	Mutagen	Main Mutations	Possible TP53 fingerprint in
*N*7-G	AFB1	GC > TA	Hepatocellular carcinoma
*N*2-G	BPDE	GC > TA	Lung cancers, smokers
*O*6-G	*N*-Methyl-*N*-nitrosourea	GC > AT	Oral, oesophageal cancers?
*O*6-G	NNK	GC > AT	Lung cancers ?
C8-G	1-Nitrosopyrene	GC > AT, GC > TA	?
C8-G	4-Aminobiphenyl	GC > TA	Bladder cancer
C8-G	2-AAF	GC > TA	Bladder cancers?
C8-G	PhIP	GC > TA	?
8-oxo-G	Oxidative agents	GC > TA	Many cancers incl lung
1,*N*2-G	Malondialdehyde	GC > TA, GC > AT	?
*N*6-A	Stryene oxide	AT > CG	?
*N*6-A	Benzo[*c*]phenantrene diol epoxide	AT > TA, AT > GC	Lung, oesophageal cancers?
*N*6-A	BPDE	AT > GC	Lung cancers
3,*N*4-C	Vinyl chloride	GC > AT	Angiosarcoma of the liver
5-OH-C, 5-OH-U, uridine glycol	Oxidative agents	GC > AT	?
N3-U	Propylene oxide	GC > AT	?
Pyrimidine dimers	UV	CC > TT tandem, GC > AT	Non-melanoma skin cancers
Apurinic	Depurinating agents	GC > TA, AT > TA	?

The second line of defense mechanisms against DNA damage by carcinogens consists in the elaborate DNA repair systems that cells have evolved to eliminate DNA adducts from the genome (Hanawalt et al. 2003; Hoeijmakers 2001). For example, the nucleotide excision repair (NER) pathway eliminates intra- and interstrand DNA crosslinks, as well as bulky DNA adducts. In contrast, bases damaged by the attachment of small chemical groups (oxidized or methylated bases) or bases fragmented by ionizing radiation or chemical oxidation are repaired through the base-excision repair (BER) pathway. A specialized, direct repair system acts through the enzyme O^6-methylguanine DNA methyltransferase, which repairs the common, miscoding methylated base O^6-methylguanine. These repair processes are complex and involve steps of damage removal (e.g. by DNA glycosylases in BER) followed by base incorporation reactions mediated by several polymerases. Furthermore, some DNA lesions that induce mutations are not repaired at the same rate on the transcribed (TS) and non-transcribed (NTS) DNA strand. For example, DNA repair experiments analyzing bulky adducts in *TP53* have shown that the non-transcribed strand is repaired more slowly than the transcribed strand, and that repair at major damage hotspots may be slower than at other positions (Denissenko et al. 1998b; Tornaletti et al. 1994). These findings support the proposal that initial DNA adduct levels, sequence-dependent adduct formation and DNA strand as well as position bias in repair contribute to the mutational spectrum.

In the best cases, effective damage removal results in the rapid elimination of DNA adducts. Incomplete or imperfect removal of damage may lead to the persistence of lesions. This phenomenon has multiple implications for cell proliferation and survival. In particular, presence of unrepaired DNA damage may trigger p53 protein stabilization and activation of a set of responses that permanently delete damaged cells from the pool of proliferative cells, either by permanent cell-cycle arrest, terminal differentiation or senescence, or by induction of apoptosis (Pluquet et al. 2001). If unrepaired damage persists during the steps of DNA reconstruction or during replication, it may cause replicative DNA polymerases to stop at the site of a lesion. This mechanism induces a third layer of protective responses against the formation of permanent damage into DNA. Polymerase stop may induce arrest of DNA replication, and cell death (again, through a set of pathways involving p53) (Bregman et al. 2000). Alternatively, the polymerases may bypass the altered base, with the possibility of base misincorporation. To that effect, cells have evolved specialized DNA polymerases that are able to bypass various types of DNA damage (Kunkel 2004; Livneh 2001). Mutations may arise when DNA adducts are bypassed incorrectly by a DNA polymerase.

The weight of spontaneous mutagenesis

There is no doubt that many mutations arise spontaneously in all organisms, and there are well-charted mechanisms by which they occur in mammalian cells. However, it is difficult to estimate which mutation arises spontaneously in humans *in vivo*, and how their frequencies vary according to gene function, cell type or differentiation state. Current knowledge is based on analysis of point mutations in human lymphocytes at the *HPRT* locus or in other genes that allow for phenotypic selection (Antonarakis et al. 2002; Dogliotti et al. 1998), as well as computational analysis of large series of genes to estimate and compare spontaneous mutation rates in mammalian genomes (Cooper et al. 2004; Kumar et al. 2002).

Based on estimates of the rate at which neutral mutations accumulate in the coding regions of the genome among mammalian lineages, the average mammalian genome mutation rate is in the range of $2.2 \ 10^{-9}$ per base pair per year (Kumar et al. 2002). This rate appears to be largely similar among different genes. However, the rate of formation of mutations at different base pairs is extremely variable. Spontaneous mutations show a strong bias towards transitions (purine to purine or pyrimidine to pyrimidine) as compared to transversions. This bias is due to the hypermutability of the CpG dinucleotides, which can mutate at a rate 10 times higher than other nucleotides according to a mechanism that is fundamentally different from those affecting other base motifs (Holliday et al. 1993; Jones et al. 1992). Briefly, a notable proportion of cytosines at CpG dinucleotides are methylated at position 5' in the genome of normal cells. The 5'methylcytosine (5mC) is significantly less stable than cytosine, and undergoes spontaneous deamination into thymine at a rate 5 times higher than the unmethylated base. This purely endogenous mutagenic process can be enhanced by oxygen and nitrogen radicals (Hussain et al. 2003). Thus, it is conceivable that the rate of formation of transitions at CpG dinucleotides can be influenced by at least four major factors: the existence of gene- or population-specific DNA methylation patterns, genetic or epigenetic variations in methyltransferase activities, exposure to enhancers such as Nitric Oxide (NO) and efficiency of repair mechanisms (Hussain et al. 2003; Schmutte et al. 1996).

In the HGMD database, G:C to A:T transitions at CpG account for 23.8% of all mutations. In *TP53*, CpG mutations represent, overall, 25% of all reported mutations, with variations from about 15% in lung cancers of smokers to close to 50% in adenocarcinomas of the gastro-digestive tract (Olivier et al. 2004). Studies by Pfeifer and collaborators have shown that there are no significant differences from one human tissue to another in the pattern of methylated cytosines at *TP53* mutation hotspots (Tornaletti et al.

1995). On the other hand, in colon cancer, a significant correlation is observed between levels of expression of the NO-generating enzyme NO-Synthase 2 (*NOS2*) and occurrence of CpG mutations (Ambs et al. 1999). This observation provides a paradigm for interpreting the high prevalence of CpG mutations in cancers such as gastric cancer (in relation with inflammation resulting from chronic infection by *H. pylori*) and adenocarcinoma of the esophagus (which often arise from an inflammatory, metaplastic lesion, the Barrett's mucosa) (Hussain et al. 2003; Olivier et al. 2004).

Another powerful possible mechanism of spontaneous mutagenesis is described as the slipped-mispairing model (Antonarakis et al. 2002). According to this model, nucleotide misincorporation may result from transient misalignment of the primer to the template due to the looping out of a base (or a short stretch of bases) from the template. This phenomenon may preferentially occur within runs of consecutive identical bases or in regions containing repetitive DNA sequences. The net result is misincorporation of a new base identical to either one of the bases flanking the mutation site. In a study of the data compiled in the IARC database, Greenblatt et al. (1996) have shown that almost all these deletions or insertions arise within one or more of the following sequence environment: monotonic base runs, adjacent or non-adjacent repeats of short tandem sequences, palindromes, and runs of purines or pyrimidines. Moreover, increased length of monotic runs correlates with increased frequency of insertion/deletion events. Thus, the polymerase slippage/misalignment model explains the vast majority of the deletions and insertions that occur in *TP53* (Greenblatt et al. 1996).

The p53 protein filter

It is an intrinsic characteristic of any mutational reporter assay that detectable mutations are confined to those defects that produce a functionally altered gene product. In fact, a large proportion of the mutations that may occur in any gene are functionally silent. The way this "protein filter" selects among mutations depends not only upon the position of the mutation in the coding sequence but also upon its type. Due to the nature of the genetic code, transitions more often result into synonymous changes than transversions. Moreover, when both types of mutations generate a change into the protein sequence, the amino-acid changes resulting from transversions are generally more severe than those resulting from transitions (Rosenberg et al. 2003). In cancers, *TP53* mutations are distributed in all coding exons, with a strong predominance in exons 5 to 8, encoding the DNA-binding domain of the protein. After initial reports that mutations cluster in the central portion of the coding sequence, many studies were limited to exons 5 to 8, resulting in an over-representation of these mutations in databases. Nonetheless, when taking into account only studies that have screened the entire coding sequence, 80% (2212/2779) of the mutations are located within exons 5-8 (Olivier et al. 2002). This observation clearly identifies sequence-specific DNA binding as the main activity targeted by mutations in cancer. Using a yeast-based functional assay to score mutations according to the ability of p53 to activate the transcription of a reporter gene, Flaman et al. (1994) have determined the number of sites where mutations could be detected following random mutagenesis in a 1182 base pairs segment encompassing exons 5-8. They found inactivating mutations at 542 sites (46% of all sites) in the open reading frame, remarkably close to the actual number of sites identified as containing mutations in human tumors (573) (Flaman et al. 1994).

In the DNA-binding domain, missense mutations have been reported at all residues, but with striking variations in prevalence. Thanks to the elucidation of the crystal structure of the core domain of p53 in complex with DNA, there is a good understanding of the structural basis of many of these mutations (Cho et al. 1994). About 30% of them fall at 6 hotspot codons (175, 245, 248, 249, 273, 282). Most of the mutations at these positions are transitions occurring at CpG dinucleotides, with the notable exception of codon 249 (discussed below). However, several of these codons are also frequent targets for transversions occurring at guanines adjacent to methylated cystosines (e.g. codons 248 and 273). This observation indicates that frequent mutations at these codons are not only due to a highly mutable sequence context but also to a particularly strong biological selection. Five of the 6 hotspots codons correspond to arginine residues (175, 248, 249, 273 282) involved in protein-DNA interactions, either by direct contact with DNA (residues 248 and 273) or by stabilization of the DNA-binding surface (residues 175, 249 and 282). On the basis of the position and structural role of these residues, it is easy to understand how their substitution may disrupt DNA binding. However, how each particular substitution alters the protein structure is still difficult to predict. There are several common substitutions at codon 248 and 273, resulting from either transitions or transversions, and there is experimental evidence that the corresponding mutant proteins may differ in their biological properties (Kato et al. 2003; Ory et al. 1994). On the other hand, the most common substitution in cancer, arginine to histidine at codon 175, is particularly difficult to predict on a purely structural basis. This substitution is relatively conservative in terms of size, shape, charge and ability to donate hydrogen bonds. However, the resulting protein has totally lost DNA-binding capacity and is generally considered as a particularly strong p53 mutant. In structural terms, the arginine to histidine substitution would appear less drastic than arginine to proline, that introduces a torsion of the protein backbone. Yet the latter substitution is rarely found in cancer, and the resulting protein conserves partial DNA-binding activities in biological assays (Kato et al. 2003). Structural studies have suggested that the most significant effect of the histidine substitution may be to prevent the correct binding of zinc to a cysteine encoded by codon 176, thus disrupting protein folding (Martin et al. 2002).

In contrast with these hotspots, most other codons are rarely mutated. Among the most rarely mutated codons (less than 5 reports in the database), codon 123 (ATC, encoding threonine) is well conserved in evolution. Experimental substitution of threonine 123 to alanine has been shown to activate, rather than suppress, DNA-binding activity. It is thus possible that mutations at specific residues are counterselected because they do not favor cancer progression. Alternatively, rare mutations may represent unselected events similar to silent mutations, which represent up to 4.4% of the mutations in the database (Strauss 1997).

Despite the fact that there is a strong structural rationale for mutations in the DNA-binding domain, it is intriguing that major functional domains and regulatory sites located in the N-terminus and C-terminus of the protein are apparently rarely mutated in cancer. For example, only 1% of the reported mutations falls within the transactivation domain (TA), which also contains the binding site for Mdm2, the main regulator of p53 stability. The C-terminus, which participates in the regulation of DNA-binding activity, contains only 0.3% of all reported mutations. Moreover, mutations at regulatory sites, such as Ser 15 and 37 (phosphorylation by ATM/ATR and/or Chk2), Phe 19, Leu 22, Trp 23 and Leu 26 (interactions with Mdm2 and transcription machinery), Ser 315 (phosphorylation by cyclin-dependent kinases), Ser 376 and 378 (interactions with 14-3-3 sigma), Lys 370, 372, 373, 381 and 382 (sites of acetylation) and Ser 392 (phosphorylation by CKII), are so rare that they may be regarded as purely accidental (Olivier et al. 2004). This observation suggests that mutation of any of these residues is not sufficient in itself to significantly inactivate p53. In the case of residues of the TA domain, studies have shown that mutation of at least two of the residues was required to inactivate p53 transcriptional activity (Lin et al. 1994).

In most cancer cells, *TP53* mutations are present on only one allele, the other being either wild-type or lost. This observation has led to the assumption that some mutants may behave in a dominant-negative fashion. Expreimentally, various mutants can dominantly inhibit wild-type p53 when co-expressed in yeast or mammalian cells (Inga et al. 1997). On the other hand, a few mutants show gain-of-function properties that are dependent upon the nature of the mutation as well as on the cellular context (Dittmer et al. 1993). Recent studies have identified that some forms of mutant p53 may interact with the products of *TP63* and *TP73*, two genes that show strong homology with *TP53* and play complex, overlapping roles in apoptosis, differenciation and morphogenesis (Strano et al. 2000; Strano et al. 2002). The capacity of mutant p53 to complex with p73 appears to be influenced by a common polymorphism at codon 72 in *TP53* (Bergamaschi et al. 2003). However, since cancer tissues often contain a non-negligible proportion of normal cells containing two wild-type alleles, most studies have not been able to discriminate whether cancer cells carrying the mutation retain or not a wild-type allele.

Recently, the use of sensitive techniques for mutation detection combined with tissue microdissection has allowed direct approaches for revealing the presence of occasional mutants within a large population of wild-type p53 sequences. In particular, cell patches containing mutant *TP53* identical to cancer cells have been found in histologically normal skin and in the epithelium of the aerodigestive tract (Jonason et al. 1996; Mandard et al. 2000; Ren et al. 1996a; Ren et al. 1996b; Ren et al. 1997; Waridel et al. 1997). This observation is compatible with the theory of "field carcinogenesis", that predicts that large fields of cells within an exposed tissue may undergo genetic changes, but that only a few of them acquire the specific combination or sequence of mutations leading to clonal expansion and cancer. Thus, acquisition of a *TP53* mutation, even at a hotspot codon, may not be sufficient to drive cancer progression, and only a small proportion of cells that have acquired a mutation actually progress into malignancy.

CARCINOGEN FINGERPRINTS IN *TP53*

A number of studies have identified potential carcinogen fingerprints in *TP53* in lung, liver, skin, bladder, oral, oesophageal cancers and in angiosarcoma of the liver (reviewed in (Hainaut et al. 2000; Harris 1996)). In this paragraph we discuss the three best characterized cases, illustrating the interplay between mutagenesis, repair and selection in the occurrence of mutation patterns (Figure 1).

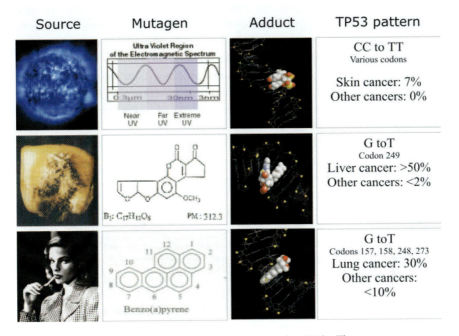

Figure 1. Textbook examples of mutagen fingerprints in TP53. The sources, mutagens, adducts and typical TP53 mutations are shown for three major cancer/risk factor associations. Top: solar UV and skin cancer (left : UV solar radiations, picture from Soho satellite); Midlle: aflatoxin and hepatocellular carcinoma (left: maize contaminated with moulds of Aspergillus sp.); Bottom: cigarette smoking and lung cancer (left: Laureen Bacall in "to have or to have not"). Right: TP53 mutation prevalence in cancers associated with the specific risk factor as compared with cancers, not associated with this risk factor.

UV radiation and carcinomas in sun-exposed tissues

There is strong experimental and epidemiological evidence linking UV radiation to the development of skin cancers (Holmquist et al. 1997; Montesano et al. 1997). *TP53* mutations are common in skin cancers that arise from transformed keratinocytes (squamous and basal cell carcinomas) but are very rare in melanoma (less than 5%), in which the p53 protein is suspected to be inactivated by other mechanisms. The *TP53* mutation spectrum in non-melanoma skin cancers (NMSC) shows a high frequency of C to T transitions (56 % of all mutations), including tandem CC to TT transitions (6% of all mutations), a type of mutation never found in tumors not related to sunlight. The positions of *TP53* mutations in skin cancer also show striking differences with other types of cancers, with hotspots at codons 177-179 and at codon 278 (Inga et al. 1998; Ziegler et al. 1994).

In terms of sequence context, many of the C to T transitions in skin cancers occur within a pyrimidine-cytosine-guanine trinucleotide sequence containing 5-methylcytosine as the central nucleotide (5'-PymCG-3'). This observation is consistent with the role of a common UV photoproduct, cyclobutane pyrimidine dimer, as the main promutagenic lesion. In a mouse transgene methylated at most CpG sequences, experimental irradiation with simulated sunlight generates a high frequency of dipyrimidine structures containing 5-methylcytosine (32% of all mutation) (Lee et al. 2003). There is evidence that absorption of near-UV by 5-methylcytosine is 5 to 10-fold higher than by cytosine. Methylation of cytosines at the 5-position also influences deamination and the resulting mutagenesis events. This particular mutagenic pathway may become prevalent under conditions of inefficient DNA repair and slow proliferation of cells in the human epidermis (Lee et al. 2003). Human DNA polymerase eta modulates susceptibility to skin cancer by promoting DNA synthesis past sunlight-induced cyclobutane pyrimidine dimers that escape nucleotide excision repair (NER). Errors made by Pol eta during dimer bypass could contribute to mutagenesis and to skin carcinogenesis in normal subjects and, particularly, in individuals with *Xeroderma pigmentosum* (XP) who are defective for NER and accumulate cyclobutane pyrimidine dimers in the skin. In skin tumors from *XP* patients, CC to TT transitions represent almost 50% of all *TP53* mutations (Dumaz et al. 1993). Moreover, repair does not occur with the same accuracy at all base positions in the *TP53* sequence. There is evidence that the preferential accumulation of mutations at codons 177, 196 and 278 in skin cancers is due to the slower repair of UV-induced lesions at these positions (Tornaletti et al. 1993; Tornaletti et al. 1994).

The clonality of *TP53* mutations in sunlight-induced skin cancers demonstrates that *TP53* mutations have an important causal role in the early steps of skin carcinogenesis. However, mutations at dipyrimidines identical to those found in skin cancers can be detected in normal skin cells of sun-exposed individuals, without affecting their morphology, and apparently, their proliferation and differentiation patterns (Jonason et al. 1996; Ren et al. 1996a; Ren et al. 1997). Mutation of *TP53* alone is thus not sufficient for initiation of skin carcinogenesis. It should be noted that although CC to TT transitions are never observed in internal organs, they can occur in other squamous tissues than skin under conditions of intense exposure to sunlight. For example, a very high prevalence of CC to TT transitions (over 50%) has been reported in a cohort of patients with squamous cell carcinoma of the conjunctiva from Uganda (Ateenyi-Agaba et al. 2004a). Current epidemiology data suggest a multiplicative model for the interaction between two main risk factors, exposure to solar light and infection by HIV, suggesting that deregulation of immune response may contribute to the development or progression of cancer of the conjunctiva. In these cancers, *TP53* mutations are frequently associated with the presence of Human Papilloma Viruses of the EV (*Epidermodysplasia verruciformis*) type (Ateenyi-Agaba et al. 2004b). Unlike their mucosal counterparts, that are aetiologically involved in cervical cancer by inactivating the p53 protein, the EV types of HPV may cooperate with *TP53* mutations to induce transformation of cells that have acquired mutations as a result of intense exposure to sunlight.

Hepatitis infections, aflatoxin and hepatocellular carcinoma (HCC)

The incidence rates of hepatocellular carcinoma (HCC) show large geographic variations, globally reflecting the prevalence of two main aetiologic factors, hepatitis B (HBV) and/or C (HCV) virus infection and exposure to high levels of aflatoxin in the diet (Chen et al. 1997). The highest incidence rates are observed in regions where most of the population is exposed to both factors, such as in parts of eastern Asia and in sub-Saharan Africa (Parkin et al. 2001). These high incidences are consistent with the fact that HBV chronicity and exposure to aflatoxin have a multiplicative effect of risk for HCC (Ross et al. 1992). The contributions of viral factors to the pathogenesis of HCC are multi-factorial and depend on virus genotype, on the presence of viral variants with enhanced tumorigenic properties, and on co-infections by different viruses (Kao et al. 2002).

HCC pathogenesis involves accumulation of genetic alterations in genes such as *p16/INK4a* (deletions, hypermethylation; in up to 50% of the cases), *β-Catenin* (mostly mutations in exon 3, 10-20%) *cyclins A* and *D* (overexpression, 20-30%), and *M6P/IGF2R* (mutations, 18-33%) (Ozturk 1999). Depending on aetiology and geographic area, mutations in *TP53* show striking differences in prevalence and pattern. In Europe and the US, where alcohol is a major risk factor in addition to viral infections, mutations occur in about 25% of HCC and show as much diversity in their type and codon position as in most other epithelial cancers. However, in high incidence areas such as Mozambique, Senegal, The Gambia (Africa) and Qidong county (China), *TP53* is mutated in over 50% of the cases and the vast majority of these mutations are a single missense, hotspot mutation at codon 249, AG<u>G</u> to AG<u>T</u>, resulting in the substitution of arginine into serine (249[ser]). This mutation is uncommon in regions where aflatoxin is not present at significant levels in the diet. In areas of intermediate exposure to aflatoxin, as for example in Thailand, the prevalence of the 249[ser] mutation is intermediate between high- and low-incidence areas. Thus, there is a dose-dependent relationship between exposure to aflatoxin, incidence of HCC and prevalence of 249[ser] mutation (Montesano et al. 1997). Mutations at codon 249 are detectable at a low prevalence in a number of other cancers, including lung cancers. However, in most cases, the mutation is a transversion affecting the second base of the codon, AGG to ATG, generating the substitution of the arginine into methionine (Olivier et al. 2002).

Aflatoxins are toxic and carcinogenic metabolites produced by several varieties of molds, mainly *Aspergillus flavus* and *Aspergillus parasiticum*. These molds contaminate a wide range of traditional agricultural products in countries with hot, humid climates, including maize, peanuts and cottonseeds. The toxins are present at significant levels in crops at the time of harvest but their concentration further increases under poor conditions of long-term food storage, in particular during the rain season. Thus, in these regions, most inhabitants of rural areas are highly exposed to aflatoxins, with seasonal variations reflecting the consumption of stored versus fresh foodstuff. Population-based surveys have demonstrated the presence of serum aflatoxin-albumin adducts in over 95% of the normal population in The Gambia, West Africa (Wild et al. 1990). Exposure starts in the perinatal period, through *in utero* transfer and breast-feeding, and continues throughout life, mainly from consumption of peanuts. Time patterns of aflatoxin-albumin adduct levels correlate with the seasonal availability of peanuts. Exposure to aflatoxins is, in principle, largely preventable but so far effective measures have been difficult to implement on a large scale in poor, rural countries (Wild et al. 2000).

There is strong experimental evidence that aflatoxins are potent hepatocarcinogens in rodents. In humans, there are good ecological correlations between the risk of HCC and the presence of biomarkers of aflatoxin exposure in serum or in urine (Turner et al. 2002). The most significant carcinogenic aflatoxin is B1 (AFB1), which is the most abundant in the diet. AFB1 is metabolized in the liver by several CYP450 enzymes (mainly 1A2 and 3A4) to a reactive AFB1-8,9-*exo*-epoxide (Mace et al. 1997). This metabolite generates a primary DNA adduct (8-9, dihydro-8-(N7-guanyl)-9-hydroxyaflatoxin; AFB1-N7-Gua), naturally converted to two secondary lesions, an apurinic (AP) site and a stable, AFB1-formamidopyrimidine (AFB1-FAPY) adduct (Smela et al. 2001). The latter is considered as the most mutagenic lesion (Smela et al. 2002). Adducts do not form equally at all guanines in *TP53* and, experimentally, promutagenic lesions have been demonstrated at several positions throughout the coding sequence of the DNA-binding domain (Puisieux et al. 1991). The sequence context of codon 249 (AGGCC) represents a site of intermediate affinity for the formation of AFB1-induced lesions. Other codons in *TP53*, including some codons that are "hotspots" in many cancers (codon 245, 248 and 273), have a similar or even greater affinity for AFB1 than codon 249 (Denissenko et al. 1998a). Thus, the selectivity for 249^{ser} in HCC cannot be solely explained by the preferential formation of adducts at this position and other factors must play a role to select this particular mutation as the major carcinogenic one in liver cells exposed to aflatoxins.

There is evidence that imperfect DNA repair may increase the risk of mutagenesis and carcinogenesis induced by AFB1. Higher levels of AFB1-DNA adducts have been detected in the placenta of healthy women from Tawain carrying the Gln399 allele of *XRCC1*, an enzyme involved in base excision repair, causing slower repair and persistence of DNA adducts (Lunn et al. 1999). Slow excision repair due to the interference of the HBx antigen of HBV with the host's repair system has been proposed as a critical factor (Jia et al. 1999). However, a faster adduct removal rate by the DNA-repair system was observed at codon 249 than at other codons (Denissenko et al. 1998a). Together, these results suggests that deficient DNA repair does not explain the high prevalence of 249^{ser} in HCC, and that biological selection may play a role to facilitate the clonal expansion of cells carrying 249^{ser} during the development of HCC.

The functional basis for biological selection of 249^{ser} in liver cells remains unknown. Transfection studies have generated only limited evidence for a specific, pro-oncogenic effect of 249^{ser} in liver cancer cells (Ponchel et al. 1994). It should be noted that liver differs from many other tissues in its mechanisms of response to genotoxic stress. The liver cell has special functions in accumulating many endogenous and exogenous reactive compounds and is equipped with an extremely efficient detoxification machinery. Recent studies suggest that the role of *TP53* in controlling life-and-death responses after DNA damage may be less important in the liver than in many other tissues, and that TP63 or TP73 may also be involved in the normal response to genotoxic stress (Petitjean et al. 2005). Thus, it is possible that accumulation of 249^{ser} corresponds to the selection of a mutant with highly unusual biochemical properties. NMR studies have shown that 249^{ser} has a residual capacity to fold into the wild-type p53 conformation (Friedler et al. 2004), suggesting that it may partially retain some of the suppressive functions of p53. It is possible that the specific, biological advantage conferred by this mutant may correspond to the capacity to maintain chronically infected liver cells walking on the tight rope between life and death, thus enhancing their chances to acquire genetic alterations leading to cancer.

Tobacco smoke and lung cancer

Cigarette smoking causes over 1 million deaths per year worldwide and is involved in over 30% of all cancer deaths in developed countries (Parkin et al. 2001). The historical prospective study on British doctors, in which Doll and collaborators followed up a cohort of about 35000 subjects for 50 years, showed that among the men born around 1920, prolonged cigarette smoking from early adult life tripled the age-specific mortality rates, with lung cancer as the main cause of mortality by cancer (Doll et al., 2004). The mainstream smoke emerging from the mouthpiece of a cigarette is an aerosol containing about 10^{10} particles/mL and 4800 compounds, including nitrogen oxides, isoprene, butadiene, benzene, styrene, formaldehyde, acetaldehyde, acrolein, furan, polycyclic aromatic hydrocarbons (PAH), N-nitrosamines, aromatic amines, and metals (Pfeifer et al. 2002).

All major histological types of lung cancers accumulate genetic alterations including allelic losses (LOH) at multiple loci on chromosome 9p, mutations in oncogenes such as *KRAS* and tumor suppressor genes such as *TP53* and *p16/INK4a*, epigenetic gene silencing through promoter hypermethylation (*p16/INK4a, APC, RASSF1*) and aberrant expression of genes involved in the control of cell proliferation (*RB1*) or apoptosis (*Bcl2*) (for review see (Gazdar et al. 2004)). Although many of these genetic changes occur independently of histological type, their frequency and timing of occurrence with respect to cancer progression are different in small cell lung carcinomas (SCLC), that originate from epithelial cells with neuro-endocrine features, and non-small cell lung carcinomas (NSCLC), that originate from bronchial or alveolar epithelial cells. Furthermore, a number of genetic and epigenetic differences have been identified between squamous cell carcinoma (SCC), that arises from bronchial epithelial cells through a squamous metaplasia/dysplasia process, and adenocarcinoma (ADC), that derives from alveolar or bronchiolar epithelial cells. However, irrespective of histological type, all forms of lung cancers are strongly associated with tobacco smoke.

TP53 mutations occur at high prevalence in all types of invasive lung cancer lesions of smokers (50-80%). Since the early nineties, it is recognized that lung cancers of smokers had a higher proportion of transversions, particularly G:C to T:A in *TP53* than most other cancer types not directly related to tobacco smoking. However, the majority of the mutations described to date have been reported in smokers and the data on mutations in non-smokers are still limited in size and heterogeneous with respect to assessment of past or indirect exposure to tobacco smoke. Overall, analysis of the IARC database shows an excess of G:C to T:A transversions in smokers (30%) versus non-smokers (13%) (Hainaut et al. 2001; Pfeifer et al. 2002). There are, however, important differences between the two categories: most smokers in the database are men with squamous cell carcinomas, whereas many non-smokers are women with adenocarcinomas. This difference is likely to reflect different habits of tobacco use rather than intrinsic differences between genders in susceptibility to carcinogenesis by tobacco smoke. In a recent study, we have analysed 131 tumors of subjects with precise assessment of current, past and indirect exposures to tobacco (Le Calvez et al., submitted). The prevalence of *TP53* mutation in these cancers showed a clear, linear dose-response relationship with tobacco consumption, cancers of heavy smokers having over 10 times higher risk of containing a mutation than cancers of never-smokers. Whereas the prevalence of G:C to T:A transversions in smokers was the same as in the IARC database (29%), a much lower prevalence was found in never-smokers (4% versus 13% in the database). Former smokers showed an intermediate prevalence of these transversions. Thus, the difference between smokers and non-smokers is stronger than inferred from the database, most probably due to the misclassification of former-smokers into non-smokers in the database. Two other mutation types also showed a difference in relation with tobacco use: A:T to G:C transitions, which were more prevalent in ever- than in never-smokers, and G:C to A:T transitions, which were more prevalent in never- and former-smokers than in current-smokers. Both G to T and A to G mutations showed a bias towards occurence on the NTS, as predicted from experimental data that show preferential DNA repair of Benzo(a)pyrene diol epoxide (BPDE) adducts in the transcribed DNA strand of *TP53* (Denissenko et al. 1998b). Thus, the above observation supports the hypothesis that these mutations are caused by the formation of bulky carcinogen adducts.

There is a remarkable concordance between the distribution of mutations in lung cancers of smokers and the experimental spectrum of DNA-adducts induced in *TP53* by PAH metabolites. Using primary bronchial cells exposed *in vitro* to PAH diol-epoxides, Pfeifer and collaborators found preferential adduct formation at guanines located in codons 156, 157, 158, 245, 248 and 273 (Denissenko et al. 1996). These guanines are adjacent to frequently methylated cytosines within methylated CpG dinucleotides. The same dinucleotides, when present within CpG-methylated mutational reporter genes, is the preferential target of G to T transversions in cells exposed to benzo[a]pyrene-7,8-diol-9,10-epoxide, the model PAH compound (Yoon et al. 2001). The mutation may arise when DNA polymerase eta bypasses a template containing a (+)- *trans-anti-* benzo[*a*]pyrene-N^2-dG adduct (BPDE-N^2-dG), and predominantly incorporates an adenine (Zhang et al. 2002). These experimental "adduction hotspots" correspond almost perfectly to the positions of G:C to T:A transversions in lung cancers of smokers, except for codon 156. In this codon, the G:C to T:A transversion results in a silent substitution that leaves p53 protein function unaltered and is therefore not picked up by biological selection. It is interesting to note that codons 248 and 273 are common mutational hotspots for G:C to A:T transitions in cancers not related to tobacco smoke such as breast, colon and brain cancers (Hainaut and Pfeifer 2001).

Predisposing genetic factors ranging from moderate to low penetrance genetic polymorphisms influence inter-individual differences and may contribute to the *TP53* mutation pattern in lung cancer (Vineis et al. 1995). It is likely that, in heavy smokers, the exposure to tobacco mutagens is so high that it overcomes any modulatory effect of genetic susceptibility in carcinogen metabolism or in DNA repair efficiency. Nevertheless, genetic susceptibility is likely to play an important role in shaping the *TP53* mutation pattern, in particular in subjects with low, short-term, or indirect exposure to tobacco smoke. Data incriminating PAH as main *TP53* mutagens in lung cancers are further strengthened by recent studies on never-smokers residents of Xuan Wei County, China, exposed to high levels of PAH throughout life from the smoky coal they burn for cooking and heating, who show a high rate of lung cancers (He et al. 1991). Over 70% of the mutations found in these cancers are G:C to T:A transversions, often occurring at similar positions as those occurring in lung cancers of smokers (DeMarini et al. 2001).

CONCLUSION

Over the past 15 years, studies on *TP53* mutation patterns have contributed to create a conceptual bridge between chemical carcinogenesis, epidemiology and molecular epidemiology. The concept is now firmly established that *TP53* mutation patterns can help to identify causes of cancer. However, the three examples discussed in this chapter show that there is no single, unique way to detect and interpret mutation patterns. In these examples, *TP53* mutation data merely confirm conclusions firmly rooted into prior epidemiological and carcinogenic evidence. We should expect that the task of interpreting mutation patterns will become more difficult in cancers for which there are less clear epidemiological clues for identifying potential risk factors.

The availability of rapid, reproducible and sensitive methods for *TP53* mutation detection should facilitate large-scale studies with the appropriate design and statistical power to identify meaningful patterns. So far, many studies have been based on retrospective, consecutive clinical series with only limited information of sufficient quality on risk factors and exposures.

An ideal type of design for future studies is case-case comparisons, as exemplified by the recent study by Dai et al. (2004) on the inverse correlation between *TP53* mutations and infection by HPV16 in oral cancers (Dai et al. 2004). The question of a causal role of HPV16 in squamous oral cancers has been raised in many studies, but none of them had sufficient power to provide a convincing answer, in particular because other important risk factors such as tobacco and alcohol consumption are strong confounders in the interpretation of mutation patterns. By designing a study on a limited number of perfectly matched oral cancers cases differing only by the presence of HPV16, Dai et al. (2004) have been able to demonstrate an inverse relationship comparable to the one that has been well established in cervical cancer. This type of study design may be applicable to many other tumor types and risk factors.

One of the main conclusions from studies on *TP53* mutation patterns is that, in many cases, it may be possible to search for low levels of mutant DNA at specific positions to assess the risk of exposure in healthy individuals. For example, in heavy smokers, transversions at codon 157 has been found in non-pathological tissues using high-sensitivity methods (Hussain et al., 2001). In the close future, these methods may become applicable in the context of early cancer detection in high risk-population, thus helping to set up targeted prevention strategies. Beyond clinical applications for treatment, this may prove, in the future, the main contribution of *TP53* research to decreasing the burden of death by cancer worldwide.

ACKNOWLEDGMENTS

The authors thank the French "Ligue contre le Cancer" for support in a form of a fellowship to H.S. F.L. is supported by a Special Training Award of IARC. Work on the IARC *TP53* mutation database is supported by funds from NIEHS, USA, and by the EU program "mutant p53".

REFERENCES

Ambs S., Bennett W.P., Merriam W.G., Ogunfusika M.O., Oser S.M., Harrington A.M., Shields P.G., Felley-Bosco E., Hussain S.P., Harris C.C. Relationship between p53 mutations and inducible nitric oxide synthase expression in human colorectal cancer.J Natl Cancer Inst. 1999; 91:86-88.

Antonarakis S.E., Krawczak M., Cooper D.N. The nature and mechanisms of human gene mutations.In: The genetic basis of human cancer. 2002; Vogelstein, B and Kinzler, W, eds, 2nd edition:7-41.

Ateenyi-Agaba C., Dai M., le Calvez F., Katongole-Mbidde E., Smet A., Tommasino M., Franceschi S., Hainaut P., Weiderpass E. *TP53* mutations in squamous-cell carcinomas of the conjunctiva: evidence for UV-induced mutagenesis.Mutagenesis. 2004a; 19:399-401.

Ateenyi-Agaba C., Weiderpass E., Smet A., Dong W., Dai M., Kahwa B., Wabinga H., Katongole-Mbidde E., Franceschi S., Tommasino M. Epidermodysplasia verruciformis human papillomavirus types and carcinoma of the conjunctiva: a pilot study.Br J Cancer. 2004b; 90:1777-1779.

Bergamaschi D., Gasco M., Hiller L., Sullivan A., Syed N., Trigiante G., Yulug I., Merlano M., Numico G., Comino A., Attard M., Reelfs O., Gusterson B., Bell A.K., Heath V., Tavassoli M., Farrell P.J., Smith P., Lu X., Crook T. p53 polymorphism influences response in cancer chemotherapy via modulation of p73-dependent apoptosis.Cancer Cell. 2003; 3:387-402.

Bregman D.B., Pestell R.G., Kidd V.J. Cell cycle regulation and RNA polymerase II.Front Biosci. 2000; 5:D244-D257.

Burchell B., Coughtrie M.W. Genetic and environmental factors associated with variation of human xenobiotic glucuronidation and sulfation.Environ Health Perspect. 1997; 105 Suppl 4:739-747.

Chen C.J., Yu M.W., Liaw Y.F. Epidemiological characteristics and risk factors of hepatocellular carcinoma.J Gastroenterol Hepatol. 1997; 12:S294-S308.

Cho Y., Gorina S., Jeffrey P.D., Pavletich N.P. Crystal structure of a p53 tumor suppressor-DNA complex: understanding tumorigenic mutations [see comments].Science. 1994; 265:346-355.

Cooper G.M., Brudno M., Stone E.A., Dubchak I., Batzoglou S., Sidow A. Characterization of evolutionary rates and constraints in three Mammalian genomes.Genome Res. 2004; 14:539-548.

Dai M., Clifford G.M., le Calvez F., Castellsague X., Snijders P.J., Pawlita M., Herrero R., Hainaut P., Franceschi S. Human papillomavirus type 16 and *TP53* mutation in oral cancer: matched analysis of the IARC multicenter study.Cancer Res. 2004; 64:468-471.

DeMarini D.M., Landi S., Tian D., Hanley N.M., Li X., Hu F., Roop B.C., Mass M.J., Keohavong P., Gao W., Olivier M., Hainaut P., Mumford J.L. Lung tumor KRAS and *TP53* mutations in nonsmokers reflect exposure to PAH-rich coal combustion emissions.Cancer Res. 2001; 61:6679-6681.

Denissenko M.F., Koudriakova T.B., Smith L., O'Connor T.R., Riggs A.D., Pfeifer G.P. The p53 codon 249 mutational hotspot in hepatocellular carcinoma is not related to selective formation or persistence of aflatoxin B1 adducts.Oncogene. 1998a; 17:3007-3014.

Denissenko M.F., Pao A., Pfeifer G.P., Tang M. Slow repair of bulky DNA adducts along the nontranscribed strand of the human p53 gene may explain the strand bias of transversion mutations in cancers.Oncogene. 1998b; 16:1241-1247.

Denissenko M.F., Pao A., Tang M., Pfeifer G.P. Preferential formation of benzo[a]pyrene adducts at lung cancer mutational hotspots in P53.Science. 1996; 274:430-432.

Dittmer D., Pati S., Zambetti G., Chu S., Teresky A.K., Moore M., Finlay C., Levine A.J. Gain of function mutations in p53.Nat Genet. 1993; 4:42-46.

Dogliotti E., Hainaut P., Hernandez T., D'Errico M., DeMarini D.M. Mutation spectra resulting from carcinogenic exposure: from model systems to cancer-related genes.Recent Results Cancer Res. 1998; 154:97-124.

Doll R., Peto R., Boreham J., Sutherland I. Mortality in relation to smoking: 50 years' observations on male British doctors.BMJ. 2004; 328:1507-1519

Dumaz N., Brougard C., Sarasin A., Daya-Grosjean L. Specific UV-induced mutation spectrum in the p53 gene of skin tumors from DNA-repair-defiicent xeroderma pigmentosum patients.Proc Natl Acad Sci U S A. 1993; 90:10529-10533.

Essigmann J.M., Wood M.L. The relationship between the chemical structures and mutagenic specificities of the DNA lesions formed by chemical and physical mutagens.Toxicol Lett. 1993; 67:29-39.

Flaman J.M., Frebourg T., Moreau V., Charbonnier F., Martin C., Chappuis P., Sappino A.P., Limacher I.M., Bron L., Benhattar J. A simple p53 functional assay for screening cell lines, blood, and tumors.Proc Natl Acad Sci U S A. 1995; 92:3963-3967.

Flaman J.M., Frebourg T., Moreau V., Charbonnier F., Martin C., Ishioka C., Friend S.H., Iggo R. A rapid PCR fidelity assay.Nucleic Acids Res. 1994; 22:3259-3260.

Friedler A., DeDecker B.S., Freund S.M., Blair C., Rudiger S., Fersht A.R. Structural distortion of p53 by the mutation R249S and its rescue by a designed peptide: implications for "mutant conformation".J Mol Biol. 2004; 336:187-196.

Gazdar A., Franklin W.A., Brambilla E., Hainaut P., Yokota J., Harris C.C. Genetic and molecular alterations.in : Pathology and genetics: Tumours of the lung, pleura, thymus and heart, Travis W D , Brambilla, E , Müller-Hermelink, K and Harris, C C , eds. 2004; IARC Press, publisher:

Greenblatt M.S., Bennett W.P., Hollstein M., Harris C.C. Mutations in the p53 tumor suppressor gene: clues to cancer etiology and molecular pathogenesis.Cancer Res. 1994; 54:4855-4878.

Greenblatt M.S., Grollman A.P., Harris C.C. Deletions and insertions in the p53 tumor suppressor gene in human cancers: confirmation of the DNA polymerase slippage/misalignment model.Cancer Res. 1996; 56:2130-2136.

Guengerich F.P. Metabolism of chemical carcinogens.Carcinogenesis. 2000; 21:345-351.

Guengerich F.P. Cytochrome P450 oxidations in the generation of reactive electrophiles: epoxidation and related reactions.Arch Biochem Biophys. 2003; 409:59-71.

Hainaut P., Hollstein M. p53 and human cancer: the first ten thousand mutations.Adv Cancer Res. 2000; 77:82-137-

Hainaut P., Pfeifer G.P. Patterns of p53 G-->T transversions in lung cancers reflect the primary mutagenic signature of DNA-damage by tobacco smoke.Carcinogenesis. 2001; 22:367-374.

Hanahan D., Weinberg R.A. The hallmarks of cancer.Cell. 2000; 100:57-70.

Hanawalt P.C., Ford J.M., Lloyd D.R. Functional characterization of global genomic DNA repair and its implications for cancer.Mutat Res. 2003; 544:107-114.

Harris C.C. The 1995 Walter Hubert Lecture--molecular epidemiology of human cancer: insights from the mutational analysis of the p53 tumour-suppressor gene.Br J Cancer. 1996; 73:261-269.

He X.Z., Chen W., Liu Z.Y., Chapman R.S. An epidemiological study of lung cancer in Xuan Wei County, China: current progress. Case-control study on lung cancer and cooking fuel.Environ Health Perspect. 1991; 94:9-13.

Hemminki K., Thilly W.G. Implications of results of molecular epidemiology on DNA adducts, their repair and mutations for mechanisms of human cancer.IARC Sci Publ. 2004; 217-235.

Hernandez-Boussard T., Montesano R., Hainaut P. Sources of bias in the detection and reporting of p53 mutations in human cancer: analysis of the IARC p53 mutation database.Genet Anal. 1999; 14:229-233.

Hoeijmakers J.H. Genome maintenance mechanisms for preventing cancer.Nature. 2001; 411:366-374.

Holliday R., Grigg G.W. DNA methylation and mutation.Mutat Res. 1993; 285:61-67.

Hollstein M., Hergenhahn M., Yang Q., Bartsch H., Wang Z.Q., Hainaut P. New approaches to understanding p53 gene tumor mutation spectra [see comments].Mutat Res. 1999; 431:199-209.

Hollstein M., Sidransky D., Vogelstein B., Harris C.C. p53 mutations in human cancers.Science. 1991; 253:49-53.

Holmquist G.P., Gao S. Somatic mutation theory, DNA repair rates, and the molecular epidemiology of p53 mutations.Mutat Res. 1997; 386:69-101.

Hussain S.P., Amstad P., Raja K., Sawyer M., Hofseth L., Shields P.G., Hewer A., Phillips D.H., Ryberg D., Haugen A., Harris C.C. Mutability of p53 hotspot codons to benzo(a)pyrene diol epoxide (BPDE) and the frequency of p53 mutations in nontumorous human lung.Cancer Res. 2001; 61:6350-6355.

Hussain S.P., Hofseth L.J., Harris C.C. Radical causes of cancer.Nat Rev Cancer. 2003; 3:276-285.

Inga A., Cresta S., Monti P., Aprile A., Scott G., Abbondandolo A., Iggo R., Fronza G. Simple identification of dominant p53 mutants by a yeast functional assay.Carcinogenesis. 1997; 18:2019-2021.

Inga A., Scott G., Monti P., Aprile A., Abbondandolo A., Burns P.A., Fronza G. Ultraviolet-light induced p53 mutational spectrum in yeast is indistinguishable from p53 mutations in human skin cancer.Carcinogenesis. 1998; 19:741-746.

Jia L., Wang X.W., Harris C.C. Hepatitis B virus X protein inhibits nucleotide excision repair.Int J Cancer. 1999; 80:875-879.

Jonason A.S., Kunala S., Price G.J., Restifo R.J., Spinelli H.M., Persing J.A., Leffell D.J., Tarone R.E., Brash D.E. Frequent clones of p53-mutated keratinocytes in normal human skin [see comments].Proc Natl Acad Sci U S A. 1996; 93:14025-14029.

Jones P.A., Rideout W.M., Shen J.C., Spruck C.H., Tsai Y.C. Methylation, mutation and cancer.Bioessays. 1992; 14:33-36.

Kao J.H., Chen D.S. Global control of hepatitis B virus infection.Lancet Infect Dis. 2002; 2:395-403.

Kato S., Han S.Y., Liu W., Otsuka K., Shibata H., Kanamaru R., Ishioka C. Understanding the function-structure and function-mutation relationships of p53 tumor suppressor protein by high-resolution missense mutation analysis.Proc Natl Acad Sci U S A. 2003; 100:8424-8429.

Kumar S., Subramanian S. Mutation rates in mammalian genomes.Proc Natl Acad Sci U S A. 2002; 99:803-808.

Kunkel T.A. DNA replication fidelity.J Biol Chem. 2004; 279:16895-16898.

Lee D.H., Pfeifer G.P. Deamination of 5-methylcytosines within cyclobutane pyrimidine dimers is an important component of UVB mutagenesis.J Biol Chem. 2003; 278:10314-10321.

Lin J., Chen J., Elenbaas B., Levine A.J. Several hydrophobic amino acids in the p53 amino-terminal domain are required for transcriptional activation, binding to mdm-2 and the adenovirus 5 E1B 55-kD protein.Genes Dev. 1994; 8:1235-1246.

Livneh Z. DNA damage control by novel DNA polymerases: translesion replication and mutagenesis.J Biol Chem. 2001; 276:25639-25642.

Loeb L.A. Cancer cells exhibit a mutator phenotype.Adv Cancer Res. 1998; 72:25-56.

Lunn R.M., Langlois R.G., Hsieh L.L., Thompson C.L., Bell D.A. XRCC1 polymorphisms: effects on aflatoxin B1-DNA adducts and glycophorin A variant frequency.Cancer Res. 1999; 59:2557-2561.

Mace K., Aguilar F., Wang J.S., Vautravers P., Gomez-Lechon M., Gonzalez F.J., Groopman J., Harris C.C., Pfeifer A.M. Aflatoxin B1-induced DNA adduct formation and p53 mutations in CYP450-expressing human liver cell lines.Carcinogenesis. 1997; 18:1291-1297.

Mandard A.M., Hainaut P., Hollstein M. Genetic steps in the development of squamous cell carcinoma of the esophagus.Mutat Res. 2000; 462:335-342.

Martin A.C., Facchiano A.M., Cuff A.L., Hernandez-Boussard T., Olivier M., Hainaut P., Thornton J.M. Integrating mutation data and structural analysis of the *TP53* tumor-suppressor protein.Hum Mutat. 2002; 19:149-164.

Miller J.A. Recent studies on the metabolic activation of chemical carcinogens.Cancer Res. 1994; 54:1879s-1881s.

Montesano R., Hainaut P., Hall J. The use of biomarkers to study pathogenesis and mechanisms of cancer: oesophagus and skin cancer as models.IARC Sci Publ. 1997; 291-301.

Olivier M., Eeles R., Hollstein M., Khan M.A., Harris C.C., Hainaut P. The IARC *TP53* database: new online mutation analysis and recommendations to users.Hum Mutat. 2002; 19:607-614.

Olivier M., Hussain S.P., Caron d.F., Hainaut P., Harris C.C. *TP53* mutation spectra and load: a tool for generating hypotheses on the etiology of cancer.IARC Sci Publ. 2004; 247-270.

Ory K., Legros Y., Auguin C., Soussi T. Analysis of the most representative tumour-derived p53 mutants reveals that changes in protein conformation are not correlated with loss of transactivation or inhibition of cell proliferation.EMBO J. 1994; 13:3496-3504.

Ozturk M. Genetic aspects of hepatocellular carcinogenesis.Semin Liver Dis. 1999; 19:235-242.

Parkin D.M., Bray F., Ferlay J., Pisani P. Estimating the world cancer burden: Globocan 2000.Int J Cancer. 2001; 94:153-156.

Petitjean A., Cavard C., Shi H., Tribollet V., Hainaut P., Caron de Fromentel C. The expression of TA and DNp63 are regulated by different mechanisms in liver cells. 2005; in press:

Pfeifer G.P., Denissenko M.F., Olivier M., Tretyakova N., Hecht S.S., Hainaut P. Tobacco smoke carcinogens, DNA damage and p53 mutations in smoking-associated cancers.Oncogene. 2002; 21:7435-7451.

Pluquet O., Hainaut P. Genotoxic and non-genotoxic pathways of p53 induction.Cancer Lett. 2001; 174:1-15.

Ponchel F., Puisieux A., Tabone E., Michot J.P., Froschl G., Morel A.P., Frebourg T., Fontaniere B., Oberhammer F., Ozturk M. Hepatocarcinoma-specific mutant p53-249ser induces mitotic activity but has no effect on transforming growth factor beta 1-mediated apoptosis.Cancer Res. 1994; 54:2064-2068.

Puisieux A., Lim S., Groopman J., Ozturk M. Selective targeting of p53 gene mutational hotspots in human cancers by etiologically defined carcinogens.Cancer Res. 1991; 51:6185-6189.

Ren Z.P., Ahmadian A., Ponten F., Nister M., Berg C., Lundeberg J., Uhlen M., Ponten J. Benign clonal keratinocyte patches with p53 mutations show no genetic link to synchronous squamous cell precancer or cancer in human skin.Am J Pathol. 1997; 150:1791-1803.

Ren Z.P., Hedrum A., Ponten F., Nister M., Ahmadian A., Lundeberg J., Uhlen M., Ponten J. Human epidermal cancer and accompanying precursors have identical p53 mutations different from p53 mutations in adjacent areas of clonally expanded non-neoplastic keratinocytes.Oncogene. 1996a; 12:765-773.

Ren Z.P., Ponten F., Nister M., Ponten J. Two distinct p53 immunohistochemical patterns in human squamous-cell skin cancer, precursors and normal epidermis.Int J Cancer. 1996b; 69:174-179.

Rosenberg M.S., Subramanian S., Kumar S. Patterns of transitional mutation biases within and among mammalian genomes.Mol Biol Evol. 2003; 20:988-993.

Ross R.K., Yuan J.M., Yu M.C., Wogan G.N., Qian G.S., Tu J.T., Groopman J.D., Gao Y.T., Henderson B.E. Urinary aflatoxin biomarkers and risk of hepatocellular carcinoma.Lancet. 1992; 339:943-946.

Schmutte C., Yang A.S., Nguyen T.T., Beart R.W., Jones P.A. Mechanisms for the involvement of DNA methylation in colon carcinogenesis.Cancer Res. 1996; 56:2375-2381.

Smela M.E., Currier S.S., Bailey E.A., Essigmann J.M. The chemistry and biology of aflatoxin B(1): from mutational spectrometry to carcinogenesis.Carcinogenesis. 2001; 22:535-545.

Smela M.E., Hamm M.L., Henderson P.T., Harris C.M., Harris T.M., Essigmann J.M. The aflatoxin B(1) formamidopyrimidine adduct plays a major role in causing the types of mutations observed in human hepatocellular carcinoma.Proc Natl Acad Sci U S A. 2002; 99:6655-6660.

Stenson P.D., Ball E.V., Mort M., Phillips A.D., Shiel J.A., Thomas N.S., Abeysinghe S., Krawczak M., Cooper D.N. Human Gene Mutation Database (HGMD): 2003 update.Hum Mutat. 2003; 21:577-581.

Strano S., Fontemaggi G., Costanzo A., Rizzo M.G., Monti O., Baccarini A., Del Sal G., Levrero M., Sacchi A., Oren M., Blandino G. Physical interaction with human tumor-derived p53 mutants inhibits p63 activities.J Biol Chem. 2002; 277:18817-18826.

Strano S., Munarriz E., Rossi M., Cristofanelli B., Shaul Y., Castagnoli L., Levine A.J., Sacchi A., Cesareni G., Oren M., Blandino G. Physical and functional interaction between p53 mutants and different isoforms of p73.J Biol Chem. 2000; 275:29503-29512.

Strauss B.S. Silent and multiple mutations in p53 and the question of the hypermutability of tumors [published erratum appears in Carcinogenesis 1998 Jan;19(1):237].Carcinogenesis. 1997; 18:1445-1452.

Tornaletti S., Pfeifer G.P. Slow repair of pyrimidine dimers at p53 mutation hotspots in skin cancer [see comments].Science. 1994; 263:1436-1438.

Tornaletti S., Pfeifer G.P. Complete and tissue-independent methylation of CpG sites in the p53 gene: implications for mutations in human cancers.Oncogene. 1995; 10:1493-1499.

Tornaletti S., Rozek D., Pfeifer G.P. The distribution of UV photoproducts along the human p53 gene and its relation to mutations in skin cancer [published erratum appears in Oncogene 1993 Dec;8(12):3469].Oncogene. 1993; 8:2051-2057.

Turner P.C., Sylla A., Diallo M.S., Castegnaro J.J., Hall A.J., Wild C.P. The role of aflatoxins and hepatitis viruses in the etiopathogenesis of hepatocellular carcinoma: A basis for primary prevention in Guinea-Conakry, West Africa.J Gastroenterol Hepatol. 2002; 17 Suppl:S441-S448.

Vineis P., Caporaso N. Tobacco and cancer: epidemiology and the laboratory.Environ Health Perspect. 1995; 103:156-160.

Vrieling H., van Zeeland A.A., Mullenders L.H. Transcription coupled repair and its impact on mutagenesis.Mutat Res. 1998; 400:135-142.

Waridel F., Estreicher A., Bron L., Flaman J.M., Fontolliet C., Monnier P., Frebourg T., Iggo R. Field cancerisation and polyclonal p53 mutation in the upper aero- digestive tract.Oncogene. 1997; 14:163-169.

Wild C.P., Hall A.J. Primary prevention of hepatocellular carcinoma in developing countries.Mutat Res. 2000; 462:381-393.

Wild C.P., Jiang Y.Z., Allen S.J., Jansen L.A., Hall A.J., Montesano R. Aflatoxin-albumin adducts in human sera from different regions of the world.Carcinogenesis. 1990; 11:2271-2274.

Wogan G.N., Hecht S.S., Felton J.S., Conney A.H., Loeb L.A. Environmental and chemical carcinogenesis.Semin Cancer Biol. 2004; 14:473-486.

Yoon J.H., Smith L.E., Feng Z., Tang M., Lee C.S., Pfeifer G.P. Methylated CpG dinucleotides are the preferential targets for G-to-T transversion mutations induced by benzo[a]pyrene diol epoxide in mammalian cells: similarities with the p53 mutation spectrum in smoking-associated lung cancers.Cancer Res. 2001; 61:7110-7117.

Zhang Y., Wu X., Guo D., Rechkoblit O., Geacintov N.E., Wang Z. Two-step error-prone bypass of the (+)- and (-)-trans-anti-BPDE-N2-dG adducts by human DNA polymerases eta and kappa.Mutat Res. 2002; 510:23-35.

Ziegler A., Jonason A.S., Leffell D.J., Simon J.A., Sharma H.W., Kimmelman J., Remington L., Jacks T., Brash D.E. Sunburn and p53 in the onset of skin cancer [see comments].Nature. 1994; 372:773-776.

Chapter 14

PROGNOSTIC AND PREDICTIVE VALUE OF *TP53* MUTATIONS IN HUMAN CANCER

Magali Olivier*, Pierre Hainaut* and Anne-Lise Børresen-Dale[&]
* *International Agency for Research on Cancer (CIRC/IARC), Lyon, France,* [&] *Department of Genetics, The Norwegian Radium Hospital, Oslo, Norway.*

INTRODUCTION

Finding reliable molecular markers for early diagnosis, prognosis and prediction of response to treatment is a major challenge for cancer management. A marker of prognosis provides information on the risk of relapse and death independently of treatment, whereas a predicitve marker provides information on the potential benefit of a specific treatment (Lonning, 2003). An early diagnostic marker helps to identify lesions at high risk of malignant transformation. Clinical stage, tumor size and morphological grade are the most reliable factors of prognosis. Among numerous molecular markers that have been tested most recently, only a few are used in clinical practice. In breast cancer for example, estrogen and progesterone receptors are used routinely as predictive markers for tumor response to anti-hormone therapy. However, about 30% of patients with positive receptor status (expected to benefit from anti-hormone treatment) will face a therapeutic failure, showing the limitations of these markers.

The tumor suppressor gene *TP53* plays a key role in many cellular pathways controlling cell proliferation, cell survival and genomic integrity (see other Chapters). It acts in response to various forms of cellular stresses to mediate antiproliferative processes. Disrupting its function promotes checkpoints defects, genomic instability and inappropriate survival, leading to the uncontrolled proliferation of damaged cells. The proliferative

P. Hainaut and K.G. Wiman (eds.), 25 Years of p53 Research, 321-338.
© 2005 *Springer. Printed in the Netherlands.*

advantage given by its inactivation and the fact that it is ubiquituously expressed explain why it is frequently found mutated in almost every type of cancers (Hainaut and Hollstein, 2000). In addition to its tumor suppressor function, *TP53* also contributes to the anti-neoplastic effects of radio- and chemotherapeutic agents. It has been shown, in various experimental *in vitro* systems as well as in mouse models, that cell cycle arrest or apoptosis induced by radiotherapy and various chemotherapeutic drugs depend on an intact *TP53* pathway (Lowe et al., 1994; O'Connor et al., 1997). These results raise the hypothesis that *TP53* could be a key player in defining tumor-sensitivity to a broad range of anti-cancer treatments in cancer patients. Moreover, the presence of a *TP53* mutation could be one of the underlying causes of drug resistance, which is the major cause of treatment failure and cancer death.

TP53 may thus be a potential marker for malignant transformation, tumor aggressiveness and treatment outcome in a broad range of cancers. Many studies have investigated, in various clinical settings, the predictive value of *TP53* mutation status for tumor response to treatment and patient outcome. Despite these efforts, no consensus has been reached and *TP53* mutation analysis is not yet used in clinical practice. In this chapter we review the pro and con of the use of *TP53* as a biomarker and propose which area of oncology could benefit from its use. We will also discuss study design and methodology issues for the development of a biomarker such as *TP53* and its implementation in clinical settings.

TP53 MUTATION STATUS AND CLINICAL OUTCOME

Although more than 500 studies have investigated the value of *TP53* status as a prognostic and/or predictive marker in various types of cancer, results have often been contradictory. Several reasons can explain this apparent confusion, from differences in study design, heterogeneity in the cohorts to methodology used to assess *TP53* status. Studies that have used gene sequencing to assess *TP53* status, have used different pre-screening methods (none, SSCP, DGGE, CDGE, TTGE, TGGE or IHC), that have different sensitivity. Moreover, a number of studies have only analyzed the central part of the protein. Although this region contains 80% of the mutations, this restricted analysis can lead to the mis-classification of 10 to 20% of the cases. However, the main reason certainly resides in the fact that the majority of studies (about 400/500) have used immuno-histochemistry (IHC) to assess *TP53* status. Earlier observations have shown that the majority of *TP53* mutations are missense mutations that accumulate in cancer cells and thus can be detected by IHC. However, it is now admitted that IHC is not suitable as a screening method for mutations since not all types of mutations are detected. Sequencing of the complete coding sequence of *TP53* shows that 10 to 25% of the mutations are truncating mutations (nonsense, frameshift or splice site mutations) that are not detected by IHC since they do not lead to a stable protein. Moreover, some cases of IHC positive cells do not carry a mutation but may result from the accumulation of the wild-type protein in response cellular stress signals. Finally, IHC studies have used different antibodies, different labeling procedures and different cut-off value for positive cases. Hence, the use of IHC leads to an unacceptable number of mis-classified cases and to a greater inter-study variability.

When only studies that have used gene sequencing to assess *TP53* mutation status are taken into account, *TP53* appears to be of prognostic value in a variety of cancers. A comprehensive list of such studies is provided on the IARC *TP53* database web site (http://www.iarc.fr/p53/). This information is summarized in Table 1, where the number of studies reporting association or lack of association between the presence of a mutation and either poor or good prognosis (patient survival and/or tumor response to treatment) is indicated. This summary table shows that association with poor prognosis has repeatedly been reported for breast, bladder, head and neck and hematological cancers. Results for colorectal, lung and esophageal cancers are more heterogeneous, but a majority of studies found an association with poor prognosis. For ovarian cancer, results remain contradictory. In brain tumors, the majority of studies show a lack of association with prognosis and two studies report an association with good prognosis. Overall, 65/93 studies found that *TP53* is a statistically significant factor of poor prognosis in various cancers. It should be noted that the majority of studies (20/27) reporting no association between *TP53* status and survival have been done on cohorts of less than 100 patients, which give insufficient statistical power to detect moderate differences in survival. Response to adjuvant chemotherapy or radiotherapy is a major determinant of patient outcome. Among 19 studies that have specifically investigated the association between tumor response to treatment and *TP53* mutation status, 14 have found that the presence of a mutation was associated with a poor response to various chemotherapy or radiotherapy regimens in breast, head and neck, hematological, colorectal, ovarian, esophageal cancers and soft tissue sarcomas. These observations are in agreement with experimental data showing a key role for *TP53* in the anti-proliferative response induced by various chemotherapeutic agents. Interestingly, one study on ovarian cancer patients showed that *TP53* status was predictive of response to treatment in patients treated with cyclophosphamide and cisplatin but not in patients treated with a paclitaxel/cisplatin regimen (Smith-Sorensen et al., 1998).

Table 1. TP53 mutation and cancer prognosis

TUMOR SITE	*Studies reporting that the presence of TP53 mutation is:		
	Significantly associated with bad prognosis	Significantly associated with good prognosis	No significant association
BLADDER	3	-	-
BRAIN	1	2	5
BREAST	17	-	3
COLORECTUM	11	-	5
ESOPHAGUS	3	1	1
HEAD&NECK	7	-	-
HEMATOL.	6	-	-
LIVER	1	-	-
LUNG	8	-	4
LYMPH NODES	1	-	-
OVARY	3	1	4
PANCREAS	1	1	-
PROSTATE	-	-	1
RENAL PELVIS	1	-	-
SINUSES	1	-	-
SOFT TISSUES	1	-	-
Total	65	5	23

*The number of studies is indicated. Data from the IARC *TP53* Database (R9, July 2004), which includes only published studies where *TP53* mutation has been analyzed by gene sequencing. Only studies with cohorts of more than 30 patients have been included in this table. The prognostic parameters investigated are patient survival and/or tumor response to treatment.

In a study on 63 advanced breast cancer patients treated with doxorubicin in a neo-adjuvant setting, a strong correlation between lack of response and presence of *TP53* mutation was observed. The same was seen for 35 breast cancer patients treated with FUMI (5 fluorouracil and Mitomycin C) in a neo-adjuvant setting (Geisler et al., 2003). These observations are in agreement with the fact that DNA-damaging agents have been shown to induce p53-dependent apoptosis whereas paclitaxel (a microtubule stabilizing agent) effects are expected to be independent of p53 function. In 2000, Berns *et al* reported that patients with *TP53* mutation showed the lowest response to tamoxifen on a series of 243 breast cancer patients. This effect was observed in ER positive patients only, suggesting that ER-dependent response to anti-hormone therapy may also depend on an intact *TP53* pathway (Berns et al., 2000).

It is not clear from the available studies whether the prognostic value of *TP53* for the overall survival of patient depends on the administration of adjuvant treatment, or if it has also a value for patients receiving only surgery. In one study of colorectal tumors where patients treated with surgery alone were included, the presence of mutations in specific regions was correlated with a shorter survival (Borresen-Dale et al., 1998). In another study on colorectal cancer, the survival of patients treated with surgery only was compared with patients receiving adjuvant chemotherapy in addition to surgery. It showed that survival was strongly correlated with the presence of *TP53* mutations in the entire cohort. However, when only patients undergoing a radical resection were considered, *TP53* mutation status was no longer of prognostic significance (Tortola et al., 1999). In esophageal cancer, *TP53* alterations (*TP53* mutation plus positive immunostaining) but not *TP53* mutation alone, was found to be significantly associated with shorter overall and disease-free survival in 91 patients treated by surgery only (Casson et al., 2003). Although these results remain to be confirmed and extended, they suggest that *TP53* mutation has a weak prognostic value, if any, in patients treated with surgery only. The capacity of *TP53* to mediate tumor response to chemotherapeutic drugs may thus be the main mechanism explaining its prognostic value.

TP53 MUTATION TYPE AND CLINICAL OUTCOME

There is now much *in vitro* experimental evidence showing that different types of *TP53* mutations have different functional consequences. Unlike other tumor suppressor genes that are inactivated by insertions or deletions leading to an absence of protein expression, most *TP53* mutations are missense mutations that lead to the over-expression of a mutant protein. More than 1800 different missense mutations have been reported in human cancer and functional assays have shown that mutant proteins show a great variability in their functional activities (see *TP53* Function Database, http://www.iarc.fr/P53). WT p53 function relies mainly on the capacity to transactivate target genes through binding to specific response elements. Loss of function (LOF) is the main consequences of missense mutations, however some mutants also exert dominant-negative effects (DNE) or show gain of function (GOF) properties. DNE corresponds to the capacity of the mutant protein to complex with the product of the remaining wild-type allele to inactivate its function. DNE results in the total abrogation of p53 protein function, even if there is still a wild-type protein expressed in the cell (Milner et al., 1991). GOF corresponds to the acquisition of novel properties by mutant p53 that do not depend upon the presence of wild-type p53

(Cadwell and Zambetti, 2001). All hotspot mutations known so far lead to a loss of specific trans-activation capacity, but the degree of LOF vary between mutants. Missense mutations outside the central DNA-binding domain more often retain transcriptional activity on a variety of promoters than mutations within the DNA-binding domain (Kato et al., 2003; Resnick and Inga, 2003). In addition to various degree of LOF, some mutant proteins exert various degrees of DNE (see *TP53* Function Database, http://www.iarc.fr/P53). It has been proposed more than 10 years ago that mutant p53 may exert pro-oncogenic effects, and that mutation was turning p53 into some kind of oncogene (Lane and Benchimol, 1990; Oren, 1992). There is now good evidence that mutant p53 can promote cancer through a GOF mechanism, such as promotion of gene amplification, or resistance to drug-induced apoptosis (reviewed in (Sigal and Rotter, 2000)). Several mutants have been shown to transactivate or potentiate the transactivation of genes such as *MDR1*, (Dittmer et al., 1993), *EGFR, c-MYC, PCNA, IGF-II* or *VEGF* (see *TP53* Function Database, http://www.iarc.fr/P53). These genes are not transactivated by the wild-type p53 protein and do not necessary possess a p53 binding-site. Mutant p53 proteins can also interact with a network of proteins that differ from wt p53. For example, some mutant p53 can form stable complexes with the products of other members of the *TP53* gene family, p63 and p73, blocking their transactivation capacity (see other chapters).

These observations suggest that different mutations may have different biological consequences *in vivo* and have led several investigators to explore whether tumor progression and tumor response to therapy may depend on the nature and localization of *TP53* mutations. In 1995, a study on colorectal cancer and another on breast cancer, have found that mutations affecting regions involved in zinc binding were of worse prognosis than others in term of survival (Borresen et al., 1995; Goh et al., 1995). Table 2 gives a summary of all studies that have found an association between the presence of specific *TP53* mutations and poor prognosis. It shows that, in various cancers, mutations affecting residues involved in zinc binding and DNA contacts (L2 and L3 loops in the DNA-binding domain) are associated with a worse prognosis than others. In breast cancer for example, it has been shown that mutations disrupting the zinc binding domain were associated with primary resistance to doxorubicin and were predicitive of an early relapse (Aas et al., 1996; Geisler et al., 2001). Similar findings were observed in another small cohort of advanced breast cancer patients treated with FUMI (5-fluorouracil and mitomycin-C) (Geisler et al., 2003). These mutations have also been found to be associated with a poor response to tamoxifen in a cohort of breast cancer patients (Berns et al., 2000), and with a shorter survival in lung and head and neck cancers (see Table 2). Functional assessments of the most common missense mutations falling within the L2-L3 loops show loss of transactivation activity towards most p53-target genes, resulting in defects in p53-dependent responses such as cell-cycle arrest or apoptosis (Aurelio et al., 2000; Ory et al., 1994). These properties fit well with the observations in cancer patients and are in agreement with a major role for *TP53* in the anti-proliferative response induced by radio- and chemotherapeutic agents.

Table 2. Specific *TP53* mutations associated with poor prognosis in cancer.

Tumor site	Country	Mutation frequency	Region associated with poor prognosis	References
BREAST	Australia	178/1037 (17%)	Exon 4	(Powell et al., 2000)
BREAST	Austria	42/205 (20%)	L2/L3 loops	(Kucera et al., 1999)
BREAST	Brazil	33/242 (14%)	DNA/Zn binding	(Nagai et al., 2003)
BREAST	Denmark	74/315 (23%)	DNA/Zn binding	(Alsner et al., 2000)
BREAST	Japan	30/76 (40%)	DNA contact	(Takahashi et al., 2000)
BREAST	Norway	?/119 (?%)	L2/L3 loops	(Borresen et al., 1995)
BREAST	Norway	26/90 (29%)	L2/L3 loops	(Aas et al., 1996; Geisler et al., 2001)
BREAST	Norway	18/35 (51%)	L2/L3 loops	(Geisler et al., 2003)
BREAST	Sweden	69/315 (22%)	Conserved regions II and V	(Bergh et al., 1995)
BREAST	Sweden	21/123 (17%)	Zn binding	(Gentile et al., 1999)
BREAST	The Netherlands	53/177 (30%)	DNA contact	(Berns et al., 1998)
COLON	USA	665/1464 (45%)	Codon 245	(Samowitz et al., 2002)
COLORECTUM	Norway	102/222 (46%)	L3 loop	(Borresen-Dale et al., 1998)
COLORECTUM	Singapore	109/192 (57%)	Non-conserved regions, codon 175	(Goh et al., 1995)
COLORECTUM	Sweden	99/189 (52%)	Non-conserved regions	(Kressner et al., 1999)
ESOPHAGUS	Japan	78/138 (56%)	L2/L3 loops	(Kihara et al., 2000)
HEAD AND	France	40/105 (38%)	DNA contact	(Temam et

Tumor site	Country	Mutation frequency	Region associated with poor prognosis	References
NECK				al., 2000)
HEAD AND NECK	Germany	39/86 (45%)	DNA contact	(Erber et al., 1998)
HEAD AND NECK	Italy	40/70 (57%)	Missense mutations on codon72 Arg allele	(Bergamaschi et al., 2003)
HEAD AND NECK	Japan	51/121 (42%)	Conserved regions and DNA/Zn binding	(Yamazaki et al., 2003)
LUNG	Europe	34/151 (22%)	Null mutations	(de Anta et al., 1997)
LUNG	Japan	75/204 (37%)	Exon 8	(Huang et al., 1998)
LUNG	Japan	65/144 (45%)	Null mutations	(Hashimoto et al., 1999)
LUNG	Japan	49/103 (47%)	Missense mutations	(Tomizawa et al., 1999)
LUNG	Norway	83/148 (56%)	L2/L3 loops	(Skaug et al., 2000)
LUNG	USA	107/188 (57%)	truncating/structural/ DNA contact mutants	(Ahrendt et al., 2003)
OVARY	Germany/ USA	99/178 (57%)	Conserved domains	(Reles et al., 2001)
OVARY	Norway		Missense on Arg ??	(Wang et al., 2004)
OVARY	USA	125/267 (47%)	Null mutations	(Rose et al., 2003)
SOFT TISSUE	Germany	15/145 (10%)	Non-frameshift mutations	(Taubert et al., 1996)

Data from the IARC *TP53* Database (R9, July 2004).

A question that remains to be fully elucidated is how the GOF activities observed for certain mutant proteins specifically affect tumor response to treatment, tumor aggressiveness and patient outcome. Among the various GOF described for mutant p53 proteins, the capacity to interact with the *TP53* family members, p63 and p73, provides interesting clues. There are now several examples showing that TAp73 is induced by various chemotherapeutic drugs, and that this activation results in the selective activation of apoptosis-related target genes (reviewed in (Gasco and Crook, 2003)). Cell assays have shown that some tumor-derived p53 mutant are able to bind to and inhibit TAp73 transactivation function (Di Como et al., 1999;

Monti et al., 2003; Strano et al., 2000). Such mutant p53 proteins are thus expected to confer a drug-resistant phenotype to tumors, due to the combined loss of p53 anti-proliferative activities and inhibition of p73 pro-apoptotic function. This hypothesis is substanciated by a recent study on advanced head and neck cancer, which showed that the efficiency with which p53 mutants inhibit TAp73-dependent apoptosis was related to efficacy of cisplatin-based chemo-radio-therapy (Bergamaschi et al., 2003).

A *TP53* polymorphism at codon 72, encoding an Arg or a Pro, has been shown to affect some *TP53* activities *in vitro*. The Arg72 variant is more susceptible to degradation by HPV E6 protein (Storey et al., 1998) and is more potent in inducing apoptosis (Dumont et al., 2003). Inhibition of TAp73 by some p53 mutants is enhanced if they expressed the arginine rather than the proline allele (Marin et al., 2000). In the study on advanced head and neck cancer by Bergamaschi *et al*, tumors carrying a p53 mutant capable of TAp73 inhibition and expressing the arginine allele had lower response rates than those expressing the same mutant with a proline allele (Bergamaschi et al., 2003). These results suggest that the two polymorphic variants of p53 are functionally distinct and that *TP53* codon72 polymorphism may influence individual responsiveness to cancer therapy.

TP53 MUTATION FOR EARLY DETECTION AND FOLLOW-UP

Although *TP53* mutations are found in almost any types of cancers, the timing of occurrence of the mutation during cancer progression is extremely variable from one cancer to another. In the classical model of stepwise progression of colorectal cancers, Fearon and Vogelstein have identified that *TP53* mutation and loss of alleles preferentially occur at the transition between late adenoma and carcinoma *in situ*, that is, at a relatively late stage in the histopathological development of these lesions (Fearon and Vogelstein, 1990). Similar findings have been reported in many common cancers, including breast and prostate cancers, although *TP53* mutations have been seen in atypical hyperplasia and DSCIS of the breast (Chitemerere et al., 1996).

In contrast, *TP53* mutation has been reported to occur at an early stage in many types of cancer that are directly caused by exogenous carcinogens. It is the case for lung cancers of smokers, non-melanoma skin cancers after exposure to UV irradiation, head and neck squamous cell carcinoma and esophageal cancers. In these cancers, *TP53* mutations are often detectable in hyperplastic and dysplastic lesions, as well as in non-involved, apparently normal tissues surrounding the tumor (Hussain et al., 2001; Mandard et al.,

2000). Moreover, the position of *TP53* mutation in the temporal sequence of events leading to cancer is not always constant for similar types of cancer. For example, in hepatocellular carcinoma (HCC), *TP53* mutations are late events in most cancers occurring in the Western population, but are very early events in most cases from West Africa and South-east Asia (Montesano et al., 1997). In these regions, HCC occurs as a consequence of exposure to aflatoxins (hepatocarcinogen contaminant of diet) and HBV. In this context, *TP53* mutations are detectable in cirrhotic liver before the onset of cancer (Livni et al., 1995). Another example is colon cancer. Apart from the well-characterised "late" involvement of *TP53* in polypoid carcinomas, there is evidence that *TP53* mutation can occur at an early stage in serrated carcinoma (Hawkins et al., 2000). *TP53* mutations have also been found in non-tumorous colonic tissue from inflamed regions in patients with ulcerative colitis (UC) (Hussain et al., 2000). In this case, they may result from an endogenous carcinogenic stress (reactive oxygen species and nitrogen species produced by the inflammatory micro-environment due to UC). It should be emphasized that the timing of occurrence of *TP53* mutations is not well established for a majority of cancers. The terms "early" or "late" event are based on the frequency of mutations observed at different tumor stages. It can't be excluded that a late event may correspond to a mutation acquired at an early stage in a tumor detected at an advanced pathological stage, the mutation providing a growth advantage and leading to the rapid development of the tumor. Therefore, screening for *TP53* mutations in dysplastic and early lesions may help identify those lesions that are at a high risk for rapid malignant evolution.

 TP53 mutations have been detected in circulating free DNA in the plasma of cancer patients, in feces of patients with colorectal cancer, in the saliva of oral cancer patients, in urine of bladder cancer patients, in sputum of lung cancer patients and in other body fluids. In many instances, these mutations were identical to the ones found in the primary tumour tissue of the patient, thus clearly establishing their tumoral origin. *TP53* mutation screening in these surrogate materials may thus be of potential use in the early detection of malignant lesions in individual at risk for these cancers, and may also be used to detect early relapse during post-treatment follow-up of patients with defined mutations in the primary lesion. In a recent study, we have analysed *TP53* and *KRAS* mutations in free DNA extracted from the plasma of healthy subjects who later developed lung, bladder, larynx, pharynx, oral cancers or leukemias. *TP53* mutations were detected in 9/374 (2.4%) and *KRAS* mutations in 11/1025 (1.0%) subjects. Six *TP53* positive (OR: 3.3; 95% CI: 0.8-13.4) and 3 *KRAS*-positive (OR: 1.0; 95% CI: 0.3-3.4) subjects developed cancer. Thus, *TP53* mutations in plasma of healthy subjects may be associated with subsequent occurrence of some types of cancers. However, further work is needed to evaluate the reliability and sensitivity of these approaches.

Circulating p53 auto-antibodies have been found in 1 to 50% of cancer patients with solid tumors (see (Soussi, 2000) for review) (Caron et al., 1987; Crawford et al., 1982). They are directed against the N- and C-terminal domains of the protein although the mutations are mainly located in the DNA-binding domain. It has been shown that they are due to a self-immunization process linked to the strong immunogenicity of the p53 mutant protein and that they correlate with the presence of missense mutations and the accumulation of p53 mutant protein in the tumor. They have been associated with high-grade tumors and poor survival in several cancers such as breast, colon, stomach and head and neck. In rare instances, these antibodies have been found in blood samples collected months to years before cancer diagnosis. Monitoring these antibodies could thus serve as early detection markers as well as markers of relapse. However, large inter-individual differences are observed suggesting that the capacity to elicit this humoral response is dependent on the biological and genetic background of the patients. Further studies are thus required to assess the significance of these antibodies in term of specificity and sensitivity for their use in clinical practice.

PERSPECTIVES: BRINGING P53 INTO CLINICAL PRACTICE

The examples discussed above show that *TP53* is a potential useful biomarker for early detection and follow-up of cancer, prediction of patient outcome and response to treatment in several types of cancer. However, its use in clinical practice still requires validation studies to precisely define in which conditions it presents a real advantage over currently available markers. In many studies, the presence of a *TP53* mutation has been found to be associated with classical clinico-pathological predictors of poor survival, ie large tumor size, positive node status, high histological grade, and low hormone receptor contents. It was also found to be associated with markers of increased cell proliferation such as high mitotic frequency and high expression of Ki-67, which are also of prognostic value. Although, in multivariate analysis, *TP53* mutation was often found to be an independent factor of prognosis in various cancer types (http://www.iarc.fr/p53/), the clinical parameters included in the multivariate models vary between studies, rendering results difficult to compare. Further studies should be carried out to specifically address this issue. These studies should be conducted on a large scale and designed with the same standards as drug trials, with well-characterized cancer cases and well-documented treatment regimens and clinical information.

One of the difficulties of conducting large validation studies is the absence of validated, high-throughput screening technology. So far, mutation analysis still relies on the sequencing of portions of the *TP53* gene to determine the exact nature and position of the mutation. Tumor DNA always contains a proportion of wild-type material due to the presence of wild-type alleles in cancer cells, or of non-cancer cells in the original tissue specimens. Therefore, DNA sequencing needs to be highly sensitive in order to detect mutant DNA against a background of wild-type material. Such sensitivity is not always achieved by standard, automated direct sequencing methods. Thus, *TP53* mutation analysis remains an expensive and labor-intensive work. The most common techniques used for *TP53* mutation analysis include PCR-based assays like single strand conformational polymorphism (SSCP), denaturing gradient gel electrophoresis (DGGE) and its variations, and DNA sequencing. Another method based on yeast functional assays was developed to detect *TP53* mutations (Ishioka et al., 1995; Moshinsky and Wogan, 1997; Scharer and Iggo, 1992a; Scharer and Iggo, 1992b). In this assay, the loss of DNA binding and transcriptional transactivation function in mutant p53 is detected by the colony color of yeast. A very elegant and reliable assay, SOMA (short oligonucleotide mass assay), has also been developed that involves PCR and mass spectrometry (Laken et al., 1998). This assay is very reliable and enables simultaneous analysis of both strands of the gene. However, it can only detect one specific mutation at a time, which is not suitable for screening the entire TP53 gene. More recently, microarray-based methods have been described. The microarray developed by Affymetrix (Santa Clara, CA) is based on direct hybridization of *TP53* DNA fragments on immobilized oligonucleotides. This array shows good specificity but its sensitivity is still limited and its application in large-scale studies needs to be further evaluated (Ahrendt et al., 1999; Wikman et al., 2000a; Wikman et al., 2000b). Another, commercial array is currently developed by Asper technologies (Tartu, Estonia), based on APEX (Arrayed Primer Extension). APEX uses solid-phase primer extension to incorporate fluorescent terminators into fixed oligos, thus requiring less oligos than methods based on specific hybridization. This method is extremely specific and sensitive and is currently being evaluated for scaling-up (Tonisson et al., 2002). These new array technologies should allow the scaling-up of *TP53* mutation screening in a close future.

The diversity of the type and functional consequences of *TP53* mutations is another issue that needs to be taken into account when analyzing the predictive value of *TP53* mutations. This is a difficult point since a clear knowledge of the biological impact of each specific mutation is still lacking for the majority of reported mutations. The impact of *TP53* polymorphisms (in particular R/P at codon 72) on the activity of mutant proteins need also to be further explored. It is of note that available studies have used various way of classifying mutation, none of which fully reflecting the biological reality. For example, mutations have often been grouped according to their codon position, despite experimental evidence showing that different amino-acid substitutions at a given codon have different functional impact (Ryan and Vousden, 1998). Gene expression profiling is another area that should help decifering the functional impact of p53 mutants. A recent study on breast cancer has identified subclasses of breast tumors based on gene expression profiles (Sorlie et al., 2001). These tumor subcalsses had different prognosis and had different frequency of *TP53* mutation. Studies on a larger scale are needed to determine if specific types of mutants are associated with specific gene expression profiles.

Future studies should address all these issues to determine to which extent the identification of a *TP53* mutation may help clinicians in the diagnosis of cancer and in the selection of the appropriate therapeutic approach. The availibility of public databases that integrate mutation data with clinical and pathological annotations as well as with functional annotations on mutant proteins will be necessary to estimate the clinical impact of specific *TP53* mutations. In the early nineties, the rapid accumulation of data on the occurrence of *TP53* mutations raised high expectations for clinical exploitation. However, over 10 years later, these expectations are still waiting to materialize into clinical practice. Recent findings show us that, in this area as in many others, the clinical reality is more complex than initially suspected and that interpretation of *TP53* mutations cannot be restricted to a "yes/no" answer. However, with the accumulation of knowledge on the specific properties of mutant proteins, we are closing down to the stage when mutation analysis will become standard practice in molecular pathology.

REFERENCES

Aas T, Borresen AL, Geisler S, Smith-Sorensen B, Johnsen H, Varhaug JE, Akslen LA, Lonning PE. 1996. Specific p53 mutations are associated with de novo resistance to doxorubicin in breast cancer patients. Nat Med 2:811-814.

Ahrendt SA, Halachmi S, Chow JT, Wu L, Halachmi N, Yang SC, Wehage S, Jen J, Sidransky D. 1999. Rapid p53 sequence analysis in primary lung cancer using an oligonucleotide probe array. Proc Natl Acad Sci U S A 96:7382-7387.

Ahrendt SA, Hu Y, Buta M, McDermott MP, Benoit N, Yang SC, Wu L, Sidransky D. 2003. p53 mutations and survival in stage I non-small-cell lung cancer: results of a prospective study. J Natl Cancer Inst 95:961-970.

Alsner J, Yilmaz M, Guldberg P, Hansen LL, Overgaard J. 2000. Heterogeneity in the clinical phenotype of TP53 mutations in breast cancer patients. Clin Cancer Res 6:3923-3931.

Aurelio ON, Kong XT, Gupta S, Stanbridge EJ. 2000. p53 mutants have selective dominant-negative effects on apoptosis but not growth arrest in human cancer cell lines. Mol Cell Biol 20:770-778.

Bergamaschi D, Gasco M, Hiller L, Sullivan A, Syed N, Trigiante G, Yulug I, Merlano M, Numico G, Comino A, Attard M, Reelfs O, Gusterson B, Bell AK, Heath V, Tavassoli M, Farrell PJ, Smith P, Lu X, Crook T. 2003. p53 polymorphism influences response in cancer chemotherapy via modulation of p73-dependent apoptosis. Cancer Cell 3:387-402.

Bergh J, Norberg T, Sjogren S, Lindgren A, Holmberg L. 1995. Complete sequencing of the p53 gene provides prognostic information in breast cancer patients, particularly in relation to adjuvant systemic therapy and radiotherapy. Nat Med 1:1029-1034.

Berns EM, Foekens JA, Vossen R, Look MP, Devilee P, Henzen-Logmans SC, van Staveren IL, van Putten WL, Inganas M, Meijer-van Gelder ME, Cornelisse C, Claassen CJ, Portengen H, Bakker B, Klijn JG. 2000. Complete sequencing of TP53 predicts poor response to systemic therapy of advanced breast cancer. Cancer Res 60:2155-2162.

Berns EM, van Staveren IL, Look MP, Smid M, Klijn JG, Foekens JA. 1998. Mutations in residues of TP53 that directly contact DNA predict poor outcome in human primary breast cancer. Br J Cancer 77:1130-1136.

Borresen-Dale AL, Lothe RA, Meling GI, Hainaut P, Rognum TO, Skovlund E. 1998. TP53 and long-term prognosis in colorectal cancer: mutations in the L3 zinc-binding domain predict poor survival. Clin Cancer Res 4:203-210.

Borresen AL, Andersen TI, Eyfjord JE, Cornelis RS, Thorlacius S, Borg A, Johansson U, Theillet C, Scherneck S, Hartman S, . 1995. TP53 mutations and breast cancer prognosis: particularly poor survival rates for cases with mutations in the zinc-binding domains. Genes Chromosomes Cancer 14:71-75.

Cadwell C and Zambetti GP. 2001. The effects of wild-type p53 tumor suppressor activity and mutant p53 gain-of-function on cell growth. Gene 277:15-30.

Caron dF, May-Levin F, Mouriesse H, Lemerle J, Chandrasekaran K, May P. 1987. Presence of circulating antibodies against cellular protein p53 in a notable proportion of children with B-cell lymphoma. Int J Cancer 39:185-189.

Casson AG, Evans SC, Gillis A, Porter GA, Veugelers P, Darnton SJ, Guernsey DL, Hainaut P. 2003. Clinical implications of p53 tumor suppressor gene mutation and protein expression in esophageal adenocarcinomas: results of a ten-year prospective study. J Thorac Cardiovasc Surg 125:1121-1131.

Chitemerere M, Andersen TI, Holm R, Karlsen F, Borresen AL, Nesland JM. 1996. TP53 alterations in atypical ductal hyperplasia and ductal carcinoma in situ of the breast. Breast Cancer Res Treat 41:103-109.

Crawford LV, Pim DC, Bulbrook RD. 1982. Detection of antibodies against the cellular protein p53 in sera from patients with breast cancer. Int J Cancer 30:403-408.

de Anta JM, Jassem E, Rosell R, Martinez-Roca M, Jassem J, Martinez-Lopez E, Monzo M, Sanchez-Hernandez JJ, Moreno I, Sanchez-Cespedes M. 1997. TP53 mutational pattern in Spanish and Polish non-small cell lung cancer patients: null mutations are associated with poor prognosis. Oncogene 15:2951-2958.

Di Como CJ, Gaiddon C, Prives C. 1999. p73 function is inhibited by tumor-derived p53 mutants in mammalian cells. Mol Cell Biol 19:1438-1449.

Dittmer D, Pati S, Zambetti G, Chu S, Teresky AK, Moore M, Finlay C, Levine AJ. 1993. Gain of function mutations in p53. Nat Genet 4:42-46.

Dumont P, Leu JI, Della PA, III, George DL, Murphy M. 2003. The codon 72 polymorphic variants of p53 have markedly different apoptotic potential. Nat Genet 33:357-365.

Erber R, Conradt C, Homann N, Enders C, Finckh M, Dietz A, Weidauer H, Bosch FX. 1998. TP53 DNA contact mutations are selectively associated with allelic loss and have a strong clinical impact in head and neck cancer. Oncogene 16:1671-1679.

Fearon ER and Vogelstein B. 1990. A genetic model for colorectal tumorigenesis. Cell 61:759-868.

Gasco M and Crook T. 2003. p53 family members and chemoresistance in cancer: what we know and what we need to know. Drug Resist Updat 6:323-328.

Geisler S, Borresen-Dale AL, Johnsen H, Aas T, Geisler J, Akslen LA, Anker G, Lonning PE. 2003. TP53 gene mutations predict the response to neoadjuvant treatment with 5-fluorouracil and mitomycin in locally advanced breast cancer. Clin Cancer Res 9:5582-5588.

Geisler S, Lonning PE, Aas T, Johnsen H, Fluge O, Haugen DF, Lillehaug JR, Akslen LA, Borresen-Dale AL. 2001. Influence of TP53 gene alterations and c-erbB-2 expression on the response to treatment with doxorubicin in locally advanced breast cancer. Cancer Res 61:2505-2512.

Gentile M, Jungestrom MB, Olsen KE, Soderkvist P, Wingren S. 1999. p53 and survival in early onset breast cancer: analysis of gene mutations, loss of heterozygosity and protein accumulation. Eur J Cancer 35:1202-1207.

Goh HS, Yao J, Smith DR. 1995. p53 point mutation and survival in colorectal cancer patients. Cancer Res 55:5217-5221.

Hainaut P and Hollstein M. 2000. p53 and human cancer: the first ten thousand mutations. Adv Cancer Res 77:81-137.

Hashimoto T, Tokuchi Y, Hayashi M, Kobayashi Y, Nishida K, Hayashi S, Ishikawa Y, Tsuchiya S, Nakagawa K, Hayashi J, Tsuchiya E. 1999. p53 null mutations undetected by immunohistochemical staining predict a poor outcome with early-stage non-small cell lung carcinomas. Cancer Res 59:5572-5577.

Hawkins NJ, Gorman P, Tomlinson IP, Bullpitt P, Ward RL. 2000. Colorectal carcinomas arising in the hyperplastic polyposis syndrome progress through the chromosomal instability pathway. Am J Pathol 157:385-392.

Huang C, Taki T, Adachi M, Konishi T, Higashiyama M, Miyake M. 1998. Mutations in exon 7 and 8 of p53 as poor prognostic factors in patients with non-small cell lung cancer. Oncogene 16:2469-2477.

Hussain SP, Amstad P, Raja K, Ambs S, Nagashima M, Bennett WP, Shields PG, Ham AJ, Swenberg JA, Marrogi AJ, Harris CC. 2000. Increased p53 mutation load in noncancerous colon tissue from ulcerative colitis: a cancer-prone chronic inflammatory disease. Cancer Res 60:3333-3337.

Hussain SP, Amstad P, Raja K, Sawyer M, Hofseth L, Shields PG, Hewer A, Phillips DH, Ryberg D, Haugen A, Harris CC. 2001. Mutability of p53 hotspot codons to benzo(a)pyrene diol epoxide (BPDE) and the frequency of p53 mutations in nontumorous human lung. Cancer Res 61:6350-6355.

Ishioka C, Englert C, Winge P, Yan YX, Engelstein M, Friend SH. 1995. Mutational analysis of the carboxy-terminal portion of p53 using both yeast and mammalian cell assays in vivo. Oncogene 10:1485-1492.

Kato S, Han SY, Liu W, Otsuka K, Shibata H, Kanamaru R, Ishioka C. 2003. Understanding the function-structure and function-mutation relationships of p53 tumor suppressor protein by high-resolution missense mutation analysis. Proc Natl Acad Sci U S A 100:8424-8429.

Kihara C, Seki T, Furukawa Y, Yamana H, Kimura Y, van Schaardenburgh P, Hirata K, Nakamura Y. 2000. Mutations in zinc-binding domains of p53 as a prognostic marker of esophageal-cancer patients. Jpn J Cancer Res 91:190-198.

Kressner U, Inganas M, Byding S, Blikstad I, Pahlman L, Glimelius B, Lindmark G. 1999. Prognostic value of p53 genetic changes in colorectal cancer. J Clin Oncol 17:593-599.

Kucera E, Speiser P, Gnant M, Szabo L, Samonigg H, Hausmaninger H, Mittlbock M, Fridrik M, Seifert M, Kubista E, Reiner A, Zeillinger R, Jakesz R. 1999. Prognostic significance of mutations in the p53 gene, particularly in the zinc-binding domains, in lymph node- and steroid receptor positive breast cancer patients. Austrian Breast Cancer Study Group. Eur J Cancer 35:398-405.

Laken SJ, Jackson PE, Kinzler KW, Vogelstein B, Strickland PT, Groopman JD, Friesen MD. 1998. Genotyping by mass spectrometric analysis of short DNA fragments. Nat Biotechnol 16:1352-1356.

Lane DP and Benchimol S. 1990. p53: oncogene or anti-oncogene? Genes Dev 4:1-8.

Livni N, Eid A, Ilan Y, Rivkind A, Rosenmann E, Blendis LM, Shouval D, Galun E. 1995. p53 expression in patients with cirrhosis with and without hepatocellular carcinoma. Cancer 75:2420-2426.

Lonning PE. 2003. Study of suboptimum treatment response: lessons from breast cancer. Lancet Oncol 4:177-185.

Lowe SW, Bodis S, McClatchey A, Remington L, Ruley HE, Fisher DE, Housman DE, Jacks T. 1994. p53 status and the efficacy of cancer therapy in vivo. Science 266:807-810.

Mandard AM, Hainaut P, Hollstein M. 2000. Genetic steps in the development of squamous cell carcinoma of the esophagus. Mutat Res 462:335-342.

Marin MC, Jost CA, Brooks LA, Irwin MS, O'Nions J, Tidy JA, James N, McGregor JM, Harwood CA, Yulug IG, Vousden KH, Allday MJ, Gusterson B, Ikawa S, Hinds PW, Crook T, Kaelin WG, Jr. 2000. A common polymorphism acts as an intragenic modifier of mutant p53 behaviour. Nat Genet 25:47-54.

Milner J, Medcalf EA, Cook AC. 1991. Tumor suppressor p53: analysis of wild-type and mutant p53 complexes. Mol Cell Biol 11:12-19.

Montesano R, Hainaut P, Hall J. 1997. The use of biomarkers to study pathogenesis and mechanisms of cancer: oesophagus and skin cancer as models. IARC Sci Publ :291-301.

Monti P, Campomenosi P, Ciribilli Y, Iannone R, Aprile A, Inga A, Tada M, Menichini P, Abbondandolo A, Fronza G. 2003. Characterization of the p53 mutants ability to inhibit p73 beta transactivation using a yeast-based functional assay. Oncogene 22:5252-5260.

Moshinsky DJ and Wogan GN. 1997. UV-induced mutagenesis of human p53 in a vector replicated in Saccharomyces cerevisiae. Proc Natl Acad Sci U S A 94:2266-2271.

Nagai MA, Schaer BH, Zago MA, Araujo SW, Jr., Nishimoto IN, Salaorni S, Guerreiro Costa LN, Silva AM, Caldas Oliveira AG, Mourao NM, Brentani MM. 2003. TP53 mutations in primary breast carcinomas from white and African-Brazilian patients. Int J Oncol 23:189-196.

O'Connor PM, Jackman J, Bae I, Myers TG, Fan S, Mutoh M, Scudiero DA, Monks A, Sausville EA, Weinstein JN, Friend S, Fornace AJ, Jr., Kohn KW. 1997. Characterization of the p53 tumor suppressor pathway in cell lines of the National Cancer Institute anticancer drug screen and correlations with the growth-inhibitory potency of 123 anticancer agents. Cancer Res 57:4285-4300.

Oren M. 1992. p53: the ultimate tumor suppressor gene? FASEB J 6:3169-3176.

Ory K, Legros Y, Auguin C, Soussi T. 1994. Analysis of the most representative tumour-derived p53 mutants reveals that changes in protein conformation are not correlated with loss of transactivation or inhibition of cell proliferation. EMBO J 13:3496-3504.

Powell B, Soong R, Iacopetta B, Seshadri R, Smith DR. 2000. Prognostic significance of mutations to different structural and functional regions of the p53 gene in breast cancer. Clin Cancer Res 6:443-451.

Reles A, Wen WH, Schmider A, Gee C, Runnebaum IB, Kilian U, Jones LA, el Naggar A, Minguillon C, Schonborn I, Reich O, Kreienberg R, Lichtenegger W, Press MF. 2001. Correlation of p53 mutations with resistance to platinum-based chemotherapy and shortened survival in ovarian cancer. Clin Cancer Res 7:2984-2997.

Resnick MA and Inga A. 2003. Functional mutants of the sequence-specific transcription factor p53 and implications for master genes of diversity. Proc Natl Acad Sci U S A 100:9934-9939.

Rose SL, Robertson AD, Goodheart MJ, Smith BJ, DeYoung BR, Buller RE. 2003. The impact of p53 protein core domain structural alteration on ovarian cancer survival. Clin Cancer Res 9:4139-4144.

Ryan KM and Vousden KH. 1998. Characterization of structural p53 mutants which show selective defects in apoptosis but not cell cycle arrest. Mol Cell Biol 18:3692-3698.

Samowitz WS, Curtin K, Ma KN, Edwards S, Schaffer D, Leppert MF, Slattery ML. 2002. Prognostic significance of p53 mutations in colon cancer at the population level. Int J Cancer 99:597-602.

Scharer E and Iggo R. 1992a. Mammalian p53 can function as a transcription factor in yeast. Nucleic Acids Res 20:1539-1545.

Sigal A and Rotter V. 2000. Oncogenic mutations of the p53 tumor suppressor: the demons of the guardian of the genome. Cancer Res 60:6788-6793.

Skaug V, Ryberg D, Kure EH, Arab MO, Stangeland L, Myking AO, Haugen A. 2000. p53 mutations in defined structural and functional domains are related to poor clinical outcome in non-small cell lung cancer patients. Clin Cancer Res 6:1031-1037.

Smith-Sorensen B, Kaern J, Holm R, Dorum A, Trope C, Borresen-Dale AL. 1998. Therapy effect of either paclitaxel or cyclophosphamide combination treatment in patients with epithelial ovarian cancer and relation to TP53 gene status. Br J Cancer 78:375-381.

Sorlie T, Perou CM, Tibshirani R, Aas T, Geisler S, Johnsen H, Hastie T, Eisen MB, van de RM, Jeffrey SS, Thorsen T, Quist H, Matese JC, Brown PO, Botstein D, Eystein LP, Borresen-Dale AL. 2001. Gene expression patterns of breast carcinomas distinguish tumor subclasses with clinical implications. Proc Natl Acad Sci U S A 98:10869-10874.

Soussi T. 2000. p53 Antibodies in the sera of patients with various types of cancer: a review. Cancer Res 60:1777-1788.

Storey A, Thomas M, Kalita A, Harwood C, Gardiol D, Mantovani F, Breuer J, Leigh IM, Matlashewski G, Banks L. 1998. Role of a p53 polymorphism in the development of human papillomavirus-associated cancer. Nature 393:229-234.

Strano S, Munarriz E, Rossi M, Cristofanelli B, Shaul Y, Castagnoli L, Levine AJ, Sacchi A, Cesareni G, Oren M, Blandino G. 2000. Physical and functional interaction between p53 mutants and different isoforms of p73. J Biol Chem 275:29503-29512.

Takahashi M, Tonoki H, Tada M, Kashiwazaki H, Furuuchi K, Hamada J, Fujioka Y, Sato Y, Takahashi H, Todo S, Sakuragi N, Moriuchi T. 2000. Distinct prognostic values of p53 mutations and loss of estrogen receptor and their cumulative effect in primary breast cancers. Int J Cancer 89:92-99.

Taubert H, Meye A, Wurl P. 1996. Prognosis is correlated with p53 mutation type for soft tissue sarcoma patients. Cancer Res 56:4134-4136.

Temam S, Flahault A, Perie S, Monceaux G, Coulet F, Callard P, Bernaudin JF, St Guily JL, Fouret P. 2000. p53 gene status as a predictor of tumor response to induction chemotherapy of patients with locoregionally advanced squamous cell carcinomas of the head and neck. J Clin Oncol 18:385-394.

Tomizawa Y, Kohno T, Fujita T, Kiyama M, Saito R, Noguchi M, Matsuno Y, Hirohashi S, Yamaguchi N, Nakajima T, Yokota J. 1999. Correlation between the status of the p53 gene and survival in patients with stage I non-small cell lung carcinoma [In Process Citation]. Oncogene 18:1007-1014.

Tonisson N, Zernant J, Kurg A, Pavel H, Slavin G, Roomere H, Meiel A, Hainaut P, Metspalu A. 2002. Evaluating the arrayed primer extension resequencing assay of TP53 tumor suppressor gene. Proc Natl Acad Sci U S A 99:5503-5508.

Tortola S, Marcuello E, Gonzalez I, Reyes G, Arribas R, Aiza G, Sancho FJ, Peinado MA, Capella G. 1999. p53 and K-ras gene mutations correlate with tumor aggressiveness but are not of routine prognostic value in colorectal cancer. J Clin Oncol 17:1375-1381.

Wang Y, Kringen P, Kristensen GB, Holm R, Baekelandt MM, Olivier M, Skomedal H, Hainaut P, Trope CG, Abeler VM, Nesland JM, Borresen-Dale AL, Helland A. 2004. Effect of the codon 72 polymorphism (c.215G>C, p.Arg72Pro) in combination with somatic sequence variants in the TP53 gene on survival in patients with advanced ovarian carcinoma. Hum Mutat 24:21-34.

Wikman FP, Lu ML, Thykjaer T, Olesen SH, Andersen LD, Cordon-Cardo C, Orntoft TF. 2000a. Evaluation of the performance of a p53 sequencing microarray chip using 140 previously sequenced bladder tumor samples. Clin Chem 46:1555-1561.

Yamazaki Y, Chiba I, Hirai A, Sugiura C, Notani K, Kashiwazaki H, Tei K, Totsuka Y, Fukuda H. 2003. Specific p53 mutations predict poor prognosis in oral squamous cell carcinoma. Oral Oncol 39:163-169.

Chapter 15

P53 LINKS TUMOR DEVELOPMENT TO CANCER THERAPY

Michael T. Hemann and Scott W. Lowe
Cold Spring Harbor Laboratory, Cold Spring Harbor, NY, USA

INTRODUCTION

Anti-cancer therapy operates on the assumption that the genetic pathways disrupted during tumorigenesis are distinct from those that mediate drug sensitivity. The main objective of this therapy is to present tumor cells with obstacles unrelated to the process of cellular transformation or to exploit vulnerabilities created by tumor development, such as uncontrolled DNA synthesis, checkpoint abnormalities, or an addiction to an oncogenic signal. Cytotoxic therapies, for example, rely on the introduction of DNA damage or the inhibition of chromosome segregation. These lesions, when introduced at high levels, elicit a DNA damage response presumably distinct from any encountered during the early stages of tumor development.

The frequent failure of these established therapies suggests that, in fact, the genetics of tumor suppression are inextricably linked to the genetics of treatment response. The single greatest culprit responsible for this phenomenon is the tumor suppressor p53. In this chapter, we will highlight results from our lab and others that contrast the distinct functions of p53 in preventing cellular transformation and tumorigenesis verses those important for therapeutic response. In doing so, we hope to emphasize the importance of tumor genetics on tumor therapy and provide a framework for understanding the unique role of p53 loss on tumor biology.

339

P. Hainaut and K.G. Wiman (eds.), 25 Years of p53 Research, 339-351.
© 2005 *Springer. Printed in the Netherlands.*

P53 AND TUMORIGENESIS

Tumor suppressors act to maintain tissue homeostasis, that is, to control the number and behavior of cells in a particular tissue within an organism (Hussain and Harris, 1998). To do so, they typically regulate processes that prevent aberrant proliferation (Vogelstein et al., 2000; Vousden and Lu, 2002). p53, the most extensively studied tumor suppressor, acts in response to diverse forms of cellular stress to mediate a variety of anti-proliferative processes. Hence, p53 is activated by DNA damage, hypoxia, or aberrant oncogene expression to promote cell-cycle checkpoints, DNA repair, cellular senescence, and apoptosis (Figure 1). As a consequence, disruption of p53 function promotes checkpoint defects, cellular immortalization, genomic instability, and inappropriate survival, allowing the continued proliferation and evolution of damaged cells. Any one or combination of these defects may contribute to the selective advantage p53 mutant cells have during tumor initiation or progression.

One of the most extensively studied areas in p53 research concerns its ability to induce apoptosis. The first hint that p53 could control apoptosis came from studies by Oren and co-workers who reintroduced p53 into a p53-deficient myeloid leukemia cell line (Yonish-Rouach et al., 1991). Here, p53 induced apoptosis in a manner that could be countered by a pro-survival cytokine. Subsequently, evidence that endogenous p53 could control apoptosis was obtained from studies using thymocytes from p53 knockout mice, which revealed that p53 was required for DNA damage-induced apoptosis, but not cell death induced by several other stimuli (Clarke et al., 1993; Lowe et al., 1993a).

The mechanism by which DNA damage activates p53 to promote apoptosis involves kinases such as ATM and perhaps the chk kinases (Iliakis et al., 2003). These kinases appear to act by phosphorylating p53 leading to its dissociation from its negative regulator Mdm2 (Prives, 1998). Although upstream signaling to p53 following DNA damage is often thought to be well understood, studies involving cells and mice where ATM and chk2 have been disrupted suggest a more complicated story. For example, ATM-deficient mice do not show the same spectrum of tumors as mice lacking p53, and ATM deficiency synergizes with p53 loss in promoting lymphoma development (Barlow et al., 1996; Donehower et al., 1992; Xu et al., 1998). Also, chk2, which is important for DNA damage induced apoptosis in murine thymocytes, is dispensable for apoptosis in some tumors and cell lines (Ahn et al., 2003; Jallepalli et al., 2003). Thus, DNA damage signaling to p53 likely represents a nonlinear pathway that varies significantly with species and cell type.

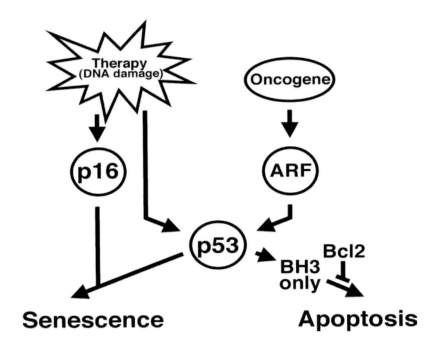

Figure 1. The p53 tumor suppressor network.

p53 is also an essential mediator of cell death in response to aberrant oncogene expression. Early work with the adenoviral oncogene E1A, as well as c-myc, showed that over-expression of either of these oncogenes significantly sensitized cells to apoptosis (Debbas and White, 1993; Hermeking and Eick, 1994; Lowe and Ruley, 1993; Rao et al., 1992). This apoptosis correlated with a significant increase in cellular p53, and, in fact, oncogene-induced cell death is substantially reduced in p53 deficient cells. Importantly, loss of p53-dependent apoptosis in response to oncogenes dramatically increases the susceptibility of cells to oncogenic transformation, such that primary p53-deficient fibroblasts are extremely susceptible to transformation by E1A and myc. Presumably, oncogenic signaling is coupled to an antiproliferative response as part of a tumor suppressive mechanism that eliminates aberrantly proliferating cells. Indeed, studies by Van Dyke, and later others, indicate that disruption of p53-dependent apoptosis correlates with tumor progression in oncogene-expressing transgenic mice (Symonds et al., 1994).

Oncogenic activation of p53 involves the tumor suppressor ARF (de Stanchina et al., 1998; Zindy et al., 1998). In fact, destabilization of p53 mediated by ARF loss can recapitulate the effects of p53 loss on oncogene-induced apoptosis in at least some settings. Interestingly, while ARF-/- murine embryonic fibroblasts (MEFs) are resistant to oncogene-induced cell death, ARF loss does not completely phenocopy p53 loss. For example, unlike p53-/- MEFs, ARF-/- MEFs are genomically stable and sensitive to DNA damage-induced cell death (Kamijo et al., 1997; Stott et al., 1998). Presumably, ARF acts in parallel with DNA damaging signal to activate p53 – indeed, ARF is not induced by DNA damage nor is it required for a DNA damage response.

Additionally, ARF may not always be essential for oncogene signaling to p53. First, ARF loss does not always mimic p53 loss in repressing apoptosis or promoting tumorigenesis in response to an oncogenic insult (Tolbert et al., 2002; Tsai et al., 2002). Second, some studies suggest that suggest that oncogenes like myc and E2F can induce DNA damage, which is required for their pro-apoptotic effects (Rogoff et al., 2002; Vafa et al., 2002). However, these latter results have been performed largely in cell culture and, to date, there is little evidence that oncogenes cooperate with loss of DNA damage response genes during the course of tumorigenesis. In fact, ATM loss has no effect on oncogene induced tumor suppression in a choroids plexus carcinoma model (Liao et al., 1999). Therefore, p53 is activated through parallel signaling pathways, the relative importance of which may depend on other factors.

As indicated above, the importance of oncogene-induced apoptosis as a mechanism of tumor suppression has been evaluated in a number of murine tumorigenesis models. In our lab, these studies have focused on a mouse model of B cell lymphomagenesis, the Eμ-myc transgenic mouse (Adams et al., 1985). This mouse over-expresses c-myc under the control of an immunoglobulin enhancer, a molecular lesion similar to that seen in human Burkitt's lymphoma. Due to the constitutive high expression of c-myc, B cells from these mice show elevated levels of both proliferation and apoptosis. B cell lymphomagenesis occurs after a 3-6 month latency in these mice and is frequently accompanied by mutations in either p53 or ARF (Eischen et al., 1999). Consistent with this result, crosses between Eμ-myc mice and germline p53 and ARF knock-out mice result in progeny with a vastly accelerated rate of B cell lymphomagenesis (Eischen et al., 1999; Schmitt et al., 1999).

The ability of ARF loss to accelerate Eμ-myc lymphomagenesis suggested that some p53 effector functions, including mitotic and DNA damage checkpoint responses, are not integral components of tumor suppression in B cells. This idea is further supported by experiments showing that over-expression of the anti-apoptotic protein bcl2 promotes lymphomagenesis as aggressively as p53 or ARF deficiency (Schmitt et al.,

2002a). Thus, the disruption of a single p53 effector function –apoptosis – was sufficient to facilitate myc-induced lymphomagenesis. This result was further supported by the *in vivo* suppression of known p53 target genes in hematopoietic stem cells (HSCs) derived from Eμ-myc transgenic mice. Surprisingly, the elimination of PUMA, a pro-apoptotic BH3-only protein and p53 target, recapitulates many of the effects of p53 loss on the development of Eμ-myc lymphomagenesis (Hemann et al, 2004). Hence, in this setting, loss of a single pro-apoptotic effector can phenocopy the effects of p53 mutations in tumor suppression.

p53 mutations are often associated with gross chromosomal instability and, in principle, this instability could fuel the evolution of malignant tumors. However, despite the fact that loss of ARF or overexpression of Bcl2 accelerates myc-induced lymphomagenesis to a similar degree as p53 loss, only p53-/- lymphomas show significant aneuploidy (Schmitt et al., 2002a). These results imply that, in this context, aneuploidy is a consequence of p53 deficiency rather than an essential mediator of tumor development. Studies examining p53 hypomorphic states are also consistent with this view(Hemann et al., 2003). Specifically, using stable RNAi technology and gene transfer into Eμ-myc hematopoietic cells, we have shown that short hairpin RNAs targeting p53 can suppress p53 to varying degrees. Interestingly, at levels of p53 suppression sufficient to compromise oncogene-induced apoptosis, the resulting tumors do not display the gross aneuploidy seen in p53-deficient cells. Nevertheless, stable suppression of p53 produces aggressive lymphomas that resembled those produced by complete p53 deficiency. Again, gross genomic instability appears dispensable for B cell lymphomagenesis. This conclusion is not limited to myc-induced lymphomas, since p53 can contribute to the progression of epithelial cancers without producing gross genomic instability (Lu et al., 2001).

The specific p53 effector functions required to suppress B cell lymphomagenesis are not, however, shared by all malignancies. Germline knock-in mice expressing a tumor-derived p53 mutant identified a distinct set of p53 effector functions essential for T cell tumor suppression (Liu et al., 2004). In this case, the point mutation renders p53 defective for oncogene-induced apoptosis but still capable of mediating p53-dependent cell cycle arrest. Knock-in mice expressing this p53 allele developed B cell lymphomas and histiocytic sarcomas, but not the T cell lymphomas commonly seen in p53 deficient mice. Thus, the ability of p53 to induce cell cycle arrest, perhaps in addition to p53-dependant apoptosis, limits T cell lymphomagenesis *in vivo*.

Together, these observations highlight three important aspects of p53 biology in early tumor development. First, the p53 effector functions involved in tumor suppression are highly context-dependent. While mediating oncogene-induced apoptosis is the single critical role of p53 in B cell tumor suppression, p53-mediated cell cycle arrest or the maintenance of genomic stability is required to suppress other malignancies. Moreover, the essential tumor suppressive functions of p53 may vary with distinct oncogenic insults. For example, whereas c-Myc activates p53 to promote apoptosis, oncogenic Ras can activate p53 to induce a senescence-like arrest (Serrano et al., 1997). Hence, during *ras*-initiated tumorigenesis, disruption of senescence may provide the immediate advantage to cells acquiring *p53* mutations whereas apoptotic defects may be byproducts of p53 loss.

Second, a myriad of genetic lesions can substitute for p53 loss during the initial stages of tumorigenesis. In B cell lymphomagenesis, ARF loss, PUMA suppression, mdm2 over-expression and bcl2 all phenocopy the effect of p53 loss (Schmitt, 1999; Eischen, 1999; Schmitt, 2002; Fridman, 2003; Hemann et al, 2004). One might imagine that in T cell lymphomagenesis the combined effects of an apoptotic defect and the loss of a cell cycle inhibitor, like p21, could recapitulate the effects of p53 loss. In other words, there are many different ways to significantly attenuate p53 function early in tumor development without directly mutating p53. It is tempting to speculate that, early in tumorigenesis, retention of some p53 functions may even be beneficial tumor development. If oncogene-induced apoptosis is the single hurdle a pre-neoplastic B cell needs to overcome, then the maintenance of genomic stability and the retention of cell cycle checkpoints may, in fact, benefit tumor proliferation. In fact, situations in which p53 exerts a pro-neoplastic effect have been documented. For example, p53 loss reduces the incidence of chemically-induced skin papillomas in mice (Kemp et al., 1993).

Finally, while disruption of apoptosis may be sufficient to promote tumor initiation, checkpoint defects and genomic instability may be important for subsequent tumor development. In the Eμ-myc model, although Bcl2 and shPUMA tumors appear at the same time or earlier than p53 deficient lymphomas, they progress to a lethal stage more slowly (Hemann et al, 2004). Furthermore, despite the similar effect of ARF loss and Bcl2 overexpression on myc-induced lymphomagenesis, these tumors show remarkably different responses to therapy. Therefore, while some p53 functions are not selected against early in tumorigenesis, they may provide added capabilities later on that allow the tumor to progress to a lethal stage (see below).

P53 AND THERAPY

Chemotherapy remains the primary treatment for systemic malignancies. However, some tumors are inherently insensitive to chemotherapeutic agents and others acquire resistance upon relapse. Most conventional agents damage cellular components, often DNA, and for years it was assumed that this damage was directly responsible for their anti-tumor effect (Johnstone et al., 2002). Consequently, drug resistance was thought to arise primarily from changes that prevented the drug-target interaction, including over-expression of drug efflux pumps (e.g. P-glycoprotein) or intracellular detoxifiers (e.g. glutathione) (Johnstone et al., 2000; Volm, 1998). It is now clear that drug-induced damage is not invariably lethal, but can instead initiate a series of post-damage responses including apoptosis, cell-cycle checkpoints, mitotic catastrophe, and cellular senescence (Johnstone et al., 2002; Schwartz and Rotter, 1998). Importantly, cells may interpret a drug-induced insult in the same way that a physiological insult, such as hypoxia or growth factor deprivation, is interpreted.

Given the significance that p53 plays in the response to physiological insults, considerable work has been done to investigate the role p53 inactivation in therapeutic response. In fact, mutations in *p53* or in the p53 pathway can produce multi-drug resistance in vitro and in vivo, and reintroduction of wild type *p53* into *p53*-null tumor cells can re-establish chemosensitivity (Wallace-Brodeur and Lowe, 1999). However, *p53* status is not a universal predictor of treatment response, in part because not all drugs absolutely require p53 for their apoptotic function (Herr and Debatin, 2001) and, in some settings, p53 loss can enhance drug-induced cell death (Bunz et al., 1999). Still, loss of p53 function correlates with multidrug resistance in many tumor types (Wallace-Brodeur and Lowe, 1999).

Further support for the essential role of p53 in mediating therapeutic responses came from early studies on the p53 null mouse. While thymocytes from wild-type mice undergo cell cycle arrest and apoptosis in response to gamma irradiation, p53-/- thymocytes show very little response to DNA damage (Lowe et al., 1993b). This complete resistance to DNA damage is not seen in cells deficient for modifiers of p53 function, like ARF or p21-deficient fibroblasts (Kamijo et al., 1997; Komarova et al., 2000). Thus, the specific absence of p53, and not single components or effector arms of the p53 pathway, most effectively disable some anti-tumor therapies.

The importance of these unique properties of p53 is again illustrated by chemotherapy studies in the Eμ-myc mouse. As indicated above, while ARF and p53 loss produce pathologically indistinguishable B cell lymphomas, tumors arising in these genetic backgrounds show very significant differences in response to therapy (Schmitt et al., 2002b). Eμ-myc ARF-/-

lymphomas undergo massive apoptosis in response to the established cytotoxic drugs adriamycin and cyclophosphamide. Mice bearing Eμ-myc ARF-/- lymphomas achieve long-lasting remissions and are frequently cured following therapy. Conversely, Eμ-myc p53-/- tumors are resistant to chemotherapy-induced apoptosis, and the mice bearing these tumors rapidly progress to end-stage disease following treatment. Presumably this reflects the ability of ARF loss to disable p53 function during tumorigenesis but not following a DNA damage response (see above).

Treatment of Eμ-myc bcl2 tumors reveals an additional component of the p53-mediated chemotherapeutic response (Schmitt et al., 2002b). The strong apoptotic block conferred by bcl2 over-expression prevents bcl2 expressing lymphoma cells from undergoing apoptosis in response to cyclophosphamide. However, these tumors still show a dramatically different response to treatment than the Eμ-myc p53-/- lymphomas. Specifically, the Eμ-myc bcl2 tumors undergo a stable therapy-induced cell cycle arrest. This arrest shows many of the hallmarks of cellular senescence, including p16 up-regulation and senescence-associated β-galactosidase staining. Due to tumor stasis, mice bearing Eμ-myc bcl2 tumors show significantly extended post-treatment survival relative to mice bearing Eμ-myc p53-/- lymphomas. These results suggest that p53 nullizygosity confers upon tumor cells the unique ability to bypass both DNA damage-induced apoptosis as well as DNA damage-induced senescence.

This work suggests that p53 provides a link between cancer genetics and cancer therapy (Johnstone et al., 2002). For example, p53-dependent apoptosis provides a brake against tumor development and contributes to the action of anticancer agents; hence, disruption of apoptosis during tumor development can simultaneously select for drug resistant cells (Schmitt and Lowe, 1999). Presumably, these interrelationships provide one explanation for intrinsic drug resistance – i.e. resistance with no preceding drug exposure. While the abrogation of certain p53 effector functions is not required during tumor evolution, p53 loss can give rise to acquired capabilities that become relevant only under therapy.

Work on tumor development and treatment response in the Eμ-myc lymphoma model has also yielded the following explanation for the unique role of p53 in tumor biology: the p53 effector functions impaired early in tumorigenesis are context-dependent, varying with cell type and oncogenic insult, and can be phenocopied by loss of specific modifiers of p53 function. However, the p53 effector functions lost as a consequence of tumor evolution or anti-tumor therapy are context-independent and involve the primary mutation or deletion of p53. In other words, many genetic alterations can substitute for p53 loss early in tumorigenesis, but p53 mutation is inevitably selected for in most advanced or chemo-resistant malignancies. This paradigm predicts that many tumors without p53 mutations at early diagnosis may acquire these mutations by the time they prove fatal.

The relative abundance of p53 mutations, versus specific pathway alterations, in end-stage and chemo-resistant cancer may result from the surprising extent to which hypomorphic states of p53 can effectively carry out certain p53 effector functions. For example, while ARF loss and mdm2 expression can significantly decrease cellular levels of p53, the remaining p53 is capable of maintaining genomic stability and an essentially normal response to DNA damage (Fridman et al., 2003; Schmitt et al., 2002b). Similarly, p53 tumor-derived point mutants defective for most p53 functions also maintain genomic stability and a normal DNA damage response (Lozano and Liu, 1998). Finally, cells with p53 levels that are reduced by >95% by stable RNAi still respond to DNA damage-inducing chemotherapeutic agents and maintain a normal karyotype {Hemann, 2003 #302; and J.T. Zilfou, unpublished results}. Thus, inhibiting the ability of p53 to mediate cell cycle arrest in response to genetic instability may necessitate the mutation or deletion of p53.

P53 AND THE TREATMENT OF HUMAN CANCERS

While much of this chapter has focused on the role of p53 in murine tumorigenesis and therapy, it is clear that striking similarities exist between the relative importance of p53 in human and mouse malignancies. p53 is mutated in the majority of human end-stage malignancies and correlates with poor prognosis in both solid tumors and lymphoid malignancies (Greenblatt et al., 1994; Nigro et al., 1989; Soussi and Beroud, 2001). Additionally, as is the case is with the Eµ-myc mouse, p53 alterations are associated with drug resistance in human cancers (Ferreira et al., 1999; Vogelstein et al., 2000). This unique importance of p53 in human malignancies has prompted significant effort into developing therapies that specifically target tumor cells with mutant p53. These therapies include: 1) the direct viral delivery of p53 into tumor cells (Horowitz, 1999), 2) the infection of tumor cells with an adenovirus specifically engineered to kill p53-deficient cells (Heise et al., 1997), and 3) the use of small molecules that re-establish wild-type p53 function to mutant p53 proteins (Bullock and Fersht, 2001).

While these strategies certainly hold promise, the successful treatment of p53-mutated tumors may involve understanding and targeting the specific tumor-suppressive functions of p53 that are relevant in specific tumor contexts. For example, if apoptosis is the essential p53 effector function abrogated during a tumor's development, then specifically targeting inhibitors of apoptosis downstream of p53 may recapitulate the effects of p53 reintroduction. Further understanding of which p53 function(s) are key to the evolution of different tumor types may ultimately identify activities required for tumor maintenance and suggest additional targets for therapeutic intervention.

ACKNOWLEDGEMENTS

The authors would like to thank J. Zilfou and R. Dickins for helpful comments and editorial advice. M.T.H. is supported by the Helen Hay Whitney Foundation; S.W.L. is supported by grants from the National Cancer Institute, the Leukemia and Lymphoma Foundation, and an AACR-NFCR research professorship.

REFERENCES

Adams, J. M., Harris, A. W., Pinkert, C. A., Corcoran, L. M., Alexander, W. S., Cory, S., Palmiter, R. D., and Brinster, R. L. (1985). The c-myc oncogene driven by immunoglobulin enhancers induces lymphoid malignancy in transgenic mice. Nature *318*, 533-538.

Ahn, J., Urist, M., and Prives, C. (2003). Questioning the role of checkpoint kinase 2 in the p53 DNA damage response. J Biol Chem *278*, 20480-20489.

Barlow, C., Hirotsune, S., Paylor, R., Liyanage, M., Eckhaus, M., Collins, F., Shiloh, Y., Crawley, J. N., Ried, T., Tagle, D., and Wynshaw-Boris, A. (1996). Atm-deficient mice: a paradigm of ataxia telangiectasia. Cell *86*, 159-171.

Bullock, A. N., and Fersht, A. R. (2001). Rescuing the function of mutant p53. Nat Rev Cancer *1*, 68-76.

Bunz, F., Hwang, P. M., Torrance, C., Waldman, T., Zhang, Y., Dillehay, L., Williams, J., Lengauer, C., Kinzler, K. W., and Vogelstein, B. (1999). Disruption of p53 in human cancer cells alters the responses to therapeutic agents. J Clin Invest *104*, 263-269.

Clarke, A. R., Purdie, C. A., Harrison, D. J., Morris, R. G., Bird, C. C., Hooper, M. L., and Wyllie, A. H. (1993). Thymocyte apoptosis induced by p53-dependent and independent pathways. Nature *362*, 849-852.

de Stanchina, E., McCurrach, M. E., Zindy, F., Shieh, S. Y., Ferbeyre, G., Samuelson, A. V., Prives, C., Roussel, M. F., Sherr, C. J., and Lowe, S. W. (1998). E1A signaling to p53 involves the p19(ARF) tumor suppressor. Genes Dev *12*, 2434-2442.

Debbas, M., and White, E. (1993). Wild-type p53 mediates apoptosis by E1A, which is inhibited by E1B. Genes Dev *7*, 546-554.

Donehower, L. A., Harvey, M., Slagle, B. L., McArthur, M. J., Montgomery, C. A., Jr., Butel, J. S., and Bradley, A. (1992). Mice deficient for p53 are developmentally normal but susceptible to spontaneous tumours. Nature *356*, 215-221.

Eischen, C. M., Weber, J. D., Roussel, M. F., Sherr, C. J., and Cleveland, J. L. (1999). Disruption of the ARF-Mdm2-p53 tumor suppressor pathway in Myc-induced lymphomagenesis. Genes Dev *13*, 2658-2669.

Ferreira, C. G., Tolis, C., and Giaccone, G. (1999). p53 and chemosensitivity. Ann Oncol *10*, 1011-1021.

Fridman, J. S., Hernando, E., Hemann, M. T., de Stanchina, E., Cordon-Cardo, C., and Lowe, S. W. (2003). Tumor promotion by Mdm2 splice variants unable to bind p53. Cancer Res *63*, 5703-5706.

Greenblatt, M. S., Bennett, W. P., Hollstein, M., and Harris, C. C. (1994). Mutations in the p53 tumor suppressor gene: clues to cancer etiology and molecular pathogenesis. Cancer Res *54*, 4855-4878.

Heise, C., Sampson-Johannes, A., Williams, A., McCormick, F., Von Hoff, D. D., and Kirn, D. H. (1997). ONYX-015, an E1B gene-attenuated adenovirus, causes tumor-specific cytolysis and antitumoral efficacy that can be augmented by standard chemotherapeutic agents. Nat Med *3*, 639-645.

Hemann, M. T., Fridman, J. S., Zilfou, J. T., Hernando, E., Paddison, P. J., Cordon-Cardo, C., Hannon, G. J., and Lowe, S. W. (2003). An epi-allelic series of p53 hypomorphs created by stable RNAi produces distinct tumor phenotypes in vivo. Nat Genet *33*, 396-400.

Hemann MT, Zilfou JT, Zhao Z, Burgess DJ, Hannon GJ, Lowe SW (2004). Suppression of tumorigenesis by the p53 target PUMA. PNAS 101, 9333-9338.

Hermeking, H., and Eick, D. (1994). Mediation of c-Myc-induced apoptosis by p53. Science *265*, 2091-2093.

Herr, I., and Debatin, K. M. (2001). Cellular stress response and apoptosis in cancer therapy. Blood *98*, 2603-2614.

Horowitz, J. (1999). Adenovirus-mediated p53 gene therapy: overview of preclinical studies and potential clinical applications. Curr Opin Mol Ther *1*, 500-509.

Hussain, S. P., and Harris, C. C. (1998). Molecular epidemiology of human cancer: contribution of mutation spectra studies of tumor suppressor genes. Cancer Res *58*, 4023-4037.

Iliakis, G., Wang, Y., Guan, J., and Wang, H. (2003). DNA damage checkpoint control in cells exposed to ionizing radiation. Oncogene *22*, 5834-5847.

Jallepalli, P. V., Lengauer, C., Vogelstein, B., and Bunz, F. (2003). The Chk2 tumor suppressor is not required for p53 responses in human cancer cells. J Biol Chem *278*, 20475-20479.

Johnstone, R. W., Ruefli, A. A., and Lowe, S. W. (2002). Apoptosis: a link between cancer genetics and chemotherapy. Cell *108*, 153-164.

Johnstone, R. W., Ruefli, A. A., and Smyth, M. J. (2000). Multiple physiological functions for multidrug transporter P-glycoprotein? Trends Biochem Sci *25*, 1-6.

Kamijo, T., Zindy, F., Roussel, M. F., Quelle, D. E., Downing, J. R., Ashmun, R. A., Grosveld, G., and Sherr, C. J. (1997). Tumor suppression at the mouse INK4a locus mediated by the alternative reading frame product p19ARF. Cell *91*, 649-659.

Kemp, C. J., Donehower, L. A., Bradley, A., and Balmain, A. (1993). Reduction of p53 gene dosage does not increase initiation or promotion but enhances malignant progression of chemically induced skin tumors. Cell *74*, 813-822.

Komarova, E. A., Christov, K., Faerman, A. I., and Gudkov, A. V. (2000). Different impact of p53 and p21 on the radiation response of mouse tissues. Oncogene *19*, 3791-3798.

Liao, M. J., Yin, C., Barlow, C., Wynshaw-Boris, A., and van Dyke, T. (1999). Atm is dispensable for p53 apoptosis and tumor suppression triggered by cell cycle dysfunction. Mol Cell Biol *19*, 3095-3102.

Liu, G., Parant, J. M., Lang, G., Chau, P., Chavez-Reyes, A., El-Naggar, A. K., Multani, A., Chang, S., and Lozano, G. (2004). Chromosome stability, in the absence of apoptosis, is critical for suppression of tumorigenesis in Trp53 mutant mice. Nat Genet *36*, 63-68.

Lowe, S. W., and Ruley, H. E. (1993). Stabilization of the p53 tumor suppressor is induced by adenovirus 5 E1A and accompanies apoptosis. Genes Dev *7*, 535-545.

Lowe, S. W., Ruley, H. E., Jacks, T., and Housman, D. E. (1993a). p53-dependent apoptosis modulates the cytotoxicity of anticancer agents. Cell *74*, 957-967.

Lowe, S. W., Schmitt, E. M., Smith, S. W., Osborne, B. A., and Jacks, T. (1993b). p53 is required for radiation-induced apoptosis in mouse thymocytes. Nature *362*, 847-849.

Lozano, G., and Liu, G. (1998). Mouse models dissect the role of p53 in cancer and development. Semin Cancer Biol *8*, 337-344.

Lu, X., Magrane, G., Yin, C., Louis, D. N., Gray, J., and Van Dyke, T. (2001). Selective inactivation of p53 facilitates mouse epithelial tumor progression without chromosomal instability. Mol Cell Biol *21*, 6017-6030.

Nigro, J. M., Baker, S. J., Preisinger, A. C., Jessup, J. M., Hostetter, R., Cleary, K., Bigner, S. H., Davidson, N., Baylin, S., Devilee, P., and et al. (1989). Mutations in the p53 gene occur in diverse human tumour types. Nature *342*, 705-708.

Prives, C. (1998). Signaling to p53: breaking the MDM2-p53 circuit. Cell *95*, 5-8.

Rao, L., Debbas, M., Sabbatini, P., Hockenbery, D., Korsmeyer, S., and White, E. (1992). The adenovirus E1A proteins induce apoptosis, which is inhibited by the E1B 19-kDa and Bcl-2 proteins. Proc Natl Acad Sci U S A *89*, 7742-7746.

Rogoff, H. A., Pickering, M. T., Debatis, M. E., Jones, S., and Kowalik, T. F. (2002). E2F1 induces phosphorylation of p53 that is coincident with p53 accumulation and apoptosis. Mol Cell Biol *22*, 5308-5318.

Schmitt, C. A., Fridman, J. S., Yang, M., Baranov, E., Hoffman, R. M., and Lowe, S W (2002a). Dissecting p53 tumor suppressor functions in vivo. Cancer Cell *1*, 289-298.

Schmitt, C. A., Fridman, J. S., Yang, M., Lee, S., Baranov, E., Hoffman, R. M., and Lowe, S. W. (2002b). A senescence program controlled by p53 and p16INK4a contributes to the outcome of cancer therapy. Cell *109*, 335-346.

Schmitt, C. A., and Lowe, S. W. (1999). Apoptosis and therapy. J Pathol *187*, 127-137.

Schmitt, C. A., McCurrach, M. E., de Stanchina, E., Wallace-Brodeur, R. R., and Lowe, S. W. (1999). INK4a/ARF mutations accelerate lymphomagenesis and promote chemoresistance by disabling p53. Genes Dev *13*, 2670-2677.

Schwartz, D., and Rotter, V. (1998). p53-dependent cell cycle control: response to genotoxic stress. Semin Cancer Biol *8*, 325-336.

Serrano, M., Lin, A. W., McCurrach, M. E., Beach, D., and Lowe, S. W. (1997). Oncogenic ras provokes premature cell senescence associated with accumulation of p53 and p16INK4a. Cell *88*, 593-602.

Soussi, T., and Beroud, C. (2001). Assessing TP53 status in human tumours to evaluate clinical outcome. Nat Rev Cancer *1*, 233-240.

Stott, F. J., Bates, S., James, M. C., McConnell, B. B., Starborg, M., Brookes, S., Palmero, I., Ryan, K., Hara, E., Vousden, K. H., and Peters, G. (1998). The alternative product from the human CDKN2A locus, p14(ARF), participates in a regulatory feedback loop with p53 and MDM2. Embo J *17*, 5001-5014.

Symonds, H., Krall, L., Remington, L., Saenz-Robles, M., Lowe, S., Jacks, T., and Van Dyke, T. (1994). p53-dependent apoptosis suppresses tumor growth and progression in vivo. Cell *78*, 703-711.

Tolbert, D., Lu, X., Yin, C., Tantama, M., and Van Dyke, T. (2002). p19(ARF) is dispensable for oncogenic stress-induced p53-mediated apoptosis and tumor suppression in vivo. Mol Cell Biol *22*, 370-377.

Tsai, K. Y., MacPherson, D., Rubinson, D. A., Crowley, D., and Jacks, T. (2002). ARF is not required for apoptosis in Rb mutant mouse embryos. Curr Biol *12*, 159-163.

Vafa, O., Wade, M., Kern, S., Beeche, M., Pandita, T. K., Hampton, G. M., and Wahl, G. M. (2002). c-Myc can induce DNA damage, increase reactive oxygen species, and mitigate p53 function: a mechanism for oncogene-induced genetic instability. Mol Cell *9*, 1031-1044.

Vogelstein, B., Lane, D., and Levine, A. J. (2000). Surfing the p53 network. Nature *408*, 307-310.

Volm, M. (1998). Multidrug resistance and its reversal. Anticancer Res *18*, 2905-2917.

Vousden, K. H., and Lu, X. (2002). Live or let die: the cell's response to p53. Nat Rev Cancer *2*, 594-604.

Wallace-Brodeur, R. R., and Lowe, S. W. (1999). Clinical implications of p53 mutations. Cell Mol Life Sci *55*, 64-75.

Xu, Y., Yang, E. M., Brugarolas, J., Jacks, T., and Baltimore, D. (1998). Involvement of p53 and p21 in cellular defects and tumorigenesis in Atm-/- mice. Mol Cell Biol *18*, 4385-4390.

Yonish-Rouach, E., Resnitzky, D., Lotem, J., Sachs, L., Kimchi, A., and Oren, M. (1991). Wild-type p53 induces apoptosis of myeloid leukaemic cells that is inhibited by interleukin-6. Nature *352*, 345-347.

Zindy, F., Eischen, C. M., Randle, D. H., Kamijo, T., Cleveland, J. L., Sherr, C. J., and Roussel, M. F. (1998). Myc signaling via the ARF tumor suppressor regulates p53-dependent apoptosis and immortalization. Genes Dev *12*, 2424-2433.

Chapter 16

NOVEL P53-BASED THERAPIES: STRATEGIES AND FUTURE PROSPECTS

Sonia Lain and David Lane
Department of Surgery and Molecular Oncology. University of Dundee. Ninewells Hospital. Dundee, Scotland, UK

INTRODUCTION

Introducing functional p53 into tumours using adenoviral vectors is leading to success in clinical trials (Edelman et al., 2003; http://www.introgen.com/infotp.html; http://www.sibiono.com) and p53 gene therapy gained regulatory approval in China in 2003 where it has been on sale since January 2004. Renewed optimism around the use of p53 gene therapy and increased understanding of p53 function suggest that many more potent and selective variants of p53 may be developed. Reactivating mutant p53 or exploiting specific properties of tumour cells carrying mutations in p53 is a greater challenge. As described elsewhere in this book, tackling this problem has led to a variety of very exciting discoveries. Here we have focused on the current approaches to activate p53 in those cancers that retain wild type p53.

Whereas more than 50% of solid tumours occurring in adults bear p53 mutations, the rate of p53 mutations is significantly lower in haematological malignant diseases, childhood cancer and malignancies associated with viral infections (Krug et al., 2002; Toren et a., 1996; Mantovani and Banks, 2001). This may be the key to the much better prognosis of children with cancer. However, the genotoxic effects of many current therapies is especially important to bear in mind when considering the treatment of young patients (Wallace et al., 2001; Schwartz, 1999). Hence, searching for novel non-genotoxic activators of the p53 response is thought to be essential

P. Hainaut and K.G. Wiman (eds.), 25 Years of p53 Research, 353-376.

in improving the treatment of those cancers in which p53 function is not abolished by mutation.

p53's Tumour Suppressor Function

The tumour suppressor activity of wild type p53 is thought to be mainly due to its ability to act as a transcription factor and induce the expression of large number of proteins (Vousden and Lu, 2002; Bourdon et al., 2002). Some of these proteins are involved in causing the inhibition of cell proliferation by promoting cell cycle arrest at different checkpoints and others can cause the induction of cell death by apoptosis. The discovery of p53 alterations that modulate its effects on different promoters and the definition of cofactors that specifically promote p53-induced apoptosis suggest that the cell-cycle arrest and apoptotic activities of p53 can be separated. These findings are thoroughly discussed in an insightful review by Vousden and Lu (2002).

While most functions of p53 involve its activity as a transcription factor, there is growing evidence on its transcription-independent activities. p53-dependent and bax-dependent apoptosis can be triggered in the absence of macromolecular synthesis and even in the absence of a nucleus (Chipuk et al., 2003). Reported data suggest that cytoplasmic p53 functions analogously to a subset of pro-apoptotic Bcl-2 proteins that permeabilise mitochondria and switch-on the apoptotic process (Chipuk et al., 2004). This implies that the induction of the pro-apoptotic transcription-independent function of p53 could also be exploited in therapy. The p53 protein has also been proposed to play a transcription independent role in the promotion and regulation of DNA repair and again modulating this function could be exploited therapeutically (Kumari et al., 2004).

Inhibition of p53's Function in Tumours Expressing Wild Type p53

In many of the tumours that encode wild type p53, its levels and activity are thought to be hampered by virtue of alterations in other cellular factors or the expression of viral oncogenes. Clear examples include the overexpression of the Mdm2 protein (Freedman et al., 1999), defects in the expression of the p14ARF tumour suppressor (Sherr, 2001), mutations in kinases such as ATM or Chk2 (Dasika et al., 1999; Bartek et al., 2001), chromosome translocations involving the PML or the nucleophosmin proteins (Salomoni and Pandolfi, 2002; Colombo et al., 2002) and the expression of alternative spliced forms of p53 (J.C. Bourdon, unpublished data). Additionally, infection with malignant viruses leads to effective downregulation of p53 levels and activity (zur Hausen, 1999a; 1999b; Lain and Lane, 2003). In HPV-positive cervical cancers, the activity of p53 is rarely abolished by mutation. Instead p53 levels are effectively decreased by the action of the oncoviral protein E6 and the cellular E3 ubiquitin ligase E6AP (reviewed in Mantovani and Banks, 2001). HPV infects the basal cells of the epithelia and HPV DNA has been found in 90% of cervical cancer and 50% of vulvar cancer (Jastreboff and Cymet, 2001). Cervical cancer is the second most common cause of cancer-related death in women world-wide, in some developing countries accounting for the highest cancer mortality (Stern et al., 2001).

Regulation of p53 degradation

If p53's primary function is to stop cell proliferation leading to a cell cycle halt or to induce cell death by apoptosis, its levels and activity must be tightly regulated. The mechanisms that control p53 gene transcription remain largely unknown and many reports have indicated that p53 levels are mainly regulated at the post-transcriptional level. However, recent work suggests *p53* promoter activation by oestrogen (Qin et al., 2002) as well as by p53 itself and NFkB (Benoit et al., 2000) and the existence of a promoter element that is involved in basal *p53* gene expression and the stress response (Noda et al., 2000). The level of *p53* gene transcription must be very important in setting the threshold of the p53 response as mice that have only a single copy of the gene show a cell autonomous haplo-insufficiency in their apoptotic response (Lowe et al., 1993; Clarke et al., 1993).

In normal non-stressed cells without mutations in the *p53* gene p53 has a very short half life. Until recently, this was thought to be due to the following autoregulatory feedback loop mechanism in which the Mdm2 protein plays a key role. Wild type p53 acts as a transcriptional activator of the *Mdm2* gene (Barak et al., 1993). In turn, Mdm2, which itself has a very short half-life due its autoubiquitination activity, has the ability to interact with p53 and to function as an E3 ubiquitin ligase that promotes the conjugation of p53 to polyubiquitin (Honda et al., 1997). This conjugation to ubiquitin serves as a tag that effectively targets p53 for degradation by the proteasome (Haupt et al., 1997; Kubbutat et al., 1997). In this way, in normal non-stressed cells, p53 levels would be maintained at a very low level and cells could be allowed to proliferate. Contributing to its 'anti-p53' function, Mdm2 is also thought to impair the transcriptional activity of p53 by masking the transactivation domain of this tumour suppressor (Chen et al., 1996). More recently, alternative E3 ubiquitin ligases for p53 have been identified. These include Pirh2 and COP1, both of which, like Mdm2, are transcriptional targets (Leng et al 2003; Dornan et al 2004). As with p53 the level of Mdm2 transcription plays a critical role in setting the sensitivity of the p53 pathway and the recent description of a common polymorphism in the human Mdm2 promoter that appears to affect tumour susceptibilty in man has brought this into sharp focus (Bond et al., 2004).

Supporting Mdm2's crucial role in the regulation of p53 activity, Mdm2 knockout mice are rescued from embryonic lethality by deletion of p53 (Montes de Oca Luna et al.,1995). A similar requirement has also been observed when MdmX (Mdm4) mice were developed, suggesting the importance of the MdmX protein in regulating the levels and activity of p53 (Parant et al., 2001; Badciong and Haas, 2002).

Tumour cells that encode mutant p53 show high levels of stable p53, although these forms are still susceptible to degradation by Mdm2 (Midgley et al., 1997). According to the current dogma, the explanation for this observation is that p53 mutation abolishes its ability to act as a transcriptional factor on the *Mdm2* gene. Hence, the levels of Mdm2 in cells carrying mutations in p53 are not sufficient to degrade p53 efficiently. This is why the detection of high levels of p53 in tumours can be used as an indicator for the existence of mutations in the *p53* gene.

The following observations have challenged the role of p53-dependent *Mdm2* transcription in the maintenance of low p53 levels in normal tissues. On the one hand, p53 status does not regulate the levels of expression of Mdm2 mRNA in non-stressed cells (Mendrysa et al., 2000). Second, and contrary to expectations, mice expressing mutant p53 alone have low levels of p53 in normal tissues (Olive et al., 2004; G. Lozano, unpublished data).

p53 activation of Mdm2 mRNA does occur but only when cells are subjected to stress (Mendrysa et al., 2000). Hence, under stress conditions, wild type p53 would promote its own degradation by increasing *Mdm2* gene transcription, whereas mutant p53 would not be able to do so and would accumulate. Possibly reflecting a stress situation, p53 levels are high in tumours appearing in mice expressing mutant p53 only and in cultured MEFs derived from these mice (Olive et al., 2004; Lang et al., 2004).

Taken together, these observations imply that the basal (p53-independent) expression of the *Mdm2* gene, despite the intrinsic instability of Mdm2, is sufficient to reduce p53 levels effectively in non-stressed cells.

p53 Can Be Activated Without Causing DNA Damage

The elucidation of the mechanisms by which p53 is activated when cells are subjected to stress has been an area of intense research, with special emphasis on the analysis of the phosphoryation and acetylation status of p53 and Mdm2 (Ljungman, 2000; Lakin and Jackson, 1999; Maya et al., 2001; Goldberg et al., 2002, Blattner et al., 2002; Zhou and Hung, 2002; Vousden and Lu, 2002; Prives and Manley, 2001). Other modifications of p53 and Mdm2 that may affect p53 stability and activity include SUMO conjugation, neddylation and methylation (Rodriguez et al., 1999; Xirodimas et al., 2004; Chuikov et al., 2004).

It is well known that p53's levels and transcriptional activity increase in cells that are irradiated or treated with DNA damaging agents. Telomere erosion during the ageing process and defects in the DNA structure due to mistakes during the processes of DNA replication and recombination are thought to trigger the p53 response by mechanisms that are likely to be similar to the response to DNA damaging agents (Vaziri and Benchimol, 1996).

p53 activity can also be increased by a variety of general stresses that do not, or at least not directly, involve DNA damage. These include hypoxia, serum starvation, heat, cold, pH, ribonucleotide depletion, glycerol and inhibition of nuclear export (reviewed in Ohnishi and Ohnishi 2001; Pluquet and Hinaut, 2001; Lain et al., 1999; Linke et al., 1996). Pioneering clinical trials support the value of some of these observations (Martin et al., 2001; Ohnishi and Ohnishi, 2001). Oncogenic signals and certain viral oncoproteins can also increase the levels of p53 (Sherr, 2001). This activation, at least in some situations, is mediated by an increase in the levels of expression of the ARF tumour suppressor, an antagonist of the Mdm2 function (see below).

WAYS TO IMPROVE PRESENT TREATMENTS

Since Mdm2 is an important inhibitor of p53 function in cells, a likely way to increase p53 levels in a non-genotoxic way is by impairing its effects on p53. In the last years, a lot of effort has been put in dissecting the p53/Mdm2 pathway. This work has given rise to a variety of ideas on how to reduce Mdm2 activity. In some cases, the aim is to specifically impair Mdm2's function on p53. Others involve the manipulation of general cellular processes. Activation of p53's transcriptional function is achieved in several cases. In others, p53 levels are increased but p53-dependent transcriptional activity is not detected. Some of these approaches are highly developed and specific small molecules are available.

Several of these ideas may also be applicable in the future to inhibit the function of other p53-targetting E3 ubiquitin ligases such as Pirh2 (Leng et al., 2004) or COP1 (Dornan et al., 2004) and also to the effect of the HPV infection on p53 levels.

Inhibiting the Interaction Between p53 and Mdm2

The p53-Mdm2 interaction sites map to the N-terminal of transactivation domain of p53 and to the N-terminal domain of Mdm2 (Picksley et al., 1994; Chen et al., 1993; Kussie et al., 1996). Targetting the physical interaction between p53 and Mdm2 has long been considered as the most direct strategy to activate p53. This has led to the identification of antibodies and peptides that validate this approach. Microinjection of the 3G5 antibody that specifically recognises the p53 binding site in Mdm2, increases p53 levels and transcriptional activity (Bottger et al., 1997). Introduction of peptides that interact with Mdm2 and prevent its association with p53 (Bottger et al., 1997; Chene et al., 2000, 2002; Garcia-Echevarria et al., 2001) or the expression of such peptides in a scaffold protein, also leads to an increase in p53 levels and activity (Bottger et al., 1997).

Computer-aided design has led to the the synthesis of a non-peptidic polycyclic molecule with affinity for Mdm2 and p53-activating effect (Zhao et al., 2002). Screening of microbial extracts identified chlorofusin (*Fusiparum sp.*) as an inhibitor of the p53-Mdm2 interaction that binds to the N-terminus of Mdm2 (Duncan et al., 2001, 2003). A detailed characterisation of the specificity of these compounds and their effects on the p53 pathway and tumour cell growth is still required.

The most potent and specific inhibitors of the p53-Mdm2 interaction published so far are the nutlins (Vassilev et al., 2004). These *cis*-imidazoline derivatives bind to the N-terminal hydrophobic pocket in Mdm2 and displace p53 with IC_{50} values in the sub-micromolar range. These compounds are cell permeable, impair the interaction between p53 and Mdm2 in vivo, induce the stabilisation of p53 and activate its transcriptional function at micromolar concentrations. In tumour cells, activation of p53 by nutlins is followed by cell cycle inhibition and apoptosis. Importantly, the effect of this familiy of compounds does not involve the induction of the DNA-damage response as assessed by monitoring p53 Serine 15 phosphorylation. Nutlins have a poor effect on cells lacking functional p53 and cause a reversible cell cycle inhibitory function in normal human fibroblasts. Oral administration of nutlins also achieves 90% inhibition of tumour growth in mouse xenograft models of p53 wild-type human tumours without noticeable toxicity in mice. Clinical studies are now in progress.

Screening of a small chemical library for inducers of p53-dependent death has recently led to the identification of a small molecule named RITA (reactivation of p53 and induction of tumor cell apoptosis) (Issaeva et al., 2004). This molecule does not bind to Mdm2, but instead binds to the amino terminus of p53 preventing p53-Mdm2 interaction in vitro and in vivo, and affecting p53 interaction with several negative regulators. RITA induces p53 accumulation in tumour cells, expression of p53 target genes, massive apoptosis in tumour cells lines expressing wild-type p53 and shows p53-dependent antitumor effect in vivo. Suggesting that RITA has limited effects in at least some normal tissues, this molecule only suppresses the growth of human normal fibroblasts and lymphoblasts upon oncogene expression.

Decreasing Mdm2 Synthesis

Synthesis of Mdm2 can be specifically decreased by antisense oligonucleotides (Chen et al., 1998) or small interefering RNAs (Saville et al., 2004). This can lead to an increase in the levels and transcriptional activity of p53 (Saville et al., 2004). Furthermore, intraperitoneal administration of anti-Mdm2 antisense oligonucleotides results in a reduction in xenograft tumour size and to a synergistic enhancement in the sensitivity of these tumours to conventional chemotherapeutic drugs (Wang et al., 1999; Tortora et al., 2000).

Inhibition of Transcription

RNA synthesis inhibitors blocking the phosphorylation of the carboxyl terminal domain (CTD) of RNA polymerase II, such as DRB, lead to increased p53 levels and paradoxically, also raise the transcription of p53-dependent genes (reviewed in Ljungman and Lane, 2004). Importantly, the p53 accumulated by this type of transcription inhibition is not phosphorylated at Serine 15, suggesting that the accumulation of p53 by DRB is not due to the activation of the DNA damage response pathway by DRB (Ljungman et al., 2001). Another way to explain the increase in p53 levels by DRB is that inhibition of transcription severely affects the levels of proteins that, like Mdm2, are very unstable. In fact, DRB treatment leads to a rapid reduction in Mdm2 levels. However, against this hypothesis, overexpression of Mdm2 is not sufficient to overcome the effect of transcription inhibition on p53 (O'Hagan and Ljungman, 2004). Other mechanisms such as the negative effect of mRNA synthesis inhibition on the export of proteins from the nucleus may cause the accumulation of p53 by DRB (Ljungman and Lane, 2004).

Instead, inhibitors of the elongation step in mRNA synthesis do lead to the accumulation of p53 phosphorylated at Serine 15, suggesting that inhibition of the elongation step in RNA polymerase II-dependent transcription may activate the transcription coupled repair response (TCR) that leads to p53 phosphorylation and activation (Ljungman and Lane, 2004).

Specific inhibitors of eukaryotic RNA polymerases I and III are not available. Suggesting that RNA polymerase I activity could signal to p53, the injection of an antibody agaist the RNA polymerase I specific factor UBF, increases p53 (Rubbi et al., 2003). Whether this leads to a reduction in RNA polymerase I activity and to an increase in the transcriptional activity of p53 has not yet be established. Nevertheless, these experiments are interesting due to the large number of small molecule activators of p53 that have a dramatic effect on the structure of the nucleolus (Rubbi et al., 2003). Disrupting nucleolar stucture has been shown to potentiate the effect of regulators of the p53/Mdm2 pathway such as the ribosomal protein L11 (Lohrum et al., 2003).

Inhibiting Ubiquitination of p53 by Mdm2

Mdm2's RING finger domain catalyses the conjugation of ubiquitin to p53 and to Mdm2 itself. This activity is necessary for the degradation of p53 as well as for the maintenance of low levels of Mdm2 (Fang et al., 2000). The observation that the accumulation of ubiquitinated forms of p53 and of Mdm2 autoubiquitination are separable is of remarkable importance (Fang et al., 2000; Xirodimas et al 2001a). If p53 and Mdm2 are ubiquitinated by distinct mechanisms, it may be possible to specifically inhibit the in the degradation of p53 complexes without the drawback of accumulating Mdm2 simultaneously. As mentioned before, binding of Mdm2 to p53 is sufficient to impair p53-dependent transcription (Chen et al., 1996).

Three chemically distinct types of compounds specifically inhibit p53 ubiquitination by Mdm2 without affecting Mdm2's ubiquitination in vitro (Lai et al., 2002). These studies show that these compounds have a reversible effect, do not compete with preconjugated-ubiquitin or p53 for Mdm2 binding and suggest a common binding site on Mdm2 for all of these inhibitors. Whether any these compounds can promote p53 stabilisation and activation in vivo remains to be investigated.

Ubiquitination of proteins occurs through the sequential actions of three enzymes. Initially ubiquitin is activated by the ubiquitin-activating enzyme (E1). Ubiquitin is then transferred from the E1 to a ubiquitin-conjugating enzyme (E2). A ubiquitin ligase (E3) then facilitates transfer of ubiquitin from an E2 to the substrate. Various E2s have been tested for their role in promoting Mdm2-mediated ubiquitination of p53 (Saville et al., 2004). Of the E2s tested only UbcH5A, -B, and -C and E2-25K support Mdm2-mediated ubiquitination of p53 in vitro. The same E2s also support Mdm2 auto-ubiquitination. According to these results, UbcH5B/C are likely to be physiological E2s for Mdm2 and contribute to the maintenance of low levels of p53 and Mdm2 in unstressed cells. Knockdown of UbcH5B/C with specific siRNAs causes accumulation of Mdm2 and p53 in unstressed cells and inhibits p53 ubiquitination and degradation. However, despite up-regulating the level of nuclear p53, UbcH5B/C knockdown is not sufficient to increase in p53-dependent transcription activity. As suggested by the authors (Saville et al., 2004), one explanation for the lack of p53 activation upon UbcH5B/C knockdown is that the concomitant accumulation of Mdm2 would block p53's transcription activating domain.

UbcH5B/C knockdown may not be sufficient to activate p53 efficiently, but its stabilising effect on p53 levels could contribute to the effect of stress on p53 function. UbcH5B/C knockdown was tested in combination with two therapeutic drugs doxorubicin and actinomycin D (Saville et al., 2004). UbcH5B/C knockdown does not sensitize p53 to activation by these two agents. However, it is interesting to note that the levels of UbcH5B/C are already reduced by doxorubicin and actinomycin D in cells, and this is likely to mask the effect of an additional decrease in UbcH5B/C levels by siRNA.

Destabilising Mdm2

E3 ubiquitin ligases catalyse the conjugation of ubiquitin to specific protein substrates promoting their degradation by the proteasome. An interesting and frequent property of these enzymes is that like Mdm2, they are able to ubiquitinate themselves and are therefore, highly unstable proteins. Studying the factors that positively or negatively affect Mdm2's stability rather than its synthesis is more likely to lead to novel small molecules that have a desired effect on tumours expressing wild type p53.

In response to DNA damage, p53/Mdm2 levels oscillate (Lev Bar-Or et al., 2000; Lahav et al., 2004). A delay in the accumulation of Mdm2 is necessary, albeit not sufficient, to explain this type of behaviour. There is clear evidence showing that Mdm2 stability is significantly decreased in response to DNA damage in p53 positive cells (Stommel et al., 2004). Furthermore, a decrease in Mdm2's stability in response to DNA damage coincides with and is necessary for the detection of an increase in p53 levels and transcriptional activity (Stommel et al., 2004).

HAUSP is a deubiquitinating enzyme first reported to bind and deubiquitinate p53 (Li et al., 2002). However, HAUSP expression also has a stabilising effect on Mdm2 and can lead to increased Mdm2 levels in the presence and absence of p53 (Li et al., 2004). Therefore, potentiating the effect of HAUSP, even if it reduces the ubiquitination of p53, does not lead to the activation of p53 (R. Berkson, unpublished data).

A remarkable finding is that p53 levels are clearly increased in HAUSP-knockout cells (Cummins et al., 2004). This suggests that that inhibiting the activity of HAUSP, unlike its activation or overexpression, could lead decrease the stability of Mdm2 and therefore have a positive effect on p53 levels and activity.

As previously discussed, only under stress situations, p53 regulates its own levels by increasing the synthesis (transcription) of its E3 ubiquitin ligase Mdm2 (Mendrysa et al., 2000) and p53's transcriptional activity is not required to maintain low p53 levels in normal tissues (Oliver et al., 2004). This implies that basal expression of Mdm2 is sufficient to destabilise p53 effectively in non-stress conditions. However, given the intrinsic instability of Mdm2, this is difficult to assume. The effect of a substrate on the fate of its E3 ubiquitin ligase is poorly understood. In the presence of their substrate, E3 ubiquitin ligases could be degraded or spared. Only if the ligase is spared from degradation, that is if it is stabilised by the binding of the substrate, one molecule of E3 ligase could be utilised to ubiquitinate and degrade a larger number of substrate units. In this situation, the E3 ubiquitin ligase would act in a highly efficient way. The fact that transcriptionally inactive p53 can stabilise its E3 ubiquitin ligase Mdm2 (Peng et al., 2001), may be key to explain the effectiveness of basal levels of Mdm2 expression. According to this model, p53 would promote its self-destruction by sparing Mdm2 from degradation. Modulating this transcriptionally independent effect of p53 over Mdm2 could provide new approaches for therapeutic intervention.

Inhibiting the Activity of the Proteasome

Proteasome inhibitors very effectively increase p53 levels and encouraging results are being obtained in clinical trials with the proteasome inhibitor PS-341 (Adams, 2002). However, whether proteasome inhibitors are effective at inducing p53-dependent transcription has not been fully established. In our hands, accumulation of p53 by proteasome inhibitors does not lead to p53-dependent transcription. Aside from the possible effects of this type of approach on multiple cellular processes, including general protein expression, proteasome inhibitors lead to the accumulation of high levels of Mdm2. Whether accumulation of p53 by proteasome inhibition can result in the activation of the transcription-independent apoptotic function of p53 (Chipuk et al., 2003) is not yet established. Our results indicate that these agents do not cause cell death in a p53-dependent way (Lain et al., 1999a; R. Berkson, unpublished data).

Accumulating p53 in the nucleus

It is generally accepted that p53 shuttles between the nucleus and the cytoplasm (reviewed in Lain et al., 1999b). An attractive approach to enhance p53 function is to induce its accumulation in the nucleus, where it can act as a transcriptional activator. The nuclear export inhibitor leptomycin B (LMB) is a specific inhibitor of the crm1 exportin, a nuclear factor that mediates the export from the nucleus of proteins that contain an HIV Rev-type of nuclear export signal (NES). LMB is an extremely potent activator of p53-dependent transcription even at sub-nanomolar concentrations (Lain et al., 1999a; Menendez et al., 2003), is very effective at increasing p53 levels in the nucleus (Freedman et al., 1998; Lain et al., 1999a) and protects p53 from Mdm2-mediated ubiquitination and degradation (Lain et al., 1999a; Xirodimas et al. 2001b). Subsequent work suggested that Mdm2 and p53 have one and two nuclear export signals, respectively (reviewed in Lain ad Lane 2003). Nevertheless, there is, as yet, no report of an interaction between any of these proteins and crm1. Therefore, it is still possible that at least part of the effect of LMB on p53 may be indirect.

Whatever the exact mechanism(s) leading to not only the accumulation, but more importantly, to the activation of the p53 transcriptional activity by LMB, our experience screening diverse compound libraries has led us to conclude that LMB is one of the most potent known activators of p53. LMB can increase p53 activity at nanomolar concentrations, which are below the concentrations required for nutlins (Menendez et al., 2003; Vassilev et al., 2004).

The effects of LMB have been tested on normal cells in monolayer culture (human primary dermal fibroblasts) and on a range of tumour cells. LMB induces the transcriptional activity of p53 in primary dermal fibroblasts, but has a relatively mild and reversible growth inhibitory effect on these cells even at micromolar concentrations (Smart et al., 1999). Instead, even at nanomolar concentrations, LMB is very potent at promoting the death of cells derived from neuroblastomas (Smart et al., 1999), one of the most frequent solid tumours occurring in children. IC_{50} values for LMB on other tumour cells in culture are also in the nanomolar range. In neuroblasoma cells, which usually express wild type p53, p53-levels and activity were increased and LMB's toxicity was to some extent dependent on the transcriptional activation of p53 (Smart et al., 1999). LMB is also able to raise p53 levels and induce its transcriptional activity and growth inhibitory functions in HPV-positive cells derived from cervix cancers. As shown in our lab (Hietanen et al., 2000), LMB increases p53 levels and activity, reduces the levels of HPV E6/E7 mRNA and significantly potentiates the apoptotic effect of low concentrations of actinomycin D in cells derived from cervical cancers. Furthermore, by introducing the dominant negative form of p53 in these cells, the killing effect of LMB on its own and the drug combination (LMB+Actinomycin D) at nanomolar concentrations was shown to be substantially p53-dependent. This study has two implications.

First it shows that HPV cells are highly sensitive to the activation of p53, and second, it suggests that it is possible to activate p53 in these cells in a non-genotoxic way.

As could be expected, inhibition of crm1-mediated nuclear export affects a variety of proteins involved in different cellular processes. Here we mention some of the observations on the effect of crm1-mediated export inhibition on proteins regulating cell proliferation and survival. Aside from p53, these include tumour suppressors such as INI1 (integrase interactor1) (Craig et al., 2002) and BRCA1 (Rodriguez and Henderson, 2000). Another well established tumour suppressor, APC (adenomatous polyposis coli), requires its nuclear export sequence for its ability to induce the degradation of β-catenin (Rosin-Arbesfeld et al., 2000). Nevertheless, an APC mutant deficient for nuclear export can sequester β-catenin and this is sufficient to inhibit its transcriptional activity (Neufeld et al., 2000). Concerted distribution of cyclins is necessary for normal cell cycle progression. In the case of cyclin B1, which is sensitive to LMB (Hagting et al., 1998), overexpression of a mutant that preferentially accumulates in the nucleus is sufficient to trigger apoptosis (Porter et al., 2003). In contrast, the activity of the NFκB survival- associated transcription factor, is negatively regulated by LMB. In particular, LMB promotes the accumulation in the nucleus of the NFκB inhibitor Iκβ where it is protected from proteasome-dependent degradation (Rodriguez et al., 1999). A marked decrease in c-myc expression in tumour cells is also observed (F. Murray-Zmijewsky, unpublished results).

It has been known for years that the pharmacokinetic behaviour of LMB is difficult to study. This together with the secondary effects observed when the drug was injected intravenously in a phase I trial stopped tests in humans almost a decade ago (Newlands et al., 1996). We consider that the promising properties of LMB in cell culture systems described over the last years should lead to the revision of the therapeutic application of LMB or more suitable derivatives in specific tumour types.

Inhibiting Mdm2 Phosphorylation

There is evidence suggesting that Mdm2 function can be impaired by small molecule kinase inhibitors. PKB-mediated phosphorylation of Mdm2 on Serines 166 and 188 correlates with inhibition of Mdm2 auto-ubiquitination and stabilisation (Feng et al., 2004). Moreover, mouse embryonic fibroblasts lacking PKBα displayed reduced Mdm2 protein levels with a concomitant increase of p53 and p21(Cip1), resulting in strongly elevated apoptosis after UV irradiation. Whether the stabilisation of Mdm2 by PKB phosphorylation is related to alterations in the subcellular localisation of Mdm2 (Mayo and Donner, 2001) needs to be investigated further. From a therapeutic point of view, it is interesting to note that the small molecule PI-3 kinase inhibitor LY294002 leads to the accumlation of p53, and at least in some cell types to the degradation of Mdm2 (Feng et al., 2004).

Mimicking p14ARF Activity

The ARF tumour suppressor (p14ARF in human, p19ARF in mouse) encoded by the INK4α/ARF locus (Quelle et al., 1995) is a small protein that inhibits degradation of the p53 tumour suppressor mediated by Mdm2 (Zhang et al., 1998; Pomerantz et al., 1998). Several non-exclusive models have been proposed to explain this effect of p14ARF. These involve inhibition of p53 export from the nucleus (Zhang and Xiong, 1999, Tao and Levine, 1999), sequestration of Mdm2 by ARF in the nucleolar compartment (Weber et al., 1999) and direct inhibition of the ubiquitin E3 ligase activity of Mdm2 on p53 (Honda and Yasuda, 1999; Midgley at al., 2000; Xirodimas et al., 2001a). The effect of ARF on cell proliferation is very likely to be primarily dependent on its ability to activate p53 (Carnero et al., 2000). This does not imply that ARF does not have effects on other proteins and pathways (Sugimoto et al., 2003), but it is clear that the consequences of these on cell proliferation in the absence of p53, are weaker (Carnero et al., 2000; Weber et al., 2000).

One of the most intriguing properties of ARF is that its sequence is very poorly conserved between species and only the exon-1β encoded region, which is essential for most of ARF's functions (Weber et al., 1999; Midgley et al., 2000; Lohrum et al., 2000; Xirodimas et al., 2001a; Weber et al., 2000; Sugimoto et al., 2003), is expressed in chicken (Kim et al., 2003). Even in this region, only six residues are strictly conserved between species. Also, the nucleolar localisation sequences in ARF (RXRR or RRXR), although present in all known ARF sequences, are positioned in different regions of the molecule. p14ARF is also characterised by a high tendency to form high molecular weight complexes, a property that is enhanced in the presence of Mdm2 (Bothner et al., 2001) and oxidative conditions (Menendez et al., 2003). Another property of this small protein is that its levels of expression are very low in normal tissues (Zindy et al., 2003). Interestingly, the p19ARF promoter is not active in most normal tissues, but is activated in tumours (Zindy et al., 2003). High levels of the protein are only found in early passage mouse fibroblasts and in the nucleolar compartment of cultured p53-negative tumour cells (Weber, 1999; Stott et al., 1998). Altogether, these three properties of ARF (poor sequence conservation, tendency to aggregate and low levels) contribute to the difficulties in the evaluation of the physiological relevance of the multiple functions suggested for ARF.

Mimicking the function of p14ARF has been proposed as a means to activate p53 function and to develop non-genotoxic therapies. Small peptides corresponding to the amino terminus of p14ARF can promote p53 stabilisation (Midgley et al., 2000; Lohrum et al., 2000). One consideration in this type of approach is that, aside from increasing and activating p53, ectopic ARF also increases the levels of Mdm2 in a p53-independent way (Xirodimas et al., 2001a). A consequence of this is that the effect of p14ARF on p53 may be limited. In fact, high levels of p14ARF are less effective than low levels of p14ARF at increasing p53-dependent transcription (our unpublished data).

Whether an increase in Mdm2 levels by ARF is biologically relevant, that is whether p14ARF limits its own positive effect on p53 at physiological levels and whether this occurs not only as consequence of the enhancement of p53's transcriptional activity on the *Mdm2* promoter but also through the ability of ARF to stabilise Mdm2 are interesting questions in basic research that are not yet answered. Nevertheless, it is clear that from a therapeutic point of view, the effect of ARF on Mdm2 levels has to be considered in approaches based on agents that mimic the ARF function. Impairing th stabilising effect of p14ARF on Mdm2 levels, without impairing its ability to stabilise p53, if possible, would potentiate its ability to activate p53.

Modulating p53/Mdm2 Acetylation

CBP/p300-mediated acetyl-transferase activity catalyses p53 acetylation and activates p53 in response to stress (reviewed in Gu et al., 2004). This p53 modification is impaired by Mdm2 (Ito et al., 2001). Interestingly, Mdm2's RING finger domain is also a target for acetylation in vitro and in vivo and acetylation of Mdm2 reduces its ability to degrade p53 (Wang et al., 2004). Two additional factors modulate p53's acetylation status: PID/MTA2, a p53-interacting protein that induces p53 deacetylation by recruiting the histone deacetylase HDAC1 complex (Luo et al., 2000) and Sir2alpha/SIRT1, a NAD-dependent histone deacetylase that impairs p53 transcriptional activity (Luo et al., 2001; Vaziri et al., 2001).

Histone deacetylase (HDAC) inhibitors are currently evaluated in clinical trials for the treatment of cancer. These small molecules cause growth arrest at the G1 and/or G2/M phases, and induce differentiation and/or apoptosis in a wide variety of tumour cells. Suberoylanilide hydroxamic acid (SAHA) is a novel small molecule histone deacetylase inhibitor with high potency in inducing differentiation and apoptosis in tumour cells (Vrana et al.,1999; Huang et al., 2000). The role played by p53 in HDAC inhibitor-induced cell death is still controversial. For example, conditional expression of wild-type p53 does not influence SAHA actions, but markedly potentiates the apoptotic effects of another HDAC inhibitor, HMBA (Huang et al., 2000). The growth arrest at G1 phase by HDAC inhibitors is thought to be highly dependent on the upregulation of p21/WAF1, but it is clear that this upregulation occurs in a p53-independent way (Vrana et al.,1999; Huang et al., 2000). Gadd45 causes cell cycle arrest at the G2/M phase transition and participates in genotoxic stress-induced apoptosis. Gadd45 is also induced by a typical HDAC inhibitor, trichostatin A (TSA), but also in a p53-independent manner (Hirose et al., 2003). Nevertheless, it is possible that induction of cell death caused by HDAC inhibitors occurs via the induction of p53-dependent proapoptotic genes like PIG3 and NOXA (Terui et al., 2003).

HDAC are induced by hypoxia and promote angiogenesis through inhibition of hypoxia-responsive tumor suppressor genes (Kim et al., 2001). Accordingly, the HDAC inhibitor trichostatin (TSA) blocks angiogenesis in vitro and in vivo. In these systems, TSA upregulates p53 and von Hippel-Lindau expression and downregulates hypoxia-inducible factor-1alpha and vascular endothelial growth factor.

Discovering Novel Small Molecules and Pathways Activating p53

Screening small molecule libraries for novel activators of p53 using cell-based assays can be used to discover lead compounds with potential therapeutic application as well as to acquire information about novel pathways regulating p53 levels and function (Berkson et al., in press). One critical point in this type of approach is to establish as unequivocally as possible that the activity of a given compound is not mediated by the effects of the molecule or its metabolic derivatives on the integrity of the genome. Aiding this analysis, there are now a variety of simple assays that assess the possible genotoxic effect of compounds. These include the analysis of the phosphorylation status of p53, phosphorylation of histone H2AX, RAD51 clustering, comet assays, inhibition by caffeine and effects on chromosome composition and centrosome number. It could be envisaged that chances of finding small molecules that affect p53 or p53-modulating proteins directly using cell based screens are small and that molecules with a mechanism of action like the one proposed for RITA are the exception. However, even though a small molecule activator of p53 does not specifically activate this tumour suppressor pathway, that is even if p53 activation is just a contributing factor, there are a series of tests available to find whether a given compound has the desired effects on tumours cells and normal cells.

The main challenge when screening small molecules in cell based assays is defining the target(s) of the drug. In some cases medicinal chemistry expertise may help to circumvent this problem. As an example of this, we have identified a small molecule that activates p53 in a non-genotoxic way and has more severe effects on tumour cell lines than on normal fibroblasts (Berkson et al., in press). The comparison of the structure of this compound with that of other better characterised molecules, has led to the identification of a family of molecules that includes several natural compounds. When these were tested, they all were found to activate p53 activity (Berkson et al., in press and unpublished data). One of these compounds is sangivamycin, a natural product with relatively well defined mechanisms of action. siRNA libraries and genetic screens, aside from being invaluable tools in the discovery of new factors modulating p53 (Berns et al., 2004; Hannon et al., 1999) may also help to define the molecular pathways affected by novel p53-activating drugs or conferring resistance to them.

CONCLUDING REMARKS

The discovery of non-genotoxic p53 activators may provide substitutes for the drugs used in the clinic or at least provide agents that can effectively sensitise tumour cells to the effects of current therapies. However, this type of approach inevitably raises the issue of specificity. That is, is it possible to effectively kill tumour by inducing the p53 response without inducing intolerable levels of cell death or ageing in normal tissues? Are the p53-induced ageing-like effects irreversible? Additionally, if p53 elicits a cytostatic response in a given tumour, will this protect malignant cells from the killing effect of therapeutic agents? These questions are very difficult to answer in a general way and will possibly depend on the characteristics of each type of cancer. The results from the clinical trials currently carried out with the nutlins, which do not cause severe damage to normal tissues in animal models (Vassilev et al., 2004), will unequivocally test the applicability of approaches described here in systemic cancer chemotherapy.

REFERENCES

Adams J. Development of the proteasome inhibitor PS-341. 2002. *Oncologist* 7: 9-16.
Badciong J.C. and Haas A.L. MdmX Is a RING Finger Ubiquitin Ligase Capable of Synergistically Enhancing Mdm2 Ubiquitination. 2002. *J Biol Chem* 277: 49668-75.
Barak Y., Juven T., Haffner R. and Oren M. 1993. Mdm2 expression is induced by wild type p53 activity. *EMBO J* 12: 461-68.

Bartek J. and Lukas J. 2001. Mammalian G1- and S-phase checkpoints in response to DNA damage. *Curr Opin Cell Biol* 13: 738-47.

Benoit V., Hellin AC., Huygen S., Gielen J., Bours V. and Merville MP. 2000. Additive effect between NF-kappaB subunits and p53 protein for transcriptional activation of human p53 promoter. *Oncogene* 19: 4787-94.

Berkson R.G, Hollick J.J, Westwood N.J., Woods J.A., Lane D.P. and Lain S. A pilot screening programme for small molecule activators of p53. *Int J Cancer,* in press.

Berns K., Hijmans E.M., Mullenders J., Brummelkamp T.R., Velds A., Heimerikx M., Kerkhoven R.M., Madiredjo M., Nijkamp W., Weigelt B., Agami R., Ge W., Cavet G., Linsley P.S., Beijersbergen R.L. and Bernards R. 2004. A large-scale RNAi screen in human cells identifies new components of the p53 pathway. *Nature* 428: 431-7.

Blattner C., Hay T., Meek D.W. and Lane D.P. 2002. Hypophosphorylation of Mdm2 augments p53 stability. *Mol Cell Biol* 22: 6170-82.

Bond G.L., Hu W., Bond E.E., Robins H., Lutzker S.G., Arva N.C., Bargonetti J., Bartel F., Taubert H., Wuerl P., Onel K., Yip L., Hwang S.J., Strong L.C., Lozano G. and Levine A.J. 2004. A Single Nucleotide Polymorphism in the MDM2 Promoter Attenuates the p53 Tumor Suppressor Pathway and Accelerates Tumor Formation in Humans. *Cell* 119: 591-602.

Bothner B., Lewis W.S., DiGiammarino E.L., Weber J.D., Bothner S.J. and Kriwacki RW. 2001. Defining the molecular basis of Arf and Hdm2 interactions. *J Mol Biol.* 314: 263-77.

Bottger A., Bottger V., Sparks A., Liu W.L., Howard S.F. and Lane D.P. 1997. Design of a synthetic Mdm2-binding mini protein that activates the p53 response in vivo. *Curr Biol* 7: 860-9.

Bourdon J.C., Renzing J., Robertson P.L., Fernandes K.N. and Lane D.P. 2002. Scotin, a novel p53-inducible proapoptotic protein located in the ER and the nuclear membrane. *J Cell Biol* 158: 235-46.

Bullock A.N. and Fersht A.R. 2001. Rescuing the function of mutant p53. *Nat Rev Cancer* 1: 68-76.

Carnero A., Hudson J.D., Price C.M. and Beach D.H. 2000. p16INK4A and p19ARF act in overlapping pathways in cellular immortalization. *Nat Cell Biol.* 2: 148-55.

Chen L., Agrawal S., Zhou W., Zhang R. and Chen J. 1998. Synergistic activation of p53 by inhibition of MDM2 expression and DNA damage. *Proc Natl Acad Sci U S A* 95: 195-200.

Chen J., Marechal V. and Levine A.J. 1993. Mapping of the p53 and mdm-2 interaction domains. *Mol Cell Biol.* 13: 4107-14.

Chen J., Wu X., Lin J. and Levine A.J. 1996. mdm-2 inhibits the G1 arrest and apoptosis functions of the p53 tumor suppressor protein. *Mol Cell Biol.* 16: 2445-52.

Chene P., Fuchs J., Carena I., Furet P. and Garcia-Echeverria C. 2002. Study of the cytotoxic effect of a peptidic inhibitor of the p53-hdm2 interaction in tumor cells. *FEBS Lett* 529: 293-7.

Chene P., Fuchs J., Bohn J., Garcia-Echeverria C., Furet P. and Fabbro D. 2000. A small synthetic peptide, which inhibits the p53-hdm2 interaction, stimulates the p53 pathway in tumour cell lines. *J Mol Biol* 299: 245-53.

Chipuk J.E., Kuwana T., Bouchier-Hayes L., Droin N.M., Newmeyer D.D., Schuler M. and Green DR. 2004. Direct activation of Bax by p53 mediates mitochondrial membrane permeabilization and apoptosis. *Science* 303: 1010-4.

Chipuk J.E., Maurer U., Green D.R. and Schuler M. 2003. Pharmacologic activation of p53 elicits Bax-dependent apoptosis in the absence of transcription. *Cancer Cell* 4: 371-81.

Chuikov S., Kurash J.K., Wilson J.R., Xiao B., Justin N., Ivanov G.S., McKinney K., Tempst P., Prives C., Gamblin S.J., Barlev N.A. and Reinberg D. 2004. Regulation of p53 activity through lysine methylation. *Nature* 432: 353-60.

Clarke A.R., Purdie C.A., Harrison D.J., Morris R.G., Bird C.C., Hooper M.L. and Wyllie A.H. Thymocyte apoptosis induced by p53-dependent and independent pathways. *Nature* 362: 849-52.

Colombo E., Marine J.C., Danovi D., Falini B. and Pelicci PG. 2002. Nucleophosmin regulates the stability and transcriptional activity of p53. *Nat Cell Biol* 4: 529-33.

Craig E., Zhang Z.K., Davies K.P. and Kalpana G.V. 2002. A masked NES in INI1/hSNF5 mediates hCRM1-dependent nuclear export: implications for tumorigenesis. *EMBO J.* 21: 31-42.

Cummins J.M., Rago C., Kohli M., Kinzler K.W., Lengauer C. and Vogelstein B. 2004. Tumour suppression: disruption of HAUSP gene stabilizes p53. *Nature* 428: 1 p following 486.

Dasika G.K., Lin S.C., Zhao S., Sung P., Tomkinson A. and Lee E.Y. 1999. DNA damage-induced cell cycle checkpoints and DNA strand break repair in development and tumorigenesis. *Oncogene* 18: 7883-99.

Dornan D., Wertz I., Shimizu H., Arnott D., Frantz G.D., Dowd P., O'Rourke K., Koeppen H. and Dixit V.M. 2004. The ubiquitin ligase COP1 is a critical negative regulator of p53. *Nature* 429: 86-92.

Duncan S.J., Cooper M.A. and Williams D.H. 2003. Binding of an inhibitor of the p53/MDM2 interaction to MDM2. *Chem Commun (Camb).* 3: 316-7.

Duncan S.J., Gruschow S., Williams D.H., McNicholas C., Purewal R., Hajek M., Gerlitz M., Martin S., Wrigley S.K. and Moore M. 2001. Isolation and structure elucidation of Chlorofusin, a novel p53-MDM2 antagonist from a Fusarium sp. *J Am Chem Soc.* 123: 554-60.

Edelman J., Edelman J. and Nemunaitis J. 2003. Adenoviral p53 gene therapy in squamous cell cancer of the head and neck region. *Curr Opin Mol Ther* 5: 611-7.

Fang S., Jensen J.P., Ludwig R.L., Vousden K.H. and Weissman A.M. 2000. Mdm2 is a RING finger-dependent ubiquitin protein ligase for itself and p53. *J Biol Chem.* 275: 8945-51.

Feng J., Tamaskovic R., Yang Z., Brazil D.P., Merlo A., Hess D. and Hemmings B.A. 2004. Stabilization of Mdm2 via decreased ubiquitination is mediated by protein kinase B/Akt-dependent phosphorylation. *J Biol Chem.* 279: 35510-7.

Freedman D.A. and Levine A.J. 1998. Nuclear export is required for degradation of endogenous p53 by MDM2 and human papillomavirus E6. *Mol Cell Biol.* 18: 7288-93.

Freedman D.A., Wu L. and Levine A.J. 1999. Functions of the MDM2 oncoprotein. *Cell Mol Life Sci* 55: 96-107.

Garcia-Echeverria C., Chene P., Blommers M.J. and Furet P. 2000. Discovery of potent antagonists of the interaction between human double minute 2 and tumor suppressor p53. *J Med Chem.* 43: 3205-8.

Goldberg Z., Vogt Sionov R., Berger M., Zwang Y., Perets R., Van Etten R.A., Oren M., Taya Y. and Haupt Y. 2002. Tyrosine phosphorylation of Mdm2 by c-Abl: implications for p53 regulation. *EMBO J* 21: 3715-27.

Gu W., Luo J., Brooks C.L., Nikolaev A.Y. and Li M. 2. Dynamics of the p53 acetylation pathway. 2004. *Novartis Found Symp.* 259:197-205.

Hagting A, Karlsson C, Clute P, Jackman M, Pines J. 1998. MPF localization is controlled by nuclear export. *EMBO J.* 17: 4127-38.

Hannon G.J., Sun P., Carnero A., Xie L.Y., Maestro R., Conklin D.S. and Beach D. 1999. MaRX: an approach to genetics in mammalian cells. *Science* 283: 1129-30.

Haupt Y., Maya R., Kazaz A. and Oren M. 1997. Mdm2 promotes the rapid degradation of p53. *Nature* 387: 296-9.

Hietanen S., Lain S., Krausz E., Blattner C. and Lane DP. 2000. Activation of p53 in cervical carcinoma cells by small molecules. *Proc Natl Acad Sci USA* 97: 8501-6.

Hirose T, Sowa Y., Takahashi S., Saito S., Yasuda C., Shindo N., Furuichi K. and Sakai T. 2003. p53-independent induction of Gadd45 by histone deacetylase inhibitor: coordinate regulation by transcription factors Oct-1 and NF-Y. Oncogene 22: 7762-73.

Honda R., Tanaka H. and Yasuda H. 1997. Oncoprotein MDM2 is a ubiquitin ligase E3 for tumor suppressor p53. *FEBS Lett* 420: 25-7.

Honda R. and Yasuda H. 1999. Association of p19(ARF) with Mdm2 inhibits ubiquitin ligase activity of Mdm2 for tumor suppressor p53. *EMBO J* 18: 22-7.

Huang L., Sowa Y., Sakai T. and Pardee A.B. 2000. Activation of the p21WAF1/CIP1 promoter independent of p53 by the histone deacetylase inhibitor suberoylanilide hydroxamic acid (SAHA) through the Sp1 sites. *Oncogene* 19: 5712-9.

Issaeva N., Bozko P., Enge M., Protopopova M., Verhoef L.G., Masucci M., Pramanik A. and Selivanova G. 2004. Small molecule RITA binds to p53, blocks p53-HDM-2 interaction and activates p53 function in tumors. *Nat Med.* 10: 1321-8.

Ito A., Lai C.H., Zhao X., Saito S., Hamilton M.H., Appella E. and Yao T.P. 2001. p300/CBP-mediated p53 acetylation is commonly induced by p53-activating agents and inhibited by MDM2. *EMBO J* 20:1331-40.

Jastreboff A.M. and Cymet T. 2002. Role of the human papilloma virus in the development of cervical intraepithelial neoplasia and malignancy. *Postgrad Med J.*, 78: 225-8.

Kim M.S., Kwon H.J., Lee Y.M., Baek J.H., Jang J.E., Lee S.W., Moon E.J., Kim H.S., Lee S.K., Chung H.Y., Kim C.W. and Kim K.W. 2001. Histone deacetylases induce angiogenesis by negative regulation of tumor suppressor genes. *Nat Med.* 4: 437-43.

Kim S.H., Mitchell M., Fujii H., Llanos S. and Peters G. 2003. Absence of p16INK4a and truncation of ARF tumor suppressors in chickens. *Proc Natl Acad Sci U S A* 100: 211-6.

Krug U., Ganser A. and Koeffler HP. 2002. Tumor suppressor genes in normal and malignant hematopoiesis. *Oncogene* 21: 3475-95.

Kubbutat MH., Jones SN. and Vousden KH. 1997. Regulation of p53 stability by Mdm2. *Nature* 387: 299-303.

Kumari A., Schultz N. and Helleday T. 2004. protects from replication-associated DNA double-strand breaks in mammalian cells. *Oncogene* 23: 2324-9.

Kussie P.H., Gorina S., Marechal V., Elenbaas B., Moreau J., Levine AJ. and Pavletich N.P. 1996. Structure of the MDM2 oncoprotein bound to the p53 tumor suppressor transactivation domain. *Science* 274: 948-53.

Lahav G., Rosenfeld N., Sigal A., Geva-Zatorsky N., Levine A.J., Elowitz M.B. and Alon U. 2004. Dynamics of the p53-Mdm2 feedback loop in individual cells. *Nat Genet* 36: 147-50.

Lai Z., Yang T., Kim Y.B., Sielecki T.M., Diamond M.A., Strack P., Rolfe M., Caligiuri M., Benfield P.A., Auger K.R. and Copeland R.A. 2002. Differentiation of Hdm2-mediated p53 ubiquitination and Hdm2 autoubiquitination activity by small molecular weight inhibitors. *Proc Natl Acad Sci U S A* 99: 14734-9.

Lain S., Midgley C., Sparks A., Lane E.B. and Lane D.P. 1999a. An inhibitor of nuclear export activates the p53 response and induces the localization of HDM2 and p53 to U1A-positive nuclear bodies associated with the PODs. *Exp Cell Res* 248: 457-72.

Lain S., Xirodimas D. and Lane D.P. 1999b. Accumulating active p53 in the nucleus by inhibition of nuclear export: a novel strategy to promote the p53 tumor suppressor function. *Exp Cell Res* 253: 315-24.

Lakin N.D. and Jackson SP. 1999. Regulation of p53 in response to DNA damage. *Oncogene* 18: 7644-55.

Lang G.A., Iwakuma T., Suh Y-A., Liu G., Rao V.A., Parant J.M., Valentin-Vega Y.A., Terzian T., Caldwell L.C., Strong L.C., El-Naggar A.K., and Lozano G. 2004. Gain of Function of a p53 Hot Spot Mutation in a Mouse Model of Li-Fraumeni Syndrome. *Cell* 119: 861-72.

Leng R.P., Lin Y., Ma W., Wu H., Lemmers B., Chung S, Parant J.M., Lozano G., Hakem R. and Benchimol S. 2003. Pirh2, a p53-induced ubiquitin-protein ligase, promotes p53 degradation. *Cell* 112: 779-91

Lev Bar-Or R., Maya R., Segel LA., Alon U., Levine A.J. and Oren M. 2000. Generation of oscillations by the p53-Mdm2 feedback loop: a theoretical and experimental study. *Proc Natl Acad Sci U S A* 97: 11250-55.

Li M., Brooks CL., Kon N. and Gu W. 2004. A dynamic role of HAUSP in the p53-Mdm2 pathway. *Mol Cell* 13: 879-86.

Li M., Chen D., Shiloh A., Luo J., Nikolaev A.Y., Qin J. and Gu W. 2002. Deubiquitination of p53 by HAUSP is an important pathway for p53 stabilization. *Nature* 416: 648-53.

Linke S.P., Clarkin K.C., Di Leonardo A., Tsou A. and Wahl G.M. 1996. A reversible, p53-dependent G0/G1 cell cycle arrest induced by ribonucleotide depletion in the absence of detectable DNA damage. *Genes Dev* 10: 934-47.

Ljungman M. 2000. Dial 9-1-1 for p53: mechanisms of p53 activation by cellular stress. *Neoplasia* 2: 208-25.

Ljungman M. and Lane D.P. 2004. Transcription - guarding the genome by sensing DNA damage. *Nat Rev Cancer* 4: 727-37.

Ljungman M., O'Hagan H.M. and Paulsen M.T. 2001. Induction of ser15 and lys382 modifications of p53 by blockage of transcription elongation. *Oncogene* 20: 5964-71.

Lohrum M.A., Ashcroft M., Kubbutat M.H. and Vousden K.H. 2000. Contribution of two independent MDM2-binding domains in p14(ARF) to p53 stabilization. *Curr Biol.* 10: 539-42.

Lohrum M.A., Ludwig R.L., Kubbutat M.H., Hanlon M. and Vousden K.H. 2003. Regulation of HDM2 activity by the ribosomal protein L11. *Cancer Cell.* 3: 577-87.

Lowe S.W., Schmitt E.M., Smith S.W., Osborne B.A. and Jacks T. (1993) p53 is required for radiation-induced apoptosis in mouse thymocytes. *Nature* 362: 847-9.

Luo J., Nikolaev A.Y., Imai S., Chen D., Su F., Shiloh A., Guarente L. and Gu W. 2001. Negative control of p53 by Sir2alpha promotes cell survival under stress. *Cell* 107: 137-48.

Luo J., Su F., Chen D., Shiloh A. and Gu W. 2000. Deacetylation of p53 modulates its effect on cell growth and apoptosis. *Nature* 408: 377-81.

Mantovani F. and Banks L. 2001. The human papillomavirus E6 protein and its contribution to malignant progression. *Oncogene* 20: 7874-87.

Martin D.S., Spriggs D. and Koutcher J.A. 2001. A concomitant ATP-depleting strategy markedly enhances anticancer agent activity. *Apoptosis* 6: 125-31.

Maya R., Balass M., Kim S.T., Shkedy D., Leal J.F., Shifman O., Moas M., Buschmann T., Ronai Z., Shiloh Y., Kastan M.B., Katzir E. and Oren M. 2001. ATM-dependent phosphorylation of Mdm2 on serine 395: role in p53 activation by DNA damage. *Genes Dev* 15: 1067-77.

Mayo L.D. and Donner D.B. 2001. A phosphatidylinositol 3-kinase/Akt pathway promotes translocation of Mdm2 from the cytoplasm to the nucleus. *Proc Natl Acad Sci U S A* 98: 11598-603.

Mendrysa S.M. and Perry M.E. 2000. The p53 Tumor Suppressor Protein Does Not Regulate Expression of Its Own Inhibitor, MDM2, Except under Conditions of Stress. *Mol Cell Biol* 20: 2023-30.

Menendez S., Khan Z., Coomber D.W., Lane D.P., Higgins M., Koufali M.M. and Lain S. 2003. Oligomerization of the human ARF tumor suppressor and its response to oxidative stress. *J Biol Chem*. 278: 18720-9.

Montes de Oca Luna R., Wagner DS., Lozano G. 1995. Rescue of early embryonic lethality in Mdm2-deficient mice by deletion of p53. *Nature* 378: 203-6.

Midgley C.A., Desterro J.M., Saville M.K., Howard S., Sparks A., Hay R.T. and Lane D.P. 2000. An N-terminal p14ARF peptide blocks Mdm2-dependent ubiquitination in vitro and can activate p53 in vivo. *Oncogene* 19: 2312-23.

Midgley C.A. and Lane D.P. 1997. p53 protein stability in tumour cells is not determined by mutation but is dependent on Mdm2 binding. *Oncogene* 15: 1179-89.

Neufeld K.L., Zhang F., Cullen B.R. and White R.L. 2000. APC-mediated downregulation of beta-catenin activity involves nuclear sequestration and nuclear export. *EMBO Rep*. 1: 519-23.

Newlands E.S., Rustin G.J. and Brampton MH. 1996. Phase I trial of elactocin. *Br J Cancer* 74: 648-9.

Noda A., Toma-Aiba Y. and Fujiwara Y. 2000. A unique, short sequence determines p53 gene basal and UV-inducible expression in normal human cells. *Oncogene* 19: 21-31.

O'Hagan H.M. and Ljungman M. 2004. Nuclear accumulation of p53 following inhibition of transcription is not due to diminished levels of MDM2. *Oncogene*. 23: 5505-12.

Ohnishi K. and Ohnishi T. 2001. Heat-induced p53-dependent signal transduction and its role in hyperthermic cancer therapy. *Int J Hyperthermia* 17: 415-27.

Olive K.P., Tuveson D.A., Ruhe Z.C., Yin B., Willis N.A., Bronson R.T., Crowley D., and Tyler Jacks. 2004. Mutant *p53* Gain of Function in Two Mouse Models of Li-Fraumeni Syndrome. *Cell* 119: 847-60.

Parant J., Chavez-Reyes A., Little N.A., Yan W., Reinke V., Jochemsen A.G. and Lozano G. 2001. Rescue of embryonic lethality in Mdm4-null mice by loss of Trp53 suggests a nonoverlapping pathway with MDM2 to regulate p53. *Nat Genet* 29: 92-5.

Peng Y., Chen L., Li C., Lu W., Agrawal S. and Chen J. 2001. Stabilization of the MDM2 oncoprotein by mutant p53. *J Biol Chem* 276: 6874-8.

Picksley S.M., Vojtesek B., Sparks A. and Lane D.P. 1994 Immunochemical analysis of the interaction of p53 with MDM2;--fine mapping of the MDM2 binding site on p53 using synthetic peptides. *Oncogene* 9: 2523-9.

Pluquet O. and Hainaut P. 2001. Genotoxic and non-genotoxic pathways of p53 induction. *Cancer Lett* 174: 1-15.

Pomerantz J., Schreiber-Agus N., Liegeois N.J., Silverman A., Alland L., Chin L., Potes J., Chen K., Orlow I., Lee H.W., Cordon-Cardo C. and DePinho R.A. 1998. The Ink4a tumor suppressor gene product, p19Arf, interacts with MDM2 and neutralizes MDM2's inhibition of p53. *Cell* 92: 713-23.

Porter LA., Cukier I.H. and Lee J.M. 2003. Nuclear localization of cyclin B1 regulates DNA damage-induced apoptosis. *Blood* 101: 1928-33.

Prives C. and Manley J.L. 2001. Why is p53 acetylated? *Cell* 107: 815-8.

Qin C., Nguyen T., Stewart J., Samudio I., Burghardt R. and Safe S. 2002. Estrogen Up-Regulation of p53 Gene Expression in MCF-7 Breast Cancer Cells Is Mediated by Calmodulin Kinase IV-Dependent Activation of a Nuclear Factor kappaB/CCAAT-Binding Transcription Factor-1 Complex. *Mol Endocrinol* 16: 1793-1809.

Quelle D.E., Zindy F., Ashmun R.A. and Sherr C.J. 1995. Alternative reading frames of the INK4a tumor suppressor gene encode two unrelated proteins capable of inducing cell cycle arrest. *Cell* 83: 993-1000.

Rodriguez J.A. and Henderson B.R. 2000. Identification of a functional nuclear export sequence in BRCA1. *J Biol Chem* 275: 38589-96.

Rodriguez M.S., Desterro J.M., Lain S., Midgley C.A., Lane D.P. and Hay R.T. 1999. SUMO-1 modification activates the transcriptional response of p53. *EMBO J* 18: 6455-61.

Rodriguez M.S., Thompson J., Hay R.T. and Dargemont C. 1999. Nuclear retention of IkappaBalpha protects it from signal-induced degradation and inhibits nuclear factor kappaB transcriptional activation. *J Biol Chem* 274: 9108-15.

Rosin-Arbesfeld R., Townsley F. and Bienz M. 2000. The APC tumour suppressor has a nuclear export function. *Nature* 406: 1009-12.

Rubbi C.P. and Milner J. 2003. Disruption of the nucleolus mediates stabilization of p53 in response to DNA damage and other stresses. *EMBO J* 22: 6068-77.

Salomoni P, Pandolfi P.P. 2002. The role of PML in tumor suppression. *Cell* 108: 165-70.

Saville M.K., Sparks A., Xirodimas D.P., Wardrop J., Stevenson L.F., Bourdon J.C., Woods Y.L. and Lane D.P. 2004. Regulation of p53 by the ubiquitin-conjugating enzymes UbcH5B/C in vivo. *J Biol Chem* 279: 42169-81.

Schwartz C.L. 1999. Long-term survivors of childhood cancer: the late effects of therapy. *Oncologist* 4: 45-54.

Shaw P., Freeman J., Bovey R. and Iggo R. 1996. Regulation of specific DNA binding by p53: evidence for a role for O-glycosylation and charged residues at the carboxy-terminus. *Oncogene* 12: 921-30.

Sherr C.J. 2001. The INK4a/ARF network in tumour suppression. *Nat Rev Mol Cell Biol* 2: 731-7.

Smart P., Lane E.B., Lane D.P., Midgley C., Vojtesek B. and Lain S. 1999. Effects on normal fibroblasts and neuroblastoma cells of the activation of the p53 response by the nuclear export inhibitor leptomycin B. *Oncogene* 18: 7378-86.

Stern P.L., Faulkner R., Veranes E.C. and Davidson E.J. 2001. The role of human papillomavirus vaccines in cervical neoplasia. *Best Pract Res Clin Obstet Gynaecol* 15: 783-99.

Stott F.J., Bates S., James M.C., McConnell B.B., Starborg M., Brookes S., Palmero I., Ryan K., Hara E., Vousden K.H. and Peters G. 1998. The alternative product from the human CDKN2A locus, p14(ARF), participates in a regulatory feedback loop with p53 and MDM2.*EMBO J* 17: 5001-14.

Soussi T., Dehouche K. and Béroud C. 2000. p53 Website and analysis of p53 gene mutations in human cancer: Forging a link between epidemiology and carcinogenesis. *Hum Mutat* 15: 105-13.

Stommel J.M. and Wahl G.M. 2004. Accelerated MDM2 auto-degradation induced by DNA-damage kinases is required for p53 activation. *EMBO J* 23: 1547-1556.

Sugimoto M., Kuo M.L., Roussel M.F. and Sherr C.J. 2003. Nucleolar Arf tumor suppressor inhibits ribosomal RNA processing. *Mol Cell* 11: 415-24.

Tao W. and Levine A.J. 1999. P19(ARF) stabilizes p53 by blocking nucleo-cytoplasmic shuttling of Mdm2. *Proc Natl Acad Sci U S A* 96: 6937-41.

Terui T., Murakami K., Takimoto R., Takahashi M., Takada K., Murakami T., Minami S., Matsunaga T., Takayama T., Kato J. and Niitsu Y. 2003. Induction of PIG3 and NOXA through acetylation of p53 at 320 and 373 lysine residues as a mechanism for apoptotic cell death by histone deacetylase inhibitors. *Cancer Res* 63: 8948-54.

Toren A., Amariglio N. and Rechavi G. 1996. Curable and non-curable malignancies: lessons from paediatric cancer. *Med Oncol* 13: 15-21.

Tortora G., Caputo R., Damiano V., Bianco R., Chen J., Agrawal S., Bianco A.R. and Ciardiello F. 2000. A novel MDM2 anti-sense oligonucleotide has anti-tumor activity and potentiates cytotoxic drugs acting by different mechanisms in human colon cancer. *Int J Cancer* 88: 804-9.

Vassilev L.T., Vu B.T., Graves B., Carvajal D., Podlaski F., Filipovic Z., Kong N., Kammlott U., Lukacs C., Klein C., Fotouhi N. and Liu EA. 2004. In vivo activation of the p53 pathway by small-molecule antagonists of MDM2. *Science* 303: 844-8.

Vaziri H. and Benchimol S. 1996. From telomere loss to p53 induction and activation of a DNA-damage pathway at senescence: the telomere loss/DNA damage model of cell aging. *Exp Gerontol* 31: 295-301.

Vaziri H., Dessain S.K., Ng Eaton E., Imai SI., Frye R.A., Pandita T.K., Guarente L. and Weinberg R.A. 2001. hSIR2(SIRT1) functions as an NAD-dependent p53 deacetylase. *Cell* 107: 149-59.

Vousden K.H. and Lu, X. 2002. Live or let die: the cell's response to p53. *Nature Reviews Cancer* 2: 594 –604.

Vrana J.A., Decker R.H., Johnson C.R., Wang Z., Jarvis W.D., Richon V.M., Ehinger M., Fisher P.B. and Grant S. 1999. Induction of apoptosis in U937 human leukemia cells by suberoylanilide hydroxamic acid (SAHA) proceeds through pathways that are regulated by Bcl-2/Bcl-XL, c-Jun, and p21CIP1, but independent of p53. *Oncogene* 18: 7016-25.

Wallace W.H., Blacklay A., Eiser C., Davies H., Hawkins M., Levitt G.A. and Jenney M.E. 2001. Developing strategies for long term follow up of survivors of childhood cancer. *BMJ* 323: 271-4.

Wang H., Zeng X., Oliver P., Le LP., Chen J., Chen L., Zhou W., Agrawal S. and Zhang R. 1999. MDM2 oncogene as a target for cancer therapy: An antisense approach. *Int J Oncol* 15: 653-60.

Wang X., Taplick J., Geva N. and Oren M. 2004. Inhibition of p53 degradation by Mdm2 acetylation. *FEBS Lett.* 2004 561: 195-201.

Weber J.D., Jeffers J.R., Rehg J.E., Randle D.H., Lozano G., Roussel M.F., Sherr C.J. and Zambetti G.P. 2000. p53-independent functions of the p19(ARF) tumor suppressor. *Genes Dev* 14: 2358-65.

Weber J.D., Taylor L.J., Roussel M.F., Sherr C.J. and Bar-Sagi D. 1999. Nucleolar Arf sequesters Mdm2 and activates p53. *Nat Cell Biol* 1: 20-6.

Xirodimas D., Saville M.K., Bourdon J.C., Hay R.T. and Lane D.P. 2004. Mdm2-mediated NEDD8 conjugation of p53 inhibits its transcriptional activity. *Cell* 118: 83-97.

Xirodimas D., Saville M.K., Edling C., Lane D.P. and Lain S. 2001a. Different effects of p14ARF on the levels of ubiquitinated p53 and Mdm2 in vivo. *Oncogene* 20: 4972-83.

Xirodimas D.P., Stephen C.W. and Lane D.P. 2001b. Cocompartmentalization of p53 and Mdm2 is a major determinant for Mdm2-mediated degradation of p53. *Exp Cell Res.* 270: 66-77.

Yoshida M. 2001. Multiple viral strategies of HTLV-1 for dysregulation of cell growth control. *Annu Rev Immunol* 19: 475-96.

Zhang Y and Xiong Y. 1999. Mutations in human ARF exon 2 disrupt its nucleolar localization and impair its ability to block nuclear export of MDM2 and p53. Mol *Cell* 3: 579-91.

Zhang Y., Xiong Y. and Yarbrough W.G. 1998. ARF promotes MDM2 degradation and stabilizes p53: ARF-INK4a locus deletion impairs both the Rb and p53 tumor suppression pathways. *Cell* 92: 725-34.

Zhao J., Wang M., Chen J., Luo A., Wang X., Wu M., Yin D. and Liu Z. 2002. The initial evaluation of non-peptidic small-molecule HDM2 inhibitors based on p53-HDM2 complex structure. *Cancer Lett* 183: 69-77.

Zhou BP. And Hung MC. 2002 Novel targets of Akt, p21(Cipl/WAF1), and MDM2. *Semin Oncol* 29(3 Suppl 11): 62-70.

Zindy F., Williams R.T., Baudino T.A., Rehg J.E., Skapek S.X., Cleveland J.L., Roussel M.F. and Sherr C.J. 2003. Arf tumor suppressor promoter monitors latent oncogenic signals in vivo. *Proc Natl Acad Sci U S A* 100: 15930-5.

zur Hausen H. Viruses in human cancers. 1999a. *Eur J Cancer* 35: 1174-81.

zur Hausen H. Viruses in human cancers. 1999b. *Eur J Cancer* 35: 1878-85.

Chapter 17

WILD-TYPE P53 CONFORMATION, STRUCTURAL CONSEQUENCES OF P53 MUTATIONS, AND MECHANISMS OF MUTANT P53 RESCUE

Andreas C. Joerger, Assaf Friedler and Alan R. Fersht
MRC Centre for Protein Engineering, Hills Road, Cambridge, UK

The tumor suppressor p53 is a transcription factor that is at the center of a network of interactions that affect the cell cycle and apoptosis (Vogelstein et al. 2000; Ryan et al. 2001). The protein is induced by a variety of stresses that include oncogene activation and DNA damage caused by chemotherapy and radiotherapy. On induction, it activates a variety of genes whose products lead to G1 and G2 cell cycle arrest and apoptosis (Vogelstein et al. 2000; Ryan et al. 2001). It is such an effective tumor suppressor that it is inactivated in virtually all cancers; in about 50% of cancers p53 is directly inactivated by mutation and in the remainder its activity is lost by perturbations of its associated pathways and interactions (Hainaut and Hollstein 2000). Reactivating mutant p53 is an important target in the development of novel therapies for cancer (Lane and Lain 2002; Lane and Hupp 2003). To understand how p53 is inactivated, it is necessary to understand its structure and how it responds to mutation. Such knowledge will provide a basis for the rational design of novel therapeutics that may reverse the effects of mutation. In this chapter, we survey the structure of the protein, the effects of mutation and how they may be reversed.

THE DOMAIN STRUCTURE OF HUMAN P53

Human p53 is a 393-residue protein, which is active as a homotetramer. Each chain has a complex domain structure that is organized into several functional and structural entities (Figure 1A) (Vousden and Lu 2002;

P. Hainaut and K.G. Wiman (eds.), 25 Years of p53 Research, 377-397.
© 2005 *Springer. Printed in the Netherlands.*

Courtois et al. 2004): The N-terminal part of the protein comprises the transactivation domain (residues 1-63). It contains regions forming protein-protein interactions with regulatory proteins such as MDM2 (residues 19-26 of p53), which triggers ubiquitination and ultimately degradation of the protein in the proteasome (Momand et al. 2000; Michael and Oren 2002; Michael and Oren 2003), and p300/CBP, which is responsible for acetylation of residues in the C-terminal part of p53 (Grossman 2001). This domain is followed by a proline-rich region (residues 64-92), which binds SH3 domains and was suggested to have a regulatory role (Walker and Levine 1996; Muller-Tiemann et al. 1998). The central (core) domain is responsible for specific DNA-binding (residues 94-292) and is the most important independent domain of p53 (Cho et al. 1994). The C-terminal part of the protein includes a tetramerization domain (residues 326-355) (Clore et al. 1994) and the negative autoregulatory domain at the extreme C-terminus (363-393), which contains acetylation and phosphorylation sites and regulates the DNA-binding activity of p53 (Prives and Manley 2001). Further elements comprise a highly conserved leucine-rich nuclear export signal within the tetramerization domain (residues 340-350) (Stommel et al. 1999) and a nuclear localization signal, which is located between the core domain and tetramerization domain (residues 302-322) (Shaulsky et al. 1990). Most cancer-associated mutations map to the DNA-binding core domain (Hainaut and Hollstein 2000), and six hot-spot mutation sites stand out as most frequently associated with human cancer (Arg175, Gly245, Arg248, Arg249, Arg273 and Arg282).

The crystal structures have been solved of both the DNA-binding (Cho et al. 1994) and the tetramerization domains (Jeffrey et al. 1995). NMR studies have revealed a very similar structure and tetrameric assembly of the tetramerization domain in solution (Clore et al. 1994; Clore et al. 1995). The tetramer exhibits D_2-symmetry and can be best described as a dimer of dimers (Figure 1B/C). Other domains such as the N-terminal transactivation domain are thought to be mainly unstructured in the free protein but at least partially structured when interacting with regulatory proteins. Crystallographic studies have shown that a 15-residue transactivation domain peptide of p53 binds to a hydrophobic cleft of the N-terminal domain of MDM2 as an amphipathic α-helix (Kussie et al. 1996). A similar result was found for a peptide derived from the C-terminal negative regulatory domain of p53 (residues 367–388). NMR spectroscopy revealed that this peptide, which has no regular structure in its native form, becomes helical upon binding to S100B (Rustandi et al. 2000). How the various domains of p53 are assembled in the full-length tetramer is, however, still not known.

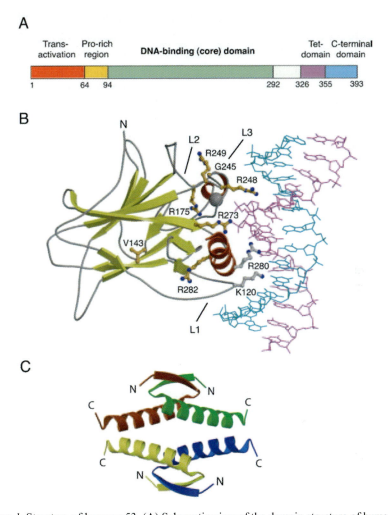

Figure 1. Structure of human p53. (A) Schematic view of the domain structure of human p53: The 393-residue protein contains an N-terminal transactivation domain, a proline-rich region, a DNA-binding (core) domain, a tetramerization domain (Tet-domain) and a C-terminal autoregulatory domain (see text for further details). (B) Structure of the DNA-binding (core) domain in complex with consensus DNA (PDB code 1TSR, molecule B). A β-sandwich provides the basic scaffold for a loop-sheet-helix motif and two large loops (L2 and L3) tethered by a zinc ion, which interact with the major and minor groove of the DNA, respectively. The zinc ion is shown as a gray sphere. Selected residues are shown as ball-and-stick models. Among these are the six hot-spot sites Arg175, Gly245, Arg248, Arg249, Arg273 and Arg282, which are most frequently mutated in human cancer (colored in orange). (C) Ribbon plot of the tetramerization (Tet-domain) domain structure (PDB code 1C26). Individual subunits within the tetramer are shown in different colors. Two monomers consisting of a β-strand and an α-helix are connected via an antiparallel β-sheet and an antiparallel helix-helix interface to form a dimer (eg. the red and green chain). Two of these dimers are assembled into a tetramer via a parallel helix-helix interface. The figures were generated using MOLSCRIPT (Kraulis 1991) and RASTER3D (Merritt and Bacon 1997).

THE CRYSTAL STRUCTURE OF THE DNA-BINDING DOMAIN

The crystal structure of the human p53 core domain has been solved at 2.2 Å resolution in complex with a consensus DNA sequence (Cho et al. 1994). Two of the three molecules in the asymmetric unit of the crystal bind to the same face of the DNA duplex, forming a head-to-tail dimer with weak protein-protein interactions between the two molecules, whereas the third molecule makes no significant contact with the DNA. Overall the structures of the three molecules are very similar, which suggests that DNA binding induces only minor structural changes within the core domain. Additional confirmation comes from the structure of p53 core domain from mouse, which has been solved in the absence of DNA (Zhao et al. 2001). The core domain structure consists of two anti-parallel β-sheets of four and five twisted strands, forming a β-sandwich with a "Greek key" topology (Figure 1B). This compact, barrel-like structure provides the basic scaffold for two large loops, L2 (residues 163-195) and L3 (residues 236-251), and a loop-sheet-helix motif at the same end of the β-sandwich. The conformation of the two large loops L2 and L3 is stabilized by a zinc ion, which is tetrahedrally coordinated by three cysteines and a histidine (Cys176, His179, Cys238 and Cys242). The loop-sheet-helix motif and the two large loops form an extended surface, rich in basic amino acids, which binds DNA. In the complex with consensus DNA, these structural elements make an extensive number of contacts with the major groove and minor groove of the DNA. Parts of the helix and the L1 loop of the loop-sheet-helix motif bind to the major groove. This involves specific interactions of Lys120 and Arg280 with bases of the pentameric consensus sequence. Arg248 from loop L3 protrudes into the adjacent minor groove where its positively charged guanidinium group is in close contact with the negatively charged DNA backbone. Interestingly, parts of this DNA-binding surface were found to overlap with the binding sites for the 53BP1 and 53BP2 proteins. 53BP1 functions as a DNA damage checkpoint protein (DiTullio et al. 2002), whereas 53BP2 forms the C-terminal half of a larger protein (ASPP2), which specifically enhances p53 induced apoptosis (Samuels-Lev et al. 2001). Crystal structures of complexes with the p53 core domain reveal that in both cases the contacting residues are concentrated on the L3 loop (Gorina and Pavletich 1996; Derbyshire et al. 2002; Joo et al. 2002).

The crystal structure of the core domain provided a framework for understanding how the common cancer-associated mutations inactivate p53. It rationalized observations that these mutations can be subdivided into two classes. The first class of mutants are "contact mutants" e.g. the cancer hot-spot mutations R248Q and R273H. These mutations directly affect DNA-binding, because an essential DNA-contacting residue is lost. In contrast, the second class of mutants, the "structural mutants", are thought to cause structural perturbations within the core domain and thus indirectly affect DNA binding. The hot-spot mutations R175H, G245S, R249S and R282W belong to this second group.

THE CONSEQUENCES AND EFFECTS OF P53 MUTATIONS AT THE MOLECULAR LEVEL

Several experimental techniques are used to study the effect of mutation on p53. The effects on the thermodynamic stability of the protein are analyzed using fluorescence spectroscopy and calorimetry, while the structural effects can be studied mainly by NMR and X-ray crystallography. NMR is especially suitable to studying the dynamic behavior of p53 mutants (Bullock et al. 1997; Wong et al. 1999; Bullock et al. 2000; Bullock and Fersht 2001).

Effect of mutations on stability

First we must define what is meant by stability as "stability" is often used in a confusing way in the p53-related literature. An increase in the cellular level of p53 is often qualitatively referred to as "stabilization of p53", regardless of whether the protein is denatured or not. This "stability" is often just a consequence of whether or not the p53 protein is degraded via MDM2, and so inactive, unstable, mutants of p53 accumulate because the mdm2 gene is not activated by p53. But, the correct quantitative definition of p53 stabilization refers to its thermodynamic properties: stabilization of p53 means an increase in the thermodynamic stability of the protein, derived from its folding-unfolding equilibrium (i.e. in the ΔG for the denaturation of a mutant compared with that of wild type). Here, when we refer to p53 stabilization, it is always the thermodynamic stabilization of the protein.

The core domain of p53 protein has a low intrinsic stability and its melting temperature, 42-44°C, is only slightly above body temperature (Figure 2) (Bullock et al. 1997; Bullock et al. 2000). The full-length protein melts at a similar temperature. Its low stability may be a consequence of the need to regulate the level of p53 by degradation by the proteasome. But, the consequence of the intrinsic low stability is that destabilizing mutations in the protein have a detrimental effect on the protein function, and slightly destabilized proteins melt at below body temperature (Figure 2) (Bullock et al. 1997; Bullock et al. 2000).

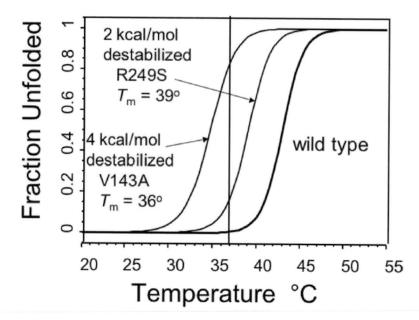

Figure 2. The denaturation curves of wild-type and mutant core domains of p53. Urea denaturation of wild-type and mutant p53 core domain at pH = 7.2 as monitored by normalized fluorescence emission at 356 nm. Data taken from ref.(Bullock et al. 1997).

The mutations can be divided into three classes (Figure 3): (I) DNA contact mutations, which have little or no effect on protein stability, but impair function due to the loss of key residues that mediate specific DNA binding. (II) Mutations that cause local distortion of the structure, and are moderately destabilizing (by <2 kcal/mol compared to the wild-type protein). Most of the non-contact mutations in the DNA-binding interface fall into this category, e.g. the loop L3 mutations G245S and R249S. (III) Mutations that result in global unfolding of the protein, and are highly destabilizing (by >3 kcal / mol compared to the wild type). All the β-sandwich mutants fall into this category. The destabilizing mutations inactivate p53 by lowering the melting temperature of its core domain to below body temperature, making it denature under physiological conditions (Bullock and Fersht 2001).

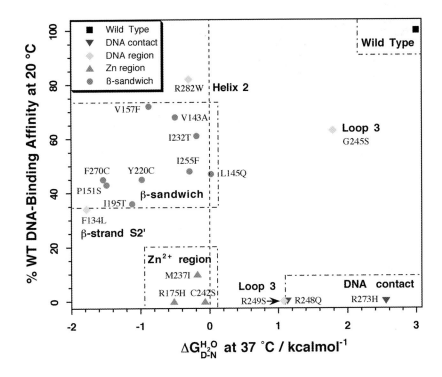

Figure 3. The different classes of mutations in p53, grouped according to stability (x-axis) and DNA binding properties. Mutant phenotypes correlate with the site of mutation as shown by a plot of stability (estimated at 37 °C) against DNA-binding affinity at 20 °C (a temperature at which all mutants are folded). A free energy of unfolding in water, ΔG, of 0 kcal/mol (shown by a dashed line) corresponds to 50% denatured protein. Data taken from ref. (Bullock and Fersht 2001).

STRUCTURAL EFFECTS OF DESTABILIZING MUTATIONS IN P53 CORE DOMAIN

HSQC NMR spectroscopy gives information about the structural changes that occur in the p53 core domain upon mutation. The chemical shift changes between the wild type and the mutant protein for each residue are taken as an indication for a structural change around that residue upon mutation, though the magnitude of the changes is not defined (Wong et al. 1999; Friedler et al. 2004). Different hot-spot mutants show different structural changes. Mutations in loop L3 (G245S, R248Q and R249S), situated in the DNA-contact region, result in structural changes in the loop L3 and L2 regions, which affect the whole DNA-binding interface but not the β-sandwich. The contact mutation R273H results mainly in changes around the mutation site. The β-sandwich mutation V143A (see Figure 1B) results in chemical shift changes across all of the β-sandwich and the DNA-binding surface (Wong et al. 1999; Friedler et al. 2004).

REVERSING THE STRUCTURAL EFFECTS OF P53 MUTATIONS WITH SMALL MOLECULES

Many mutations in p53 lead to its misfolding and aggregating. This is an important example of a more general phenomenon, where misfolding of proteins leads to disease. Many other human diseases, such as amyloidoses, cystic fibrosis and lysosomal storage diseases are caused by protein misfolding, which occurs mainly due to mutations that disrupt the three-dimensional structure of the protein (Dobson 1999; Fan et al. 1999; Morello et al. 2000; Bullock and Fersht 2001; Fan 2003). A novel therapeutic strategy for such diseases is to develop small molecules that will assist in refolding of the proteins, which will result in their reactivation. Such molecules are termed "chemical chaperones" (Morello et al. 2000; Fan 2003).

Refolding of p53 mutants to their correct native structure should lead to their reactivation (Bullock and Fersht 2001). Destabilized p53 mutants undergo a folding – unfolding equilibrium, which involves numerous states ranging from native and native-like structures, through distorted structures, to globally unfolded states (Bullock et al. 1997; Bullock et al. 2000; Bullock and Fersht 2001). A small molecule chemical chaperone that binds the native, but not the denatured state, would result in a shift of the equilibrium towards the native state leading to thermodynamic stabilization and to restoration of activity (Bullock and Fersht 2001; Friedler et al. 2002) (Figure 4A). Once a chemical chaperone is bound preferentially to the native state, the equilibrium is shifted towards this state.

Kinetic instability of p53 core domain mutants

The thermodynamically unstable p53 core domain mutants are also kinetically unstable, and denature within minutes at 37 °C (Friedler et al. 2003). The half-life ($t_{1/2}$) of unfolding of wild-type p53 core domain is 9 min. The hot-spot mutant G245S, which is moderately destabilized, has $t_{1/2} =$ 4.6 min. The highly destabilized mutant I195T has a $t_{1/2}$ of less than one minute. After unfolding, the denatured proteins aggregate, the rate increasing with higher concentration of protein (Friedler et al. 2003).

The kinetic instability and rapid unfolding rates of oncogenic p53 mutants define the suitable rescue strategy for these mutants. In principle, the binding of a genuine ligand of p53, e.g. DNA or a p53-binding protein will also shift the conformational equilibrium in favor of a wild-type-like structure. However, because of their short half-lives at body temperature, many p53 mutants denature very rapidly after biosynthesis and never get the chance to bind their natural ligand. This kinetic instability may thus be the underlying reason for their inactivation. p53-stabilizing chemical chaperones have to act immediately after biosynthesis, before the protein unfolds, and keep it folded until it enters the nucleus and binds its sequence-specific target DNA or other natural ligands, which then take over (Figure 4B) (Friedler et al. 2003).

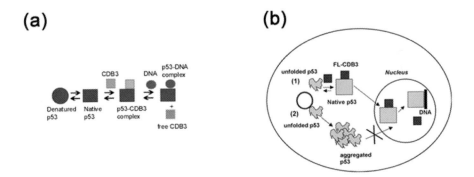

Figure 4. Mechanism of action of FL-CDB3, a chemical chaperone that refolds destabilized p53 core domain: (A) Chaperone strategy for rescue of p53: A schematic model of the proposed mechanism of action for FL-CDB3. See text for details. Taken from ref. (Friedler et al. 2002). (B) A model for action of p53 stabilizing drugs. Newly synthesized mutant p53 is released from the ribosome (circle) in its unfolded form. Then, two pathways are possible: (1) In presence of a stabilizing drug such as FL-CDB3 (small square) the protein refolds to its native conformation (large square), is transported into the nucleus as a complex with the drug, and then binds its natural ligand, e.g. its target DNA. (2) In the absence of the drug, the unfolded protein aggregates and can not be transported into the nucleus and bind DNA. Taken from ref. (Friedler et al. 2003).

SEARCH FOR SMALL MOLECULE CHEMICAL CHAPERONES THAT STABILIZE P53 CORE DOMAIN

Potential p53-stabilizing small molecule drugs may be found by random screening or by rational design. Various p53-activating compounds have been discovered using cell-based assay screening (Selivanova et al. 1997;

Selivanova et al. 1998; Selivanova et al. 1999; Bykov et al. 2002; Bykov et al. 2003; Peng et al. 2003). Biophysical studies show that these do not bind p53 core domain and so function by different mechanisms (Friedler et al. 2002; Rippin et al. 2002; Issaeva et al. 2003; Wang et al. 2003). The compound CP-31398 (Foster et al. 1999), found from random screening of the Pfizer drug library for compounds that stabilize p53, also does not bind to the core domain *in vitro* (Rippin et al. 2002), but functions *in vivo* by inhibiting ubiquitination of p53 and hence its degradative pathway (Wang et al. 2003).

The surface of p53 lacks well-defined clefts for binding small molecules. Rational design of p53-stabilizing compounds has focused so far on peptides that mimic natural proteins that bind p53 core domain (Friedler et al. 2002). These are excellent candidates to serve as chemical chaperones for p53, because they have already been optimized by nature to bind the native state of the p53 core domain. Further, being larger and more flexible than the typical small-molecule drugs, peptides could target the DNA-binding interface of p53 more efficiently even in the absence of a well-defined binding pocket.

CDB3: A CHEMICAL CHAPERONE THAT BINDS AND STABILIZES MUTANT P53 *IN VITRO* AND *IN VIVO*

The structure of the complex between p53 core and the p53 binding protein 2 (53BP2 or ASPP, Figure 5A) (Gorina and Pavletich 1996) was the first starting point in designing small binding peptides. Initial screening for binding used HSQC NMR and surface plasmon resonance (BIAcore) to identify a 9 amino acid residue peptide, CDB3 (Core Domain Binding peptide number 3), which binds to p53 core domain. CDB3 is derived from the p53-binding loop of the protein 53BP2 (residues 490-498) and its sequence is REDEDEIEW (Friedler et al. 2002).

NMR HSQC chemical shift analysis shows that CDB3 binds p53 in loop L1, helix H2 and strand S8, which are at the edge of the DNA binding site, partly overlapping it (Figure 5C) (Friedler et al. 2002). The fluorescein-labelled peptide (FL-CDB3) binds wild-type p53 core domain with a dissociation constant of 600 nM, measured by fluorescence anisotropy. It stabilizes mutant p53 core domain against chemical and thermal denaturation, (Friedler et al. 2002), and restores DNA binding of the highly destabilized mutant I195T back to wild-type level (Friedler et al. 2002). CDB3 also slows down the unfolding rate of p53 core domain (Friedler et al. 2003). FL-CDB3 is 50 times more active than the non-labeled CDB3, and was used as a lead compound for *in vivo* studies. *In vivo*, FL-CDB3 acts as a chemical chaperone and rescues the tumor suppressing function of oncogenic mutants of p53 in living cells (Issaeva et al. 2003) (for details see Selivanova et al., this book).

THE HOT-SPOT MUTANT R249S: A CASE STUDY OF RESTORING THE NATIVE CONFORMATION IN A STRUCTURAL P53 MUTANT BY FL-CDB3

The mutation R249S is one of the most common cancer-associated p53 mutations ("hot-spots"). As it is highly frequent in hepatocellular carcinoma, its rescue is an important therapeutic target (Aguilar et al. 1993). The structural effects of the R249S mutation were analyzed by a combination of HSQC and relaxation NMR techniques. R249S has a native structure that resembles that of wild type, and it does not adopt a denatured "mutant conformation" (Wong et al. 1999; Friedler et al. 2004). The structural effects of the mutation are an increased flexibility of the β–sandwich and a local distortion throughout the DNA-binding interface (Figure 5B). The R249S mutation results in an ensemble of native and native-like conformations in a dynamic equilibrium (Friedler et al. 2004).

As FL-CDB3 was designed to rescue mutants of p53 by binding specifically to their native structure, and since R249S undergoes a dynamic equilibrium, R249S is a natural target for rescue by chemical chaperones such as FL-CDB3. NMR studies show that FL-CDB3 changes the chemical shifts of R249S back towards the wild-type values and thus reverses the structural effects of mutation (Friedler et al. 2004). FL-CDB3 binds to both p53 core wild type and R249S at the same site, which includes loop L1, helix H2 and strand S8. However, there is an additional set of chemical shift changes upon FL-CDB3 binding to R249S: at the site distorted by the mutation, which is remote from the peptide binding site and situated at the other side of the DNA-binding interface (Figure 5D), including the loop L2 and helix H1 region and the loop between strand S6 and strand S7 (Friedler et al. 2004). The chemical shift values for these residues are shifted back towards wild type values, indicating a shift of the conformational equilibrium of the R249S mutant towards the wild type conformation (Figure 5E/F) (Friedler et al. 2004).

(a)   **(b)**

(c)   **(d)**

(e)   **(f)**

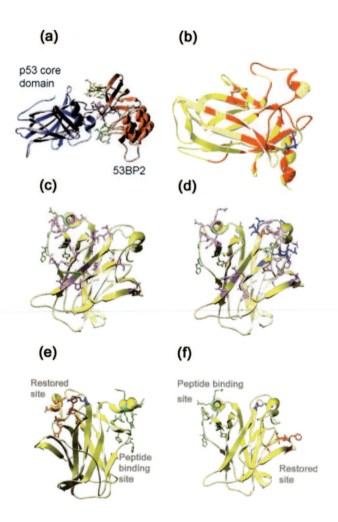

Figure 5. Reversing the structural effects of mutation in p53 core domain by FL-CDB3: (A) design of FL-CDB3. The crystal structure of the p53 core domain (blue)-53BP2 (red) complex (coordinates taken from (Gorina and Pavletich 1996)) is shown with the three p53-binding loops in 53BP2 highlighted. CDB3 (residues 490-498) is shown in purple. (B) Structural effects of the R249S mutation on p53 core domain - chemical shift analysis: Residues in R249S, which change their chemical shift compared to the wild type (0.125 ppm $< \Delta\partial$ for 15N, 0.025 ppm $< \Delta\partial$ for 1H), are color-coded red. Residues that retain their chemical shift or those that could not be assigned are color-coded yellow. The mutation site is color-coded blue. (C) Chemical shift changes ($\Delta\partial$) in wild-type p53 core domain upon binding to FL-CDB3. The FL-CDB3 binding site consists mainly of loop L1 and helix H2. Deviations above 5 times the standard deviation ($\Delta\partial > 0.25$ppm for 15N and $\Delta\partial > 0.05$ ppm for 1H) were considered significant (color-coded green). $\Delta\partial$ differences between 2.5 times and 5 times the standard deviation ($0.125 < \Delta\partial < 0.25$ ppm for 15N, $0.025 < \Delta\partial < 0.05$ ppm

for 1H) were considered as medium (color-coded purple), and $\Delta\partial$ differences below 2.5 times the standard deviation ($\Delta\partial < 0.125$ ppm for 15N and $\Delta\partial < 0.025$ ppm for 1H) were considered insignificant (color-coded yellow). See (Friedler et al. 2004) for residue numbers. (D) Chemical shift changes ($\Delta\partial$) in R249S upon binding to FL-CDB3. The peptide-binding site is the same as for wild type (shown in panel C) but additional chemical shift changes are observed at the other end of the protein. Color code is the same as in panel C. In addition, residues that disappeared in the mutant spectrum compared to the peptide-free wild type are color-coded blue and the mutation sites are color-coded red. (E) Residues in loop L2 and helix H1 of R249S whose chemical shifts are shifted towards wild type upon binding of FL-CDB3. The color code is: Green: FL-CDB3 binding residues. Red: residues in above region of R249S whose chemical shifts are reversed towards wild type conformation upon FL-CDB3 binding. Blue: mutation site. For clarity, the p53 core domain is viewed from the opposite side compared to panels C and D. (F) same as (E) for the loop between strands 6 and 7 and from a different orientation. Panels B-F are taken from (Friedler et al. 2004). The picture was generated using swissPDB viewer (Guex and Peitsch 1997).

DESIGN OF A SUPERSTABLE P53 CORE DOMAIN MUTANT

Not only is the inherently low thermodynamic stability of the wild-type p53 protein the Achilles' heel of human p53, but it also makes the protein difficult to study by biophysical and structural methods. As we have seen above, even weakly destabilizing mutations can trigger the unfolding and hence the loss of activity of human p53 core domain at physiological temperature. Conversely, it should be possible to stabilize both wild-type and mutant proteins by single amino acid substitutions that contribute to the overall stability of the structural scaffold without compromising its DNA-binding activity and hence its function *in vivo*. Such stabilized p53 variants could be better suited for many experimental purposes and could furthermore show possible ways to reactivate tumorigenic p53 mutants.

By adopting a semi-rational approach on the basis of the molecular evolution of p53, a superstable mutant of the human p53 core domain was designed (Nikolova et al. 1998). A comparison of sequences of p53 homologues from more than 20 species revealed many naturally occurring mutations. A number of the identified mutations were introduced into human p53 core domain and the effect on the thermodynamic stability of the protein was measured. Such a strategy should minimize the selection of mutations that impair activity. In addition, the N239Y substitution was included, which is not naturally occurring, but had been reported to restore activity in some of the common cancer mutants (Brachmann et al. 1998) (see below). The most stable substitutions were finally combined in a quadruple mutant. Four point mutations, M133L, V203A, N239Y and N268D, stabilize the core domain by 2.65 kcal/mol without impairing binding to *gadd45* promoter DNA (Nikolova et al. 1998). The effects of the point mutations, which are located in different regions of the core domain, are nearly additive. The main contribution to stability increase comes from the N239Y and N268D substitutions, which are of particular interest, because they are both known to act as second-site suppressors for various cancer-associated mutations (Brachmann et al. 1998). Interestingly, these mutations were also found in a study on the *in vitro* evolution of thermostable p53 variants (Matsumura and Ellington 1999).

CRYSTAL STRUCTURE OF THE SUPERSTABLE P53 CORE DOMAIN MUTANT

In order to understand the molecular basis for its increased stability, the high-resolution crystal structure (1.9 Å) of this mutant in the absence of DNA has been solved (Joerger et al. 2004). This structure reveals that the four point mutations cause only local structural changes, whereas the overall structure of the β-sandwich and the DNA-binding surface is conserved (Figure 6A). As is to be expected, the largest deviations from the wild-type structure can be found in the loop regions. This is either due to an induced fit movement upon DNA binding (loop L1) or is a direct consequence of a different crystallographic environment of the molecules in both structures and thus reflects the inherent flexibility of some of these regions (S7/S8 region).

As shown by urea-induced unfolding studies, the mutations M133L and V203A make the smallest contribution to the stability increase in the quadruple mutant (Nikolova et al. 1998). The crystal structure reveals that this is achieved by only small structural changes, mainly by local re-packing of the side chains around the mutations sites in the loop-sheet-helix motif and the S5/S6 turn, respectively. The structure provides an explanation for the large effect of the N239Y and N268D mutations on stability increase. The N239Y substitution is located in the L2 loop, which forms part of the DNA-binding surface. On the basis of the DNA-bound structure of wild type, Asn239 makes no direct contact with DNA. Tyr239 in the quadruple mutant is partly solvent exposed. The mutation induces minor changes in neighboring side chains such as Leu137 to accommodate the tyrosyl moiety (Figure 6B). Tyr239 is within a 5 Å distance of the zinc ligands Cys238, Cys242 and His179 and thus directly connects with zinc binding. Tyr239 seems to stabilize loop L3 as reflected by the reduced crystallographic thermal factors in the structure of the quadruple mutant in this region.

The N268D mutation alters the hydrogen-bonding network (Figure 6C). In wild type, the side chain amide of Asn268 on β-strand S10 forms a hydrogen bond with the backbone oxygen of Phe109 on β-strand S1. In the quadruple mutant, the side chain of Asp268 has flipped relative to the position of the asparagine in wild type. It uses both of its carboxylate oxygens to form two new hydrogen bonds with backbone nitrogen atoms of Ser269 on the same strand and Leu111 on strand S10. This links the two sheets of the β-sandwich in an energetically much more favorable way than is achieved in the wild type. Although the N239Y and N268D substitutions act on different regions of the core domain and do this in a very different way, their contribution to the thermodynamic stability of the protein is similar. Overall they seem to reduce the structural plasticity of the protein. This is evident from both the reduced relative temperature factors in certain regions of the quadruple mutant crystal structure and from NMR studies on the dynamic properties of the protein in solution (Ang H. C., Freund S.M.V. and Fersht A.R. unpublished results).

The structure of the quadruple mutant not only provides us with insights into the molecular basis for stabilizing mutations in p53, but also shows the potential practical use of this mutant for experimental purposes. So far many biophysical studies on p53 have been hampered by the low thermodynamic stability of the destabilized mutants and even the wild type itself. The superstable quadruple mutant provides a more rigid and stable structural framework, while maintaining the overall structural characteristics of the wild-type protein. We are currently studying the structural effects of the cancer hot-spot mutations in the context of the quadruple mutant. The proteins do indeed show higher stability and longer lifetimes than their respective counterparts in the wild-type context, which makes them more amenable to NMR and crystallographic studies.

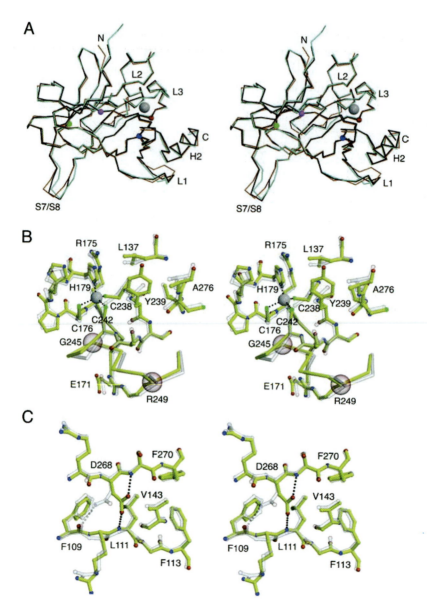

Figure 6. The crystal structure of the superstable p53 core domain quadruple mutant M133L/V203A/N239Y/N268D. (A) Stereo view of the Cα-trace of the p53 core domain quadruple mutant (PDB code 1UOL, chain A; black) superimposed on DNA-free wild type (PDB code 1TSR, chain A; blue) and DNA-bound wild type (PDB code 1TSR, chain B; orange). The zinc ion is shown as a large gray sphere. The mutation sites in the quadruple mutant are marked by small colored spheres, indicating the spatial separation of the four mutation sites. The color coding is as follows: blue, M133L; green, V203A; red, N239Y; magenta, N268D. (B) Stereo view of the region around the N239Y mutation site. The

structure of p53 core domain quadruple mutant M133L/V203A/N239Y/N268D (chain A; yellow) is superimposed on the wild-type structure (chain A; transparent light gray). The zinc ion is shown as a gray sphere. The large semi-transparent red spheres indicate the location of the two cancer hot-spot sites Gly245 and Arg249. (C) Stereo view of the region around the N268D mutation site. The structure of p53 core domain quadruple mutant M133L/V203A/N239Y/N268D (chain A; yellow) is superimposed on the wild-type structure (chain A; transparent light gray). The figures which were generated using MOLSCRIPT (Kraulis 1991) and RASTER3D (Merritt and Bacon 1997) are taken from (Joerger et al. 2004).

REVERSING THE STRUCTURAL EFFECTS OF TUMORIGENIC MUTATIONS BY SECOND-SITE SUPPRESSOR MUTATIONS

The deleterious effect of many common cancer mutations can be reversed by intragenic suppressor mutations. Using a yeast selection system in combination with mammalian reporter gene and apoptosis assays, Brachmann *et al.* (Brachmann et al. 1998) have identified several second-site suppressor mutations for the cancer hot-spot mutations G245S and R249S as well as V143A, the classical example of a temperature sensitive cancer mutant (Zhang et al. 1994). Most of these suppressor mutations are not specific for only one single cancer mutation but restore activity in a whole subset of different mutants albeit at different levels. The suppressor mutations N239Y and S240N, which are not found in any sequence of the known p53 homologues, show a very similar rescue profile. They both suppress the cancer hot-spot mutant G245S and to a lesser extent restore activity in V143A. At low temperatures N239Y was found to suppress even R175H. N268D, another of these second-site suppressor mutations, is naturally occurring in p53 of various rodents and was found to restore activity in V143A. Remarkably, H168R, which in conjunction with T123A is reported to rescue the cancer hot-spot mutant R249S, causes cancer when on its own. A more recent study reported that 16 of 30 of the most frequently occurring cancer mutants can be rescued by changes in residues 235, 239 and 240 either alone or in combination (Baroni et al. 2004).

Insights into the mechanism of rescue

The effect of the above mutants on the stability and DNA-binding activity of p53 core domain was studied to further elucidate how they rescue tumorigenic p53 mutants (Nikolova et al. 2000). Double mutant cycles show that N268D and N239Y act as "global stability" suppressors by increasing the stability of the cancer mutants G245S and V143A. The free energy changes are additive. In contrast, the suppressor mutation H168R seems to be rather specific for the R249S mutation: despite destabilizing wild type, H168R has virtually no effect on the stability of R249S, but restores its binding affinity for the *gadd45* promoter. NMR studies show that H168R reverses some of the structural changes induced by R249S (Nikolova et al. 2000).

In the wild-type and the quadruple mutant structures, the guanidinium group of Arg249 forms a salt bridge with the carboxylate group of Glu171 in the L2 loop and hydrogen bonds with the backbone oxygens of Gly245 and Met246 in the L3 loop (Figure 2B). This stabilizes the conformation of the L2/L3 region allowing for specific DNA binding. Loss of these interactions in mutant R249S results in destabilization of the protein concomitant with a conformation of the L2/L3 region which is no longer compatible with effective DNA binding. The H168R mutation has a similar effect on the p53 core domain. Owing to its close proximity, Arg168 presumably directly disrupts the salt bridge between Glu171 and Arg249 and thus compromises the structural integrity of the DNA binding surface. In the structural context of the R249S mutant, however, Arg168 seems to be able to mimic the role of Arg249 in the wild type and at least partly restores an active, wild-type like conformation in the L2/L3 region.

The crystal structure of the superstable quadruple mutant provides further insight into how the two second-site suppressor mutants N239Y and N268D rescue some of the common structurally destabilized cancer mutants. Tyr239 stabilizes the conformation of the L2/L3 loop region, in particular the L3 region harboring the DNA contact residue Arg248 (Figure 2B). This stabilizing effect seems to compensate for the structural perturbation caused by the G245S mutation resulting in a rescue of this cancer mutant. The rescue of V143A by N268D can also be readily explained. NMR studies indicate that the V143A mutation results in changes in almost all the residues of the β-sandwich (Wong et al. 1999). In wild type and the superstable quadruple mutant (Figure 6C), the side chain of Val143, which is located on β-strand S3, points towards the hydrophobic core of the β-sandwich, where it is completely buried (Cho et al. 1994; Joerger et al. 2004). Val143 lies at the very heart of a cluster of hydrophobic residues contributed by different regions of the protein. Apparently, the truncation of the methyl groups in mutant V143A is detrimental to the packing of these hydrophobic core residues. The mechanism of rescue by an N268D substitution can be pinpointed to the formation of new hydrogen bonds. The energy loss of more than 3 kcal/mol arising from the creation of a hydrophobic cavity in mutant V143A can be partly compensated for by Asp268, which bridges the two sheets of the β-sandwich via hydrogen bonds in an energetically more favorable way than could be achieved by the asparagine in wild type. Apart from an effect on the overall stability of the protein, a local effect may be crucial in rescuing V143A, too. Leu111, which via its backbone nitrogen acts as a hydrogen bond donor for Asp268, makes direct contact with one of the methyl groups of Val143 (Figure 6B). The hydrogen bond with Asp268 significantly reduces the flexibility of residues Leu111 and Phe113 in the immediate vicinity of Val143 as reflected by the *B*-factors in the crystal structure. This may directly counteract the loss of interactions by the V143A mutation by providing a more rigid structural framework around the mutation site.

An even more detailed and concise picture of possible ways to rescue mutant p53 will most likely emerge from additional crystal structures of cancer mutants in combination with their respective second-site suppressors or in the context of the superstable quadruple mutant.

CONCLUDING REMARKS

Over the past 25 years, studies carried out using molecular biology in combination with protein engineering experiments and structural studies by both NMR and protein crystallography have resulted in a wealth of information about the structure and function of human p53. We now have a much clearer picture of how human p53 is inactivated by cancer mutations. In particular the effects of the cancer hot-spot mutations in the DNA-binding core domain of human p53 are well understood. They either remove essential DNA contacts or destabilize the protein both locally and globally. This provides us with the foundations for tackling the ultimate goal – finding a cure for cancer. It may still be a long way to go until we can hold an effective drug in our hands, but the initial results from second-site suppressor mutations and screenings for generic molecules that stabilize the native state of human p53 provide us with a promising starting point.

REFERENCES

Aguilar, F., Hussain, S. P. & Cerutti, P. (1993) Proc Natl Acad Sci U S A 90, 8586-90.

Baroni, T. E., Wang, T., Qian, H., Dearth, L. R., Truong, L. N., Zeng, J., Denes, A. E., Chen, S. W. & Brachmann, R. K. (2004) Proc Natl Acad Sci U S A 101, 4930-5.

Brachmann, R. K., Yu, K., Eby, Y., Pavletich, N. P. & Boeke, J. D. (1998) Embo J 17, 1847-59.

Bullock, A. N. & Fersht, A. R. (2001) Nature cancer reviews 1, 68-76.

Bullock, A. N., Henckel, J. & Fersht, A. R. (2000) Oncogene 19, 1245-56.

Bullock, A. N., Henckel, J., DeDecker, B. S., Johnson, C. M., Nikolova, P. V., Proctor, M. R., Lane, D. P. & Fersht, A. R. (1997) Proc Natl Acad Sci U S A 94, 14338-42.

Bykov, V. J., Issaeva, N., Shilov, A., Hultcrantz, M., Pugacheva, E., Chumakov, P., Bergman, J., Wiman, K. G. & Selivanova, G. (2002) Nat Med 8, 282-8.

Bykov, V. J., Selivanova, G. & Wiman, K. G. (2003) Eur J Cancer 39, 1828-34.

Cho, Y., Gorina, S., Jeffrey, P. D. & Pavletich, N. P. (1994) Science 265, 346-355.

Clore, G. M., Ernst, J., Clubb, R., Omichinski, J. G., Kennedy, W. M., Sakaguchi, K., Appella, E. & Gronenborn, A. M. (1995) Nat Struct Biol 2, 321-33.

Clore, G. M., Omichinski, J. G., Sakaguchi, K., Zambrano, N., Sakamoto, H., Appella, E. & Gronenborn, A. M. (1994) Science 265, 386-91.

Courtois, S., de Fromentel, C. C. & Hainaut, P. (2004) Oncogene 23, 631-8.

Derbyshire, D. J., Basu, B. P., Serpell, L. C., Joo, W. S., Date, T., Iwabuchi, K. & Doherty, A. J. (2002) Embo J 21, 3863-3872.

DiTullio, R. A., Jr., Mochan, T. A., Venere, M., Bartkova, J., Sehested, M., Bartek, J. & Halazonetis, T. D. (2002) Nat Cell Biol 4, 998-1002.

Dobson, C. M. (1999) Trends Biochem Sci 24, 329-32.

Fan, J. Q. (2003) Trends Pharmacol Sci 24, 355-60.

Fan, J. Q., Ishii, S., Asano, N. & Suzuki, Y. (1999) Nat Med 5, 112-5.

Foster, B. A., Coffey, H. A., Morin, M. J. & Rastinejad, F. (1999) Science 286, 2507-10.

Friedler, A., DeDecker, B. S., Freund, S. M., Blair, C., Rüdiger, S. & Fersht, A. R. (2004) J Mol Biol 336, 187-96.

Friedler, A., Hansson, L. O., Veprintsev, D. B., Freund, S. M., Rippin, T. M., Nikolova, P. V., Proctor, M. R., Rüdiger, S. & Fersht, A. R. (2002) Proc Natl Acad Sci U S A 99, 937-42.

Friedler, A., Veprintsev, D. B., Hansson, L. O. & Fersht, A. R. (2003) J Biol Chem 278, 24108-12.

Gorina, S. & Pavletich, N. P. (1996) Science 274, 1001-5.

Grossman, S. R. (2001) Eur J Biochem 268, 2773-8.

Guex, N. & Peitsch, M. C. (1997) Electrophoresis 18, 2714-23.

Hainaut, P. & Hollstein, M. (2000) Adv Cancer Res 77, 81-137.

Issaeva, N., Friedler, A., Bozko, P., Wiman, K. G., Fersht, A. R. & Selivanova, G. (2003) Proc Natl Acad Sci U S A 100, 13303-7.

Jeffrey, P. D., Gorina, S. & Pavletich, N. P. (1995) Science 267, 1498-1502.

Joerger, A. C., Allen, M. D. & Fersht, A. R. (2004) J Biol Chem 279, 1291-6.

Joo, W. S., Jeffrey, P. D., Cantor, S. B., Finnin, M. S., Livingston, D. M. & Pavletich, N. P. (2002) Genes Dev 16, 583-93.

Kraulis, P. J. (1991) J. Appl. Crystallogr. 24, 946-950.

Kussie, P. H., Gorina, S., Marechal, V., Elenbaas, B., Moreau, J., Levine, A. J. & Pavletich, N. P. (1996) Science 274, 948-953.

Lane, D. P. & Hupp, T. R. (2003) Drug Discov Today 8, 347-55.

Lane, D. P. & Lain, S. (2002) Trends Mol Med 8, S38-42.

Matsumura, I. & Ellington, A. D. (1999) Protein Sci 8, 731-40.

Merritt, E. A. & Bacon, D. J. (1997) Methods Enzymol. 277, 505-524.

Michael, D. & Oren, M. (2002) Curr Opin Genet Dev 12, 53-9.

Michael, D. & Oren, M. (2003) Semin Cancer Biol 13, 49-58.

Momand, J., Wu, H. H. & Dasgupta, G. (2000) Gene 242, 15-29.

Morello, J. P., Petaja-Repo, U. E., Bichet, D. G. & Bouvier, M. (2000) Trends Pharmacol Sci 21, 466-9.

Muller-Tiemann, B. F., Halazonetis, T. D. & Elting, J. J. (1998) Proc Natl Acad Sci U S A 95, 6079-84.

Nikolova, P. V., Henckel, J., Lane, D. P. & Fersht, A. R. (1998) Proc Natl Acad Sci U S A 95, 14675-80.

Nikolova, P. V., Wong, K. B., DeDecker, B., Henckel, J. & Fersht, A. R. (2000) Embo J 19, 370-8.

Peng, Y., Li, C., Chen, L., Sebti, S. & Chen, J. (2003) Oncogene 22, 4478-87.

Prives, C. & Manley, J. L. (2001) Cell 107, 815-8.

Rippin, T. M., Bykov, V. J., Freund, S. M., Selivanova, G., Wiman, K. G. & Fersht, A. R. (2002) Oncogene 21, 2119-29.

Rustandi, R. R., Baldisseri, D. M. & Weber, D. J. (2000) Nat Struct Biol 7, 570-4.

Ryan, K. M., Phillips, A. C. & Vousden, K. H. (2001) Curr Opin Cell Biol 13, 332-7.

Samuels-Lev, Y., O'Connor, D. J., Bergamaschi, D., Trigiante, G., Hsieh, J. K., Zhong, S., Campargue, I., Naumovski, L., Crook, T. & Lu, X. (2001) Mol Cell 8, 781-94.

Selivanova, G., Iotsova, V., Okan, I., Fritsche, M., Strom, M., Groner, B., Grafstrom, R. C. & Wiman, K. G. (1997) Nat Med 3, 632-8.

Selivanova, G., Kawasaki, T., Ryabchenko, L. & Wiman, K. G. (1998) Semin Cancer Biol 8, 369-78.

Selivanova, G., Ryabchenko, L., Jansson, E., Iotsova, V. & Wiman, K. G. (1999) Mol Cell Biol 19, 3395-402.

Shaulsky, G., Goldfinger, N., Ben-Zeev, A. & Rotter, V. (1990) Mol. Cell. Biol. 10, 6565-6577.

Stommel, J. M., Marchenko, N. D., Jimenez, G. S., Moll, U. M., Hope, T. J. & Wahl, G. M. (1999) EMBO J 18, 1660-1672.

Vogelstein, B., Lane, D. & Levine, A. J. (2000) Nature 408, 307-10.

Vousden, K. H. & Lu, X. (2002) Nat Rev Cancer 2, 594-604.

Walker, K. K. & Levine, A. J. (1996) Proc Natl Acad Sci U S A 93, 15335-40.

Wang, W., Takimoto, R., Rastinejad, F. & El-Deiry, W. S. (2003) Mol Cell Biol 23, 2171-81.

Wong, K. B., DeDecker, B. S., Freund, S. M., Proctor, M. R., Bycroft, M. & Fersht, A. R. (1999) Proc Natl Acad Sci U S A 96, 8438-42.

Zhang, W., Guo, X. Y., Hu, G. Y., Liu, W. B., Shay, J. W. & Deisseroth, A. B. (1994) Embo J 13, 2535-44.

Zhao, K., Chai, X., Johnston, K., Clements, A. & Marmorstein, R. (2001) J Biol Chem 276, 12120-7.

Chapter 18

MUTANT P53 REACTIVATION AS A NOVEL STRATEGY FOR CANCER THERAPY

Galina Selivanova, Vladimir J.N. Bykov and Klas G. Wiman
Dept. of Oncology-Pathology, Cancer Center Karolinska (CCK) and the Microbiology & Tumor Biology Center (MTC), Karolinska Institute Stockholm, Sweden

THE TARGET: ABUNDANTLY EXPRESSED POINT MUTANT P53

Inactivation of the p53 tumor suppressor by point mutation occurs in a large fraction of human tumors, including almost all tumor types (see p53 Mutation database at http://www.iarc.fr/p53). A majority of p53 mutations are missense mutations that give rise to the expression of mutant p53 proteins with one amino acid substitution. This pattern of mutation stands in sharp contrast to those of most other tumor suppressor genes, e.g. the Rb and p16 genes, which are frequently inactivated by homozygous deletion, smaller deletions or promoter methylation that either results in complete lack of expression of the protein, or expression of a truncated unstable protein. This suggests that p53 mutation not only serves to inactivate p53 but that expression of mutant p53 itself may provide a selective advantage to tumor cells and promote tumor growth. First, point mutant p53 proteins may act in a dominant negative manner, i.e. inhibit the activity of a wild type allele present in the same cell through hetero-oligomerization that forces wild type p53 to adopt a mutant conformation (Milner and Medcalf 1991). In addition, mutant p53 proteins may have acquired novel activites that could support the growth of tumors. These so called gain-of-function (GOF) activities of mutant p53 could involve promiscuous DNA binding and illegitimate activation of target genes, such as the c-Myc oncogene, the multidrug

P. Hainaut and K.G. Wiman (eds.), 25 Years of p53 Research, 399-419.

resistance gene (MDR1), VEGF, and the dUTPase gene ((Frazier et al. 1998; Pugacheva et al. 2002; Tsang et al. 2003); www.iarc.fr/p53), whose activation could contribute to tumor development. Moreover, mutant p53 could enhance cell cycle progression and/or cell survival through novel interactions with cellular protein partners, as examplified by the binding of mutant p53 to p73 and other p53 family members (Di Como et al. 1999; Marin and Kaelin 2000; Strano et al. 2002; Monti et al. 2003; Strano and Blandino 2003). Importantly, interaction of mutant p53 with p73 prevents p73-induced apoptosis in response to DNA damage. The proposed mutant p53 gain-of-function activities need to be studied further in order to assess their contribution to tumor development at various stages.

Mutant p53 is not only retained by tumor cells but usually also expressed at high levels. The reasons for this become evident upon a closer look at normal regulation of p53 expression. Since p53 is a potent inhibitor of cell growth and inducer of cell death, its activity must be tightly controlled to ensure normal cell proliferation and survival. This is achieved through several mechanisms, including regulation of p53 protein stability and activity, and regulation at the level of transcription, translation and subcellular localization. Rapid p53 induction in response to stress conditions occurs mainly via post-translational mechanisms, leading to the stabilization and activation of the p53 protein (reviewed in (Ryan et al. 2001)). The MDM2 protein, which can both inhibit p53's transcriptional activity and target p53 for degradation, is a key p53 regulator. MDM2 functions as an E3 ubiquitin ligase, the final component of the enzyme cascade that results in the conjugation of ubiquitin to proteins, targeting them for degradation by the proteasome. MDM2 ubiquitinates both p53 and itself, contributing to the rapid turnover of both proteins (reviewed in (Ryan et al. 2001)). Since MDM2 is a transcriptional target of p53 itself, it forms an autoregulatory feedback loop with p53: increased p53 activity leads to increased expression of its own negative regulator. Several additional p53 autoregulatory loops exist. The expression of p14ARF, a potent negative regulator of MDM2, is repressed by p53 (Sherr 2000). On the other hand, p53 activates the expression of a positive regulator of MDM2, cyclin G1 (Okamoto et al. 2002). Furthermore, p53 induces expression of the Wip1 gene which encodes a phosphatase that inhibits the p53-activating p38 MAP kinase by dephosphorylation (Takekawa et al. 2000), and expression of two novel E3 ubiquitin ligases, Pirh2 and COP1, which mediate p53 degradation (Leng et al. 2003; Dornan et al. 2004). Thus, several pathways that downregulate p53 protein expression are dependent on p53 transcriptional activity. Due to the inability of mutant p53 to activate transcription of these genes, pathways responsible for p53 degradation are disrupted in tumor cells expressing mutant p53.

Furthermore, mutant p53 may be posttranslationally modified in tumors ((Minamoto et al. 2001; Melnikova et al. 2003); reviewed by Bode and Dong (Bode and Dong 2004). This is because stress signalling in tumor cells activates various p53-modifying enzymes. DNA damage induces N-terminal p53 phosphorylation at several residues. The ATM/ATR, DNA-PK and PI-3 kinases phosphorylate Ser-15, Chk1/Chk2 phosphorylate Ser-20, and p38 MAP kinase and homeodomain-interacting protein kinase-2 (HIPK2) phosphorylate Ser-46. N-terminal p53 phosphorylation (reviewed in (Xu 2003)) augments the interaction with the p300/CBP acetyltransferases, establishing a phosphorylation/ acetylation cascade that results in enhanced p53 activity. The acetyltransferases p300/CBP and pCAF bind within the amino terminus of p53 and mediate acetylation of the lysines in the carboxy terminus, facilitating the function of p53 as transcriptional activator (Gu and Roeder 1997). Phosphorylation at Ser-15 and Ser-46 facilitates p53 transactivation function, binding to p300/CBP, and induction of apoptosis (Lambert et al. 1998; Bulavin et al. 1999), but does not directly influence the interaction with HDM2 (Dumaz and Meek 1999). Constitutive activation of DNA damage checkpoints in at least some tumors suggest that activated DNA damage signalling pathways can induce posttranslational modifications of mutant p53 in tumors (DiTullio et al. 2002).

Oncogenic stress may also promote posttranslational modification of mutant p53 in tumor cells. p14ARF, which blocks p53-MDM2 interaction by direct binding to MDM2, is induced by oncogene activation (Sherr 2000). Interestingly, p14ARF can prevent the inhibition of p53 acetylation by MDM2, illustrating another role for p14ARF in p53 activation (Ito et al. 2001). Thus, enhanced expression of p14ARF observed in mutant p53-carrying tumor cells (Lindstrom et al. 2000), might contribute to posttranslational modifications of mutant p53. Moreover, both Myc and E2F1 induce p53 phosporylation that stabilizes p53 in the absence of ARF (Lindstrom and Wiman 2003). E2F1 has been shown to promote p53 phosphorylation via the ATM signalling pathway (Powers et al. 2004). In addition, oncogenic Ras induces p53 N-terminal phosphorylation by p38MAPK at Ser-33 and Ser-46 via inhibition of the Wip1(PPM1D) phosphatase (Bulavin et al. 2002).

In conclusion, loss of p53 transcriptional activity in mutant p53-expressing tumor cells relaxes the tight control of p53 protein levels, while DNA damage and oncogenic signalling trigger posttranslational modifications of p53. One can therefore envisage abundant and activated mutant p53 as a "loaded gun" (Selivanova 2001). However, the gun's trigger is locked by a mutation, preventing activation of the cellular suicide program.

CLINICAL PROBLEM POSED BY MUTANT P53-CARRYING TUMORS

Since p53 has a critical role in the cellular response to DNA damage, and since radiotherapy and most commonly used chemotherapeutic drugs directly or indirectly induce DNA damage, the p53 status of a tumor is expected to have an impact on therapy response and clinical outcome. The role of p53 status for the therapeutic response can be studied at different levels: cultured human tumor cells, animal tumor models (mouse), and cancer patients. Screening of a large set of cell lines with defined p53 status revealed that most clinically used chemotherapeutic agents were more efficient in tumor cells carrying wild type p53. However, certain drugs including colchicine derivatives and microtubule inhibitors (taxanes) were more active in a mutant p53 context or regardless of p53 status (Weinstein et al. 1997). In addition, wild type p53-dependent and independent activation of the cell cycle regulator p21 can induce cell cycle arrest that may increase cell survival in some genetic backgrounds in response to chemotherapeutic agents and gamma irradiation (Bunz et al. 1998).

Studies in mouse tumor models have shown that disruption of the p53 pathway increases resistance to DNA-damaging chemotherapeutic agents and gamma irradiation. In Emµ-myc transgenic lymphomas, p53-induced apoptosis is critical for tumor suppression after DNA damage (Schmitt 2003). Emµ-myc lymphomas derived from crosses with p53 null or INK4a null mice showed increased resistance to alkylating agents (cyclophosphamide) and radiation therapy compared to Emµ-myc lymphomas from mice with intact p53 and INK4a loci (Schmitt et al. 1999). Notably, induction of p53 or p16-dependent senescence was associated with better response of Emµ–Myc lymphomas to cyclophosphamide in vivo (Schmitt et al. 2002a; Schmitt et al. 2002b) and a senescence response in human breast cancer cells following exposure to doxorubicin (Elmore et al. 2002). Thus, apoptosis and senescence may both contribute to the p53-dependent responses to chemotherapeutic drugs.

The clinical relevance of p53 status has been addressed in numerous studies. The evidence is perhaps most persuasive in breast cancer. Several studies have demonstrated that mutant p53 status is associated with poor prognosis (http://www.iarc.fr/p53; see also for example (Bergh et al. 1995; Borresen et al. 1995). Patients with mutant p53-carrying tumors showed poor responses to treatment with anthracyclines (Geisler et al. 2001; Rahko et al. 2003). Wild type p53 status was associated with sensitivity to low dose doxorubicin treatment (Geisler et al. 2001). Moreover, the type of p53 mutation has clinical impact. Mutations that give rise to amino acid substitutions in the L2/L3 loop of the p53 core domain appear to predict a particularly poor clinical outcome in various tumor types (Soussi and Beroud 2001; Geisler et al. 2003). Nonetheless, some studies have indicated that p53 status has no value as predictive marker (Wang et al. 2004) or that mutant p53 is in fact a better predictor for complete therapeutic response (Bertheau et al. 2002). It is clear, therefore, that p53 status can have different predictive value even within the same tumor type depending on the choice of regimen, and presumably a range of other factors as well. For instance, amplification of the c-myc oncogene in wild type p53-carrying tumors was associated with improved survival in a phase III 5-fluorouracil clinical trial (Seoane et al. 2002). To a certain extent, the conflicting data on the relevance of p53 status for clinical outcome are probably due to inaccurate methods for assessing p53 status. DNA sequencing of the entire coding region is the method of choice.

MUTANT P53 REACTIVATION: AN EFFICIENT AND SELECTIVE STRATEGY FOR CANCER THERAPY

p53 is a critical regulator of cell growth and survival which responds to a wide range of stress signals. Oncogenic stress, including activation of Ras, Myc, ß-catenin, cdk4 or cyclins D/E, or inactivation of tumor suppressors p16, Rb, or APC, triggers p53 activation and p53-dependent cell death. Therefore, p53 inactivation either by mutation of the p53 gene iself or alterations of other critical regulators of the p53 pathway, e.g. MDM2 or p14ARF, is probably a necessary requirement for evading apoptosis during tumor development. The p53 pathway is thus a prime target for therapeutic intervention in tumors, and a number of possible strategies to rehabilitate or exploit this pathway have been designed, including p53 gene therapy and other approaches (reviewed in (Bykov and Wiman 2003; Lane and Hupp 2003)).

As pointed out above, p53 mutations in tumors are unique among tumor suppressor gene mutations in that the great majority of them are missense mutations that result in single amino acid substitutions in the protein. A myriad of different amino acid substitutions in the p53 core domain produce a common effect – the partial unfolding of the core domain (reviewed in (Bullock and Fersht 2001)) which disrupts its DNA binding activity and thus leads to inactivation of p53 as a tumor suppressor. This shared feature of mutant p53 proteins suggests the possibility of designing novel p53-specific drugs that can stabilize the folding of many mutant forms of p53 and thus rescue their function.

Therapies aimed at restoring the function of mutant p53 in tumors will re-establish the broken link between oncogene-induced pro-apoptotic signalling and the cell death machinery and thus trigger massive apoptosis in presensitized tumor cells (Figure 1). Moreover, for reasons that are not entirely understood, expression of exogenous p53 protein in non-malignant cells caused no detrimental effects in vitro and in vivo (Bossi et al. 2000). As discussed above, mutant p53 is frequently overexpressed and posttranslationally modified in tumor cells. These factors combined should render tumor cells highly sensitive to mutant p53 reactivation, providing a basis for efficient and selective therapy without any need for specific tumor targeting.

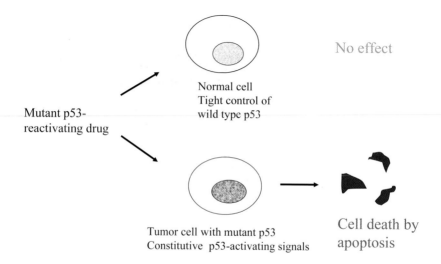

Tumor cell-specific mutant p53-dependent apoptosis

No effect

Normal cell
Tight control of
wild type p53

Mutant p53-
reactivating drug

Tumor cell with mutant p53
Constitutive p53-activating signals

Cell death by
apoptosis

Figure 1. Strategy for elimination of tumor cells by mutant p53 reactivation. Compounds that target mutant p53 and restore wild type p53 function will selectively induce apoptosis in tumor cells, due to high levels of post-translationally modified mutant p53 and sensitization of tumor cells to apoptosis as result of oncogene activation and aberrant cell growth control. In contrast, such compounds will not significantly affect normal cells that express minute levels of wild type p53 and do not carry genetic alterations that disrupt cell cycle regulation.

REACTIVATION OF MUTANT P53 BY MANIPULATING WITH THE C-TERMINUS

Ample experimental evidence suggests that the p53 carboxy-terminal domain can influence the DNA binding of the core domain. Different modifications at the C-terminus, including phosphorylation and acetylation, activate the specific binding of p53 in vitro and in cells (reviewed in (Selivanova et al. 1998)). Furthermore, the folding of the core is affected by the C-terminus. For instance, the binding of the bacterial heat shock protein DnaK or the monoclonal antibody PAb421 to the C-terminal basic region, and to a lesser extent its deletion, renders the core domain resistant to a thermal denaturation as assessed by the retention of PAb1620 epitope upon heating to 37°C (Hansen et al. 1996). Short synthetic peptides derived from the p53 C-terminus were shown to activate p53's latent DNA binding (Hupp et al. 1995; Selivanova et al. 1997). Remarkably, the DNA binding and transactivation function of several mutant p53 proteins, including the hot spot mutants His-273, Cys-273, Gln-248, Trp-282 and Ser-245, could be rescued by such manipulations with p53's C-terminus ((Halazonetis et al. 1993; Hupp et al. 1993; Niewolik et al. 1995; Selivanova et al. 1997); reviewed in (Selivanova et al. 1998)). Also, the transcriptional transactivation function of one of the most common mutant, His-273, was restored in living cells by microinjection of the PAb421 antibody or a PAb421-derived Fv fragment in tumor cells (Abarzua et al. 1995; Caron de Fromentel et al. 1999).

Using a set of 48 overlapping p53-derived 22-mer peptides, we found that one particular peptide, designated peptide 46 and representing residues 361-382 of human p53, not only increased DNA binding by wild p53 in a gel shift assay but also restored the transcriptional transactivation function to several hot spot p53 mutants in reporter assays. Overlapping peptides containing more N-terminal or C-terminal sequences, e.g. peptides 45 and 47, were inactive. Moreover, introduction of peptide 46 into human tumor cells as a fusion peptide with a membrane-penetrating peptide derived from the Drosophila Antennapedia protein resulted in massive apoptosis in a mutant p53-dependent manner (Selivanova et al. 1997). Thus, reactivation of endogenous mutant p53 can trigger apoptosis in human cells carrying mutant p53. Subsequent studies by others showed that the same peptide could rescue mutant p53 and trigger apoptosis in a range of human tumor cells (Kim et al. 1999). Our observation that C-terminal peptide could restore the DNA binding activity to isolated core domain proteins indicated that it interacts directly with the p53 domain in which tumor-derived mutations are clustered (Selivanova et al. 1999). We and others hypothesized that the C-terminal peptide reactivates mutant p53 through the stabilization of the core domain folding and/or establishment of novel DNA contacts (Abarzua et al. 1996; Selivanova et al. 1997; Selivanova et al. 1999). These findings provided a proof-of-principle of mutant p53 rescue by a small molecule added in trans and raised hopes for the development of anticancer drugs for treatment of mutant p53-carrying tumors.

The hypothesis that direct reactivation of endogenous p53 protein in tumor cells will be therapeutically beneficial has recently been addressed in vivo. Treatment of preclinical terminal peritoneal carcinomatosis and peritoneal lymphoma models with a D-isomer C-terminal p53 peptide fused to a carrier HIV-tat peptide resulted in significant increases in lifespan (greater than 6-fold) and the generation of disease-free animals (Snyder et al. 2004). Interestingly, the p53-derived C-terminal peptide activated p53 in cancer cells but not in normal cells. Thus, specific activation of endogenous p53 activity by a macromolecular agent is therapeutically effective in preclinical models of terminal human malignancy.

RATIONAL DESIGN OF A MUTANT P53-REACTIVATING PEPTIDE

Since mutant p53 protein folding is destabilized, a promising strategy to reactivate mutant p53 is to identify small molecules capable of stabilizing the proper fold of the protein and/or raising its melting temperature. In principle, p53-stabilizing small molecules may be produced by rational design or identified by random screening of chemical libraries. The first rational design of a molecule capable of stabilizing the folding of the p53 core domain was undertaken by Fersht and colleagues. The idea behind this approach is that a ligand which binds to the properly folded p53 core domain will shift the equilibrium between native and unfolded core domain conformation towards the active wild type fold. A short peptide derived from the p53-interacting protein ASPP (previously referred to as 53BP2) (Samuels-Lev et al. 2001) served as a basis for the design of such a ligand. ASPP binds to the core domain of p53 (Gorina and Pavletich 1996), augments p53-dependent transactivation and promotes p53-dependent apoptosis (Samuels-Lev et al. 2001). CDB3, a nine-residue peptide, was designed on the basis of structural information about of the p53-ASPP/53BP2 complex. NMR provided solid evidence that CDB3 and its derivative FL-CDB3 with attached fluorescein bind to the core domain and stabilize the folding of mutant p53 core domain proteins (Friedler et al. 2002). NMR analysis revealed that CDB3 binds to a site in p53 that partly overlaps with its positively charged DNA binding site. However, in vitro studies showed that CDB3 can restore the sequence-specific DNA binding to mutant p53 proteins, including the Thr-195 mutant which is highly destabilized. It is possible that CDB3 acts as a "chaperone" that binds to mutant p53 during biosynthesis before rapid unfolding of the protein; the peptide could subsequently be replaced by the tighter-binding cognate DNA.

The major challenge in designing a target-specific drug from experiments in vitro is to achieve binding to the same target in cells in a context of multiple cellular proteins. Notably, FL-CDB3 can enter tumor cells and bind p53 in the cells. A dramatic induction of the native PAb1620 conformation was observed upon FL-CDB3 treatment of cells expressing p53 mutants His-273 and His-175, along with a decrease in the fraction of the unfolded PAb240 form (Issaeva et al. 2003). A simultaneous restoration of the transcriptional function of two hot spot p53 mutants, His-273 and His-175, was observed, as demonstrated by induction of the endogenous p53 targets p21 and MDM2 and a p53-dependent lacZ reporter. The transcriptional activity of wild type p53 was also stimulated by FL-CDB3 (Issaeva et al. 2003). Interestingly, FL-CDB3 caused accumulation of both wild type and mutant p53 in cells. The biological response induced by FL-CDB3, although moderate, was p53-dependent as shown by using isogenic cell lines that differ only in p53 status. Thus, FL-CDB3, the first rationally designed p53-reactivating molecule, can serve as a lead for the further development of anti-cancer drugs aimed at rescue of p53 tumor suppressor function.

SCREENING FOR SMALL MOLECULES THAT REACTIVATE MUTANT P53

A different strategy to identify small molecules that can reactivate mutant p53 is random screening of chemical libraries. Such screening can be performed either using a test tube assay aimed at the identification of compounds that restore native conformation and/or DNA binding activity to mutant p53 proteins, or using cellular assays based on induction of a specific cellular response, e.g. apoptosis, in the presence of mutant p53 but not in its absence. Both types of screens have been carried out and have allowed the identification of molecules with mutant p53-modulating properties (Bykov et al. 2003; Bykov and Wiman 2003) (Figure 2).

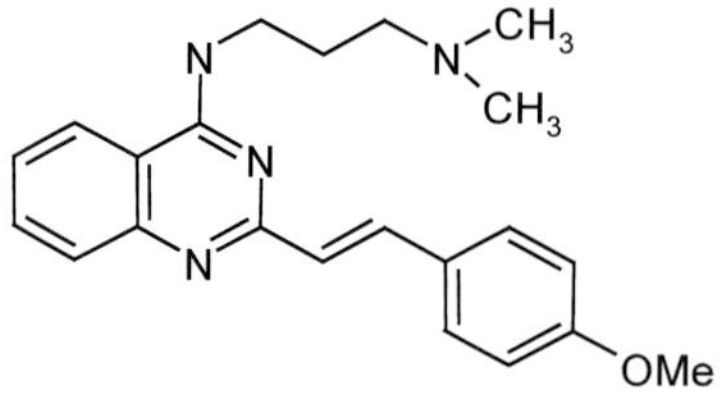

CP-31398

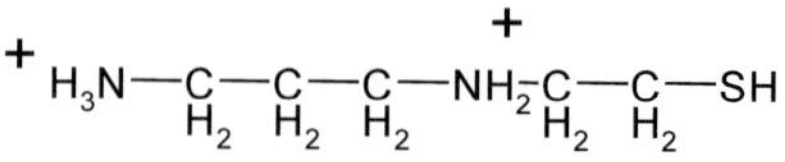

WR-1065

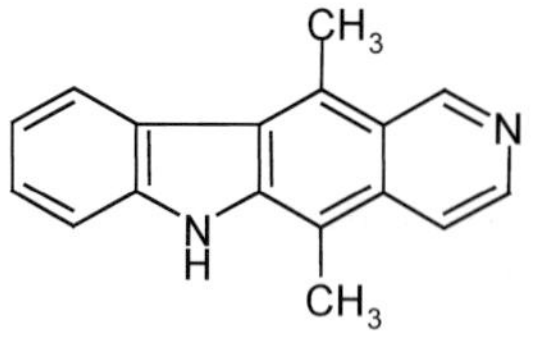

ellipticine

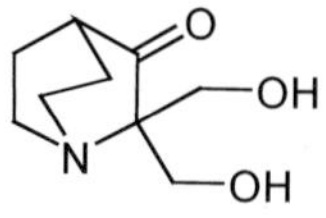

PRIMA-1

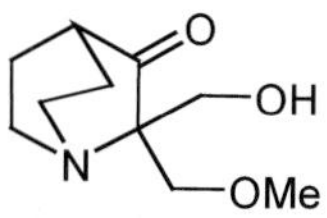

PRIMA-1MET

Figure 2. Structures of mutant p53-targeting compounds. Several low molecular weight compounds target mutant p53 and induce mutant p53-dependent cell death. The molecular structures for CP-31398 (Foster et al. 1999), WR-1065 (Tacka et al. 2002), ellipticine (Shi et al. 1998), PRIMA-1 and PRIMA-1MET (Bykov et al. 2002b; Bykov et al., 2005) are shown.

Rastinejad and colleagues screened a chemical library for compounds that could stabilize the native fold of the purified wild type p53 core domain protein after heat denaturation (Foster et al. 1999). One hit, CP-31398, stabilized mutant p53 core domains (Ala-173 and His-273) in vitro and restored both native conformation and transcriptional transactivation activity to Ala-173 mutant p53 in living cells. Treatment of mutant p53-expressing tumor cells with CP-31398 induced the p53 target gene p21. Furthermore, intraperitoneal administration of CP-31398 inhibited growth of human tumor cells carrying mutant p53 (at position 249 or 241) in nude mice without any apparent toxicity.

Subsequent studies have confirmed that CP-31398 treatment induces the p53 target genes p21, MDM2, Bax and KILLER/DR5 (Luu et al. 2002; Takimoto et al. 2002). However, the effect of CP-31398 appears to be cell line-dependent, since the induction of p53 target genes was not observed in several tumor cell lines expressing endogenous mutant p53. Induction of cell death by CP-31398 was shown to involve a caspase-dependent pathway in several cell lines expressing wild type or mutant p53, but caspase-independent pathways were also invoked in glioma cell lines (Wischhusen et al. 2003). It was proposed that CP-31398 can induce rapid p53-dependent cell death but also p53-independent cell death upon prolonged exposure. As CP-31398 may be unstable in vivo, rapid degradation of the compound might reduce its nonspecic toxic effects on tissues in vivo (Rippin et al. 2002).

The mechanism behind CP-31398-mediated mutant p53 refolding is still unclear. Structural studies did not reveal any binding of CP-31398 to the p53 core domain (Rippin et al. 2002). Experimental evidence indicates that CP-31398 can promote the correct folding of newly synthesized p53 only (Foster et al. 1999). This could account for the lack of a detectable direct interaction in the NMR studies. Interestingly, CP-31398 stabilizes wild type p53 (Luu et al. 2002; Takimoto et al. 2002). This involves inhibition of wild type p53 ubiquitination but the compound does not affect p53 phosphorylation or the interaction between p53 and MDM2 (Wang et al. 2003).

CP-31398 is an interesting molecule with activity on both mutant and wild-type p53 proteins in various protein and cellular assays, and anti-tumor activity in mice. However, it needs further optimization to reduce cellular toxicity and increase specificity for the p53 pathway. Ultimately, the anticancer efficacy of CP-31398 or a derivative thereof should be tested in clinical trials.

The aminothiol WR1065, an active metabolite of amifostine, is used clinically to attenuate side effects of chemotherapy and radiotherapy. Amifostine is an antimutagen that protects normal cells from irradiation-induced DNA damage but has no protective effect on cancer cells. Maurici and Hainaut and their colleagues found that amifostine partially rescued several mutant forms of p53 for transcriptional transactivation in a yeast-based transcription assay (Maurici et al. 2001). The same group also showed that treatment with WR1065 at micromolar concentrations could restore wild-type conformation to the Met-272 mutant p53 in an oesophageal carcinoma cell line, enhance its DNA binding, and induce expression of certain p53 target genes followed by G1 cell cycle arrest (North et al. 2002). Further studies revealed that the accumulation of wild type p53 triggered by WR1065 in MCF-7 cells occurs via activation of the JNK (c-Jun N-terminal kinase) pathway, resulting in p53 phosphorylation at Thr-81 and inhibition of proteasome-mediated degradation of p53 (Pluquet et al. 2003a). Another study demonstrated that WR1065 can reduce thiol groups in p53 and enhance p53's redox-dependent DNA binding (Pluquet et al. 2003b). The ability of WR1065 to reactivate other mutant forms of p53 should be addressed in future studies.

In order to identify low molecular compounds that would preferentially kill human tumor cells expressing mutant p53, we performed a cell-based screening of a chemical library from the National Cancer Institute (http://dtp.nci.nih.gov), using a subline of p53 null Saos-2 osteosarcoma cells carrying exogenous Tet-regulated p53 His-273 mutant p53 (Bykov et al. 2002b). A screening assay based on the biological response of two isogenic tumor cell lines which differ only in p53 status has several advantages. First, it circumvents the problem of differential uptake and metabolism of compounds by different tumor cell lines. Moreover, compounds with nonspecific cellular toxicity and compounds that cannot enter the cells will not score. Finally, it allows identification of compounds which can rescue p53 function in cells irrespective of the molecular mechanism. Since no a priori assumptions about the mechanism are made, this approach may lead to the identification of compounds with previously unknown mechanisms.

We identified PRIMA-1, a molecule that induced significant apoptosis in Saos-2-His273 cells expressing mutant p53 but not in the same cells in the absence of mutant p53 expression (Bykov et al. 2002b). PRIMA-1 (for p53 Reactivation and Induction of Massive Apoptosis) is a quinuclidine with a molecular weight of 185. PRIMA-1MET, a methylated version of PRIMA-1, has even higher potency (Figure 2). PRIMA-1 induced apoptosis in a mutant p53-dependent manner as shown using a panel of cell lines expressing different Tet-regulated p53 mutants. Furthermore, PRIMA-1 stimulated DNA binding of a wide range of mutant p53 proteins in gel shift assays. Interestingly, the Phe-176 mutant p53 which lacks Cys-176 that holds a Zn atom critical for the structural integrity of the DNA binding domain (Cho et al. 1994) was refractory to reactivation by PRIMA-1. In accordance with this finding, KRC/Y renal cell carcinoma cells were relatively resistant to PRIMA-1.

We observed that PRIMA-1 was more efficient in rescuing p53 conformation and DNA binding in cellular extracts than in preparations of recombinant p53. This might be due to the higher extent of unfolding of p53 produced in bacteria. In addition, it is possible that specific post-translational modifications such as phosphorylation and acetylation that are present in mutant p53 derived from human tumor cells but not in bacterially produced p53 are required for mutant p53 reactivation by PRIMA-1. It is also conceivable that PRIMA-1 acts through cellular proteins such as Hsp70/90 that could regulate folding of p53.

Since PRIMA-1 induces mutant p53-dependent apoptosis and stimulates DNA binding of mutant p53, we tested if it would affect expression of p53 target genes in a mutant p53-dependent manner. Indeed, PRIMA-1 treatment triggered induction of the p53 target genes p21, MDM-2, and PUMA in mutant p53-carrying human tumor cells, but not in human tumor cells carrying wild-type p53. To confirm that PRIMA-1 exerts its antitumor effect through restoration of p53's transcriptional transactivation activity, we treated human ovarian carcinoma cells (SKOV) expressing a transcription-defective p53 double mutant with a single amino acid substitution in the core domain (His-175) and two amino acid substitutions in the N-terminal transactivation domain (Leu-22/Trp-23) with PRIMA-1. These cells were relatively resistant to PRIMA-1, whereas the same cells expressing His-175 mutant p53 with an intact N-terminal transactivation domain were sensitive. This demonstrates that p53's transactivation function is important for PRIMA-1-mediated tumor cell killing.

Our in vivo studies of PRIMA-1 in SCID mice revealed a statistically significant inhibition of the growth of human xenografts carrying mutant p53. No apparent toxic effects of PRIMA-1 treatment were observed. As a continuation of these in vivo studies, it will be important to examine the effect of PRIMA-1 on spontaneously arising tumours in mutant p53 transgenic or knock-in mice.

To confirm that PRIMA-1 can suppress tumor growth in a mutant p53-dependent fashion, we undertook an analysis of the available information in the National Cancer Institute database (http://dtp.nci.nih.gov). This database contains information about the effects of a large number of compounds, including many known anti-cancer drugs, on a panel of 60 human tumor cell lines. We found that the growth-inhibiting effect of PRIMA-1 correlated to mutant p53 levels but not to the levels of wild type p53 or cell proliferation rates (Bykov et al. 2002a). In contrast, cisplatin, 5-fluorouracil, methotrexate and doxorubicin showed preferential killing or growth suppression of wild type p53-carrying tumor cell lines. This pattern of growth suppression distinguishes PRIMA-1 from currently used anticancer drugs, and supports the notion that PRIMA-1 acts through fundamentally different mechanisms.

More recently, PRIMA-1 was shown to induce p53-dependent apoptosis in the absence of transcription or de novo protein synthesis, and even in the absence of a nucleus (Chipuk et al. 2003). This suggests that PRIMA-1 may induce apoptosis in several ways, both via p53-dependent transcription and by augmenting the interaction of p53 with Bcl-2 protein family members in cytoplasm. Cooperation between the different apoptotic pathways induced by PRIMA-1 might be highly beneficial for its anti-tumor activity.

It is currently unkown whether PRIMA-1 targets mutant p53 directly or via another protein. If PRIMA-1 does bind mutant p53, identification of the site of interaction should greatly facilitate further improvement of PRIMA-1 as a mutant p53-reactivating drug, as well as the design of novel molecular scaffolds with higher potency and specificity. Identification of an indirect mechanism for mutant p53 reactivation acting through a specific cellular target would also aid further drug development. Meanwhile, screening for more potent and mutant p53-specific PRIMA analogs will be required in order to identify a candidate drug for clinical trials in cancer patients.

In addition to the molecules described above, certain other classes of compounds have been shown to preferentially kill mutant p53-carrying tumor cells. Geldanamycin was identified as a compound capable of inducing mutant p53-dependent cell death via binding to the cellular chaperone Hsp90 that unfolds mutant p53 (Blagosklonny et al. 1996; Neckers 2002). Another screen for mutant p53-dependent growth inhibition lead to the discovery of the ellipticines (Shi et al. 1998). Subsequent studies have shown that ellipticine can restore wild type conformation and DNA binding to mutant p53 and transactivation of the p53 target genes p21 and MDM2 (Peng et al. 2003). Interestingly, rapamycin, an inhibitor TOR kinase, has been shown to relieve TOR-mediated suppression of ASK1 (apoptosis-signal-regulating kinase) in the absence of functional p53, leading to activation of a c-JUN signalling cascade and subsequent apoptosis (reviewed in (Bjornsti and Houghton 2004)). Thus, compounds that induce mutant p53-dependent growth inhibition and/or apoptosis may act at different levels. Some of them may induce mutant p53 refolding, possibly through direct binding, whereas other molecules can reactivate mutant p53 or exploit the absence of wild type p53 through interactions with other cellular targets.

SYNERGY WITH CURRENTLY USED CHEMOTHERAPEUTIC DRUGS

Many anticancer drugs used in the clinic today, e.g. cisplatin or 5-fluorouracil, activate wild type p53-mediated apoptotic pathways (O'Connor et al. 1997; Bunz et al. 1999; Lai et al. 2000; Lane 2004). As a result, treatment of wild type p53-carrying tumors is generally more successful than treatment of those that lack wild type p53 (see above and http://www.iarc.fr/p53).

The frequent acquisition of resistance to chemotherapy via p53 inactivation emphasizes the need to identify novel compounds that target tumor cells deficient in wild type p53-dependent apoptosis or drugs that restore tumor suppressor function to mutant p53. Among the drugs used clinically today, only paclitaxel and its analogs preferentially target cells with impaired p53 function (O'Connor et al. 1997; Weinstein et al. 1997). Strategies aimed at mutant p53 reactivation may therefore provide improved treatment for tumors that are resistant to current anticancer drugs.

Nevertheless, the fundamental problem of acquired resistance will undoubtedly apply to mutant p53-reactivating strategies as well. In this case, resistance might arise through selection for cells with impaired mutant p53 expression due to deletion or promoter methylation. Therefore, mutant p53 reactivation could be combined with other therapeutic regimens to minimize the emergence of resistant variants. Since wild type p53-carrying tumors tend to show better response to many chemotherapeutic drugs, particularly those that damage DNA, mutant p53 rescue may increase the sensitivity of mutant p53-carrying tumor cells to such drugs. Combined treatment with CP-31398 and cisplatin, doxorubicin (adriamycin), or VP-16 resulted in an additive effect in cultured tumor cells (Takimoto et al. 2002). However, treatment with PRIMA-1 and chemotherapeutic drugs like cisplatin resulted in a synergistic apoptosis-inducing effect in cultured cells and in tumor xenografts in SCID mice (Bykov et al. 2005). Interestingly, the combination of CP-31398 and PRIMA-1 inhibited growth of mutant p53-expressing cells in a synergistic manner (Bykov et al. 2005). The exact mechanism responsible for the observed synergy has not been elucidated. As mentioned above, the growth inhibitory effect of PRIMA-1 is dependent on mutant p53 protein levels (Bykov et al. 2002a). It is possible that DNA damage induced by chemotherapeutic drugs will further increase p53 phosphorylation and thus enhance levels of mutant p53 protein, which should make the tumor cells more responsive to PRIMA-1. If so, any agent that causes a further enhancement of the levels of mutant p53 in tumor cells could potentially show synergy with PRIMA-1.

FUTURE PERSPECTIVES

Mutant p53 reactivation is clearly an attractive strategy for cancer therapy. Significant progress has been made since the first proof-of-principle study of mutant p53 reactivation by a short synthetic peptide (Selivanova et al. 1997). Elucidation of the molecular mechanism of mutant p53 refolding by different classes of small molecules will hopefully pave the way for the development of more specific, less toxic and more potent p53-reactivators. Global analyses of gene and protein expression patterns induced by p53 reactivating molecules should provide a better understanding of their molecular mechanisms of action. Furthermore, the ability of these molecules to modulate p53 levels, conformation and biological activity may make them important new tools for studying the properties of wild type and mutant p53 proteins and pathways for p53 regulation in cells.

ACKNOWLEDGEMENTS

We thank the Swedish Cancer Society, the Swedish Research Council, the EU 5th and 6th Framework Programs, the Ingabritt & Arne Lundberg Foundation, the Cancer Society of Stockholm, King Gustaf V Jubilee Fund, and the Karolinska Institute for generous support.

REFERENCES

Abarzua, P., J.E. LoSardo, M.L. Gubler, and A. Neri. 1995. Microinjection of monoclonal antibody PAb421 into human SW480 colorectal carcinoma cells restores the transcription activation function to mutant p53. *Cancer Res* 55: 3490-4.

Abarzua, P., J.E. LoSardo, M.L. Gubler, R. Spathis, Y.A. Lu, A. Felix, and A. Neri. 1996. Restoration of the transcription activation function to mutant p53 in human cancer cells. *Oncogene* 13: 2477-82.

Bergh, J., T. Norberg, S. Sjogren, A. Lindgren, and L. Holmberg. 1995. Complete sequencing of the p53 gene provides prognostic information in breast cancer patients, particularly in relation to adjuvant systemic therapy and radiotherapy. *Nat Med* 1: 1029-34.

Bertheau, P., F. Plassa, M. Espie, E. Turpin, A. de Roquancourt, M. Marty, F. Lerebours, Y. Beuzard, A. Janin, and H. de The. 2002. Effect of mutated TP53 on response of advanced breast cancers to high-dose chemotherapy. *Lancet* 360: 852-4.

Bjornsti, M.A. and P.J. Houghton. 2004. The TOR pathway: a target for cancer therapy. *Nat Rev Cancer* 4: 335-48.

Blagosklonny, M.V., J. Toretsky, S. Bohen, and L. Neckers. 1996. Mutant conformation of p53 translated in vitro or in vivo requires functional HSP90. *Proc Natl Acad Sci U S A* 93: 8379-83.

Bode, A.M. and Z. Dong. 2004. Post-translational modification of p53 in tumorigenesis. *Nat Rev Cancer* 4: 793-805.

Borresen, A.L., T.I. Andersen, J.E. Eyfjord, R.S. Cornelis, S. Thorlacius, A. Borg, U. Johansson, C. Theillet, S. Scherneck, and S. Hartman. 1995. TP53 mutations and breast cancer prognosis: particularly poor survival rates for cases with mutations in the zinc-binding domains. *Genes Chromosomes Cancer* 14: 71-5.

Bossi, G., R. Scardigli, P. Musiani, R. Martinelli, M.P. Gentileschi, S. Soddu, and A. Sacchi. 2000. Development of a murine orthotopic model of leukemia: evaluation of TP53 gene therapy efficacy. *Cancer Gene Ther* 7: 135-43.

Bulavin, D.V., O.N. Demidov, S. Saito, P. Kauraniemi, C. Phillips, S.A. Amundson, C. Ambrosino, G. Sauter, A.R. Nebreda, C.W. Anderson, A. Kallioniemi, A.J. Fornace, Jr., and E. Appella. 2002. Amplification of PPM1D in human tumors abrogates p53 tumor-suppressor activity. *Nat Genet* 31: 210-5.

Bulavin, D.V., S. Saito, M.C. Hollander, K. Sakaguchi, C.W. Anderson, E. Appella, and A.J. Fornace, Jr. 1999. Phosphorylation of human p53 by p38 kinase coordinates N-terminal phosphorylation and apoptosis in response to UV radiation. *Embo J* 18: 6845-54.

Bullock, A.N. and A.R. Fersht. 2001. Rescuing the function of mutant p53. *Nat Rev Cancer* 1: 68-76.

Bunz, F., A. Dutriaux, C. Lengauer, T. Waldman, S. Zhou, J.P. Brown, J.M. Sedivy, K.W. Kinzler, and B. Vogelstein. 1998. Requirement for p53 and p21 to sustain G2 arrest after DNA damage. *Science* 282: 1497-501.

Bunz, F., P.M. Hwang, C. Torrance, T. Waldman, Y. Zhang, L. Dillehay, J. Williams, C. Lengauer, K.W. Kinzler, and B. Vogelstein. 1999. Disruption of p53 in human cancer cells alters the responses to therapeutic agents. *J Clin Invest* 104: 263-9.

Bykov, V.J., N. Issaeva, G. Selivanova, and K.G. Wiman. 2002a. Mutant p53-dependent growth suppression distinguishes PRIMA-1 from known anticancer drugs: a statistical analysis of information in the National Cancer Institute database. *Carcinogenesis* 23: 2011-8.

Bykov, V.J., N. Issaeva, A. Shilov, M. Hultcrantz, E. Pugacheva, P. Chumakov, J. Bergman, K.G. Wiman, and G. Selivanova. 2002b. Restoration of the tumor suppressor function to mutant p53 by a low-molecular-weight compound. *Nat Med* 8: 282-8.

Bykov, V.J., G. Selivanova, and K.G. Wiman. 2003. Small molecules that reactivate mutant p53. *Eur J Cancer* 39: 1828-34.

Bykov, V.J. and K.G. Wiman. 2003. Novel cancer therapy by reactivation of the p53 apoptosis pathway. *Ann Med* 35: 458-65.

Bykov, V.J., N. Zache, H. Stridh, J. Westman, J. Bergman, G. Selivanova, and K.G. Wiman. 2005. PRIMA-1^{MET} synergizes with cisplatin to induce tumor cell apoptosis. *Oncogene*, in press.

Caron de Fromentel, C., N. Gruel, C. Venot, L. Debussche, E. Conseiller, C. Dureuil, J.L. Teillaud, B. Tocque, and L. Bracco. 1999. Restoration of transcriptional activity of p53 mutants in human tumour cells by intracellular expression of anti-p53 single chain Fv fragments. *Oncogene* 18: 551-7.

Chipuk, J.E., U. Maurer, D.R. Green, and M. Schuler. 2003. Pharmacologic activation of p53 elicits Bax-dependent apoptosis in the absence of transcription. *Cancer Cell* 4: 371-81.

Cho, Y., S. Gorina, P.D. Jeffrey, and N.P. Pavletich. 1994. Crystal structure of a p53 tumor suppressor-DNA complex: understanding tumorigenic mutations. *Science* 265: 346-55.

Di Como, C.J., C. Gaiddon, and C. Prives. 1999. p73 function is inhibited by tumor-derived p53 mutants in mammalian cells. *Mol Cell Biol* 19: 1438-49.

DiTullio, R.A., Jr., T.A. Mochan, M. Venere, J. Bartkova, M. Sehested, J. Bartek, and T.D. Halazonetis. 2002. 53BP1 functions in an ATM-dependent checkpoint pathway that is constitutively activated in human cancer. *Nat Cell Biol* 4: 998-1002.

Dornan, D., I. Wertz, H. Shimizu, D. Arnott, G.D. Frantz, P. Dowd, K. O'Rourke, H. Koeppen, and V.M. Dixit. 2004. The ubiquitin ligase COP1 is a critical negative regulator of p53. *Nature* 429: 86-92.

Dumaz, N. and D.W. Meek. 1999. Serine15 phosphorylation stimulates p53 transactivation but does not directly influence interaction with HDM2. *Embo J* 18: 7002-10.

Elmore, L.W., C.W. Rehder, X. Di, P.A. McChesney, C.K. Jackson-Cook, D.A. Gewirtz, and S.E. Holt. 2002. Adriamycin-induced senescence in breast tumor cells involves functional p53 and telomere dysfunction. *J Biol Chem* 277: 35509-15.

Foster, B.A., H.A. Coffey, M.J. Morin, and F. Rastinejad. 1999. Pharmacological rescue of mutant p53 conformation and function. *Science* 286: 2507-10.

Frazier, M.W., X. He, J. Wang, Z. Gu, J.L. Cleveland, and G.P. Zambetti. 1998. Activation of c-myc gene expression by tumor-derived p53 mutants requires a discrete C-terminal domain. *Mol Cell Biol* 18: 3735-43.

Friedler, A., L.O. Hansson, D.B. Veprintsev, S.M. Freund, T.M. Rippin, P.V. Nikolova, M.R. Proctor, S. Rudiger, and A.R. Fersht. 2002. A peptide that binds and stabilizes p53 core domain: chaperone strategy for rescue of oncogenic mutants. *Proc Natl Acad Sci U S A* 99: 937-42.

Geisler, S., A.L. Borresen-Dale, H. Johnsen, T. Aas, J. Geisler, L.A. Akslen, G. Anker, and P.E. Lonning. 2003. TP53 gene mutations predict the response to neoadjuvant treatment with 5-fluorouracil and mitomycin in locally advanced breast cancer. *Clin Cancer Res* 9: 5582-8.

Geisler, S., P.E. Lonning, T. Aas, H. Johnsen, O. Fluge, D.F. Haugen, J.R. Lillehaug, L.A. Akslen, and A.L. Borresen-Dale. 2001. Influence of TP53 gene alterations and c-erbB-2 expression on the response to treatment with doxorubicin in locally advanced breast cancer. *Cancer Res* 61: 2505-12.

Gorina, S. and N.P. Pavletich. 1996. Structure of the p53 tumor suppressor bound to the ankyrin and SH3 domains of 53BP2. *Science* 274: 1001-5.

Gu, W. and R.G. Roeder. 1997. Activation of p53 sequence-specific DNA binding by acetylation of the p53 C-terminal domain. *Cell* 90: 595-606.

Halazonetis, T.D., L.J. Davis, and A.N. Kandil. 1993. Wild-type p53 adopts a 'mutant'-like conformation when bound to DNA. *Embo J* 12: 1021-8.

Hansen, S., T.R. Hupp, and D.P. Lane. 1996. Allosteric regulation of the thermostability and DNA binding activity of human p53 by specific interacting proteins. CRC Cell Transformation Group. *J Biol Chem* 271: 3917-24.

Hupp, T.R., D.W. Meek, C.A. Midgley, and D.P. Lane. 1993. Activation of the cryptic DNA binding function of mutant forms of p53. *Nucleic Acids Res* 21: 3167-74.

Hupp, T.R., A. Sparks, and D.P. Lane. 1995. Small peptides activate the latent sequence-specific DNA binding function of p53. *Cell* 83: 237-45.

Issaeva, N., A. Friedler, P. Bozko, K.G. Wiman, A.R. Fersht, and G. Selivanova. 2003. Rescue of mutants of the tumor suppressor p53 in cancer cells by a designed peptide. *Proc Natl Acad Sci U S A* 100: 13303-7.

Ito, A., C.H. Lai, X. Zhao, S. Saito, M.H. Hamilton, E. Appella, and T.P. Yao. 2001. p300/CBP-mediated p53 acetylation is commonly induced by p53-activating agents and inhibited by MDM2. *Embo J* 20: 1331-40.

Kim, A.L., A.J. Raffo, P.W. Brandt-Rauf, M.R. Pincus, R. Monaco, P. Abarzua, and R.L. Fine. 1999. Conformational and molecular basis for induction of apoptosis by a p53 C-terminal peptide in human cancer cells. *J Biol Chem* 274: 34924-31.

Lai, S.L., R.P. Perng, and J. Hwang. 2000. p53 gene status modulates the chemosensitivity of non-small cell lung cancer cells. *J Biomed Sci* 7: 64-70.

Lambert, P.F., F. Kashanchi, M.F. Radonovich, R. Shiekhattar, and J.N. Brady. 1998. Phosphorylation of p53 Serine 15 Increases Interaction with CBP. *J. Biol. Chem.* 273: 33048-33053.

Lane, D. 2004. p53 from pathway to therapy. *Carcinogenesis* 25: 1077-1081.

Lane, D.P. and T.R. Hupp. 2003. Drug discovery and p53. *Drug Discov Today* 8: 347-55.

Leng, R.P., Y. Lin, W. Ma, H. Wu, B. Lemmers, S. Chung, J.M. Parant, G. Lozano, R. Hakem, and S. Benchimol. 2003. Pirh2, a p53-induced ubiquitin-protein ligase, promotes p53 degradation. *Cell* 112: 779-91.

Lindstrom, M.S., U. Klangby, R. Inoue, P. Pisa, K.G. Wiman, and C.E. Asker. 2000. Immunolocalization of human p14(ARF) to the granular component of the interphase nucleolus. *Exp Cell Res* 256: 400-10.

Lindstrom, M.S. and K.G. Wiman. 2003. Myc and E2F1 induce p53 through p14ARF-independent mechanisms in human fibroblasts. *Oncogene* 22: 4993-5005.

Luu, Y., J. Bush, K.J. Cheung, Jr., and G. Li. 2002. The p53 stabilizing compound CP-31398 induces apoptosis by activating the intrinsic Bax/mitochondrial/caspase-9 pathway. *Exp Cell Res* 276: 214-22.

Marin, M.C. and W.G. Kaelin, Jr. 2000. p63 and p73: old members of a new family. *Biochim Biophys Acta* 1470: M93-M100.

Maurici, D., P. Monti, P. Campomenosi, S. North, T. Frebourg, G. Fronza, and P. Hainaut. 2001. Amifostine (WR2721) restores transcriptional activity of specific p53 mutant proteins in a yeast functional assay. *Oncogene* 20: 3533-40.

Melnikova, V.O., A.B. Santamaria, S.V. Bolshakov, and H.N. Ananthaswamy. 2003. Mutant p53 is constitutively phosphorylated at Serine 15 in UV-induced mouse skin tumors: involvement of ERK1/2 MAP kinase. *Oncogene* 22: 5958-66.

Milner, J. and E.A. Medcalf. 1991. Cotranslation of activated mutant p53 with wild type drives the wild-type p53 protein into the mutant conformation. *Cell* 65: 765-74.

Minamoto, T., T. Buschmann, H. Habelhah, E. Matusevich, H. Tahara, A.L. Boerresen-Dale, C. Harris, D. Sidransky, and Z. Ronai. 2001. Distinct pattern of p53 phosphorylation in human tumors. *Oncogene* 20: 3341-7.

Monti, P., P. Campomenosi, Y. Ciribilli, R. Iannone, A. Aprile, A. Inga, M. Tada, P. Menichini, A. Abbondandolo, and G. Fronza. 2003. Characterization of the p53 mutants ability to inhibit p73 beta transactivation using a yeast-based functional assay. *Oncogene* 22: 5252-60.

Neckers, L. 2002. Hsp90 inhibitors as novel cancer chemotherapeutic agents. *Trends Mol Med* 8: S55-61.

Niewolik, D., B. Vojtesek, and J. Kovarik. 1995. p53 derived from human tumour cell lines and containing distinct point mutations can be activated to bind its consensus target sequence. *Oncogene* 10: 881-90.

North, S., O. Pluquet, D. Maurici, F. El-Ghissassi, and P. Hainaut. 2002. Restoration of wild-type conformation and activity of a temperature-sensitive mutant of p53 (p53(V272M)) by the cytoprotective aminothiol WR1065 in the esophageal cancer cell line TE-1. *Mol Carcinog* 33: 181-8.

O'Connor, P.M., J. Jackman, I. Bae, T.G. Myers, S. Fan, M. Mutoh, D.A. Scudiero, A. Monks, E.A. Sausville, J.N. Weinstein, S. Friend, A.J. Fornace, Jr., and K.W. Kohn. 1997. Characterization of the p53 tumor suppressor pathway in cell lines of the National Cancer Institute anticancer drug screen and correlations with the growth-inhibitory potency of 123 anticancer agents. *Cancer Res* 57: 4285-300.

Okamoto, K., H. Li, M.R. Jensen, T. Zhang, Y. Taya, S.S. Thorgeirsson, and C. Prives. 2002. Cyclin G recruits PP2A to dephosphorylate Mdm2. *Mol Cell* 9: 761-71.

Peng, Y., C. Li, L. Chen, S. Sebti, and J. Chen. 2003. Rescue of mutant p53 transcription function by ellipticine. *Oncogene* 22: 4478-87.

Pluquet, O., S. North, A. Bhoumik, K. Dimas, Z. Ronai, and P. Hainaut. 2003a. The cytoprotective aminothiol WR1065 activates p53 through a non-genotoxic signaling pathway involving c-Jun N-terminal kinase. *J Biol Chem* 278: 11879-87.

Pluquet, O., S. North, M.J. Richard, and P. Hainaut. 2003b. Activation of p53 by the cytoprotective aminothiol WR1065: DNA-damage-independent pathway and redox-dependent modulation of p53 DNA-binding activity. *Biochem Pharmacol* 65: 1129-37.

Powers, J.T., S. Hong, C.N. Mayhew, P.M. Rogers, E.S. Knudsen, and D.G. Johnson. 2004. E2F1 uses the ATM signaling pathway to induce p53 and Chk2 phosphorylation and apoptosis. *Mol Cancer Res* 2: 203-14.

Pugacheva, E.N., A.V. Ivanov, J.E. Kravchenko, B.P. Kopnin, A.J. Levine, and P.M. Chumakov. 2002. Novel gain of function activity of p53 mutants: activation of the dUTPase gene expression leading to resistance to 5-fluorouracil. *Oncogene* 21: 4595-600.

Rahko, E., G. Blanco, Y. Soini, R. Bloigu, and A. Jukkola. 2003. A mutant TP53 gene status is associated with a poor prognosis and anthracycline-resistance in breast cancer patients. *Eur J Cancer* 39: 447-53.

Rippin, T.M., V.J. Bykov, S.M. Freund, G. Selivanova, K.G. Wiman, and A.R. Fersht. 2002. Characterization of the p53-rescue drug CP-31398 in vitro and in living cells. *Oncogene* 21: 2119-29.

Ryan, K.M., A.C. Phillips, and K.H. Vousden. 2001. Regulation and function of the p53 tumor suppressor protein. *Curr Opin Cell Biol* 13: 332-7.

Samuels-Lev, Y., D.J. O'Connor, D. Bergamaschi, G. Trigiante, J.K. Hsieh, S. Zhong, I. Campargue, L. Naumovski, T. Crook, and X. Lu. 2001. ASPP proteins specifically stimulate the apoptotic function of p53. *Mol Cell* 8: 781-94.

Schmitt, C.A. 2003. Senescence, apoptosis and therapy--cutting the lifelines of cancer. *Nat Rev Cancer* 3: 286-95.

Schmitt, C.A., J.S. Fridman, M. Yang, E. Baranov, R.M. Hoffman, and S.W. Lowe. 2002a. Dissecting p53 tumor suppressor functions in vivo. *Cancer Cell* 1: 289-98.

Schmitt, C.A., J.S. Fridman, M. Yang, S. Lee, E. Baranov, R.M. Hoffman, and S.W. Lowe. 2002b. A senescence program controlled by p53 and p16INK4a contributes to the outcome of cancer therapy. *Cell* 109: 335-46.

Schmitt, C.A., M.E. McCurrach, E. de Stanchina, R.R. Wallace-Brodeur, and S.W. Lowe. 1999. INK4a/ARF mutations accelerate lymphomagenesis and promote chemoresistance by disabling p53. *Genes Dev* 13: 2670-7.

Selivanova, G. 2001. Mutant p53: the loaded gun. *Curr Opin Investig Drugs* 2: 1136-41.

Selivanova, G., V. Iotsova, I. Okan, M. Fritsche, M. Strom, B. Groner, R.C. Grafstrom, and K.G. Wiman. 1997. Restoration of the growth suppression function of mutant p53 by a synthetic peptide derived from the p53 C-terminal domain. *Nat Med* 3: 632-8.

Selivanova, G., T. Kawasaki, L. Ryabchenko, and K.G. Wiman. 1998. Reactivation of mutant p53: a new strategy for cancer therapy. *Semin Cancer Biol* 8: 369-78.

Selivanova, G., L. Ryabchenko, E. Jansson, V. Iotsova, and K.G. Wiman. 1999. Reactivation of mutant p53 through interaction of a C-terminal peptide with the core domain. *Mol Cell Biol* 19: 3395-402.

Seoane, J., H.V. Le, and J. Massague. 2002. Myc suppression of the p21(Cip1) Cdk inhibitor influences the outcome of the p53 response to DNA damage. *Nature* 419: 729-34.

Sherr, C.J. 2000. The Pezcoller lecture: cancer cell cycles revisited. *Cancer Res* 60: 3689-95.

Shi, L.M., Y. Fan, T.G. Myers, P.M. O'Connor, K.D. Paull, S.H. Friend, and J.N. Weinstein. 1998. Mining the NCI anticancer drug discovery databases: genetic function approximation for the QSAR study of anticancer ellipticine analogues. *J Chem Inf Comput Sci* 38: 189-99.

Snyder, E.L., B.R. Meade, C.C. Saenz, and S.F. Dowdy. 2004. Treatment of Terminal Peritoneal Carcinomatosis by a Transducible p53-Activating Peptide. *PLoS Biol* 2: E36.

Soussi, T. and C. Beroud. 2001. Assessing TP53 status in human tumours to evaluate clinical outcome. *Nat Rev Cancer* 1: 233-40.

Strano, S. and G. Blandino. 2003. p73-mediated chemosensitivity: a preferential target of oncogenic mutant p53. *Cell Cycle* 2: 348-9.

Strano, S., G. Fontemaggi, A. Costanzo, M.G. Rizzo, O. Monti, A. Baccarini, G. Del Sal, M. Levrero, A. Sacchi, M. Oren, and G. Blandino. 2002. Physical interaction with human tumor-derived p53 mutants inhibits p63 activities. *J Biol Chem* 277: 18817-26.

Tacka, K.A., J.C. Dabrowiak, J. Goodisman, and A.K. Souid. 2002. Kinetic analysis of the reactions of 4-hydroperoxycyclophosphamide and acrolein with glutathione, mesna, and WR-1065. *Drug Metab Dispos* 30: 875-82.

Takekawa, M., M. Adachi, A. Nakahata, I. Nakayama, F. Itoh, H. Tsukuda, Y. Taya, and K. Imai. 2000. p53-inducible wip1 phosphatase mediates a negative feedback regulation of p38 MAPK-p53 signaling in response to UV radiation. *Embo J* 19: 6517-26.

Takimoto, R., W. Wang, D.T. Dicker, F. Rastinejad, J. Lyssikatos, and W.S. el-Deiry. 2002. The mutant p53-conformation modifying drug, CP-31398, can induce apoptosis of human cancer cells and can stabilize wild-type p53 protein. *Cancer Biol Ther* 1: 47-55.

Tsang, W.P., S.P. Chau, K.P. Fung, S.K. Kong, and T.T. Kwok. 2003. Modulation of multidrug resistance-associated protein 1 (MRP1) by p53 mutant in Saos-2 cells. *Cancer Chemother Pharmacol* 51: 161-6.

Wang, W., R. Takimoto, F. Rastinejad, and W.S. El-Deiry. 2003. Stabilization of p53 by CP-31398 inhibits ubiquitination without altering phosphorylation at serine 15 or 20 or MDM2 binding. *Mol Cell Biol* 23: 2171-81.

Wang, Y., A. Helland, R. Holm, H. Skomedal, V.M. Abeler, H.E. Danielsen, C.G. Trope, A.L. Borresen-Dale, and G.B. Kristensen. 2004. TP53 mutations in early-stage ovarian carcinoma, relation to long-term survival. *Br J Cancer* 90: 678-85.

Weinstein, J.N., T.G. Myers, P.M. O'Connor, S.H. Friend, A.J. Fornace, Jr., K.W. Kohn, T. Fojo, S.E. Bates, L.V. Rubinstein, N.L. Anderson, J.K. Buolamwini, W.W. van Osdol, A.P. Monks, D.A. Scudiero, E.A. Sausville, D.W. Zaharevitz, B. Bunow, V.N. Viswanadhan, G.S. Johnson, R.E. Wittes, and K.D. Paull. 1997. An information-intensive approach to the molecular pharmacology of cancer. *Science* 275: 343-9.

Wischhusen, J., U. Naumann, H. Ohgaki, F. Rastinejad, and M. Weller. 2003. CP-31398, a novel p53-stabilizing agent, induces p53-dependent and p53-independent glioma cell death. *Oncogene* 22: 8233-45.

Xu, Y. 2003. Regulation of p53 responses by post-translational modifications. *Cell Death Differ* 10: 400-3.

Chapter 19

NOVEL APPROACHES TO P53-BASED THERAPY: ONYX-015

Frank McCormick
UCSF Comprehensive Cancer Center and Cancer Research Institute, 240 Sutter St, San Francisco, USA

INTRODUCTION

Cancer is caused by gain of function of proteins involved in proliferation and survival, and loss of function of proteins that regulate these processes (Hanahan and Weinberg, 2000). Strategies for treating cancer generally involve development of small molecules that block hyperactive enzymes, or take advantage of abnormal expression of protein targets on the surface of cancer cells. Developing therapies based on loss of function of tumor suppressors presents novel challenges. Loss of the protein phosphates PTEN and loss of the G1/S checkpoint protein pRB occurs frequently in cancer, and offers a number of potential drug targets. Loss of PTEN leads to hyper-activation of downstream enzymes such as AKT and mTOR (McCormick, 2004) whereas loss of pRB leads to hyper-activation of the transcription factor E2F, and increased expression of numerous potential targets, some of which have already been exploited for cancer therapy (dihydrofolate reductase and thymidylate synthase, for example, are the targets of methotrexate and 5-fluorouracil, respectively). Loss of p53, on the other hand, does not appear to offer any direct targets for intervention: in contrast to PTEN and pRB, p53 is a positive regulatory protein, whose targets are obviously lost rather than hyper-activated in cancer cells.

The challenge of cancer therapy based on p53 has been addressed in several ways. Most directly, vectors have been designed in which p53 is delivered to cancer cells by gene therapy. This approach, using a non-

P. Hainaut and K.G. Wiman (eds.), 25 Years of p53 Research, 421-429.

replicating adenovirus vectors, is currently under Phase III clinical investigation in the US (Swisher and Roth, 2002) and has already been approved in China. It has the merit that delivery of p53 may have effects that are selective for cancer cells: for example, normal cells often growth arrest when p53 is expressed, cancer cells often undergo apoptosis. Furthermore, normal cells express Mdm2, and maintain p53 at low levels through active degradation. In principle, normal cells should be capable of degrading additional p53 that is introduced ectopically. In contrast, cancer cells expressing mutant p53 often have low levels of Mdm2, since Mdm2 is a transcriptional target of p53, and high levels of the Mdm2 inhibitor p14ARF (Bates et al., 1998). Introduction of ectopic p53 into such tumor cells may therefore generate a p53 response that is more sustained because of lower Mdm2 activity and more effective because responses are pro-apoptotic rather than cytostatic. The limitation of this approach is generally assumed to be the sunstantial technical hurdle of transducing every cell in a tumor with p53. However, a bystander effect following p53 transduction has been noted (Nishizaki et al., 1999) and an inflammatory response provoked by high levels of viral proteins may also contribute to killing effects in non-infected cells (McCormick, 2001). On the other hand, gene therapy is currently restricted to local treatment, since systemic delivery is inefficient and induces neutralizing antibodies. Cancer gene therapy still requires technological development before it can be considered a candidate for mainstream cancer treatment.

Other approaches to cancer therapy based on the p53 pathway include development of drugs blocking Mdm2-p53 interaction or preventing Mdm2-mediated p53 breakdown (Vassilev et al., 2004). These approaches are obviously limited to cancer cells that retain wild type p53 but depend on hyperactive Mdm2 to keep p53 at a tolerably low level. Such approaches have not yet entered clinical evaluation but are of considerable interest. Mdm2 is an enzyme that functions as an E3 ligase for p53: as such it offers potential as a drug target that may be more amenable to medicinal chemistry than sites of interaction between with Mdm2.

We have pursued a different approach: we have developed a virus whose replication and cell killing depends on loss of p53, at least in theory. This virus is the adenovirus mutant dl1520, or ONYX-015 (Bischoff et al., 1996).

DNA TUMOR VIRUSES, RB AND P53

The history of p53 research is embedded in the field of DNA tumor virus biology. p53, of course, was first described as a protein that co-immunoprecipitates with antisera against SV40 T-antigens. At this time, the relationship between tumor antigens was dissected using 2D tryptic fingerprint analysis of proteins, and in vitro systems for translating viral mRNAs. The cellular origin of p53 was revealed by 2D tryptic fingerprint mapping of p53-like proteins from rat, hamster and mouse cells, each of which had a distinct map. Presence of p53 in immunoprecipitates was either due to immunologic cross-reactivity, or physical association. This was resolved using polyclonal or monoclonal antibodies against T-antigen, or against p53 itself. The latter antiserum enabled us to identify p53 in normal cells in the absence of SV40 for the first time (McCormick and Harlow, 1980; Milner and McCormick, 1980).

Parallel efforts to identify tumor antigens of polyomavirus were complicated by the presence of a middle-T antigen that had no counterpart in SV40, though its molecular weight made it appear suspiciously like p53. When p53 was proven to be a host protein that associates with SV40 T-antigen, and polyomavirus middle T was proven to be a splice variant from the early region of the viral genome, the hunt was on for polyomavirus proteins that interacted with p53. However, this search failed:: we now believe that this polyomavirus inactivates p53 through in indirect mechanism that disrupts the p14ARF-p53 pathway (Lomax and Fried, 2001). This mechanism appears to involve binding of polyomavirus small-t antigen to pp2A, an event that may somehow affects Mdm2/p53 interaction. However, association of adenovirus E1B 55K with p53 followed the early successes with SV40 T-antigen (Kao et al., 1990). Likewise, Howley and coworkers showed that HPV targets degradation of p53 through the cellular E6-associated protein (Scheffner et al., 1990). 25 years after the first identification of p53, we can now conclude that most, if not all, small DNA viruses expend significant genetic capital on mechanisms for blocking p53 function during infection.

These results represented a early inroads into the close relationship between viral proteins and key cellular pathways. Discovery of interactions between E1a and RB from Drs Harlow, Livingston and others represented another breakthrough in this direction. The concept that viral proteins inactivate critical cellular regulatory proteins emerged through a consensus of these pioneering discoveries. In support of this concept, it was frequently observed that mutant forms of viral proteins that fail to bind p53 or RB were defective in virus replication and transformation assays, suggesting that these viral-cell interactions are essential. However, formal proof was lacking: The possibility remained that viral proteins target unknown cellular proteins that are critical for replication, and that interactions with p53 and RB coincided with these critical interactions. Consideration of a method of testing the hypothesis formally led to the concept of ONYX-015 and equivalent viruses that are selective for RB-deficient cells; the prediction to be tested was that virus defective in proteins targeting p53 or RB should replicate at wild type levels in cells that lack p53 or RB. If so, it would be possible to conclude that p53 or RB are indeed the crucial targets of these viral proteins. Furthermore, it was already known that such mutants were defective in normal cells, and so the concept that viruses could be competent for replication in tumor cells but defective in normal cells was established (Bischoff et al., 1996).

To test this hypothesis, we obtained mutant adenoviruses in which E1a or E1B had been mutated or deleted, and tested their ability to grow in cancer cells lacking p53 or RB. We chose adenoviruses because they are relatively easy to grown in human cells in culture and a wealth of mutants defective in E1a or E1B-55K proteins had been well characterized.

ADENOVIRUS E1B 55K AND E4 ORF 6

E1B 55K binds directly to p53 at a site that overlaps with the site of Mdm2 binding (Yew et al., 19900). (Figure 1). As a result, p53 is stabilized through binding, and its transcriptional activity is repressed. However, during lytic infection of human cells with type 3 and 5 group C adenoviruses, a second viral protein, E4orf6 enters the complex and targets p53 for destruction (Dobner et al., 1996). Cellular proteins that associate with E1B 55K have been implicated in the degradation process. p53 is thus cleared from normal cells during replication (Harada et al., 2002; Querido et al., 2001). Targeted destruction of p53 during infection is obviously consistent with the hypothesis that p53 has the capacity to block virus replication, as discussed above. Such a block could be mediated by p53-dependent induction of p21, a protein that inhibits cdk2 activity and blocks entry into S-phase, or by induction of apoptosis through induction of Bax, or other p53-dependent mediators of apoptosis. We are currently attempting to determine which of these p53 functions, if any, is responsible for attenuation of adenovirus infection. Indeed, deletion of E1B strongly attenuates replication of adenoviruses in normal cells, and high levels of p53 accumulate, but direct proof that these events are causally related has not yet been obtained. A technical hurdle relates to the difficulty of working with low-passage human cells in culture and inefficient methods of manipulating these cells. It is now well established that the process of growing cells in tissue culture leads to induction of p53, through the stress of growing cells on plastic and at non-physiological oxygen tension. Established cell lines, such as NIH 3T3 cells, are generally those in which the p53 pathway is defective, either through loss of p53 itself or loss of the upstream regulator p14ARF. For these reasons, we cannot depend on established cell lines to measure the role of p53 in replication.

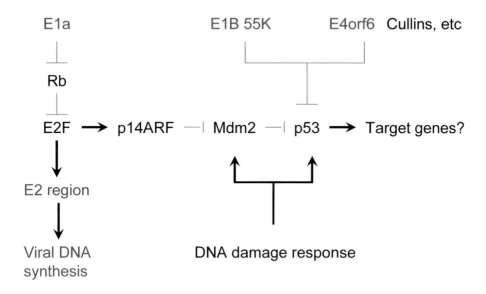

Figure 1. Viral proteins that interact with the RB and p53 pathways.

ONYX-015 and p53

Replication of ONYX-015, dl1520, fails to express E1B 55K, through a deletion and mutation in the E1B 55K gene. ONYX-015 replication in primary human epithelial cells is severely attenuated. Nuclear p53 is induced to super-physiological levels. However, in many tumor cells, ONYX-015 replicates at near wild-type levels (Heise et al., 1997; Harada and Berck, 1999). In other tumor cells attenuate replication of ONYX-015 severely. One of these cell lines, the human osteosarcoma cell U20S, is particular interest. This cell expresses wild-type p53 whose activity can be measured and indeed activated by DNA damaging reagents. This correlation led us to conclude that attenuation of ONYX-015 replication was mediated by p53. This assumption was later shown to be incorrect: elimination of p53 did not rescue ONYX-015 replication as predicted.

Other investigators, and workers at ONYX, noted a lack of correlation between ONYX-015 replication and p53 status, suggesting that the premise underlying ONYX-015 was incorrect (Rothmann et al., 1998). However, these experiments were flawed by the fact the p53 can be inactivated by a number of mechanisms in addition to genomic mutation. The most common of these is loss of the upstream regulator p14ARF (McCormick, 2000; Ries et al., 2000). Other mechanisms include Mdm2 amplification and expression of HPV E6 proteins. Lack of correlation of ONYX-015 replication with p53 mutations does not, therefore, detract from the original concept of ONYX-015.

Confusion regarding the status of p53 in tumor cells was exaggerated by the fact that wild type p53 can be activated in p14ARF-deficient tumor cells, suggesting that p53 regulatory pathways are intact in such tumors. However, the pathway that is triggered during adenovirus infection is distinct from that triggered by DNA damage. Loss of p14ARF disrupts induction of p53 during infection (see Figure 1). Ectopic expression of p14ARF in cancer cells that retain wild type p53 leads to massive p53 induction and attenuation of ONYX-015 replication.

When ONYX-015 replication is examined in p53-null cells, the importance of other unrelated functions of E1B 55K become apparent. In such cells, ONYX-015 replicates with a wide range of efficiencies (Harada and Berck, 1999). Replication is therefore dependent on the degree to which tumor cells provide other functions of E1B 55K. These functions include a role in efficient export of viral mRNAs (Babiss et al., 1985) and efficient shut-off of host protein synthesis. We are currently investigating the mechanisms by which E1B 55K regulates these functions and the ways in which permissive tumor cells provide these functions and allow efficient replication of ONYX-015.

Clinical effects of ONYX-015

Development of ONYX-015 and related concepts requires a number of important advances:

1. With respect to ONYX-015, it is clear that replication of this virus depends on factors unrelated to p53: the degree to which cancer cells provide additional E1B 55K functions varies, and the molecular basis of this needs to be fully understood so that improvements can be made. Currently, we are investigating the basis of variable replication of ONYX-015 in tumor cells and the mechanisms of restriction in normal and non-permissive cells.

2. The role of the adenovirus serotype C receptor, CAR, needs to be better understood in vivo. In cells in culture, variable expression of CAR accounts, tom some degree, for variable infectivity of cancer cells. In some cases, pharmacologic intervention can up-regulate CAR and increase infectivity: whether these observations translate into clinical improvements remains to be seen (Anders et al., 2003).

3. For systemic administration, it will be necessary to reduce or avoid production of neutralizing antibodies. This could be achieved, in principle, through inhibition of B-cell expansion with agents such as Rituxan (Reid, unpublished), or be other means.

4. Investment is necessary to facilitate large-scale production and purification of virus for commercial purposes, at costs that are acceptable.

These advances may be achieved through a combination of academic research and biotechnology, and each indeed appears feasible. However, investment in this approach to cancer therapy depends entirely on clinical evidence that the approach is viable and effective, and this must be the next chapter in the development of this platform.

Future prospects

ONYX-015 entered Phase I clinical trials in 1996. Increasing doses of virus were injected directly in tumors of the head and neck. The purpose of this study was to evaluate safety, and to establish proof of principle through biopsy of injected material. Indeed it was discovered that ONYX-015 replicates in tumors in vivo, and replication appeared to be selective for tumor tissue. Further, evidence of clinical activity was observed, and a Phase II trial was initiated. Results were published in 2000, and responses considered encouraging enough to warrant further studies and a pivotal Phase III study was opened (Khuri et al., 2000). Unfortunately, this study was not pursued due to commercial and financial issues unrelated to this discussion. Phase II studies with ONYX-015 include treatment of the pre-malignant disease oral leukoplakia (Rudin et al., 2003) and intrahepatic artery infusion for metastatic colorectal cancer (Reid et al., 2002). Results of these studies suggest clinical activity and an excellent safety profile. A number of anecdotal responses have been reported, some dramatic, but controlled studies are necessary to determine the potential benefit of these protocols objectively.

CONCLUSION

At the current time, the promising approach to p53-therapy in the near future appears to be the straightforward approach of delivering p53 to tumor cells by direct injection. A Phase III study using the Introgen vector is underway. Development of ONYX-015 is on hold, due to lack of investment in development and commercialization. The complexity of its mechanism of selectivity has diminished enthusiasm for this agent, and the lack of controlled clinical studies has led to uncertainty regarding its clinical potential. Meanwhile, small molecule approaches to blocking Mdm2 activity or restoring biological activity to certain mutant forms of p53 are under investigation, but are probably years from clinical testing. Opportunities abound for new creative approaches to treating cancer based on loss of p53: while therapies based on RB and PTEN are already undergoing clinical evaluation (CDK4 inhibitors, mTOR inhibitors, for example) we expect that 25 years of p53 research will eventually lead to clinical success, despite the more daunting challenges posed by loss of this crucial tumor suppressor..

REFERENCES

Anders M., Christian C., McMahon M., McCormick F., Korn W.M. Inhibition of the Raf/MEK/ERK pathway up-regulates expression of the coxsackievirus and adenovirus receptor in cancer cells. Cancer Res. 2003. 63: 2088-2095.

Babiss L.E., Ginsberg H.S., Darnell J.E. Jr. Adenovirus E1B proteins are required for accumulation of late viral mRNA and for effects on cellular mRNA translation and transport. Mol Cell Biol. 1985. 5: 2552-2558.

Bates S., Phillips A.C., Clark P.A., Stott F., Peters G., Ludwig R.L., Vousden K.H. p14ARF links the tumour suppressors RB and p53. Nature. 1998. 395: 124-125.

Bischoff J.R., Kirn D.H., Williams A., Heise C., Horn S., Muna M., Ng L., Nye J.A., Sampson-Johannes A., Fattaey A., McCormick F. An adenovirus mutant that replicates selectively in p53-deficient human tumor cells. Science. 1996. 274: 373-376.

Dobner T., Horikoshi N., Rubenwolf S., Shenk T. Blockage by adenovirus E4orf6 of transcriptional activation by the p53 tumor suppressor. Science. 1996. 272: 1470-1473.

Hanahan D., Weinberg R.A. The hallmarks of cancer. Cell. 2000. 100: 57-70.

Harada J.N., Berk A.J. p53-Independent and -dependent requirements for E1B-55K in adenovirus type 5 replication. J Virol. 1999. 73: 5333-5344.

Harada J.N., Shevchenko A., Pallas D.C., and Berk A.J. Analysis of the adenovirus E1B-55K-anchored proteome reveals its link to ubiquitination machinery. J Virol. 2002. 76: 9194-9206.

Heise C., Sampson-Johannes A., Williams A., McCormick F., Von Hoff D.D., Kirn D.H. ONYX-015, an E1B gene-attenuated adenovirus, causes tumor-specific cytolysis and antitumoral efficacy that can be augmented by standard chemotherapeutic agents. Nat Med. 1997. 3: 639-645.

Kao C.C., Yew P.R., Berk A.J.. Domains required for in vitro association between the cellular p53 and the adenovirus 2 E1B 55K proteins. Virology. 1990. 179: 806-814.

Khuri F.R., Nemunaitis J., Ganly I., Arseneau J., Tannock I.F., Romel L., Gore M., Ironside J., MacDougall R.H., Heise C., Randlev B., Gillenwater A.M., Bruso P., Kaye S.B., Hong W.K., Kirn D.H. a controlled trial of intratumoral ONYX-015, a selectively-replicating adenovirus, in combination with cisplatin and 5-fluorouracil in patients with recurrent head and neck cancer [see comments]. Nat Med. 2000. 6: 879-885.

Lomax M., Frie M.. Polyoma virus disrupts ARF signaling to p53. Oncogene. 2001. 20: 4951-4960.

McCormick F. ONYX-015 selectivity and the p14ARF pathway. Oncogene. 2000. 19: 6670-6672.

McCormick F. Cancer gene therapy: fringe or cutting edge?. Nat Rev Cancer. 2001. 1: 130-141.

McCormick F. Cancer: survival pathways meet their end. Nature. 2004. 428: 267-269.

McCormick F., Harlow, E. Association of a murine 53,000-dalton phosphoprotein with simian virus 40 large-T antigen in transformed cells. J Virol. 1980. 34: 213-224.

Milner J., McCormick F. Lymphocyte stimulation: concanavalin A induces the expression of a 53K protein. Cell Biol Int Rep. 1980. 4: 663-667.

Nishizaki M., Fujiwara T., Tanida T., Hizuta A., Nishimori H., Tokino T., Nakamura Y., Bouvet M., Roth J.A., Tanaka N. Recombinant adenovirus expressing wild-type p53 is antiangiogenic: a proposed mechanism for bystander effect. Clin Cancer. 1999. Res 5: 1015-1023.

Querido E., Blanchette P., Yan Q., Kamura T., Morrison M., Boivin D., Kaelin W.G., Conaway R.C., Conaway J.W., Branton P.E. Degradation of p53 by adenovirus E4orf6 and E1B55K proteins occurs via a novel mechanism involving a Cullin-containing complex. Genes Dev. 2001. 15: 3104-3117.

Reid T., Galanis E., Abbruzzese J., Sze D., Wein L.M., Andrews J., Randlev B., Heise C., Uprichard M., Hatfield M., Rome L., Rubin J., Kirn D. Hepatic arterial infusion of a replication-selective oncolytic adenovirus (dl1520): phase II viral, immunologic, and clinical endpoints. Cancer Res. 2002. 62: 6070-6079.

Ries S.J., Brandts C.H., Chung A.S., Biederer C.H., Hann B.C., Lipner E.M., McCormick F., Korn W.M. Loss of p14ARF in tumor cells facilitates replication of the adenovirus mutant dl1520 (ONYX-015). Nat Med. 2000. 6: 1128-1133.

Rothmann T., Hengstermann A., Whitaker N.J., Scheffner M. zur Hausen, H. Replication of ONYX-015, a potential anticancer adenovirus, is independent of p53 status in tumor cells. J Virol 1998. 72: 9470-9478.

Rudin C.M., Cohen E.E., Papadimitrakopoulou V.A., Silverman S. Jr., Recant W., El-Naggar A.K., Stenson K., Lippman S.M., Hong W.K., Vokes E.E. An attenuated adenovirus, ONYX-015, as mouthwash therapy for premalignant oral dysplasia. J Clin Oncol. 2003. 21: 4546-4552.

Scheffner M., Werness B.A., Huibregtse J.M., Levine A.J., Howley P.M. The E6 oncoprotein encoded by human papillomavirus types 16 and 18 promotes the degradation of p53. Cell. 1990. 63: 1129-1136.

Swisher S.G., Roth J.A. Clinical update of Ad-p53 gene therapy for lung cancer. Surg Oncol Clin N Am. 2002. 11: 521-535.

Vassilev L.T., Vu B.T., Graves B., Carvajal D., Podlaski F., Filipovic Z., Kong N., Kammlott U., Lukacs C., Klein C., Fotouhi N., Liu E.A.. In vivo activation of the p53 pathway by small-molecule antagonists of MDM2. Science. 2004. 303: 844-848.

Yew P.R., Kao C.C., Berk, A.J. Dissection of functional domains in the adenovirus 2 early 1B 55K polypeptide by suppressor-linker insertional mutagenesis. Virology. 1990. 179: 795-805.

Chapter 20

P53 AS SEEN BY AN OUTSIDER

George Klein
Microbiology and Tumor Biology Center, Karolinska Institute, Stockholm

Has the p53-field made a major impact on cancer research? Indeed it has. From its earliest beginnings, cancer research has been looking for "the" fundamental change in cancer cells, the ultimate common denominator. The idea that such a change must exist, was, if not abandoned, substantially mollified by the increasing realization, from the late 1950s, that cancer development is a multistep process, based on the individual reassortment of several unit characteristics (Foulds, 1958) or, as we now say, phenotypic traits.

p53 was named the "molecule of the year" a decade ago, nevertheless, due to the fact that it was the most commonly mutated gene in the largest variety of human tumors. Subsequently, it has been realized that the p53 pathway is inactivated at one point or another in most and perhaps all human tumors. Together with the equally frequent inactivation of the Rb pathway, the idea of a common – or, rather, a small number of common – key changes in cancer returned, although at a higher level of sophistication.

Historically, Rb and p53 were the first tumor suppressor genes that have been identified. They were followed by many others. We do not know what the total number might be. The many repeatedly occurring LOHs in the major human solid tumors that have not been accounted for, and the existence of gene clusters with tumor suppressing activity within three regions of human 3p, measuring 2.42, 1.23 and 7.8 Mb in size, has prompted us to suggest that the number of genes capable of suppressing or at least antagonizing tumor growth may be much larger than usually thought (Imreh et al., 2003).

This possibility and the added realization that both Rb and p53 are individual components of extensive regulatory networks with many participating molecules, it is appropriate to ask whether their early discovery may have biased the subsequent development of the field. Both genes have

P. Hainaut and K.G. Wiman (eds.), 25 Years of p53 Research, 431-438.
© 2005 *Springer. Printed in the Netherlands.*

been discovered through indirect and quite tortuous routes, led by accidental findings and experimental convenience. Has there been undue focus on these two genes? Is it possible that many others could have turned out to be equally, if not more important?

How can you approach such a question? The DNA tumor viruses have given us some clues. None of the oncogenic papova- , adeno- or herpesviruses have evolved to cause tumors. Lethal viruses are at selective disadvantage, compared to their low or non-pathogenetic relatives. They are only tumorigenic under experimental conditions: in vitro, in non-natural hosts or, sometimes, in special categories of natural hosts. They have, nevertheless, evolved the ability to encode specific proteins that can bind and degrade or otherwise inactivate Rb and p53. Since several of the viruses are unrelated and have different strategies, this is a true example of convergent evolution. The viral proteins responsible for these effects are also responsible for the transforming action of each virus. In most cases, the natural host of the virus responds with effective immunological responses that can keep the transformed cells under effective control. Tumor development occurs then only as an accident of immunosuppression or cytogenetic changes.

WHAT IS THE FUNCTIONAL MEANING OF RB AND P53 INACTIVATION IN THE CONTEXT OF THE VIRAL STRATEGIES?

One common denominator is the need to expand the virus carrying cell population. In addition, the integrating viruses, such as SV40 and adeno, must induce an S-phase in their target cells, in order to open the chromatin before they can integrate their proviral DNA within the cellular genome. Episomal viruses like HPV may to also have put or to keep their target cells in the cycle, in order to establish the correct episomal-chromosomal balance. However this may be, the fact remains that unrelated viruses have evolved the capacity to inactivate the Rb and p53 proteins, rather than other members of the corresponding pathways. Is the dominating position of these two key proteins the reason for their multifunctionality?

The cellular gene Bmi1 has an at least superficially similar ability to inactivate both pathways (Lund and Lohuizen, 2004). It does not act on the two target proteins directly, however, but through a common regulator gene, INK4a. One of several protein products of its multiply spliced mRNA, p16, is an upstream regulator of the Rb pathway. Another of its products, p14 ARF, a totally different protein, generated by **frameshift** and differential splicing, is one of the main regulators of the p53 pathway. The evolution of a double code that can be read in two different frames is a remarkable and perhaps unique feature for a cellular gene although it is well documented for the spatially much more confined viral genes. The independent but coordinatable functions of p14ARF and p16, known as key regulators of the cell cycle and of apoptosis, respectively, and their derivative roles as key components of the two most important tumor suppressor pathways, is even more remarkable. Apparently, the partly double-coded DNA sequence of a single gene has endowed it with functions that are usually thought in relation to more complex signaling pathways or master switch mechanisms. Bmi1 appears as a central controlling element for several functions of the complex INK4a/ARF gene.

Coordinated or at least parallel dysregulation of the Rb and p53 pathways is not only a convergent feature of DNA tumor virus strategies but is also an important characteristic of spontaneous tumor development and progression. The impairment of both pathways by genetic and/or epigenetic mechanisms has emerged as a common marker of most and perhaps all cancer cells. It is now a central feature of cancer cell biology, and one of the few generalizations that can be made. It makes sense. Dysregulation of the Rb dependent normal control of the cell cycle and the crippling of one of the most important apoptotic mechanisms by the impairment of two independent but intercommunicating pathways is an efficient way to favor the evolution of the cancer cell phenotype. Still, it has to be noted that the inactivation of both pathways in e.g. Bmi1 transgenic mice does not lead to a general tumor phenotype. They mainly develop B and T cell lymphomas. Explanations are being sought in developmental and tissue specific differences in the function of the key proteins.

Returning the main topic of this book, it may be asked why most if not all cancer cells have to inactivate the p53 pathway? This may be required not only because p53 controls the main apoptotic routes, but also, and very particularly, because it can be triggered by illegitimately activated oncogenes, such as ras, myc, E1A and E2F, via the p14/ARF regulator.

The resistance of the cancer cell to apoptosis is never complete. Even though every cancer cell becomes relatively resistant during tumor development, its responsiveness to irradiation and genotoxic chemicals reflects residual apoptotic sensitivity. The therapeutic effects of these treatments are not due to a direct action on DNA (which would require much higher doses), but to the apoptotic reaction that can be elicited even by minor DNA damage. Since there are several known p53 independent apoptotic pathways, this is no surprise in itself, but it does raise questions about the hierarchy of the different apoptotic pathways. How high can a cancer cell climb on the ladder of apoptotic resistance? The progressive "sculpting" of the malignant phenotype by sequential mutations and/or epigenetic inactivation of more and more apoptotic pathways can contribute to therapeutic resistance, but it is not clear how far this may proceed and what may be seen as the last citadel of sensitivity.

In view of its virtually universal inactivation in tumors, the p53 pathway is probably at the bottom of this "sculpting", perhaps because it is eminently efficient in nipping clones, driven to expansion by illegitimately activated oncogenes, in the bud. It may be added that the **early** impairment of p53 effector functions during tumorigenesis are context dependent, since they vary with cell type and oncogenic insult. They can be phenocopied by the loss of specific modifiers of p53 function. The p53 effector functions lost as **a consequence** of tumor evolution or antitumor therapy are context independent and involve primary mutation or deletion of p53.

Possible clues to the context dependence of the p53 impairment are approached by several articles in this volume. Hemann and Lowe point out that the disruption of apoptosis is particularly important for tumor initiation, whereas subsequent tumor development is mostly influenced by the disruption of cell cycle checkpoints and genome instability. The effects of ARF and p53 inactivation are not identical, as shown by the therapeutic response of the lymphomas that develop in different knockout mice. Emu-myc ARF-/- lymphomas respond to certain drugs with massive apoptosis, followed by long lasting remission or cure. In contrast, Emu-myc p53-/- lymphomas are resistant to chemotherapeutically induced apoptosis and progress rapidly to end stage disease. This is consistent with the view that while ARF loss disables a crucial function of p53 in relation to the tumorigenic process, it is not equally effective in abolishing the DNA damage response.

The context dependence of p53 impairment is also reflected by the tissue predilections of the oncogenic effect. Oncogene induced, p53 dependent apoptosis is apparently critical in preventing the development of B-cell derived tumors, whereas the roles of the p53 pathway in arresting the cell cycle and the maintenance of genomic stability are important for suppressing malignancies originating in other tissues. Moreover, the efficiency of oncogene induced apoptosis varies between different illegitimately activated oncogenes. Loss or silencing of other proapoptotic genes such as PUMA, or ARF, overexpression of the p53 antagonist MDM2 or of the antiapoptotic bcl-2 gene may all mimic the effects of p53 loss in given cellular contexts.

The interaction between activated oncogenes, impaired suppressor genes and the genetic background is a large but relatively unexplored field. As shown in the articles of Hemann, Lowe, and Donehower and Lozano and colleagues, INK 4a -/-, p53-/-, and INK4a/ARF double knockout mice differ with regard to tumor frequency, and, importantly, the curability of the tumors by chemotherapy. The latter is relatable to the frequency of spontaneous apoptosis. The analysis of these differences may have implications for therapeutically oriented translational research.

The article of Donehower and of Lozano et al. is also relevant in another context. They point out that p53 mutants that are oncogenic in transgenics induce a different spectrum of tumors for each mutation. Even tumors induced by different mutants in the same tissue differ with regard to their frequency, latency period, pathology, and metastatic potential. Relevant variables include the expression level of the transgene, the sensitivity of the given tissue to the dominant negative effect of the mutant and the genetic background of the transgenic hosts.

In what way can variations in the genetic background influence the probability of cancer development? Much has been learned about this question in the premolecular era, particularly during the first three decades after mid-century when mouse cancer genetics were intensely pursued. Mouse mammary carcinoma was one of the favorite objects of study. Analysis of spontaneous tumor incidence in hybrids between low and high mammary cancer strains and their intercross and breakcross hybrids by Bittner, Heston, Andervont, Mühlbock and others (reviewed in Heston, 1963) has revealed the existence of at least three functionally different genes or groups of genes that affect the probability of breast cancer development. One group influences the ability of the mouse mammary tumor virus (MMTV), a contributory factor, to replicate. The second creates a hormonal environment that is conducive to mammary tumor development not only in breeder females but also in virgins. The third group is most interesting and potentially relevant in a broader context. Its experimental detection was based on the fact that hybrids between two inbred strains of mice accept grafts from both parental strains. High and low cancer strains were crossed and the normal mammary glands of the first generation hybrid females were removed surgically. Subsequently, normal glands from both parental strains were successfully implanted into the hybrid mice and monitored for cancer development. The incidence was approximately ten times higher in the implants derived from the high, compared to the low cancer parent. Since the glands of both parental strains were now in the same host, exposed to the same virus and hormone levels, it became obvious that selective inbreeding of the mice for high or low tumor incidence has **also fixed genes that influence the probability of tumor development at the level of the target tissue.** Similar evidence was obtained in thymic lymhomas by Kaplan, lung adenomas by Kirschbaum, and adrenal tumors by Huseby and Bittner. This interesting area has not been analyzed further and responsible genes have not been identified. Still, the convergent genetic evidence from four different tumor systems clearly shows that modifier genes can influence the probability of tumor development in a tissue or cell type specific fashion.

Here a new question arises, however. Can we be sure that we are dealing with modifier **genes?** No longer. **Epigenetic** modifications of chromatin structure have emerged as a powerful source of developmental and also cancer related variation. The recent work of Feinberg and his group has shown that differences in the stringency of parental imprinting may be associated with different risk levels for colon carcinoma (Cui et al., 2003). More specifically, the loss of epigenetic silencing of the IGF2 gene, a feature of many human cancers, was also found in normal PBLs in about 10% of the normal human population, and was then associated with a 3,5-5 fold increase of colorectal adenoma risk. Experiments in a mouse model have confirmed that the loss of imprinting leads to an expanded stem cell compartment and is a modifier of neoplastic risk. Feinberg et. al. suggested that the increased number of adenomas was due to increased tumor initiation, rather than increased tumor progression. Earlier work on mice has shown the existence of genetic differences in the activity of enzymes involved in DNA methylation.

The most common changes that affect the p53 pathway in human tumors, such as mutation or deletion of p53 or ARF, or upregulation of MDM2, act at the genetic, i.e. the DNA level, although epigenetic inactivation of ARF may occur occasionally. In contrast, p16, the other main product of INK4a gene and an important regulator of the Rb pathway as already discussed, is usually inactivated by DNA methylation (Baylin et al., 1998). The reasons for these gene and also tissue related differences in the mode of genetic vs epigenetic inactivation of the same tumor suppressor gene are not understood.

Further development of the p53 and related fields will provide increased understanding of the genetic and epigenetic (including tissue specific) factors that modify the risk of tumor development. This will have to be related to the surveillance mechanisms of the normal organism against tumor development. With the exception of the protection against the growth of virus transformed cells the known mechanisms are largely non-immunological. They can act at the following levels (summarized in Klein, 2004):

a. Genetic surveillance (largely DNA repair)
b. Epigenetic surveillance (based on differences in imprinting stringency)
c. Intracellular surveillance (mainly apoptosis)
d. Intercellular surveillance (social interactions between cells)

The role of p53 and associated pathways is firmly established for a) and c).

POSSIBLE CLINICAL APPLICATIONS

Since all cancer cells are relatively apoptosis resistant, and since most therapeutic agents act by inducing apoptosis, partial or full reestablishment of apoptotic sensitivity is an obvious goal. But is it attainable?

In view of the fact that the p53 pathway is out of order in perhaps all cancer cells, attempts to rehabilitate it appear important. The modest but unequivocal success achieved by the large scale empirical testing of small molecules is encouraging. If the problem of targeting an entire tumor can be solved, and substantial effects can be achieved, treatment resistance will undoubtedly arise, as has been the case with all other forms of therapy. The versatility provided by the dynamics of tumor cell population is extraordinary. Would I not be prevented by the horror of antropomorphisms, I would call it ingenious.

The problem is not new. Resistance to initially effective treatment is a problem with chemotherapy, hormone therapy, and, most recently, molecularly targeted therapy (Glivec, Herceptin). So far, it could be best circumvented in chemotherapy, where unrelated drugs with different action mechanisms could be combined.

The idea behind combination therapy is seductively simple. If the frequency of cells resistant to drug A is 10^{-6} and resistant cells to drug B also occur with the same probability, the frequency of doubly resistant cells is 10^{-12}, beyond the realm of what is usually possible.

What can the use of additional, independently acting drugs aim at in the present case? It can not be limited to further fortification of the p53 pathway in addition to the primary drug, since this is unpredictable and potentially counterproductive. Rather, it might involve the targeting of additional, p53-independent apoptotic pathways. The multitude of these pathways and the fact that not all of them are impaired in established tumor cells, may provide potential Achilles heels.

The rehabilitation of the p53 pathway, even if only partial, and of other apoptotic pathways, may also increase the sensitivity of the tumor cells to cytotoxic drugs, as indicated by the knockout mouse experiments already discussed.

Tumor evolution is a combinatorial game. Tumor therapy will have to be combinatorial as well, making use of residual or rehabilitated apoptotic pathways and correspondingly influential cytotoxic drugs, hormones and anti hormones, growth factors and antigrowth factors, combined in a judicious way. The powerful high-throughput techniques for the massive analysis of gene expression should be helpful to work towards the orbit jump in data collection and analysis that will be required.

REFERENCES

Baylin, S.B., Herman, J.G., Graff, J.R., Verino, P.M., Issa, J-P. 1998. Alterations in DNA methylation: A fundamental aspect of neoplasia. Advances in Cancer Research. 72, 141-196.

Cui, H., Cruz-Correa, M., Giardiello, FM., Hutcheon, DF., Kafonek, DR., Brandenburg, S., Wu, Y., He, X., Powe, NR., Feinberg, AP. 2003, Loss of IGF2 imprinting: a potential marker of colorectal cancer risk. Science 299, 1753-1755.

Foulds, L., 1958. The natural history of cancer. J. Chron. Diseases 8, 2-37.

Heston, W.F. 1963. Genetics of Neoplasia. In: Methodology in Mamalian Genetics. W.J.Burdette, Editor, Holden-Day, San Francisco, pp247-248.

Imreh, S., Klein, G., Zabarovsky, E.R. 2003. Search for unknown tumor-antagonizing genes. Genes, Chromosomes and Cancer 38, 307-321.

Klein, G. 2004. Cancer, apoptosis, and nonimmune surveillance, Cell Death and Differentiation, 11, 13-17.

Lund, AH., van Lohuizen, M. 2004. Polycomb complexes and silencing mechanisms. Curr Opin Cell Biol. 3, 239-246.

INDEX

N

O

P